Methods in Molecular Biology™

Series Editor
John M. Walker
School of Life Sciences
University of Hertfordshire
Hatfield, Hertfordshire, AL10 9AB, UK

For further volumes:
http://www.springer.com/series/7651

Bacterial Cell Surfaces

Methods and Protocols

Edited by

Anne H. Delcour

Department of Biology and Biochemistry, University of Houston, Houston, Texas, USA

Humana Press

Editor
Anne H. Delcour
Department of Biology and Biochemistry
University of Houston
Houston, Texas, USA

ISSN 1064-3745 ISSN 1940-6029 (electronic)
ISBN 978-1-4939-5993-8 ISBN 978-1-62703-245-2 (eBook)
DOI 10.1007/978-1-62703-245-2
Springer New York Heidelberg Dordrecht London

Softcover re-print of the Hardcover 1st edition 2013

Cover illustration caption: Immuno-electron microscopy picture of *Corynebacterium diphtheriae* cells over-expressing SpaA pili. Black dots are gold particles conjugated to secondary antibodies recognizing SpaA pilin-specific antibodies

Printed on acid-free paper

Humana Press is a brand of Springer
Springer is part of Springer Science+Business Media (www.springer.com)

Preface

Surface and envelope processes play a fundamental role in the interplay between bacteria and their environment and in the successful adaptation of these microorganisms to a variety of ecological niches. A bacterial cell perceives and interacts with its habitat, be it a human host, a decaying leaf, or the surface of a coral, through processes that must involve its outermost barriers. In many cases, surface-exposed appendages, such as pili or adhesins, anchor the bacterium to a substratum, oftentimes a eukaryotic cell. Many of these cell-surface appendages have proven to be essential virulence factors, whose absence severely compromises or completely impedes the pathogen's ability to establish an infection. Bacteria can be surrounded by additional material, such as exopolysaccharides or capsules, and can display an extremely complex array of macromolecules in their outermost layer. The presence and the composition of these external layers are also key factors in the pathogenic process or the formation of bacterial communities inside hosts or in the environment.

But the role of the cell surface is not limited to the physical contact with the outside world. Lipid bilayers provide cellular integrity and protection against the chemical hazards of the external milieu. Cell walls play important roles in cell viability and morphogenesis. Membrane proteins such as transporters, channels, and receptors allow detection of signals and the regulated flow of organic and inorganic material across membranes. Membrane proteins of the cell envelope are also participants in the vast array of secretion systems and biosynthetic machineries utilized for the targeting, secretion, or surface presentation of external structures, toxins, and enzymes. Subcellular compartments, such as the periplasm, play important roles in the biogenesis of surface components, stress response, and transport across the envelope. Because of the complexity and variety of the macromolecular makeup of the bacterial envelope, microbiologists have been engaged for decades—and will continue to be for years to come—in deciphering its composition, understanding its biogenesis, and examining its regulation in response to external conditions.

In recent years, Molecular Microbiology is emerging as one of the cutting-edge disciplines in the biological sciences, thanks to the multidisciplinary and integrative approaches taken by investigators seeking to understand the intricacies of the microbial world and how it affects human health and the biosphere. In particular, recent advances in structural biology, proteomics, and imaging techniques, together with the traditional biochemical and genetic approaches, provide for an exciting look into the structure, function, and regulation of the bacterial cell envelope. These trends and the variety of techniques applied to the study of bacterial cell surfaces are reflected in this volume. In *Bacterial Cell Surfaces: Methods and Protocols*, the reader will find examples of traditional and innovative tools for the study of protein structure and function and enzymatic activities, the purification and analysis of macromolecules and their complexes, and the investigation of regulatory mechanisms and cell biological processes. The book is aimed at the microbiologist, biochemist, or cell biologist, whether a beginner graduate student, or a veteran experimentalist, who wishes to learn new methodologies and take advantage of the years of research and protocol optimization from the best laboratories.

Chapters in this book are contributed by highly recognized experts in a wide variety of topics that pertain to bacterial cell surfaces and the bacterial envelope. My rationale in the choice of areas was to provide a comprehensive view to the field, while at the same time, carefully selecting the leading laboratories in the various areas of specialization. For the purpose of this book, the bacterial cell surface was operationally defined as any structure or compartment external to the cytoplasmic membrane, and thus, safe for a chapter on electrophysiology of bacterial channels, the vast number of processes inherent exclusively to the plasma membrane is not addressed in this volume. After a couple of introductory chapters which provide reviews of the bacterial cell surface and a history of its microscopic investigation, the book is organized in a directional fashion with topics dealing with purely external structures at first, followed by investigations of components that are progressively more "inward" across the cell envelope, such as outer membrane, cell walls, and periplasm. Gram-negative bacteria are well represented, but a few chapters are focused on Gram-positive bacteria and Mycobacteria.

I wish to thank all the outstanding authors who have contributed chapters for their consistent cooperation and efforts; the series Editor, John Walker, for his relentless guidance and patience; and my current students for our enjoyable discussions about this book.

Houston, TX, USA ***Anne H. Delcour***

Contents

Preface . . . . *v*
Contributors . . . . *ix*

1 A Bird's Eye View of the Bacterial Landscape 1
Natividad Ruiz

2 Visualizing the Bacterial Cell Surface: An Overview 15
Harald Engelhardt

3 Purification of the Outer Membrane Usher Protein and Periplasmic Chaperone-Subunit Complexes from the P and Type 1 Pilus Systems 37
Nadine S. Henderson and David G. Thanassi

4 Experimental Manipulation of the Microbial Functional Amyloid Called Curli 53
Yizhou Zhou, Daniel R. Smith, David A. Hufnagel, and Matthew R. Chapman

5 Visualization of Gram-positive Bacterial Pili 77
Chungyu Chang, I-Hsiu Huang, Antoni P.A. Hendrickx, and Hung Ton-That

6 Single Cell Microfluidic Studies of Bacterial Motility 97
Adrien Ducret, Olivier Théodoly, and Tâm Mignot

7 Chromatographic Analysis of the *Escherichia coli* Polysialic Acid Capsule 109
Susan M. Steenbergen and Eric R. Vimr

8 Analysis of Exopolysaccharides in *Myxococcus xanthus* Using Confocal Laser Scanning Microscopy 121
Wei Hu, Renate Lux, and Wenyuan Shi

9 Assessment of Multidrug Efflux Assemblies by Surface Plasmon Resonance 133
Elena B. Tikhonova and Helen I. Zgurskaya

10 Fluorescence Microscopy and Proteomics to Investigate Subcellular Localization, Assembly, and Function of the Type II Secretion System 157
Tanya L. Johnson, Aleksandra E. Sikora, Ryszard A. Zielke, and Maria Sandkvist

11 Pore Formation by T3SS Translocators: Liposome Leakage Assay 173
Eric Faudry, Caroline Perdu, and Ina Attrée

12 Isolation of Bacterial Type IV Machine Subassemblies 187
Mayukh K. Sarkar, Seyyed I. Husnain, Simon J. Jakubowski, and Peter J. Christie

13 Production and Crystallization of Bacterial Type V Secretion Proteins 205
Hye-Jeong Yeo

14 Assembly of Bacterial Outer Membrane Proteins 223
Jan Grijpstra, Martine P. Bos, and Jan Tommassen

15 The Outer Membrane of Gram-Negative Bacteria: Lipid A Isolation and Characterization 239
Jessica V. Hankins, James A. Madsen, Brittany D. Needham, Jennifer S. Brodbelt, and M. Stephen Trent

16 Quantitative and Qualitative Preparations of Bacterial Outer Membrane Vesicles 259
Halima Chutkan, Ian MacDonald, Andrew Manning, and Meta J. Kuehn

17 In Vitro Peptidoglycan Synthesis Assay with Lipid II Substrate 273
Jacob Biboy, Nhat Khai Bui, and Waldemar Vollmer

18 Extraction of Cell Wall-Bound Teichoic Acids and Surface Proteins from *Listeria monocytogenes* 289
Filipe Carvalho, María Graciela Pucciarelli, Francisco García-del Portillo, Didier Cabanes, and Pascale Cossart

19 Subfractionation and Analysis of the Cell Envelope (Lipo) polysaccharides of *Mycobacterium tuberculosis* 309
Anna E. Grzegorzewicz and Mary Jackson

20 Protein Disulfide Bond Formation in the Periplasm: Determination of the In Vivo Redox State of Cysteine Residues 325
Katleen Denoncin, Valérie Nicolaes, Seung-Hyun Cho, Pauline Leverrier, and Jean-François Collet

21 Using Reporter Genes and the *Escherichia coli* ASKA Overexpression Library in Screens for Regulators of the Gram Negative Envelope Stress Response. . . . 337
Julia L. Wong, Stefanie L. Vogt, and Tracy L. Raivio

22 Isolation of Bacteria Envelope Proteins 359
Shu Quan, Annie Hiniker, Jean-François Collet, and James C.A. Bardwell

23 Patch Clamp Electrophysiology for the Study of Bacterial Ion Channels in Giant Spheroplasts of *E. coli* 367
Boris Martinac, Paul R. Rohde, Charles G. Cranfield, and Takeshi Nomura

24 Electrophysiological Characterization of Bacterial Pore-Forming Proteins in Planar Lipid Bilayers 381
Owen S. Mapingire, Beau Wager, and Anne H. Delcour

Index *397*

Contributors

INA ATTRÉE • *Bacterial Pathogenesis and Cellular Responses Group, Centre National de la Recherche Scientifique, Université Grenoble I, Commissariat à l'Energie Atomique, iRTSV, Grenoble, France*

JAMES C.A. BARDWELL • *Howard Hughes Medical Institute, University of Michigan, Ann Arbor, MI, USA; Department of Molecular, Cellular, and Developmental Biology, University of Michigan, Ann Arbor, MI, USA*

JACOB BIBOY • *Institute for Cell and Molecular Biosciences, The Centre for Bacterial Cell Biology, Newcastle University, Newcastle upon Tyne, UK*

MARTINE P. BOS • *Department of Medical Microbiology and Infection Control, VU University Medical Center, Amsterdam, The Netherlands*

JENNIFER S. BRODBELT • *Department of Chemistry and Biochemistry, The University of Texas at Austin, Austin, TX, USA*

NHAT KHAI BUI • *Institute for Cell and Molecular Biosciences, The Centre for Bacterial Cell Biology, Newcastle University, Newcastle upon Tyne, UK*

DIDIER CABANES • *Group of Molecular Microbiology, Institute for Molecular and Cell Biology, Porto, Portugal*

FILIPE CARVALHO • *Group of Molecular Microbiology, Institute for Molecular and Cell Biology, Porto, Portugal*

CHUNGYU CHANG • *Department of Microbiology and Molecular Genetics, University of Texas Health Science Center, Houston, TX, USA*

MATTHEW R. CHAPMAN • *Department of Molecular, Cellular, and Development Biology, University of Michigan, Ann Arbor, MI, USA*

SEUNG-HYUN CHO • *Brussels Center for Redox Biology and de Duve Institute, Université Catholique de Louvain, Brussels, Belgium; WELBIO, Brussels, Belgium*

PETER J. CHRISTIE • *Department of Microbiology and Molecular Genetics, University of Texas Medical School at Houston, Houston, TX, USA*

HALIMA CHUTKAN • *Department of Molecular Genetics and Microbiology, Duke University Medical Center, Durham, NC, USA*

JEAN-FRANÇOIS COLLET • *Brussels Center for Redox Biology and de Duve Institute, Université Catholique de Louvain, Brussels, Belgium; WELBIO, Brussels, Belgium*

PASCALE COSSART • *Unité des Interactions Bactéries-Cellules, Institut Pasteur, Paris, France*

CHARLES G. CRANFIELD • *Victor Chang Cardiac Research Institute, Darlinghurst, Australia*

ANNE H. DELCOUR • *Department of Biology and Biochemistry, University of Houston, Houston, TX, USA*

KATLEEN DENONCIN • *Brussels Center for Redox Biology and de Duve Institute, Université Catholique de Louvain, Brussels, Belgium; WELBIO, Brussels, Belgium*

ADRIEN DUCRET • *Laboratoire de Chimie Bactérienne, Institut de Microbiologie de la Méditerranée, Université Aix-Marseille, Marseille, France*
HARALD ENGELHARDT • *Max-Planck-Institut für Biochemie, Martinsried, Germany*
ERIC FAUDRY • *Bacterial Pathogenesis and Cellular Responses Group, Centre National de la Recherche Scientifique, Université Grenoble I, Commissariat à l'Energie Atomique, iRTSV, Grenoble, France*
JAN GRIJPSTRA • *Department of Molecular Microbiology, Utrecht University, Utrecht, The Netherlands; Institute of Biomembranes, Utrecht University, Utrecht, The Netherlands*
ANNA E. GRZEGORZEWICZ • *Mycobacteria Research Laboratories, Department of Microbiology, Immunology, and Pathology, Colorado State University, Fort Collins, CO, USA*
JESSICA V. HANKINS • *Section of Molecular Genetics and Microbiology, The University of Texas at Austin, Austin, TX, USA*
NADINE S. HENDERSON • *Department of Molecular Genetics and Microbiology, Stony Brook University, Stony Brook, NY, USA*
ANTONI P.A. HENDRICKX • *Department of Microbiology, University of Chicago, Chicago, IL, USA*
ANNIE HINIKER • *Department of Anatomic Pathology, University of California San Francisco, San Francisco, CA, USA*
WEI HU • *School of Dentistry, University of California, Los Angeles, CA, USA; State Key Laboratory of Microbial Technology, School of Life Science, Shandong University, Jinan, Shandong, China*
I-HSIU HUANG • *Department of Microbiology and Molecular Genetics, University of Texas Health Science Center, Houston, TX, USA*
DAVID A. HUFNAGEL • *Department of Molecular, Cellular, and Development Biology, University of Michigan, Ann Arbor, MI, USA*
SEYYED I. HUSNAIN • *Department of Biochemistry, University of Leicester, Leicester, UK*
MARY JACKSON • *Mycobacteria Research Laboratories, Department of Microbiology, Immunology, and Pathology, Colorado State University, Fort Collins, CO, USA*
SIMON J. JAKUBOWSKI • *Department of Microbiology and Molecular Genetics, University of Texas Medical School at Houston, Houston, TX, USA*
TANYA L. JOHNSON • *Department of Microbiology and Immunology, University of Michigan Medical School, Ann Arbor, MI, USA*
META J. KUEHN • *Department of Biochemistry, Duke University Medical Center, Durham, NC, USA*
PAULINE LEVERRIER • *Brussels Center for Redox Biology and de Duve Institute, Université Catholique de Louvain, Brussels, Belgium; WELBIO, Brussels, Belgium*
RENATE LUX • *School of Dentistry, University of California, Los Angeles, CA, USA*
IAN MACDONALD • *Department of Molecular Genetics and Microbiology, Duke University Medical Center, Durham, NC, USA*
JAMES A. MADSEN • *Department of Chemistry and Biochemistry, The University of Texas at Austin, Austin, TX, USA*
ANDREW MANNING • *Department of Biochemistry, Duke University Medical Center, Durham, NC, USA*

OWEN S. MAPINGIRE • *Department of Biology and Biochemistry, University of Houston, Houston, TX, USA*
BORIS MARTINAC • *Victor Chang Cardiac Research Institute, Darlinghurst, Australia; St. Vincent's Clinical School, University of New South Wales, Sydney, Australia*
TÂM MIGNOT • *Laboratoire de Chimie Bactérienne, Institut de Microbiologie de la Méditerranée, Université Aix-Marseille, Marseille, France*
BRITTANY D. NEEDHAM • *The Institute of Cellular and Molecular Biology, The University of Texas at Austin, Austin, TX, USA*
VALÉRIE NICOLAES • *Brussels Center for Redox Biology and de Duve Institute, Université Catholique de Louvain, Brussels, Belgium; WELBIO, Brussels, Belgium*
TAKESHI NOMURA • *Victor Chang Cardiac Research Institute, Darlinghurst, Australia*
CAROLINE PERDU • *Bacterial Pathogenesis and Cellular Responses Group, Centre National de la Recherche Scientifique, Université Grenoble I, Commissariat à l'Energie Atomique, iRTSV, Grenoble, France*
FRANCISCO GARCÍA-DEL PORTILLO • *Centro Nacional de Biotecnología, Madrid, Spain*
MARÍA GRACIELA PUCCIARELLI • *Departamento de Biología Molecular, Universidad Autónoma de Madrid, Madrid, Spain; Centro Nacional de Biotecnología, Universidad Autónoma de Madrid, Madrid, Spain*
SHU QUAN • *Howard Hughes Medical Institute, University of Michigan, Ann Arbor, MI, USA; Department of Molecular, Cellular, and Developmental Biology, University of Michigan, Ann Arbor, MI, USA*
TRACY L. RAIVIO • *Department of Biological Sciences, University of Alberta, Edmonton, AB, Canada*
PAUL R. ROHDE • *Victor Chang Cardiac Research Institute, Darlinghurst, Australia*
NATIVIDAD RUIZ • *Department of Microbiology, The Ohio State University, Columbus, OH, USA*
MARIA SANDKVIST • *Department of Microbiology and Immunology, University of Michigan Medical School, Ann Arbor, MI, USA*
MAYUKH K. SARKAR • *Department of Microbiology and Molecular Genetics, University of Texas Medical School at Houston, Houston, TX, USA*
WENYUAN SHI • *School of Dentistry, University of California, Los Angeles, CA, USA; Department of Microbiology, Immunology, and Molecular Genetics, University of California, Los Angeles, CA, USA*
ALEXSANDRA E. SIKORA • *Department of Pharmaceutical Sciences, Oregon State University, Corvallis, OR, USA*
DANIEL R. SMITH • *Department of Molecular, Cellular, and Development Biology, University of Michigan, Ann Arbor, MI, USA*
SUSAN M. STEENBERGEN • *Department of Pathobiology, University of Illinois at Urbana-Champaign, Urbana, IL, USA*
DAVID G. THANASSI • *Department of Molecular Genetics and Microbiology, Stony Brook University, Stony Brook, NY, USA*
OLIVIER THÉODOLY • *Adhesion & Inflammation, Université Aix-Marseille, Marseille, France*
ELENA B. TIKHONOVA • *Department of Chemistry and Biochemistry, University of Oklahoma, Norman, OK, USA*

JAN TOMMASSEN • *Department of Molecular Microbiology, Utrecht University, Utrecht, The Netherlands; Institute of Biomembranes, Utrecht University, Utrecht, The Netherlands*
HUNG TON-THAT • *Department of Microbiology and Molecular Genetics, University of Texas Health Science Center, Houston, TX, USA*
M. STEPHEN TRENT • *Section of Molecular Genetics and Microbiology, The University of Texas at Austin, Austin, TX, USA; The Institute of Cellular and Molecular Biology, The University of Texas at Austin, Austin, TX, USA*
ERIC R. VIMR • *Department of Pathobiology, University of Illinois at Urbana-Champaign, Urbana, IL, USA*
STEFANIE L. VOGT • *Department of Biological Sciences, University of Alberta, Edmonton, AB, Canada*
WALDEMAR VOLLMER • *Institute for Cell and Molecular Biosciences, The Centre for Bacterial Cell Biology, Newcastle University, Newcastle upon Tyne, UK*
BEAU WAGER • *Department of Biology and Biochemistry, University of Houston, Houston, TX, USA*
JULIA L. WONG • *Department of Biological Sciences, University of Alberta, Edmonton, AB, Canada*
HYE-JEONG YEO • *Department of Biology and Biochemistry, University of Houston, Houston, TX, USA*
HELEN I. ZGURSKAYA • *Department of Chemistry and Biochemistry, University of Oklahoma, Norman, OK, USA*
YIZHOU ZHOU • *Department of Molecular, Cellular, and Development Biology, University of Michigan, Ann Arbor, MI, USA*
RYSZARD A. ZIELKE • *Department of Pharmaceutical Sciences, Oregon State University, Corvallis, OR, USA*

Chapter 1

A Bird's Eye View of the Bacterial Landscape

Natividad Ruiz

Abstract

Bacteria interact with the environment through their cell surface. Activities as diverse as attaching to a catheter, crawling on a surface, swimming through a pond, or being preyed on by a bacteriophage depend on the composition and structure of the cell surface. The cell surface must also protect bacteria from harmful chemicals present in the environment while allowing the intake of nutrients and excretion of toxic molecules. Bacteria have evolved four main types of bacterial cell surfaces to accomplish these functions: those of the typical gram-negative and gram-positive bacteria, and those of the *Actinobacteria* and *Mollicutes.* So few types seems remarkable since bacteria are very diverse and abundant, and they can live in many different environments. However, each species has tweaked these stereotypical bacterial surfaces to best fit its needs. The result is an amazing diversity of the bacterial landscape, most of which remains unexplored. Here I give an overview of the main features of the bacterial cell surface and highlight how advances in methodology have moved forward this field of study.

Key words: Cell envelope, Outer membrane, Peptidoglycan, Periplasm, Teichoic acids, S-layer, Capsule, Adhesion

1. Introduction

We have come a long way since Antonie van Leeuwenhoek described bacterial motility when observing samples with a rudimentary microscope. Now, with a few touches on a mobile electronic device, one can download videos showing a computer animation of the bacterial flagellum architecture, operation, and biogenesis. Both van Leeuwenhoek's drawings and the computer animation from the Namba laboratory were based on empirical data and driven by curiosity and love for science. What sets them apart is the technology and methodology behind them.

The history of the research on the bacterial flagellum illustrates how our understanding of the bacterial cell surface has evolved as advances in methodology occurred. What started from a simple

Anne H. Delcour (ed.), *Bacterial Cell Surfaces: Methods and Protocols*, Methods in Molecular Biology, vol. 966, DOI 10.1007/978-1-62703-245-2_1,

observation of motility using the light microscope has led to more than 2,000 articles on the bacterial flagellum. These studies have used methodology that ranges from light microscopy to electron and atomic force microscopy, from denaturing gel electrophoresis to crystallography, from simple genetic screens to genome-wide competitive selection and microarray-based genetic footprinting. Many approaches and techniques contributed to these studies and, at times, certain findings were considered major breakthroughs; however, no single approach or technique has been the magic bullet for understanding this bacterial nanomachine.

In this chapter, I use the term *bacterial cell surface* to refer to the outermost permanent layer that surrounds a bacterium. In other words, the bacteria cell surface is the bacterial skin. As I describe below, there are other layers that can dress the bacterial cell surface, such as an S-layer or capsule, but these are typically neither permanent nor essential. We know of four main types of bacterial cell surfaces: the cell surface of typical gram-positive and gram-negative bacteria, and the surface of *Actinobacteria* and *Mollicutes*. Although each type has key architectural and compositional features that are conserved, each bacterial species has its own version of one of these types of surfaces. Moreover, in response to environmental cues, a bacterium can remodel its own surface by altering the expression or modification of certain components, such as proteins and lipopolysaccharides (LPS) (1, 2). Thus, the bacterial cell surface is a complex and dynamic structure.

Realization that bacteria might have different types of cell surfaces probably first occurred in the late 1880s. Methods for differential staining of clinical specimens were developed then and they revealed that not all bacterial species stain the same (3–5). Staining properties were therefore established as a main way to classify bacteria into gram-positive bacteria, gram-negative bacteria, and acid-fast bacteria. This was and continues to be extremely useful when diagnosing bacterial infections (4, 5). For a long time, it was thought that whether a bacterium would retain or lose a particular stain depended on the composition of its cell envelope. This assumption turned out to be correct but it took many decades to understand why, since the small size of bacteria made studying their anatomy quite difficult (5).

For decades, the number of layers and composition of the bacterial cell envelope was unknown. Many studies on the cell surface were limited to serological analyses of pathogenic organisms, such as the classification of streptococci conducted by Rebecca Lancefield (6). Therefore, high resolution electron microscopy (EM) and the development of cell fractionation protocols were major breakthroughs to advance the field of the bacterial cell envelope in the 1950s–1970s (7–11). Once layers were identified and separated, studies focused on analyzing the composition of the bacterial cell surface using biochemical methods. These studies revealed that the cell surface can be composed of three types of molecules:

proteins, polysaccharides, and lipids. However, determining the identity of specific surface proteins, polysaccharides, and lipids required the development and application of additional analytical methods, such as SDS-PAGE, nuclear magnetic resonance, and mass spectrometry (12–18).

Genetic approaches also played a central role in elucidating the composition of the bacterial cell surface. They were especially important in identifying specific envelope components, as well as the genes involved in their biosynthesis. Since genetic tools for *Escherichia coli* and *Salmonella* were developed early on, much of the early advances were made on the outer membrane of these gram-negative bacteria (19–23). Interestingly, the genetic information we have obtained from many different organisms can now be used to deduce *in silico* general features of the surface of many bacterial species whose genomes have been sequenced without having to culture them.

In the 1990s, some of the major developments in the field of bacterial cell surface came from structural and functional studies. Crystallography and electron microscopy revealed the high-resolution structure of β-barrel proteins and some surface organelles (24–33). Meanwhile, interest in molecular microbial pathogenesis drove the identification and functional analyses of many surface components with a role in bacterium–host interactions, particularly adhesins, since adhesion is crucial in pathogenesis and is a tractable phenotype (28, 34–36).

Although elucidating the structure and composition of a bacterium's surface is critical, we must also understand how the cell builds and maintains it. Moreover, the bacterial surface is dynamic; it changes in response to environmental signals. Therefore, understanding how the cell regulates these changes is also essential. Genetic and biochemical approaches have primarily advanced these fields, especially during the last 15 years with respect to the biogenesis of the gram-negative and gram-positive cell surface (26, 37–43). Describing these advances is a topic too vast to cover here, but I would like to point out that much of what we have learned in these areas has been proven to be fundamental concepts in biology that reach beyond the field of bacterial cell surface. For example, studies on the biogenesis of the gram-negative outer membrane have advanced our understanding of processes that are conserved from bacteria to humans, such as protein targeting and translocation to and across membranes, and folding of β-barrel membrane proteins (44–47). In addition, studies on the bacterial cell surface have given us great insight into two-component systems, small RNAs, stress responses, assembly of multi-protein machines, and complex regulatory systems to name a few (2, 27, 48–56).

I will next discuss the function and overall composition of the four main types of bacterial cell surfaces. I have included references to many reviews that can provide the reader a better understanding of the many topics I can only cover lightly in this chapter.

2. Functions of the Bacterial Cell Surface

The bacterial cell surface plays essential roles for the survival and behavior of bacteria. Since it delimits the cell, the bacterial cell surface serves as the ultimate protective barrier for the bacterium. The cell surface must withstand variations in physical environmental conditions, such as temperature and osmolarity, as well as the chemical stresses, such as changes in pH and toxic small molecules. The best characterized example to illustrate the barrier function of the bacterial cell surface is the gram-negative outer membrane. Gram-negative bacteria are naturally resistant to many antibiotics because these chemicals cannot readily cross their outer membrane. Resistance to large hydrophilic antibiotics stems from the size exclusion imposed by the outer membrane proteins that function as porins, while, as described below, the presence of LPS in the outer leaflet is responsible for the resistance to hydrophobic antibiotics (57). In the *Actinobacteria*, surface lipids are also crucial for serving as a barrier to the entry of many antibiotics, while in the gram positives, the peptidoglycan and teichoic acids are thought to act as a permeability barrier (58–62).

This barrier quality of the cell surface must be balanced with another important function of the cell surface, which is to allow the exchange of molecules with the environment and other cells. In other words, the cell surface must be a selective barrier. To solve this issue, the cell surface controls molecule exchange by containing specific types of lipids and transporter proteins. The latter can serve as importers and exporters, including efflux pumps that excrete toxic chemicals that somehow enter the cell (63–66). In some cases, substrates that are too large to enter the cell are first partially degraded by surface-exposed enzymes into smaller components that can be imported. For example, glycosylases on the surface of *Streptococcus pneumonia* hydrolyze complex carbohydrates so that they can be used as carbon source. Interestingly, these surface enzymes have also been proposed to expose receptors on host cells that are important for colonization (67–70).

Another essential function of the cell surface intrinsic from its outermost location is to serve as the site for physical contact between the cell and all other surfaces. These external surfaces can be abiotic, cellular, or viral, and the outcome of these interactions can vary from beneficial, such as the acquisition of antibiotic resistance genes from a bacterial neighbor and the adhesion of the microbiome to the host, to detrimental, such as the interaction with a lytic virus or contact-dependent inhibition with the wrong neighbor (71–73). These outcomes have probably driven a great deal of the diversity that exists among the cell surface of different bacterial species, as well as the remodeling that the surface of a single bacterium can undergo in response to its environment.

Lastly, the cell surface also plays an essential role in bacterial behavior. Swimming, swarming, walking, and gliding depend on surface appendages, such as flagella, pili, and "legs," as well as possibly on surface-associated exopolysaccharide or slime (74–77). Even the arrest of flagellum rotation can serve as a sense of touch to detect external surfaces and induce the production of adhesins (78). Small-molecule signals present in the environment can trigger responses in a bacterium. In order to function, these signals must either bind to a component of the bacterial surface or cross the cell surface to enter the cell. Similarly, many small molecules and proteins that bacteria use to signal to other organisms are produced in the cytoplasm, so they must somehow translocate across the cell surface. To do this, bacteria have evolved a series of specialized secretion systems that transport proteins across the cell surface into the environment and other cells (79); in addition, gram-negative bacteria can package pieces of their cell surface and periplasm as outer membrane vesicles that can deliver signals and virulence factors to other bacteria and host cells (80). Thus, much of what bacteria do and how they do it is influenced by their cell surface.

3. The Cell Surface of Gram-positive Bacteria

Bacteria that retain crystal violet after Gram staining are known as gram-positive bacteria (3, 5). Their cell envelope is composed of a cytoplasmic membrane that is surrounded by layers of the rigid polymer peptidoglycan that ultimately constitutes the cell surface. The >20 nm-thick peptidoglycan mesh also serves as a barrier and as an exoskeleton that protects the cytoplasmic membrane from the high turgor pressure in the cytoplasm (81). Recently, atomic-force microscopy has provided structural information of the architecture of the peptidoglycan of *B. subtilis* and a model has emerged where the long peptidoglycan strands are arranged into coiled cables (82). Somehow, long anionic polymers known as teichoic acids are threaded through the peptidoglycan layers so that they can be exposed to the cell surface (41, 58). These teichoic acids are mostly composed of glycerol phosphate and sugar–phosphate repeats and they can be covalently attached to peptidoglycan (wall teichoic acids) or to membrane lipids (lipoteichoic acids). Notably, teichoic acids can serve as adhesins and bacteriophage receptors (58). In gram-positive bacteria, surface-exposed proteins and organelles such as pili are covalently attached to the peptidoglycan by transpeptidases of the sortase family (39, 83). Recognition and attachment of these surface proteins requires an LPXTG motif at the carboxyl terminus of the precursor protein.

4. The Gram-negative Cell Surface

Unlike gram-positive bacteria, gram-negative bacteria do not retain crystal violet after Gram staining because they contain a peptidoglycan layer that is only ~2–8 nm thick (3, 5, 81). In addition, the structure of their cell envelope differs from that of the gram-positive envelope in that it contains two membranes known as the inner or cytoplasmic membrane and the outer membrane. The aqueous space between these membranes, known as the periplasm, is ~20 nm wide and it contains the peptidoglycan layer (40). Thus, the outer membrane constitutes the cell surface of gram-negative bacteria.

The outer membrane is anchored to the peptidoglycan layer and it is an asymmetric lipid bilayer composed of phospholipids in the inner leaflet and lipopolysaccharide or lipooligosaccharide (LPS or LOS, respectively) in the outer leaflet. These glycolipids are crucial for providing a barrier against hydrophobic compounds, since they establish strong lateral interactions that make the outer membrane a fairly impermeable gel (57). Some gram-negative bacteria, such as the spirochetes, lack LPS (57). Instead, they possess diacylglycerol-containing glycolipids that include a cholesteryl-galactolipid (84, 85).

There are two types of outer membrane proteins, lipoproteins and integral outer membrane proteins (OMPs). Lipoproteins are anchored to the outer membrane through a covalently linked lipid moiety (86, 87) and they can either face the periplasm or be surface exposed (88, 89). In *E. coli*, even the surface-exposed lipoproteins are thought to be lipid-anchored to the inner leaflet (88); in contrast, the surface-exposed lipoproteins of the *Borrelia* spirochete are anchored to the outer leaflet of the outer membrane (90). Integral OMPs span the membrane bilayer in a barrel conformation (91). Most are β-barrels but exceptions exists, such as Wza, which is a barrel composed of α-helices (92). Although most of these barrel proteins serve as channels that facilitate the entry and exit of molecules (57), some are enzymes, such as the acyltransferase PagP, the deacylase PagL, and the protease OmpT (1, 93). In addition, many of them also serve as adhesins to host surfaces and bacteriophages.

5. The Actinobacterial Cell Surface

The *Actinobacteria* are known as acid-fast bacteria because they are not decolorized easily by an acid–alcohol mixture after staining with an arylmethane dye (4). There have been many controversies regarding the classification of this group of bacteria. According to 16S RNA phylogenetic analysis, the *Actinobacteria* have been classified as GC-rich gram-positive organisms. However, genome-wide comparison reveals that they are more closely related to gram-negative bacteria (94, 95). More importantly, the structure of the cell envelope and composition of the cell surface of *Actinobacteria* is also controversial. Although these bacteria were classified as gram positives, it was clear that they were not typical. Their cell envelope was known to be impermeable to many antibiotics that normally enter the gram-positive cell envelope (96). In addition, they were known to contain a variety of complex glycolipids and mycolic acids, which have very long acyl chains of 30–90 carbons and are attached to arabinogalactan. In the early 1980s, Minnikin proposed that these bacteria possess an outer membrane (97). Thus, their cell envelope was proposed to be composed of a cytoplasmic membrane, a periplasmic space containing a peptidoglycan layer to which arabinogalactan is covalently attached to, and an outer membrane. However, difficulties with resolving envelope layers in EM images and with physically separating these layers through fractionation procedures made it difficult to reach a consensus in the scientific community (98). In 1993, X-ray diffraction studies supported the existence of an outer membrane (99), but it was not until 2008 that clear visualization of the outer membrane was possible with the use of cryo-EM of vitreous sections (100, 101).

Although it is now accepted that mycolic acids are present in the outer membrane, it is not clear how they are arranged in this bilayer (100, 101). What the actual lipid composition of the outer membrane is also remains unclear because of the difficulty in isolating pure fractions. Very recently, however, analyzing the lipid composition of the *Corynebacterium glutamicum* outer membrane has been possible using an extraction method that employs reverse surfactant micelles of an ester sodium salt (102). Analyses using mass spectrometry and thin-layer chromatography of the fraction have revealed an unusual composition which might differ from that of the *Mycobacterium* species. The inner leaflet of *C.glutamicum* contains large amounts of the phospholipid cardiolipin, which appears to be tightly associated with the peptidoglycan–arabinogalactan layer through an unknown mechanism. In contrast, the outer leaflet is composed of trehalose dimycolates, phosphatidylinositol mannosides, and highly apolar lipids. Similar studies on mycobacterial species are clearly needed.

The surface proteins of *Actinobacteria* are also poorly characterized. Biophysical and structural studies have demonstrated that the mycobacterial surface has porins that are thought to facilitate the passage of hydrophilic molecules across the outer membrane (103, 104). Only the structure of MspA has been solved so far. This structure shows that MspA is different from typical gram-negative β-barrel proteins, since octamers of MspA adopt a goblet-like conformation with a single central channel (104). The identification of more outer membrane proteins and the elucidation of their high-resolution structures are needed in order to understand whether this goblet-like conformation is conserved among the actinobacterial outer membrane proteins.

6. The Mollicute Cell Surface

Mollicutes, such as the mycoplasmas, are among the smallest culturable bacteria known (105). These bacteria possess a single membrane and lack peptidoglycan. Therefore, their cell surface is the cytoplasmic membrane, which is composed of phospholipids, choline-containing phosphoglycolipids, and proteins (106). Given that these bacteria lack a periplasmic space and a peptidoglycan layer, most surface-exposed proteins are lipoproteins anchored to the outer leaflet of the cytoplasmic membrane. These are abundant and they include the majority of the surface antigens known to undergo antigenic variation (107).

Interestingly, although certain *Mycoplasma* species lack flagella and pili, they are motile. They use an ATP-dependent mechanism of gliding motility that is poorly understood (108, 109). In *Mycoplasma pneumonia*, both gliding motility and adherence to host cells are mediated by the mycoplasma terminal organelle or button, which is an electron-dense, membrane-bound cell extension distal end (110, 111). But perhaps, the best studied motility system is that of *Mycoplasma mobilis*, which forms a membrane protrusion at a cell pole and glides over surfaces in the direction of this protrusion. A unique model for this "centipede" motility has been proposed. In this model, gliding depends on four proteins that localize to the base of the membrane protrusion and that are supported by the cytoskeleton. One of these proteins forms 50 nm-long "legs" that receive movements generated in the cytoplasm by ATP hydrolysis by the motor protein. This results in repeated binding, pull, and release of the legs along the surface (112). High-resolution structural and microscopy information are being currently used to probe and understand this novel mechanism of motility.

7. Nonessential Surface Layers

Capsules. Bacterial capsules are extracellular polysaccharides that are intimately associated with many types of bacterial cell surfaces. In most cases, the mechanism for anchoring these polysaccharides to the cell surface is unknown. Nevertheless, it is clear that capsules are often important for virulence, resistance to phage absorption, and survival of the bacterium in unfavorable environments. For reviews on bacterial capsules, I suggest these references (113–116).

S-layers. The beautiful lattice of S-layers was discovered by EM in the gram-negative *Spirillum serpens* in the early 1950s (117). Since then, they have been found in many different types of bacteria (76, 118, 119). S-layers are built from a single protein that self-polymerizes into nanolattices. In some cases, S-layer proteins are glycoproteins. Notably, the first glycoprotein discovered in bacteria was a *Clostridium* S-layer protein (120). In gram-positive bacteria, the S-layer is directly bound to the peptidoglycan (121), while in gram-negative bacteria and *Actinobacteria*, it is associated with their outer membranes (122). Surprisingly, the function of these surface layers remains largely unknown. Nevertheless, because of their self-assembly properties and their crystalline arrangements, S-layers are the subjects of many crystallographic and biophysical studies, and more recently, their utility in nanobiotechnology is being explored (123).

8. Final Thoughts

We have acquired a vast amount of knowledge on the bacterial cell surface over the past 100 years, but there is much more to learn. How biogenesis of the cell surface occurs and is coordinated with the rest of the cellular compartments, and how the cell regulates and remodels the composition of its cell surface are fundamental processes that still require much research even in the most studied bacteria. In addition, most of what we know has come from studies that focus on very few organisms, when compared to the great diversity of bacterial species. Although many fundamental concepts will be conserved, I am sure that many more surprises are yet to come. Once again, the bacterial flagellum is already showing us this (124).

Acknowledgments

I thank Emily Buttler and Rebecca Davis for their helpful suggestions in the preparation of this manuscript.

References

1. Raetz CR, Reynolds CM, Trent MS, Bishop RE (2007) Lipid A modification systems in gram-negative bacteria. Annu Rev Biochem 76:295–329
2. Guillier M, Gottesman S (2006) Remodelling of the Escherichia coli outer membrane by two small regulatory RNAs. Mol Microbiol 59:231–247
3. Gram HCJ (1884) Über die isolirte färbung der Schizomyceten in schnitt- und trocken-präparaten. Fortschr Med 2:185–189
4. Madison BM (2001) Application of stains in clinical microbiology. Biotech Histochem 76:119–125
5. Beveridge TJ (2001) Use of the gram stain in microbiology. Biotech Histochem 76:111–118

6. Lancefield RC (1933) A serological differentiation of human and other groups of hemolytic streptococci. J Exp Med 57:571–595
7. van Iterson, Ruys AC (1960) The fine structure of the Mycoplasmataceae (microorganisms of the pleuropneumonia group PPLO). 1. Mycoplasma hominis, M. fermentans and M. salivarium. J Ultrastruct Res 3:282–301
8. Glauert AM, Thornley MJ (1969) The topography of the bacterial cell wall. Annu Rev Microbiol 23:159–198
9. Miura T, Mizushima S (1968) Separation by density gradient centrifugation of two types of membranes from spheroplast membrane of Escherichia coli K12. Biochim Biophys Acta 150:159–161
10. Razin S, Hasin M, Ne'eman Z, Rottem S (1973) Isolation, chemical composition, and ultrastructural features of the cell membrane of the mycoplasma-like organism Spiroplasma citri. J Bacteriol 116:1421–1435
11. Salton MR (1974) Isolation of cell walls from Gram-positive bacteria. Methods Enzymol 31:653–667
12. Laemmli UK (1970) Cleavage of structural proteins during the assembly of the head of bacteriophage T4. Nature 227:680–685
13. Brennan PJ (1989) Structure of mycobacteria: recent developments in defining cell wall carbohydrates and proteins. Rev Infect Dis 11(Suppl 2):S420–S430
14. Chatterjee D, Hunter SW, McNeil M, Jardine I, Brennan PJ (1989) Structure and function of mycobacterial glycolipids and glycopeptidolipids. Acta Leprol 7(Suppl 1):81–84
15. Johnson WC, Silhavy TJ, Boos W (1975) Two-dimensional polyacylamide gel electrophoresis of envelope proteins of Escherichia coli. Appl Microbiol 29:405–413
16. Nouwens AS, Walsh BJ, Cordwell SJ (2003) Application of proteomics to Pseudomonas aeruginosa. Adv Biochem Eng Biotechnol 83:117–140
17. Blache D, Bruneteau M, Michel G (1980) Lipid A of lipopolysaccharide from Escherichia coli K12 CR34. Biochimie 62:191–194
18. Qureshi N, Takayama K, Ribi E (1982) Purification and structural determination of nontoxic lipid A obtained from the lipopolysaccharide of Salmonella typhimurium. J Biol Chem 257:11808–11815
19. Raetz CR, Whitfield C (2002) Lipopolysaccharide endotoxins. Annu Rev Biochem 71:635–700
20. Kadner RJ, Liggins GL (1973) Transport of vitamin B12 in Escherichia coli: genetic studies. J Bacteriol 115:514–521
21. Schnaitman CA, Klena JD (1993) Genetics of lipopolysaccharide biosynthesis in enteric bacteria. Microbiol Rev 57:655–682
22. Randall-Hazelbauer L, Schwartz M (1973) Isolation of the bacteriophage lambda receptor from Escherichia coli. J Bacteriol 116: 1436–1446
23. Raetz CR (1986) Molecular genetics of membrane phospholipid synthesis. Annu Rev Genet 20:253–295
24. Kubori T, Matsushima Y, Nakamura D, Uralil J, Lara-Tejero M, Sukhan A, Galan JE, Aizawa SI (1998) Supramolecular structure of the Salmonella typhimurium type III protein secretion system. Science 280:602–605
25. Hung DL, Hultgren SJ (1998) Pilus biogenesis via the chaperone/usher pathway: an integration of structure and function. J Struct Biol 124:201–220
26. Waksman G, Hultgren SJ (2009) Structural biology of the chaperone-usher pathway of pilus biogenesis. Nat Rev Microbiol 7:765–774
27. Erhardt M, Namba K, Hughes KT (2010) Bacterial nanomachines: the flagellum and type III injectisome. Cold Spring Harb Perspect Biol 2:a000299
28. Olsen A, Jonsson A, Normark S (1989) Fibronectin binding mediated by a novel class of surface organelles on Escherichia coli. Nature 338:652–655
29. Weiss MS, Abele U, Weckesser J, Welte W, Schiltz E, Schulz GE (1991) Molecular architecture and electrostatic properties of a bacterial porin. Science 254:1627–1630
30. Weiss MS, Kreusch A, Schiltz E, Nestel U, Welte W, Weckesser J, Schulz GE (1991) The structure of porin from Rhodobacter capsulatus at 1.8 A resolution. FEBS Lett 280:379–382
31. Weiss MS, Schulz GE (1992) Structure of porin refined at 1.8 A resolution. J Mol Biol 227:493–509
32. Weiss MS, Wacker T, Weckesser J, Welte W, Schulz GE (1990) The three-dimensional structure of porin from Rhodobacter capsulatus at 3 A resolution. FEBS Lett 267:268–272
33. Jap BK, Walian PJ, Gehring K (1991) Structural architecture of an outer membrane channel as determined by electron crystallography. Nature 350:167–170
34. Hultgren SJ, Abraham S, Caparon M, Falk P, St Geme JW 3rd, Normark S (1993) Pilus and nonpilus bacterial adhesins: assembly and function in cell recognition. Cell 73:887–901
35. de Graaf FK (1990) Genetics of adhesive fimbriae of intestinal Escherichia coli. Curr Top Microbiol Immunol 151:29–53

36. Razin S, Jacobs E (1992) Mycoplasma adhesion. J Gen Microbiol 138:407–422
37. Ruiz N, Kahne D, Silhavy TJ (2006) Advances in understanding bacterial outer-membrane biogenesis. Nat Rev Microbiol 4:57–66
38. Ruiz N, Kahne D, Silhavy TJ (2009) Transport of lipopolysaccharide across the cell envelope: the long road of discovery. Nat Rev Microbiol 7:677–683
39. Hendrickx AP, Budzik JM, Oh SY, Schneewind O (2011) Architects at the bacterial surface—sortases and the assembly of pili with isopeptide bonds. Nat Rev Microbiol 9:166–176
40. Silhavy TJ, Kahne D, Walker S (2010) The bacterial cell envelope. Cold Spring Harb Perspect Biol 2:a000414
41. Swoboda JG, Campbell J, Meredith TC, Walker S (2010) Wall teichoic acid function, biosynthesis, and inhibition. Chembiochem 11:35–45
42. Kline KA, Dodson KW, Caparon MG, Hultgren SJ (2010) A tale of two pili: assembly and function of pili in bacteria. Trends Microbiol 18:224–232
43. Ricci DP, Silhavy TJ (2012) The bam machine: a molecular cooper. Biochim Biophys Acta 1818:1067–1084
44. Benson SA, Hall MN, Silhavy TJ (1985) Genetic analysis of protein export in Escherichia coli K12. Annu Rev Biochem 54:101–134
45. Emr SD, Schwartz M, Silhavy TJ (1978) Mutations altering the cellular localization of the phage lambda receptor, an Escherichia coli outer membrane protein. Proc Natl Acad Sci U S A 75:5802–5806
46. Hagan CL, Silhavy TJ, Kahne D (2011) beta-Barrel membrane protein assembly by the Bam complex. Annu Rev Biochem 80:189–210
47. Walther DM, Rapaport D, Tommassen J (2009) Biogenesis of beta-barrel membrane proteins in bacteria and eukaryotes: evolutionary conservation and divergence. Cell Mol Life Sci 66:2789–2804
48. Majdalani N, Gottesman S (2005) The Rcs phosphorelay: a complex signal transduction system. Annu Rev Microbiol 59:379–405
49. MacRitchie DM, Buelow DR, Price NL, Raivio TL (2008) Two-component signaling and gram negative envelope stress response systems. Adv Exp Med Biol 631:80–110
50. Vogt SL, Raivio TL (2012) Just scratching the surface: an expanding view of the Cpx envelope stress response. FEMS Microbiol Lett 326:2–11
51. Jordan S, Hutchings MI, Mascher T (2008) Cell envelope stress response in Gram-positive bacteria. FEMS Microbiol Rev 32:107–146
52. Ades SE (2008) Regulation by destruction: design of the sigmaE envelope stress response. Curr Opin Microbiol 11:535–540
53. Porter SL, Wadhams GH, Armitage JP (2011) Signal processing in complex chemotaxis pathways. Nat Rev Microbiol 9:153–165
54. Lertsethtakarn P, Ottemann KM, Hendrixson DR (2011) Motility and chemotaxis in campylobacter and helicobacter. Annu Rev Microbiol 65:389–410
55. Chevance FF, Hughes KT (2008) Coordinating assembly of a bacterial macromolecular machine. Nat Rev Microbiol 6:455–465
56. Hanson BR, Neely MN (2012) Coordinate regulation of Gram-positive cell surface components. Curr Opin Microbiol 15:204–210
57. Nikaido H (2003) Molecular basis of bacterial outer membrane permeability revisited. Microbiol Mol Biol Rev 67:593–656
58. Weidenmaier C, Peschel A (2008) Teichoic acids and related cell-wall glycopolymers in Gram-positive physiology and host interactions. Nat Rev Microbiol 6:276–287
59. Jarlier V, Nikaido H (1990) Permeability barrier to hydrophilic solutes in Mycobacterium chelonei. J Bacteriol 172:1418–1423
60. Jarlier V, Nikaido H (1994) Mycobacterial cell wall: structure and role in natural resistance to antibiotics. FEMS Microbiol Lett 123:11–18
61. Lambert PA (2002) Cellular impermeability and uptake of biocides and antibiotics in Gram-positive bacteria and mycobacteria. J Appl Microbiol 92(Suppl):46S–54S
62. Scherrer R, Cabrera Beaman T, Gerhardt P (1971) Macromolecular sieving by the dormant spore of Bacillus cereus. J Bacteriol 108:868–873
63. Li XZ, Nikaido H (2004) Efflux-mediated drug resistance in bacteria. Drugs 64:159–204
64. Li XZ, Nikaido H (2009) Efflux-mediated drug resistance in bacteria: an update. Drugs 69:1555–1623
65. Jones CM, Niederweis M (2010) Role of porins in iron uptake by Mycobacterium smegmatis. J Bacteriol 192:6411–6417
66. Niederweis M (2008) Nutrient acquisition by mycobacteria. Microbiology 154:679–692
67. Camara M, Boulnois GJ, Andrew PW, Mitchell TJ (1994) A neuraminidase from Streptococcus pneumoniae has the features of a surface protein. Infect Immun 62:3688–3695

68. Burnaugh AM, Frantz LJ, King SJ (2008) Growth of Streptococcus pneumoniae on human glycoconjugates is dependent upon the sequential activity of bacterial exoglycosidases. J Bacteriol 190:221–230
69. King SJ, Hippe KR, Weiser JN (2006) Deglycosylation of human glycoconjugates by the sequential activities of exoglycosidases expressed by Streptococcus pneumoniae. Mol Microbiol 59:961–974
70. Marion C, Burnaugh AM, Woodiga SA, King SJ (2011) Sialic acid transport contributes to pneumococcal colonization. Infect Immun 79:1262–1269
71. van Hoek AH, Mevius D, Guerra B, Mullany P, Roberts AP, Aarts HJ (2011) Acquired antibiotic resistance genes: an overview. Front Microbiol 2:203
72. Nobbs AH, Jenkinson HF, Jakubovics NS (2011) Stick to your gums: mechanisms of oral microbial adherence. J Dent Res 90:1271–1278
73. Aoki SK, Pamma R, Hernday AD, Bickham JE, Braaten BA, Low DA (2005) Contact-dependent inhibition of growth in Escherichia coli. Science 309:1245–1248
74. Jarrell KF, McBride MJ (2008) The surprisingly diverse ways that prokaryotes move. Nat Rev Microbiol 6:466–476
75. Kaiser D (2007) Bacterial swarming: a re-examination of cell-movement patterns. Curr Biol 17:R561–R570
76. Engelhardt H (2007) Are S-layers exoskeletons? The basic function of protein surface layers revisited. J Struct Biol 160:115–124
77. Mauriello EM, Mignot T, Yang Z, Zusman DR (2010) Gliding motility revisited: how do the myxobacteria move without flagella? Microbiol Mol Biol Rev 74:229–249
78. Li G, Brown PJ, Tang JX, Xu J, Quardokus EM, Fuqua C, Brun YV (2012) Surface contact stimulates the just-in-time deployment of bacterial adhesins. Mol Microbiol 83:41–51
79. Desvaux M, Hebraud M, Talon R, Henderson IR (2009) Secretion and subcellular localizations of bacterial proteins: a semantic awareness issue. Trends Microbiol 17:139–145
80. Ellis TN, Kuehn MJ (2010) Virulence and immunomodulatory roles of bacterial outer membrane vesicles. Microbiol Mol Biol Rev 74:81–94
81. Vollmer W, Seligman SJ (2010) Architecture of peptidoglycan: more data and more models. Trends Microbiol 18:59–66
82. Hayhurst EJ, Kailas L, Hobbs JK, Foster SJ (2008) Cell wall peptidoglycan architecture in Bacillus subtilis. Proc Natl Acad Sci U S A 105:14603–14608
83. Mandlik A, Swierczynski A, Das A, Ton-That H (2008) Pili in Gram-positive bacteria: assembly, involvement in colonization and biofilm development. Trends Microbiol 16:33–40
84. Schroder NW, Eckert J, Stubs G, Schumann RR (2008) Immune responses induced by spirochetal outer membrane lipoproteins and glycolipids. Immunobiology 213:329–340
85. Schroder NW, Schombel U, Heine H, Gobel UB, Zahringer U, Schumann RR (2003) Acylated cholesteryl galactoside as a novel immunogenic motif in Borrelia burgdorferi sensu stricto. J Biol Chem 278:33645–33653
86. Tokuda H, Matsuyama S (2004) Sorting of lipoproteins to the outer membrane in *E. coli*. Biochim Biophys Acta 1694:IN1–IN9
87. Narita S (2011) ABC transporters involved in the biogenesis of the outer membrane in gram-negative bacteria. Biosci Biotechnol Biochem 75:1044–1054
88. Cowles CE, Li Y, Semmelhack MF, Cristea IM, Silhavy TJ (2011) The free and bound forms of Lpp occupy distinct subcellular locations in Escherichia coli. Mol Microbiol 79:1168–1181
89. Epstein EA, Reizian MA, Chapman MR (2009) Spatial clustering of the curlin secretion lipoprotein requires curli fiber assembly. J Bacteriol 191:608–615
90. Chen S, Kumru OS, Zuckert WR (2011) Determination of Borrelia surface lipoprotein anchor topology by surface proteolysis. J Bacteriol 193:6379–6383
91. Fairman JW, Noinaj N, Buchanan SK (2011) The structural biology of beta-barrel membrane proteins: a summary of recent reports. Curr Opin Struct Biol 21:523–531
92. Dong C, Beis K, Nesper J, Brunkan-Lamontagne AL, Clarke BR, Whitfield C, Naismith JH (2006) Wza the translocon for E. coli capsular polysaccharides defines a new class of membrane protein. Nature 444:226–229
93. Grodberg J, Dunn JJ (1988) ompT encodes the Escherichia coli outer membrane protease that cleaves T7 RNA polymerase during purification. J Bacteriol 170:1245–1253
94. Fu LM, Fu-Liu CS (2002) Is Mycobacterium tuberculosis a closer relative to Gram-positive or Gram-negative bacterial pathogens? Tuberculosis (Edinb) 82:85–90
95. Fu LM, Fu-Liu CS (2002) Genome comparison of Mycobacterium tuberculosis and other bacteria. OMICS 6:199–206
96. Nikaido H, Jarlier V (1991) Permeability of the mycobacterial cell wall. Res Microbiol 142:437–443

97. Minnikin DE (1982) Lipids: complex lipids, their chemistry, biosynthesis and roles. In: Ratledge C, Stanford J (eds) The biology of mycobacteria. Physiology, identification and classification. Academic, New York, pp 94–184
98. McNeil MR, Brennan PJ (1991) Structure, function and biogenesis of the cell envelope of mycobacteria in relation to bacterial physiology, pathogenesis and drug resistance; some thoughts and possibilities arising from recent structural information. Res Microbiol 142:451–463
99. Nikaido H, Kim SH, Rosenberg EY (1993) Physical organization of lipids in the cell wall of Mycobacterium chelonae. Mol Microbiol 8:1025–1030
100. Zuber B, Chami M, Houssin C, Dubochet J, Griffiths G, Daffe M (2008) Direct visualization of the outer membrane of mycobacteria and corynebacteria in their native state. J Bacteriol 190:5672–5680
101. Hoffmann C, Leis A, Niederweis M, Plitzko JM, Engelhardt H (2008) Disclosure of the mycobacterial outer membrane: cryo-electron tomography and vitreous sections reveal the lipid bilayer structure. Proc Natl Acad Sci U S A 105:3963–3967
102. Bansal-Mutalik R, Nikaido H (2011) Quantitative lipid composition of cell envelopes of Corynebacterium glutamicum elucidated through reverse micelle extraction. Proc Natl Acad Sci U S A 108:15360–15365
103. Trias J, Jarlier V, Benz R (1992) Porins in the cell wall of mycobacteria. Science 258:1479–1481
104. Faller M, Niederweis M, Schulz GE (2004) The structure of a mycobacterial outer-membrane channel. Science 303:1189–1192
105. Fraser CM, Gocayne JD, White O, Adams MD, Clayton RA, Fleischmann RD, Bult CJ, Kerlavage AR, Sutton G, Kelley JM, Fritchman RD, Weidman JF, Small KV, Sandusky M, Fuhrmann J, Nguyen D, Utterback TR, Saudek DM, Phillips CA, Merrick JM, Tomb JF, Dougherty BA, Bott KF, Hu PC, Lucier TS, Peterson SN, Smith HO, Hutchison CA 3rd, Venter JC (1995) The minimal gene complement of Mycoplasma genitalium. Science 270:397–403
106. Rottem S (2002) Choline-containing lipids in mycoplasmas. Microbes Infect 4:963–968
107. Razin S, Yogev D, Naot Y (1998) Molecular biology and pathogenicity of mycoplasmas. Microbiol Mol Biol Rev 62:1094–1156
108. Uenoyama A, Miyata M (2005) Gliding ghosts of Mycoplasma mobile. Proc Natl Acad Sci U S A 102:12754–12758
109. Jaffe JD, Miyata M, Berg HC (2004) Energetics of gliding motility in Mycoplasma mobile. J Bacteriol 186:4254–4261
110. Krause DC, Balish MF (2001) Structure, function, and assembly of the terminal organelle of Mycoplasma pneumoniae. FEMS Microbiol Lett 198:1–7
111. Biberfeld G, Biberfeld P (1970) Ultrastructural features of Mycoplasma pneumoniae. J Bacteriol 102:855–861
112. Miyata M (2010) Unique centipede mechanism of Mycoplasma gliding. Annu Rev Microbiol 64:519–537
113. Yother J (2011) Capsules of Streptococcus pneumoniae and other bacteria: paradigms for polysaccharide biosynthesis and regulation. Annu Rev Microbiol 65:563–581
114. Sambou T, Dinadayala P, Stadthagen G, Barilone N, Bordat Y, Constant P, Levillain F, Neyrolles O, Gicquel B, Lemassu A, Daffe M, Jackson M (2008) Capsular glucan and intracellular glycogen of Mycobacterium tuberculosis: biosynthesis and impact on the persistence in mice. Mol Microbiol 70:762–774
115. Cuthbertson L, Mainprize IL, Naismith JH, Whitfield C (2009) Pivotal roles of the outer membrane polysaccharide export and polysaccharide copolymerase protein families in export of extracellular polysaccharides in gram-negative bacteria. Microbiol Mol Biol Rev 73:155–177
116. Lemassu A, Ortalo-Magne A, Bardou F, Silve G, Laneelle MA, Daffe M (1996) Extracellular and surface-exposed polysaccharides of non-tuberculous mycobacteria. Microbiology 142(Pt 6):1513–1520
117. Houwink AL (1953) A macromolecular mono-layer in the cell wall of Spirillum spec. Biochim Biophys Acta 10:360–366
118. Messner P, Steiner K, Zarschler K, Schaffer C (2008) S-layer nanoglycobiology of bacteria. Carbohydr Res 343:1934–1951
119. Ristl R, Steiner K, Zarschler K, Zayni S, Messner P, Schaffer C (2011) The s-layer glycome-adding to the sugar coat of bacteria. Int J Microbiol 2011.pii:127870
120. Sleytr UB, Thorne KJ (1976) Chemical characterization of the regularly arranged surface layers of Clostridium thermosaccharolyticum and Clostridium thermohydrosulfuricum. J Bacteriol 126:377–383
121. Zhao G, Ali E, Sakka M, Kimura T, Sakka K (2006) Binding of S-layer homology modules from Clostridium thermocellum SdbA to peptidoglycans. Appl Microbiol Biotechnol 70:464–469
122. Lounatmaa K, Brander E (1989) Crystalline cell surface layer of Mycobacterium bovis BCG. J Bacteriol 171:5756–5758

123. Sleytr UB, Schuster B, Egelseer EM, Pum D, Horejs CM, Tscheliessnig R, Ilk N (2011) Nanobiotechnology with s-layer proteins as building blocks. Prog Mol Biol Transl Sci 103:277–352

124. Chen S, Beeby M, Murphy GE, Leadbetter JR, Hendrixson DR, Briegel A, Li Z, Shi J, Tocheva EI, Muller A, Dobro MJ, Jensen GJ (2011) Structural diversity of bacterial flagellar motors. EMBO J 30:2972–2981

Chapter 2

Visualizing the Bacterial Cell Surface: An Overview

Harald Engelhardt

Abstract

The ultrastructure of bacteria is only accessible by electron microscopy. Our insights into the architecture of cells and cellular compartments such as the envelope and appendages is thus dependent on the progress of preparative and imaging techniques in electron microscopy. Here, I give a short overview of the development and characteristics of methods applied for imaging (components of) the bacterial surface and refer to key investigations and exemplary results. In the beginning of electron microscopy, fixation of biological material and staining for contrast enhancement were the standard techniques. The results from freeze-etching, metal shadowing and from ultrathin-sections of plastic-embedded material shaped our view of the cellular organization of bacteria. The introduction of cryo-preparations, keeping samples in their natural environment, and three-dimensional (3D) electron microscopy of isolated protein complexes and intact cells opened the door to a new dimension and has provided insight into the native structure of macromolecules and the in situ organization of cells at molecular resolution. Cryo-electron microscopy of single particles, together with other methods of structure determination, and cellular cryo-electron tomography will provide us with a quasi-atomic model of the bacterial cell surface in the years to come.

Key words: Correlative microscopy, Cryo-electron microscopy, Cryo-electron tomography, Cryo-sectioning, Electron microscopy, Focused ion beam micromachining (FIB), Freeze-etching, Metal shadowing, Negative staining, Single particles, Template matching

1. From Staining to Visualization

Bacteria came into virtual existence in 1683 when Antoni van Leeuwenhoek detected *viva animalcula* in dental plaques by means of his self-made microscopes (1). Since then, microbiology has been intertwined with microscopy; and even in recent years, achievements of modern fluorescence microscopy and cryo-electron tomography founded a new (or revived an atrophied) discipline in microbiology, namely microbial cell biology. Investigations of the cellular organization and its compartments with new microscopies also shed light on the architecture of the microbial cell surface,

Anne H. Delcour (ed.), *Bacterial Cell Surfaces: Methods and Protocols*, Methods in Molecular Biology, vol. 966,
DOI 10.1007/978-1-62703-245-2_2, © Springer Science+Business Media New York 2013

and of the bacterial cell envelope in particular, that is still being explored. 130 years ago and almost exactly 200 years after the initial observation of bacteria, Robert Koch developed a staining method to visualize and differentiate the causing agent of tuberculosis. Paul Ehrlich who improved the method shortly after interpreted the acid-resistant staining behavior as a specific property of the cell envelope (2) and thus realized that bacteria may have different cell wall compositions. The staining introduced by Hans Christian Gram in 1884 (3) had even more impact. It distinguished between two larger groups of microorganisms and was not only taken as indicative for the cell wall permeability (4) and thickness (5, 6) a number of decades later, but also for the apparent cell wall architecture of *gram-positive* (thick peptidoglycan, no outer membrane) and *gram-negative* bacteria (thin peptidoglycan, outer membrane in addition). However, cell envelopes turned out to be more diverse and the behavior of Gram staining often led to misinterpretations of the cell wall composition. The extreme chemical variability of the cell surface (mainly of lipopolysaccharides and flagellar antigens) became apparent by serological typing of *Salmonella* strains introduced in 1926 (7) and its systematic documentation based on the Kauffmann–White scheme (8), which meanwhile lists more than 2,500 entries (9). Staining and labeling methods for light microscopy have been and still are of diagnostic value but they could hardly reveal structural details of the bacterial cell surface because of the limited resolution power of light optical instruments.

The situation changed with the invention of transmission electron microscopy (EM) in 1932 (10), the first electron microscopical images of bacteria encouraging further studies (11, 12), and the development of appropriate preparation and imaging techniques (still in progress) that provided structural insights and prompted the research on microbial cell envelopes. Early investigations still employed staining methods known from cytology of eukaryotes (13). The era of fruitful investigations began in the early 1950s (14, 15), flourished in the 1960s and later when ultrathin sectioning had become a common method, and lasts until today, now profiting from cryo-electron microscopy of vitrified and unstained biological material in a close-to-native state. Any progress in the visualization of cell envelopes, their internal structure, and of its appendages preceded technical and preparative innovations. Major developments of EM after its invention (three-dimensional (3D) EM, cryo-electron tomography [CET]) occurred in about 30-years steps, as well as methods for specimen preparation (ultrathin sectioning, cryo-sectioning, cryo-focused-ion-beam micromachining [cryo-FIB]). Other significant improvements for imaging biological material, i.e., contrast and resolution enhancements (heavy metal shadowing, negative staining, EM of unstained material, zero-loss energy filtering, imaging correction by phase-plates) showed a periodicity of about 15 years, so that we observe a minor

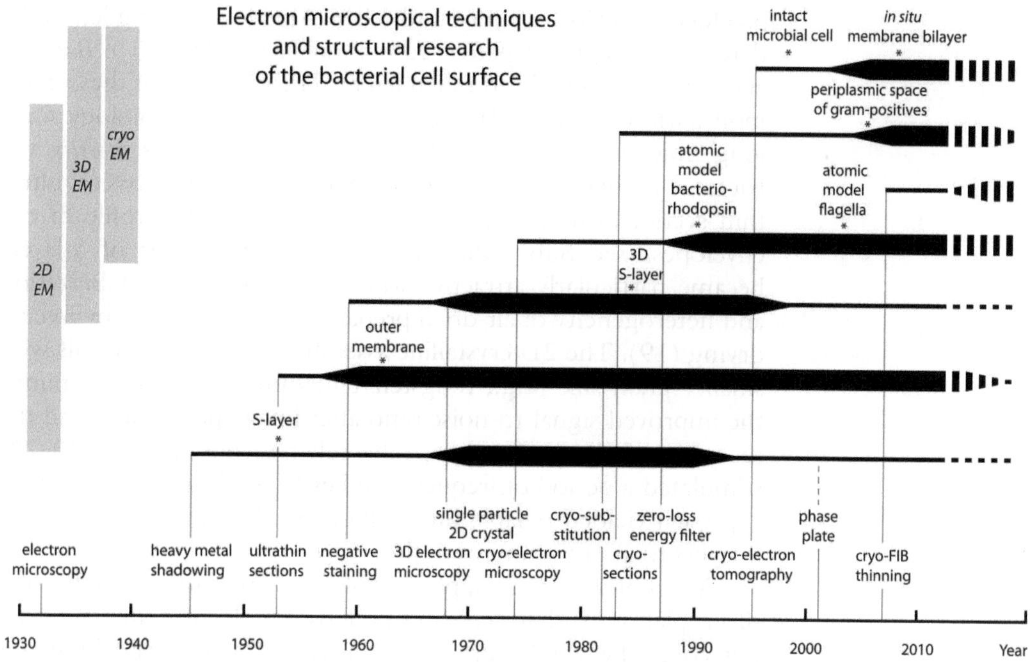

Fig. 1. Technical and preparative developments in transmission electron microscopy and their impact on structural research of microbial cell envelopes and their constituents. The *thickness* of time lines indicates the importance of methods. *Dashed lines* denote assessments for future applications. Some key discoveries in microbial surface structure research are indicated.

or major innovation step every 5–9 years that is of importance for structure research in (micro)biology (Fig. 1). The evolution of EM techniques was of course accompanied and partly driven by the immense progress of computational performance and efficient software solutions for image processing. The route led from chemically fixed, stained, and dehydrated specimens to native and vitrified preparations devoid of artificial modifications, from 2D projections to 3D reconstructions, and from low resolution images to quasi-atomic models for representative structures. Some important events of EM cell surface research are indicated in Fig. 1.

This chapter gives an overview of electron microscopical approaches, which have been of particular importance for structural research in microbiology, highlights its contributions to our understanding of the microbial cell surface, and outlines the perspectives for the near future.

2. Looking at Surfaces

2.1. Heavy-Metal Shadowing

Metal coating ("shadowing") of isolated and dried cell envelopes was one of the first systematically applied contrasting methods (16, 17) when it had become clear that biological material produced only low contrast in the EM. Metal shadowing led to the detection of regularly patterned cell wall layers (15, 18) in *Bacteria* and *Archaea* (as realized much later). The existence of these common surface layers (S-layers) was ignored in microbiology for a long time since neither *Escherichia coli* nor *Bacillus subtilis*, the bacterial model organisms, possess S-layers, and the classical ultrathin sections often failed to display the complete profile of cell envelopes (see Subheading 3.1). Metal shadowing of S-layers became particularly attractive when the structural destabilization and heterogeneity of air-dried preparations was avoided by freeze-drying (19). The 2D crystalline assemblies, the use of metals with smaller grain size (e.g., tungsten–tantalum instead of platinum), the improved signal-to-noise ratio after image processing, and the recovery of the surface shape by relief reconstruction (20–22) stimulated a period of frequent studies (Figs. 1 and 2). The latter approach yielded reliable information on the surface structure of S-layers to about 2 nm resolution (23).

A technical variant, applying only low amounts of metal (gold, silver, platinum) that decorates preferred molecular sites but does not cover the whole surface, was used to identify and highlight periodic structures in freeze-dried or freeze-etched protein assemblies (24, 25). Metal shadowing and decoration of isolated material lost its appeal when 3D electron microscopy came into play (see Subheading 4). Today, contrasting by metal coating is used for some special applications only.

2.2. Freeze-Etching and Freeze-Fracturing

Cells, frozen as a whole, freed from ice by deep-etching (sublimation above −100°C), contrasted by metal coating, and finally prepared for EM by the replica technique (26–28), show their very surface. For a long time, this was the only method to look at cells from outside. Again, it was a useful approach for identifying and characterizing the outermost layer of bacterial cells. By this way the orientation of S-layers could be determined and comparisons with the isolated layer and its 3D reconstruction revealed conformational variations (upon isolation). Although informative, such comprehensive investigations remained scarce (Fig. 2, (29)).

Freeze-fracturing—where the frozen cells are broken prior to etching and metal coating—not only displayed cell surfaces but also part of the envelope profile. Since fracture faces propagate along surfaces and hydrophobic ones in particular, often the inner or outer leaflet of membranes was visible and thus indicative of lipid membranes (30).

2.3. Scanning Electron and Scanning Probe Microscopy

Imaging of surface structures from abiotic and biological material is an original domain of scanning electron microscopy (SEM). This approach usually also employs metal coating and is limited in lateral resolution for several reasons so that molecular details of cell surfaces can hardly be investigated (31). Therefore, SEM does

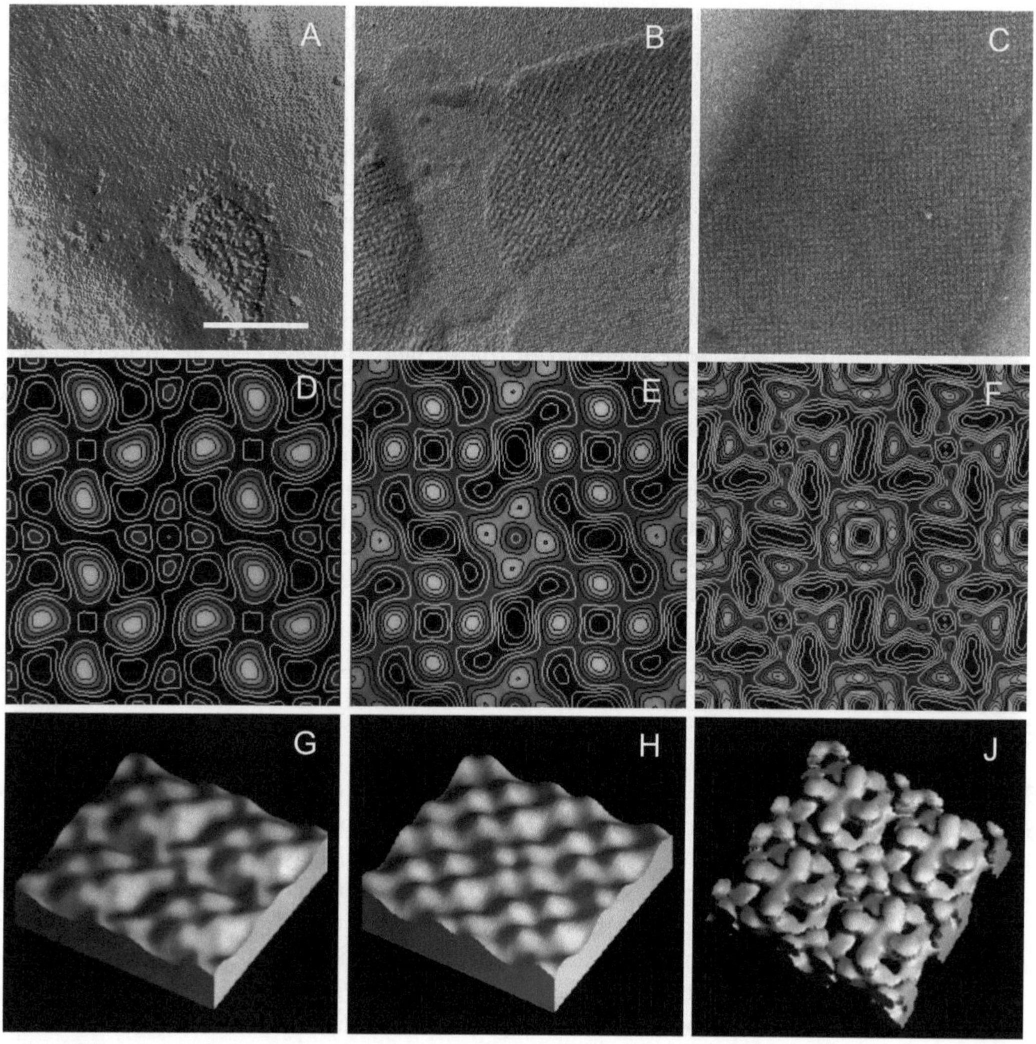

Fig. 2. Staining and reconstructions of the S-layer from *Sporosarcina ureae*. (**a**, **d**, **g**) Freeze-etched cell unidirectionally shadowed by tungsten-tantalum. (**b**, **d**, **h**) Isolated S-layer unidirectionally shadowed by tungsten-tantalum. (**c**, **f**, **j**) Isolated S-layer negatively stained with Na/K-phosphotungstate. (**d**, **e**) Surface relief reconstruction of the outer surface after averaging of unit cells; *contour lines* indicate identical levels of the relief hight (*black* areas are valleys). The central tetragonal domain has a different height in the isolated S-layer and indicates a conformational change upon detachment from the cell wall. (**f**) Averaged S-layer lattice; *black* areas indicate stain-filled gaps, *bright* areas represent protein mass. (**g**, **h**) Models of reliefs from the outer surface. (**j**) Three-dimensional (3D) (isosurface) model of the S-layer based on tilt-series data. Scale bar indicates 200 nm (valid for **a**–**c**). The averages and models (**d**–**j**) show four unit cells each (lattice constant 12.9 nm).

not play a prominent role in studies of the bacterial surface although low-voltage SEM promises interesting applications in biology (32).

In the 1980s, a novel type of microscopy was invented and introduced a fascinating approach to gain topographical and spectroscopic information from surfaces, i.e., scanning probe microscopy (SPM) (33, 34). Atomic force microscopy (AFM) of fully hydrated and uncoated protein assemblies, once more of S-layers

(35, 36), and membranous samples containing densely packed protein complexes (2D crystalline porins (37)), yielded topographical images of high signal-to-noise ratio, so that single complexes could be inspected. AFM has an impressive z-resolution down to 1 Å and a subnanometer resolution in *x–y*-direction, depending on the tip geometry and the structure of the sample (38, 39). Functionalization of the tip and particular scanning modes allow for probing various molecular features such as polarity and charges (40), hydrophobicity, elasticity or roughness (41), and can be applied to probe molecular forces for binding and unfolding (42, 43). Investigations of the surface of intact cells are possible (44); but—according to a hidden law—the larger the object under the tip the weaker the resolution of details. The preferred domain of SPM for topographical investigations is flat specimens such as membranes and densely packed proteins with moderate corrugation that firmly adhere to the supporting substrate. Similar to other approaches probing *surface* characteristics, scanning microscopies are complementary to methods for *3D structure* analysis.

3. Profile of Cell Envelopes

3.1. Ultrathin Sectioning: Looking into Microbes

Intact bacteria, either directly dried or soaked with (heavy) metals were not suited for EM because of the extremely low or high contrast, preventing imaging of details (45). The solution to this problem was thin-sectioning, a technique that was already known from preparations of tissues. However, the sections had to be ultrathin (≤100 nm) and this required the development of new microtomes (14, 46, 47). In the 1960s and 1970s the common methods for fixation (glutaraldehyde), dehydration (water-ethanol exchange), staining (OsO_4, $Pb(OH)_2$, ruthenium red), embedding (epoxid resins), and polymerization (60°C) were established and represented the basis for most of the cytological investigations in the following two decades. In the early 1960s, the existence of an outer membrane became apparent (48–50) and the characteristic difference of the cell wall thickness of gram-positive and gram-negative bacteria was established. The development and peculiarity of the architectural layers cytoplasmic membrane, peptidoglycan, outer membrane, polysaccharide capsules (slime), and sometimes other (unidentified) components were investigated and shaped our view of the composition and variability of the bacterial cell envelope (31, 51). Interestingly, S-layers were only rarely detected or recognized in ultrathin sections (52). One of the reasons was that they usually did not show up very clearly—with some remarkable exceptions (53–55)—and another one that microbiologists were not prepared to *perceive* this structure (56). A systematic investigation of the staining behavior of ultrathin sections from *Deinococcus*

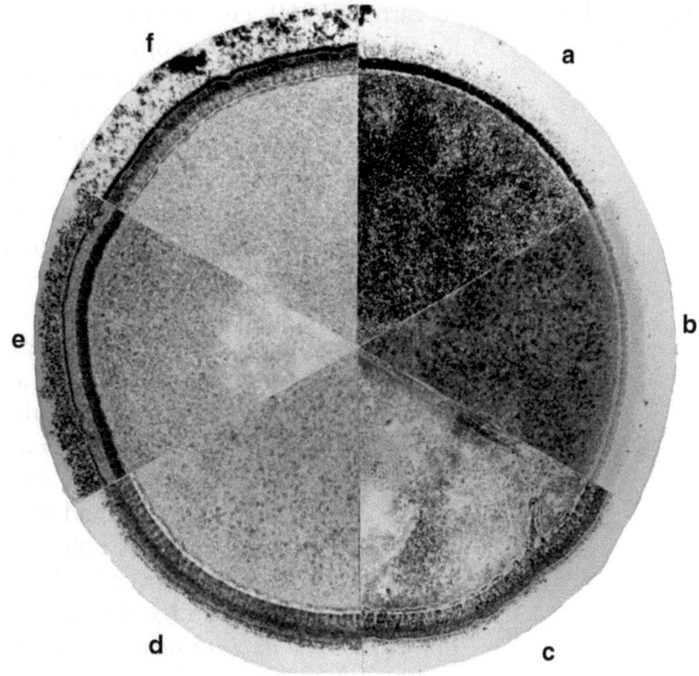

Fig. 3. Effect of different fixation and staining conditions for ultrathin sections of *Deinococcus radiodurans.* The cells were treated with (**a**) OsO_4 and PbO, (**b**) glutaraldehyde, OsO_4, and $Pb(OH)_2$, (**c**) tannin, OsO_4, and $Pb(OH)_2$, (**d**) tannin, glutaraldehyde, OsO_4, and $Pb(OH)_2$, (**e**) glutaraldehyde, OsO_4, ruthenium red, and $Pb(OH)_2$, (**f**) tannin, glutaraldehyde, OsO_4, ruthenium red, and $Pb(OH)_2$. All samples were embedded in epoxide resin and polymerized at 60°C. Most of the sections show the thick fenestrated peptidoglycan, but only the section in (**d**) indicates the existence of an outermost layer possessing a periodic structure (S-layer "HPI") (Courtesy of W. Baumeister, Martinsried).

radiodurans—possessing a thick fenestrated peptidoglycan, an outer membrane, an S-layer, and a polysaccharide capsule (57)—illustrates the impact of preparation conditions for the visualization of cell envelope components and for the interpretation of the cellular architecture (Fig. 3).

Another problem was the harsh preparation conditions that were suspected to alter the appearance of morphological details and to denature protein complexes. The structural integrity of cells was improved by cryo-substitution and new embedding materials, polymerizing in the cold (58). Hence, proteins were conserved better and could be localized by immuno-EM easier than before (59). Today, this is the standard procedure for bacteria, ideally being combined with high-pressure freezing for improved structural preservation. However, substitution of cellular water and staining, even in the cold, could still not be expected to keep the organization of macromolecules untouched. If a near-to-life situation has to be guaranteed there is no way other than freezing the samples and keeping them frozen in the microscope without any chemical treatment.

3.2. Cryo-Sectioning: Pure Nature

As early as 1960, Humberto Fernández-Morán explored methods for cryo-electron microscopy (60). Yet, it took another two decades until basic experiments encouraged the development of cryo-sectioning to a reproducible and reliable technique (61–63). As an important prerequisite, Jaques Dubochet demonstrated that quick freezing prevents water from forming ice crystals that would destroy the (molecular structure of) biological samples otherwise; water could be ***vitrified*** (64). Vitrification is possible by plunge-freezing (fast injection of samples into a cryogen such as ethane at ≈–180°C (65)) if the specimen thickness does not exceed ≈10 μm. Thicker samples require freezing at high pressure (about 2,000 bar) that lowers the freezing rate of water and suppresses the formation of cubic ice up to a freezing depth of ≈100 μm (66). Cryo-sectioning turned out to be a demanding technique that found its way into other laboratories only slowly. But ultrathin sections of native frozen-hydrated cells, and imaging based on the contrast of untreated biological material only, almost instantly succeeded to clarify open questions of the cell wall architecture in bacteria. Valério Matias and Terry Beveridge demonstrated a periplasmic space in the gram-positives *Bacillus subtilis* and *Staphylococcus aureus* (67, 68), and two other groups showed the bilayer structure of the peculiar outer membrane in mycobacteria and corynebacteria (69, 70), which was not obvious in conventional preparations (71, 72).

Cryo-sectioning is not free of artifacts. Forces acting during the cutting process have various consequences for the frozen sections (73, 74). The most severe disadvantage is compression in the cutting direction that affects the morphology of the cell envelope and the bilayer structure of membranes (Fig. 4, (69)). Although it is

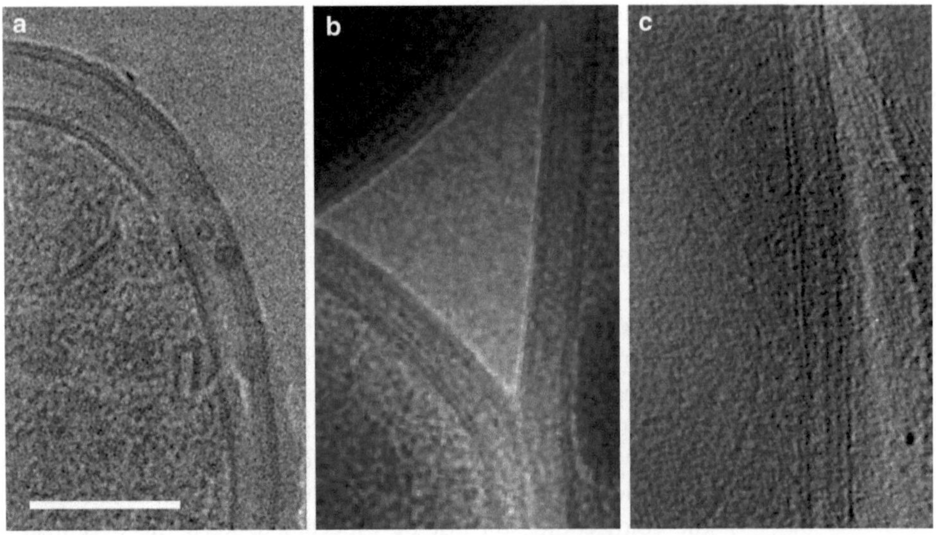

Fig. 4. Mycobacterial cell envelope in cryo-electron microscopy. (**a**) Cryo-section (thickness ≈35 nm), the membrane bilayer structure is preserved perpendicular to and compressed in the cutting direction (from the *right*). (**b**) EM projection of a cell thinned by cryo-FIB milling; the bilayer is visible in all orientations but displayed with lower contrast due to the specimen thickness (≈300 nm). (**c**) Cryo-electron tomography of an intact cell, *x*–*y*-slice from the tomogram. Bar indicates 100 nm.

possible to detect periplasmic protein complexes in less squeezed regions of the cell (75) a thinning method without compression effects was desirable for studies of microbial cell envelopes.

3.3. Cryo-Focused Ion Beam Micromachining (Cryo-FIB): Thinning Without Artifacts

Dual beam scanning electron microscopes are equipped with an electron gun and an ion gun (gallium) in addition that is used to abrase material from specimens in a controlled manner. This milling technique is wellknown from material science and was applied to plastic-embedded cells and tissues for imaging by SEM (76, 77). Michael Marko and colleagues described the first experiments with frozen-hydrated bacteria thinned by FIB milling for further investigation in the EM (78, 79). Cryo-FIB has meanwhile been optimized for 2D and 3D electron microscopy and was shown to completely avoid compression of frozen samples and membrane structures (Fig. 4, (80, 81)). Specimen thinning by FIB micromachining will likely become a key technique for the investigation of prokaryotic (and eukaryotic) cells and for cryo-electron tomography in particular (see Subheading 5).

Methods for imaging surfaces and EM of ultrathin sections essentially provide structural information in two dimensions only, either in *x*–*y*- (surfaces) or in *x*–*z*-direction (thin sections). The development of 3D techniques, covering all dimensions (*x*, *y*, *z*), thus denoted a major step in EM.

4. Macromolecules and Protein Assemblies in Three Dimensions

4.1. Negative Staining of Isolated Protein Complexes

Metal-coating by thermo-electric evaporation provided sufficient contrast for EM but was not suited to yield structural information from all orientations of isolated macromolecules and protein assemblies. The idea to embed particles in heavy metal salts such as Na/K- phosphotungstate (82) and others introduced later (see ref. 83 for a compilation) opened the new field of molecular EM. The technique of negative staining avoided drastic flattening on the specimen support upon drying and stabilized the protein complexes in the electron beam (83, 84). Negatively stained virus capsids (85) and large enzymes (86, 87) were the first molecular specimens that could be reconstructed in 3D. While the investigation of a single negatively stained protein molecule was insufficient, reconstructions including averaging procedures provoked the development of 3D electron microscopy (88) and image processing systems (89–93). Averaging exploits the structural redundancy of regular assemblies such as filamentous and 2D crystalline structures, or uses high numbers of individual molecules, imaged from different projection directions, in single particle approaches.

Negative staining not only opened the door to the third dimension in EM but also improved the resolution of molecular details. The apparent resolution limit of 1.3 nm (83) corresponds to the diameter of a globular protein domain consisting of eight amino acids (94). This was good enough to get insight into the architecture of filamentous and 2D crystalline protein assemblies. 3D reconstructions of S-layers showed common architectural principles and the variability of the outer surface (95–98) and inspired the search for individual and general functions (29, 99, 100). Studies on major outer membrane proteins in isolated cell walls and reconstituted in artificial 2D crystals clarified the arrangement of pores in trimeric porins (101, 102) before cryo-electron microscopy pushed the resolution limit further (103, 104) and X-ray crystallography succeeded to solve the atomic structure (105). Major contributions to the basic understanding of bacterial appendages concerned the helical organization of flagellins in the flagellar filament and hook (106, 107) and of spirillin in a large basal disk of the *Wolinella succinogenes* flagellar apparatus forming an Archimedian spiral (108).

Although the 3D reconstruction of negatively stained protein complexes and membranous structures was a major step in EM, and the future of the contrast-enhancing technique was regarded optimistic in the early 1990 (83), its application has several drawbacks. (1) Negative staining only offers access to the *shape* of molecules to about 1.3–1.7 nm resolution, but not to its intrinsic structure. (2) Complexes embedded in membranes cannot be stained, only their domains outside of the lipid bilayer and large pores. (3) Artificial variability of protein complexes introduced by embedding in stain, i.e., in high salt concentrations upon drying, and accompanying effects of positive staining can hardly be identified, which affects the interpretation of structural features (109). To overcome the limitations, negative staining was occasionally replaced by polyhydrated compounds (e.g., glucose, erythrose, glycerol) often conjugated with gold (aurothioglucose (110)) or other metal clusters (111). The resolution and structural preservation was considerably increased, but now the specimens became radiation sensitive and enforced low dose microscopy resulting in images of low signal-to-noise ratio. Glucose embedding was extremely successful with the 2D crystalline bacteriorhodopsin still residing in the natural lipid membrane (112). Soluble proteins required a more general approach that consequently led to cryo-electron microscopy of proteins embedded in vitreous ice without contrast-enhancing additives. Negative staining, thus, lost its importance for molecular 3D EM (Fig. 2.1) but is still esteemed for its simplicity and fast application in preparation control and for preliminary structural investigations.

4.2. Cryo-Electron Microscopy: High-Resolution Structures of Regular Assemblies and Single Particles

Frozen-hydrated samples smoothly replaced negatively stained preparations in single particle EM when the microscopes became accessible to remote control and image processing faster due to powerful computers. Although cryo-EM is much more costly and laborious, it has unbeatable advantages. (1) The image signal originates from the biological material, (2) the macromolecules are embedded in their natural environment (water) and do not experience treatments other than vitrification, (3) the resolution is not limited by any staining and can potentially reach the quasi-atomic level (113).

The low signal-to-noise ratio of images of radiation-sensitive macromolecules requires averaging of 10^3–10^6 particles, depending on the number of identical subunits per particle, the intrinsic conformational variability of protein complexes that enforces classification into structural subsets (114, 115), and the resolution desired. If proteins are organized in a regular pattern, alignment and averaging is particularly effective. This was demonstrated and exploited with the first quasi-atomic structure of a 2D crystalline membrane protein, i.e., bacteriorhodopsin (116). The next high-resolution structures were the result of the consequent application of cryo-EM with filamentous structures, where the protein units occur in a helical arrangement. The benefit of such protein filaments is that one EM image already contains projections of the repeating unit in different orientations that can be combined to a 3D electron density map (117–119). The structures of the flagellar filament (120–124) and the hook (125–127) were solved to near-atomic resolution. Some of the latter models were obtained from cryo-EM 3D-reconstructions combined with the X-ray structure of isolated (truncated) flagellin by means of molecular docking. This synergetic combination of techniques, the *hybrid approach* (128–130), has meanwhile become a common strategy to solve the atomic structure of large, flexible and very complex protein assemblies that cannot be crystallized as a whole. A number of other helical appendages (fimbriae, pili) were also investigated by cryo-EM in a similar way (131). Isolated flagellar motors and the switch complex (132–134), periplasmic and outer membrane components of, e.g., the type III (135, 136) and type IV secretion machineries (137, 138), and other periplasmic multicomponent complexes are not filamentous or helical but possess an internal rotational symmetry so that the 3D reconstruction is facilitated in the course of averaging individual particles, again leading to high-resolution models in combination with other methods of protein structure research.

The single particle approach in cryo-EM is a powerful method that will—together with X-ray crystallography—contribute to solve the structure of protein complexes and molecular machines isolated from microbial cell surfaces in the years to come.

5. Cells in Three Dimensions

5.1. Cryo-Electron Tomography (CET): The Integrative Approach

The 3D reconstruction of an individual biological structure is obtained by the tomographic approach. The specimen is tilted around an axis from ideally −90° to +90° (less in practice) and projected every 1°–3° to collect a 3D data set (tilt series). The projections are aligned and the 3D electron density map (tomogram) calculated. The basic principles are known since the late 1960s (86, 139). But only when the technical progress allowed for automatic control of electron microscopes (140, 141), electron energy filtering (142, 143), recording of data by sensitive cameras, and demanding computations, it became possible to realize the vision of reconstructing an entire cell in three dimensions. The group of Wolfgang Baumeister pioneered the development of cryo-electron tomography for frozen-hydrated cells in a close-to-life state and is continuing to improve it (65, 144–147). The major characteristics of cryo-tomograms are: (1) they contain the signatures of the entire proteome of a cell, (2) they present the positions, orientations, and interactions of macromolecular complexes in a "snapshot" of a close-to-life situation, (3) they allow to trace intracellular filaments (the cytoskeleton) and large macromolecular assemblies in 3D and to visualize the spatial organization of the cytoplasm and the cell envelope at macromolecular dimensions. Challenges are (1) the low signal-to-noise ratio of original data, (2) the resolution-limiting imaging conditions (defocus), (3) the crowded nature of intact cells with very high concentrations of macromolecules and a heterogeneous composition, making the identification of individual molecules demanding, and (4) the uniqueness of cells that prevents signal enhancement by averaging (except for redundant molecules contained in tomograms, see Subheading 5.3). A comprehensive compilation of the technical aspects of CET is found in (65). The presentation of structural details in the cellular context and in a close-to-life state provides unprecedented insight into the complete 3D organization of cells and new findings in almost all of the reconstructions of bacteria published to date. These include microbial S-layers (148), the mycobacterial outer membrane (69), extracellular membrane vesicles (149), the peptidoglycan (150), the bacterial flagellar apparatus (151–153), the membrane-bound cytoskeleton mediating cell motion in mollicutes (154), and a rapidly growing number of cytoplasmic structures; for selected reviews and articles see refs. 155–159.

It is the fate of new methods that they have to reproduce already established knowledge and to prove that they are able to confirm previous discoveries. Usually, membranes are only reconstructed as solid structures in CET (150, 152, 157), not revealing the lipid bilayer characteristics that are wellknown from ultrathin sectioning since decades. This is due to extreme underfocus conditions, producing high contrast, but limiting the resolution to about 4–6 nm. However, it is possible to adjust the imaging conditions appropriately and to reconstruct the lipid bilayer of inner and outer membranes (69). In addition, attempts have been made to correct for the focus effects (i.e., for oscillations of the contrast transfer function) by image processing and to enhance the resolution of 3D reconstructions to 2 nm or better (160, 161). Theoretically, it is possible to amend the contrast transfer function already in the

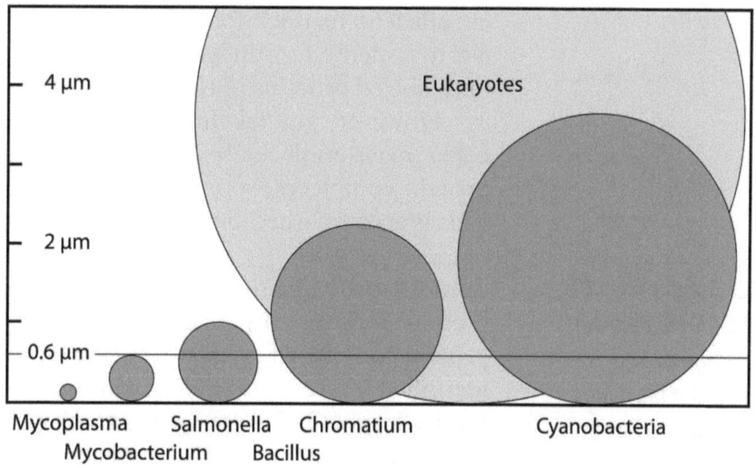

Fig. 5. Scheme of bacterial cell sizes and suitable specimen thickness for cryo-electron tomography (0.6 µm).

microscope in the course of imaging. Recent studies evaluated experimental approaches that appear to be promising and will directly extend the usable resolution for CET (162, 163).

One important limitation is the suitable specimen thickness for tomographic reconstructions. Since electrons strongly interact with atoms the thickness of biological material should ideally not exceed ≈0.6 µm (0.5–1.0 µm) with 300 keV electron microscopes, the instruments commonly used for cellular CET (65). This limits direct CET applications to an adequate group of microbes and requires thicker cells to be thinned (Fig. 5). As outlined in Subheading 3.3, the method of choice will be FIB milling since it does not only allow to abrase material without compression artifacts but also to cut out lamella from cells in arbitrary positions and orientations (81).

5.2. Correlative Cryo-Microscopy

The inspection and definition of cellular structures in tomograms is feasible by eye or automated segmentation (146, 164) if they are large and characteristic enough (e.g., membranes, filamentous aggregates, flagellar motor). However, the unambiguous identification of (smaller) molecular complexes requires specific labeling or comparison with known structures by an objective approach. Classical labeling with immunogold is possible for surface structures of intact cells that are accessible from outside. The widely applied label green fluorescent protein (GFP) is not directly detectable by electron microscopy. But it can be used to identify subregions of big cells or individual bacteria in multicellular aggregates (consortia) by cryo-fluorescence microscopy for selective thinning by FIB and/or imaging by CET. The *correlative microscopy* approach (80, 81, 165) has been developed in recent years and is advantageous for selecting appropriate cells or cellular regions in an efficient manner for CET. Recently, fusion of proteins with electron-dense ferritin was achieved and can directly be applied to find labeled proteins in tomograms (166).

However, any labeling may modify native interactions with other macromolecules; it would thus be desirable if unaltered protein complexes were detectable by their individual structures. This is accomplished by *template matching*.

5.3. Combining CET and Single Particle Averaging: The Template Matching Approach

Tomograms contain the signatures of the entire proteome being present in the reconstructed volume. Provided that the signal of a protein complex of interest is strong enough it can be detected and identified by cross-correlating tomograms with the 3D structure of the molecule (167–169). The template may originate from X-ray crystallography or single particle EM (see Subheading 4.2). Examples for soluble protein complexes in bacteria demonstrate the potential of the approach (159, 170). It needs some effort to extract the high-resolution information of tomograms and to selected sub-tomograms containing the identified protein complexes. But the combination of artifact-free cryo-preparations of cells and cell membranes in particular, appropriate imaging conditions and of image processing approaches, correcting for contrast variations due to focus conditions and the focus gradient in projections of the tilted specimen (160), promises to get insight into the structure of membrane (and other) protein complexes to 1–1.5 nm resolution (171). Template matching finds the positions and orientations of the respective protein complexes that can be extracted (sub-tomograms) and 3D averaged afterwards. If the sub-volumes include some neighboring space specifically interacting structures will be captured in an in situ situation. By this way labile or only temporarily existing but functional protein aggregations may be identified and structurally investigated.

6. Perspectives

Future investigations of bacterial structures will reside in the cold and will predominantly make use of 3D information. Cryo-electron tomography and its associated techniques will take over what electron microscopy of ultrathin sections of plastic-embedded material has been for structural research in microbiology for the last four decades. The goal is to visualize whole cells at molecular resolution and to evaluate molecular interactions in situ. The template matching approach is still in its beginning and awaits broader application in the years to come. The challenge is the reliable identification of smaller protein complexes in the crowded cytoplasm and in membranes, which would benefit from electron detectors with improved signal transformation in the higher resolution range and from more efficient correction approaches for the contrast transfer function in the course of 3D reconstruction. The introduction of commercial solutions for cryo-FIB micromachining and phase-corrected electron microscopy employing phase plates could denote a further step forward. With these improvements, the perspectives to successfully explore the largely uncharted territories of the bacterial cell envelope—the periplasm and the outer (and inner) membrane in situ—are promising.

References

1. van Leeuwenhoek A (1684) Microscopical observations about animals in the scurf of the teeth, the substance called jvorms in the nose, and the cuticula consisting of scales. Phil Transact Royal Soc 159:308
2. Ehrlich P (1882) Deut Med Wochenschr 8:269–270
3. Gram HC (1884) Über die isolierte Färbung der Schizomyceten in Schnitt- und Trockenpräparaten. Fortschr Med 2:185–189
4. Salton MRJ (1964) The bacterial cell wall. American Elsevier Publ Co, New York
5. Beveridge TJ (1990) Mechanism of gram variability in selected bacteria. J Bacteriol 172:1609–1620
6. Popescu A, Doyle RJ (1996) The gram stain after more than a century. Biotech Histochem 71:145–151
7. White PB (1926) Further studies of the *Salmonella* group. Great Britain Medical Research Council 103, p 3–160
8. Salmonella Subcommittee (1934) J Hyg 34:333–350
9. Grimont PAD, Weill FX (2007) Antigenic formulae of the *Salmonella* servoars, 9th edn. World Health Organization Collaborating Centre for Reference and Research on Salmonella, Institut Pasteur, Paris
10. Knoll M, Ruska E (1932) Das Elektronenmikroskop. Z Phys 78:318–339
11. Krause F (1937) Das magnetische Elektronenmikroskop und seine Anwendung in der Biologie. Naturwissenschaften 25:817–825
12. Marton L (1937) La microscopie électronique des objects biologiques. Bull Acad Roy Belg Cl Sci 23:672–678
13. Knaysi G (1949) Cytology of bacteria II. Bot Rev 15:106–151
14. Chapman GB, Hillier J (1953) Electron microscopy of ultra-thin sections of bacteria I. Cellular division in *Bacillus cereus*. J Bacteriol 66:362–373
15. Houwink AL (1953) A macromolecular monolayer in the cell wall of *Spirillum* spec. Biochim Biophys Acta 10:360–366
16. Müller HO (1942) Die Ausmessung der Tiefe übermikroskopischer Objekte. Kolloid-Zeitschrift 99:6–28
17. Williams RC, Wyckoff RWG (1945) Electron shadow micrography of the tobacco mosaic virus protein. Science 101:594–596
18. Houwink AL (1956) Flagella, gas vacuoles and cell-wall structure in *Halobacterium halobium*; an electron microscope study. J Gen Microbiol 15:146–150
19. Wildhaber I, Gross H, Engel A, Baumeister W (1985) The effects of air-drying and freeze-drying on the structure of a regular protein layer. Ultramicroscopy 16:411–422
20. Guckenberger R (1985) Surface reliefs derived from heavy-metal-shadowed specimens—Fourier space technique applied to periodic objects. Ultramicroscopy 16:357–370
21. Engelhardt H, Guckenberger R, Hegerl R, Baumeister W (1985) High resolution shadowing of freeze-dried bacterial photosynthetic membranes: multivariate statistical analysis and surface relief reconstruction. Ultramicroscopy 16:395–410
22. Baumeister W, Guckenberger R, Engelhardt H, Woodcock CLF (1986) Metal shadowing and decoration in electron microscopy of biological macromolecules. Ann N Y Acad Sci 483:57–76
23. Wildhaber I, Hegerl R, Barth M, Gross H, Baumeister W (1986) Three-dimensional reconstruction of a freeze-dried and metal-shadowed bacterial surface layer. Ultramicroscopy 19:57–68
24. Bachmann L, Becker R, Leupold G, Barth M, Guckenberger R, Baumeister W (1985) Decoration and shadowing of freeze-etched catalase crystals. Ultramicroscopy 16:305–320
25. Rübenkamm E, Braun N, Bachmann L, Bacher A, Brandt J, Baumeister W, Weinkauf S (1995) Quantitative evaluation of heavy metal decoration on protein molecules: contrast, specificity and resolution. Ultramicroscopy 58:337–351
26. Moor H, Mühlethaler K, Waldner H, Frey-Wyssling A (1961) A new freezing-ultramicrotome. J Biophys Biochem Cytol 10:1–13

27. Reimer L, Schulte C (1966) Elektronenmikroskopische Oberflächenabdrücke und ihr Auflösungsvermögen. Naturwissenschaften 53:489–497
28. Holt SC, Trüper HG, Takács BJ (1968) Fine structure of *Ectothiorhodospira mobils* strain 8113 thylakoids: chemical fixation and freeze-etching studies. Arch Mikrobiol 62:111–128
29. Engelhardt H (2007) Are S-layers exoskeletons? The basic function of surface protein layers revisited. J Struct Biol 160:115–124
30. Sleytr U, Adam H, Klaushofer H (1969) Die Feinstruktur der Zellwand und Cytoplasmamembran von *Clostridium nigrificans*, dargestellt mit Hilfe der Gefrierätz- und Ultradünnschnittechnik. Arch Mikrobiol 66:40–58
31. Holt SC, Beveridge TJ (1982) Electron microscopy: its development and application to microbiology. Can J Microbiol 28:1–53
32. Schatten H (2011) Low voltage high-resolution SEM (LVHRSEM) for biological structural and molecular analysis. Micron 42:175–185
33. Binnig G, Rohrer H, Gerber C, Weibel E (1982) Surface studies by scanning tunneling microscopy. Phys Rev Lett 49:57–61
34. Binnig G, Quate CF, Gerber C (1986) Atomic force microscope. Phys Rev Lett 56:930–933
35. WiegräbeW,NonnenmacherM,Guckenberger R, Wolter O (1991) Atomic force microscopy of a hydrated bacterial surface protein. J Microsc 163:79–84
36. Müller DJ, Baumeister W, Engel A (1996) Conformational change of the hexagonally packed intermediate layer of *Deinococcus radiodurans* monitored by atomic force microscopy. J Bacteriol 178:3025–3030
37. Schabert FA, Henn C, Engel A (1995) Native *Escherichia coli* OmpF porin surfaces probed by atomic force microscopy. Science 268:92–94
38. Engel A, Schoenenberger C-A, Müller DJ (1997) High resolution imaging of native biological sample surfaces using scanning probe microscopy. Curr Opin Struct Biol 7:279–284
39. Scheuring S, Müller DJ, Stahlberg H, Engel H-A, Engel A (2002) Sampling the conformational space of membrane protein surfaces with the AFM. Eur Biophys J 31:172–178
40. Philippsen A, Im W, Engel A, Schirmer T, Roux B, Müller DJ (2002) Imaging the electrostatic potential of transmembrane channels: atomic probe microscopy of OmpF porin. Biophys J 82:1667–1676
41. Knapp HF, Wiegräbe W, Heim M, Eschrich R, Guckenberger R (1995) Atomic force microscope measurements and manipulation of Langmuir-Blodgett films with modified tips. Biophys J 69:708–715
42. Gonçalves RP, Scheuring S (2006) Manipulating and imaging individual membrane proteins by AFM. Surf Interface Anal 38:1413–1418
43. Müller DJ (2008) AFM: a nanotool in membrane biology. Biochemistry 47:7986–7998
44. Anselmetti D, Hansmeier N, Kalinowski J, Martini J, Merkle T, Pamisano R, Ros R, Schmied K, Sischka A, Toensing K (2007) Analysis of subcellular surface structures, function and dynamics. Anal Bioanal Chem 387:83–89
45. Marton L (1976) Early application of electron microscopy to biology. Ultramicroscopy 1:281–296
46. Baker RF, Pease DC (1949) Sectioning of the bacterial cell for the electron microscope. Nature 163:282
47. Sjöstrand FS (1953) A new microtome for ultrathin sectioning for high resolution electron microscopy. Experientia 9:114–115
48. Kellenberger E, Ryter A (1958) Cell wall and cytoplasmic membrane of *Escherichia coli*. J Biophys Biochem Cytol 4:323–326
49. Murray RGE (1963) On the cell wall structure of *Spirillum serpens*. Can J Microbiol 9:381–392
50. Claus GW, Roth LE (1964) Fine structure of the gram-negative bacterium *Acetobacter suboxydans*. J Cell Biol 20:217–233
51. Costerton JW (1979) The role of electron microscopy in the elucidation of bacterial structure and function. Annu Rev Microbiol 33:459–479
52. Sleytr UB (1978) Regular arrays of macromolecules on bacterial cell walls: structure, chemistry, assembly, and function. Int Rev Cytol 53:1–64
53. Remsen CC, Watson SW, Trüper HG (1970) Macromolecular subunits in the walls of marine photosynthetic bacteria. J Bacteriol 103:254–257
54. Hageage GJ, Gherna RL (1971) Surface structure of *Chromatium okenii* and *Chromatium weissei*. J Bacteriol 106:687–690
55. Jeffries P, Wilkinson JF (1978) Electron microscopy of the cell wall complex of *Methylomonas albus*. Arch Microbiol 119:227–229
56. Murray RGE (1978) Form and function. In: Norris JR, Richmond MH (eds) Essays in microbiology. I Bacteria. Wiley, New York, pp 2/30–2/31
57. Emde B, Wehrli E, Baumeister W (1980) The topography of the cell wall of *Micrococcus*

radiodurans. In: Brederoo P, de Priester W (ed) Proc 7th Europ Congr Electron Microscopy, vol. 2, Leiden, 1980, p 460–461

58. Graham LL, Beveridge TJ (1990) Evaluation of freeze-substitution and conventional embedding protocols for routine electron microscopic processing of eubacteria. J Bacteriol 172:2141–2149
59. Tokunaga M, Kusamichi M, Koike H (1986) Ultrastructure of outermost layer of cell wall in *Candida albicans* observed by rapid-freezing technique. J Electron Microsc 35:237–246
60. Fernández-Morán H (1960) Low-temperature preparation techniques for electron microscopy of biological specimens based on rapid freezing with liquid Helium II. Ann N Y Acad Sci 85:689–713
61. Dubochet J, McDowall AW, Menge B, Schmid EN, Lickfeld KG (1983) Electron microscopy of frozen-hydrated bacteria. J Bacteriol 155:381–390
62. Al-Amoudi A, Chang J-J, Leforestier A, McDowall A, Salamin LM, Norlén LPO, Richter K, Blanc NS, Studer D, Dubochet J (2004) Cryo-electron microscopy of vitreous sections. EMBO J 23:3583–3588
63. Dubochet J (2012) Cryo-EM—the first thirty years. J Microsc 245:221–224
64. Dubochet J, McDowall AW (1981) Vitrification of pure water for electron microscopy. J Microsc 124:RP3–RP4
65. Plitzko JM, Baumeister W (2007) Cryoelectron tomography (CET). In: Hawkes PW, Spence JCH (eds) Science of microscopy, vol I. Springer, New York, pp 535–604
66. Shimoni E, Müller M (1998) On optimizing high-pressure freezing: from heat transfer theory to a new microbiopsy device. J Microsc 192:236–247
67. Matias VRF, Beveridge TJ (2005) Cryo-electron microscopy reveals native polymeric cell wall structure in *Bacillus subtilis* 168 and the existence of a periplasmic space. Mol Microbiol 56:240–251
68. Matias VRF, Beveridge TJ (2006) Native cell wall organization shown by cryo-electron microscopy confirms the existence of a periplasmic space in *Staphylococcus* aureaus. J Bacteriol 188:1011–1021
69. Hoffmann C, Leis A, Niederweis M, Plitzko JM, Engelhardt H (2008) Disclosure of the mycobacterial outer membrane: cryo-electron tomography and vitreous sections reveal the lipid bilayer structure. Proc Natl Acad Sci U S A 105:3963–3967
70. Zuber B, Chami M, Houssin C, Dubochet J, Griffith G, Daffé M (2008) Direct visualization of the outer membrane of mycobacteria and corynebacteria in their native state. J Bacteriol 190:5672–5680
71. Paul TR, Beveridge TJ (1992) Reevaluation of envelope profiles and cytoplasmic ultrastructure of mycobacteria processed by conventional embedding and freeze- substitution protocols. J Bacteriol 174:6508–6517
72. Bleck CKE, Merz A, Gutierrez MG, Walther P, Dubochet J, Zuber B, Griffith G (2009) Comparison of different methods for thin sections EM analysis of *Mycobacterium smegmatis*. J Microsc 237:23–38
73. Al-Amoudi A, Studer D, Dubochet J (2005) Cutting artefacts and cutting process in vitreous sections for cryo-electron microscopy. J Struct Biol 150:109–121
74. Han H-M, Zuber B, Dubochet J (2008) Compression and crevasses in vitreous sections under different cutting conditions. J Microsc 230:167–171
75. Zuber B, Haenni M, Ribeiro T, Minnig K, Lopes F, Moreillon P, Dubochet J (2006) Granular layer in the periplasmic space of gram-positive bacteria and fine structures of *Enterococcus gallinarum* and *Streptococcus gordonii* septa revealed by cryo-electron microscopy of vitreous sections. J Bacteriol 188:6652–6660
76. Ballerini M, Milani M, Batani M, Squadrini F (2001) Focused ion beam techniques for the analysis of biological samples: a revolution in ultramicroscopy? Proc SPIE 4261:92–104
77. Heymann JAW, Hayles M, Gestmann I, Giannuzzi LA, Lich B, Subramaniam S (2006) Site-specific 3D imaging of cells and tissues with a dual beam microscope. J Struct Biol 155:63–73
78. Marko M, Hsieh C, Moberlychan W, Mannella CA, Frank J (2006) Focused ion beam milling of vitreous water: prospects for an alternative to cryo-ultramicrotomy of frozen- hydrated biological samples. J Microsc 222:42–47
79. Marko M, Hsieh C, Schalek R, Frank J, Mannella C (2007) Focused-ion-beam thinning of frozen-hydrated biological specimens for cryo-electron microscopy. Nat Methods 4:2015–2017
80. Rigort A, Bäuerlein FJB, Leis A, Gruska M, Hoffmann C, Laugks T, Böhm U, Eibauer M, Gnaegi H, Baumeister W, Plitzko JM (2010) Micromachining tools and correlative approaches for cellular cryo-electron tomography. J Struct Biol 172:169–179
81. Rigort A, Bäuerlein FJB, Villa E, Eibauer M, Laugks T, Baumeister W, Plitzko JM (2012) Focused ion beam micromachining of

eukaryotic cells for cryoelectron tomography. Proc Natl Acad Sci U S A 109:4449–4454

82. Brenner S, Horne RW (1959) A negative staining method for high resolution electron microscopy of viruses. Biochim Biophys Acta 34:103–110
83. Bremer A, Henn C, Engel A, Baumeister W, Aebi U (1992) Has negative staining still a place in biomacromolecular electron microscopy? Ultramicroscopy 46:85–111
84. Anderson TF (1962) Negative staining and its use in the study of viruses and their serological reactions. In: Harris RJC (ed) Symposium of the international society for cell biology, vol 1. Academic, New York, pp 251–262
85. DeRosier DJ, Klug A (1968) Reconstruction of three-dimensional structures from electron micrographs. Nature 217:130–134
86. Hoppe W, Langer R, Knesch G, Poppe C (1968) Proteinkristallstrukturanalyse mit Elektronenstrahlen. Naturwissenschaften 55:333–336
87. Hoppe W, Gassmann J, Hunsmann N, Schramm HJ, Sturm M (1974) Three-dimensional reconstruction of individual negatively stained yeast fatty-acid synthetase molecules from tilt series in the electron microscope. Hoppe Seylers Z Physiol Chem 355:1483–1487
88. Frank J (1989) Image analysis of single molecules. Electron Microsc Rev 2:53–74
89. Hegerl R (1996) The EM program package: a platform for image processing in biological electron microscopy. J Struct Biol 116:30–34
90. van Heel M, Harauz G, Orlova EV, Schmidt R, Schatz M (1996) A new generation of the IMAGIC image processing system. J Struct Biol 116:17–24
91. Sorzano COS, Marabini R, Velázques-Muriel J, Bilbao-Acstro JR, Scheres SHW, Carazo JM, Pascual-Montano A (2004) XMIPP: a new generation of an open-source image processing package for electron microscopy. J Struct Biol 148:194–204
92. Shaik TR, Gao H, Baxter WT, Asturias FJ, Boisset N, Leith A, Frank J (2008) SPIDER image processing for single-particle reconstruction of biological macromolecules from electron micrographs. Nat Protoc 3:1941–1974
93. Korinek A, Beck F, Baumeister W, Nickell S, Plitzko JM (2011) Computer controlled cryo-electron microscopy—TOM^2 a software package for high-throughput applications. J Struct Biol 175:394–405
94. Engelhardt H (1991) Electron microscopy of microbial cell wall proteins. Surface topography, three-dimensional reconstruction, and strategies for two-dimensional crystallization. In: Latgé JP, Boucias D (eds) Fungal cell wall and immune response, NATO ASI Series H53. Springer, Berlin, pp 11–25
95. Saxton WO, Baumeister W, Hahn M (1984) Three-dimensional reconstruction of imperfect two-dimensional crystals. Ultramicroscopy 13:57–70
96. Bingle WH, Engelhardt H, Page WJ, Baumeister W (1987) Three-dimensional structure of the regular tetragonal surface layer of *Azotobacter vinelandii*. J Bacteriol 169:5008–5015
97. Engelhardt H, Cejka Z, Baumeister W (1988) Three-dimensional structure of surface layers from various *Bacillus* and *Clostridium* species. In: Sleytr UB, Messner P, Pum D, Sára M (eds) Crystalline bacterial cell surface layers. Springer, Berlin, pp 87–91
98. Tsuboi A, Engelhardt H, Santarius U, Tsukagoshi N, Udaka S, Baumeister W (1989) Three-dimensional structure of the surface protein layer (MW layer) of *Bacillus brevis* 47. J Ultrastruct Mol Struct Res 102:178–187
99. Beveridge TJ, Pouwels PH, Sára M et al (1997) Functions of S-layers. FEMS Microbiol Rev 20:99–149
100. Engelhardt H, Peters J (1998) Structural research on surface layers—a focus on stability, surface layer homology domains, and surface layer-cell wall interactions. J Struct Biol 124:276–302
101. Engel A, Massalski A, Schindler H, Dorset DL, Rosenbusch JP (1985) Porin channel triplets merge into single outlets *in Escherichia coli* outer membranes. Nature 317:643–645
102. Chalcroft JP, Engelhardt H, Baumeister W (1987) Structure of the porin from a bacterial stalk. FEBS Lett 211:53–58
103. Sass JJ, Massalski A, Beckmann E, Büldt G, Dorset D, van Heel M, Rosenbusch JP, Zeitler E, Zemlin F (1988) High-resolution cryomicroscopy of porin (OmpF). In: Bailey GW (ed) Proc 46th Ann Meet Electron Microscopy Soc Am. San Francisco Press, San Francisco, pp 146–147
104. Jap B, Downing KH, Walian PJ (1990) Structure of PhoE porin in projection at 3.5 Å resolution. J Struct Biol 103:57–63
105. Weiss MS, Abele U, Weckesser J, Welte W, Schiltz E, Schulz GE (1991) Molecular architecture and electrostatic properties of a bacterial porin. Science 254:1627–1630
106. Trachtenberg S, DeRosier DJ, Macnab RM (1987) Three-dimensional structure of the complex flagellar filament of *Rhizobium lupini*

and its relation to the structure of the plain filament. J Mol Biol 195:603–620
107. Morgan DG, Machnab RM, Francis NR, DeRosier DJ (1993) Domain organization of the subunit of the *Salmonella typhimurium* flagellar hook. J Mol Biol 229:79–84
108. Engelhardt H, Schuster SC, Baeuerlein E (1993) An Archimedian spiral: the basal disk of the *Wolinella* flagellar hook. Science 262:1046–1048
109. Woodcock CL, Baumeister W (1990) Different representations of a protein structure obtained with different negative stains. Eur J Cell Biol 51:45–52
110. Rachel R, Jakubowski U, Tietz H, Hegerl R, Baumeister W (1986) Projected structure of the surface protein of *Deinococcus radiodurans* determined to 8 Å resolution by cryomicroscopy. Ultramicroscopy 20:305–316
111. Jakubowski U, Hegerl R, Formanek H, Volker S, Santarius U, Baumeister W (1988) Three-dimensional reconstruction of the HPI-layer of *Deinococcus radiodurans* embedded in Cd-thioglycerol. Inst Phys Conf Ser 93(3):381–382
112. Henderson R, Unwin PNT (1975) Three-dimensional model of purple membrane obtained by electron microscopy. Nature 257:28–32
113. Glaeser RM, Hall RJ (2011) Reaching the information limit in cyro-EM of biological macromolecules: experimental aspects. Biophys J 100:2331–2337
114. Penczek P, Radermacher M, Frank J (1992) Three-dimensional reconstruction of single particles embedded in ice. Ultramicroscopy 40:33–52
115. Frank J, Radermacher M (1992) Three-dimensional reconstruction of single particles negatively stained or in vitreous ice. Ultramicroscopy 46:241–262
116. Henderson R, Baldwin JM, Ceska TA, Zemlin F, Beckmann E, Downing KH (1990) Model for the structure of bacteriorhodopsin based on high-resolution electron cryo- microscopy. J Mol Biol 213:899–929
117. DeRosier DJ, Moore PB (1970) Reconstruction of three-dimensional images from electron micrographs of structures with helical symmetry. J Mol Biol 52:355–369
118. Morgan DG, DeRosier D (1992) Processing images of helical structures: a new twist. Ultramicroscopy 46:263–285
119. Egelman EH (2000) A robust algorithm for the reconstruction of helical filaments using single-particle methods. Ultramicroscopy 85:225–234
120. Mimori Y, Samashita I, Murata K, Fujiyoshi Y, Yonekura K, Toyoshima C, Namba K (1995) The structure of the R-type straight flagellar filament of *Salmonella* at 9 Å resolution by electron cryomicroscopy. J Mol Biol 249:69–87
121. Morgan DG, Owen C, Melanson LA, DeRosier DJ (1995) Structure of bacterial flagellar filaments at 11 Å resolution: packing of the α-helices. J Mol Biol 249:88–110
122. Yonekura K, Maki-Yonekura S, Namba K (2003) Complete atomic model of the bacterial flagellar filament by electron cryomicroscopy. Nature 424:643–650
123. Galkin VE, Yu X, Bielnicki J, Heuser J, Ewing CP, Guerry P, Egelman EH (2008) Divergence of quaternary structures among bacterial flagellar filaments. Science 320:382–385
124. Maki-Yonekura S, Yonekura K, Namba K (2010) Conformational change of flagellin for polymorphic supercoiling of the flagellar filament. Nat Struct Mol Biol 17:417–422
125. Samatey FA, Natsunami H, Imada K, Gagashima S, Shaikh T, Thomas DR, Chen JZ, DeRosier DJ, Kitao A, Namba K (2004) Structure of the bacterial flagellar hook and implication for the molecular universal joint mechanism. Nature 431:1062–1068
126. Shaik TR, Thomas DR, Chen JZ, Samatey FA, Matsunami H, Imada K, Namba K, DeRosier DJ (2005) A partial atomic structure for the flagellar hook of *Salmonella typhimurium*. Proc Natl Acad Sci U S A 102:1023–1028
127. Fujii T, Kato T, Namba K (2009) Specific arrangement of α-helical coiled coils in the core domain of the bacterial flagellar hook for the universal joint function. Structure 17:1485–1493
128. Steven AC, Baumeister W (2008) The future is hybrid. J Struct Biol 163:186–195
129. Wriggers W (2010) Using *Situs* for the integration of multi-resolution structures. Biophys Rev 2:21–27
130. Trabuco LG, Schreiner E, Gumbart J, Hsin J, Villa E, Schulten K (2011) Applications of the molecular dynamics flexible fitting method. J Struct Biol 173:420–427
131. Fujii T, Cheung M, Blanco A, Kato T, Blocker AJ, Namba K (2012) Structure of a type III secretion needle at 7-Å resolution provides insights into its assembly and signaling mechanisms. Proc Natl Acad Sci U S A 109:4461–4466
132. Young HS, Dang H, Lai Y, DeRosier DJ, Khan S (2003) Variable symmetry in *Salmonella typhimurium* flagellar motors. Biophys J 84:571–577

133. Thomas DR, Francis NR, Xu C, DeRosier DJ (2006) The three-dimensional structure of the flagellar rotor from a clockwise-locked mutant of *Salmonella enterica* serovar typhimurium. J Bacteriol 188:7039–7048
134. Minamino T, Imada K, Namba K (2008) Molecular motors of the bacterial flagella. Curr Opin Struct Biol 18:693–701
135. Schraidt O, Lefebre MD, Brunner MJ, Schmied WH, Schmidt A, Radics J, Mechtler K, Galán JE, Marlovits TC (2010) Topology and organization of the *Salmonella typhimurium* type III secretion needle complex components. PLoS Pathog 6:e1000824
136. Schraidt O, Marlovits TC (2011) Three-dimensional model of *Salmonella's* needle complex at subnanometer resolution. Science 331:1192–1195
137. Fronzes R, Schäfer E, Wang L, Saibil HR, Orlova EV, Waksman G (2009) Structure of a type IV secretion system core complex. Science 323:266–268
138. Chandran V, Fronzes R, Duquerroy S, Cronin N, Navaza J, Waksman G (2009) Structure of the outer membrane complex of a type IV secretion system. Nature 462:1011–1015
139. Hart RG (1968) Electron microscopy of unstained biological material: the polytropic montage. Science 159:1464–1467
140. Dierksen K, Typke D, Hegerl R, Koster AJ, Baumeister W (1992) Towards automatic electron tomography. Ultramicroscopy 40:71–87
141. Dierksen K, Typke D, Hegerl R, Baumeister W (1993) Towards automatic electron tomography II. Implementation of autofocus and low-dose procedures. Ultramicroscopy 49:109–120
142. Grimm R, Koster J, Ziese U, Typke D, Baumeister W (1996) Zero-loss energy filtering under low-down conditions using a post-column energy filter. J Microsc 183:60–68
143. Grimm R, Typke D, Baumeister W (1998) Improving image quality by zero-loss energy filtering: quantitative assessment by means of image cross-correlation. J Microsc 190:339–349
144. Medalia O, Weber I, Frangaiks AS, Nicastro D, Gerisch G, Baumeister W (2002) Macromolecular architecture in eukaryotic cells visualized by cryoelectron tomography. Science 298:1209–1213
145. Baumeister W (2002) Electron tomography: towards visualizing the molecular organization of the cytoskeleton. Curr Opin Struct Biol 12:679–684
146. Lucic V, Förster F, Baumeister W (2005) Structural studies by electron tomography: from cells to molecules. Annu Rev Biochem 74:833–865
147. Leis A, Rockel B, Andrees L, Baumeister W (2009) Visualizing cells at the nanoscale. Trends Biochem Sci 34:60–70
148. Nickell S, Hegerl R, Baumeister W, Rachel R (2003) *Pyrodictium* cannulae enter the periplasmic space but do not enter the cytoplasm, as revealed by cryo-electron tomography. J Struct Biol 141:34–42
149. Schrempf H, Koebsch I, Walter S, Engelhardt H, Meschke H (2011) Extracellular *Streptomyces* veiscles: amphorae for survival and defence. Microb Biotechnol 4:286–299
150. Gan L, Chen S, Jensen GJ (2008) Molecular organization of Gram-negative peptidoglycan. Proc Natl Acad Sci U S A 105:18953–18957
151. Murphy GE, Leadbetter JR, Jensen GJ (2006) In situ structure of the complete *Treponema primitia* flagellar motor. Nature 442: 1062–1064
152. Izard J, Hsieh C-E, Limberger R, Mannella CA, Marko M (2008) Native cellular architecture of *Treponema denticola* revealed by cryo-electron tomography. J Struct Biol 163:10–17
153. Chen S, Beeby M, Murphy GE, Leadbetter JR, Hendrixon DR, Briegel A, Li Z, Shi J, Tocheva EI, Müller A, Dobro MM, Jensen GJ (2011) Structural diversity of bacterial flagellar motors. EMBO J 30:2972–2981
154. Kürner J, Frangakis AS, Baumeister W (2005) Cryo-electron tomography reveals the cytoskeletal structure of *Spiroplasma melliferum*. Science 307:436–438
155. Jensen GJ, Briegel A (2007) How electron cryotomography is opening a new window onto prokaryotic ultrastructure. Curr Opin Struct Biol 17:260–267
156. Morris DM, Jensen GJ (2008) Toward a biomechanical understanding of whole bacterial cells. Annu Rev Biochem 77:583–613
157. Ting CS, Hsieh C, Sundararaman S, Mannella C, Marko M (2007) Cryo-electron tomography reveals the comparative three-dimensional architecture of *Prochlorococcus*, a globally important marine cyanobacterium. J Bacteriol 289:4485–4493
158. Beck M, Malmström JA, Lange V, Schmidt A, Deutsch EW, Aebersold R (2009) Visual proteomics of the human pathogen Leptospira interrogans. Nat Methods 6:817–823
159. Ortiz JO, Brandt F, Matias VRF, Sennels L, Rappsilber J, Scheres SHW, Eibauer M, Hartl FU, Baumeister W (2010) Structure of

hibernating ribosomes studied by cryoelectron tomography in vitro and in situ. J Cell Biol 190:613–621

160. Fernández JJ, Li S, Crowther RA (2006) CTF determination and correction in electron cryotomography. Ultramicroscopy 106: 587–596
161. Xiong Q, Morphew MK, Schwartz CL, Hoenger AH, Mastronarde DN (2009) CTF determination and correction for low dose tomographic tilt series. J Struct Biol 168: 378–387
162. Danev R, Nagayama K (2001) Transmission electron microscopy with zernike phase plate. Ultramicroscopy 88:243–252
163. Danev R, Kanamaru S, Marko M, Nagayama K (2010) Zernike phase contrast cryo- electron tomography. J Struct Biol 171:174–181
164. Rigort A, Günther D, Hegerl R, Baum D, Weber B, Prohaska S, Medalia O, Baumeister W, Hege H-C (2012) Automated segmentation of electron tomograms for a quantitative description of actin filament networks. J Struct Biol 177:135–144
165. Sartori A, Gatz R, Beck F, Rigort A, Baumeister W, Plitzko JM (2007) Correlative microscopy: bridging the gap between fluorescence light microscopy and cryo-electron tomography. J Struct Biol 160:135–145
166. Wang Q, Mercogliano CP, Löwe J (2011) A ferritin-based label for cellular electron cryotomography. Structure 19:147–154
167. Frangakis AS, Böhm J, Förster F, Nickell S, Nicastro D, Typke D, Hegerl R, Baumeister W (2002) Identification of macromolecular complexes in cryoelectron tomograms of phantom cells. Proc Natl Acad Sci U S A 99:14153–14158
168. Frangakis AS, Förster F (2004) Computational exploration of structural information from cryo-electron tomograms. Curr Opin Struct Biol 14:325–331
169. Nickell S, Kofler C, Leis AP, Baumeister W (2006) A visual approach to proteomics. Nat Rev Mol Cell Biol 7:225–230
170. Ortiz JO, Förster F, Kürner J, Linaroudis AA, Baumeister W (2006) Mapping 70S ribosomes in intact cells by cryoelectron tomography and pattern recognition. J Struct Biol 156:334–341
171. Eibauer M (2011) Korrektur der transferfunktionen CTF und MTF in der Kryoelektronentomographie. Thesis, Technical University of Munich

Chapter 3

Purification of the Outer Membrane Usher Protein and Periplasmic Chaperone-Subunit Complexes from the P and Type 1 Pilus Systems

Nadine S. Henderson and David G. Thanassi

Abstract

Understanding molecular mechanisms of protein secretion by bacteria requires the purification of secretion machinery components and the isolation of complexes between the secretion machinery and substrate proteins. Here, we describe methods for the purification of proteins from the chaperone/usher pathway, which is a conserved secretion pathway dedicated to the assembly of polymeric surface fibers termed pili or fimbriae in gram-negative bacteria. Specifically, we describe the isolation of the PapC and FimD usher proteins from the bacterial outer membrane, and the purification of PapD-PapG and FimC-FimH chaperone-subunit complexes from the periplasm. These Pap and Fim proteins belong to the P and type 1 pilus systems of uropathogenic *Escherichia coli*, respectively.

Key words: Pili, Fimbriae, Usher, Chaperone-subunit complex, Outer membrane, Periplasm, Protein purification

1. Introduction

The chaperone/usher (CU) pathway is a conserved secretion system present in gram-negative bacteria dedicated to the assembly of polymeric surface fibers termed pili or fimbriae (1–3). In the CU pathway, a periplasmic chaperone binds pilus subunit proteins as they enter the periplasm via the Sec general secretion machinery and delivers the subunits to the outer membrane (OM) usher protein. The chaperone allows proper folding of pilus subunits and maintains the subunits in an assembly-competent state (4, 5). The usher provides a multifunctional platform for pilus biogenesis, catalyzing the organized assembly of subunits into the pilus fiber and

Anne H. Delcour (ed.), *Bacterial Cell Surfaces: Methods and Protocols*, Methods in Molecular Biology, vol. 966,
DOI 10.1007/978-1-62703-245-2_3,

Providing a channel for secretion of the fiber to the cell surface (6, 7). The usher exists in the OM as a dimeric complex, although the monomer appears to be the functional unit for pilus assembly (6). Many pili assembled by the CU pathway are composite fibers composed of multiple different subunit proteins. The tip of the pilus fiber contains an adhesive protein, termed the adhesin, which confers function to the pilus by binding to specific host cell receptors. The assembly machinery and subunit proteins of the CU pathway have been purified, and structures have been generated for a number of individual proteins as well as for complexes corresponding to pilus assembly intermediates. As a result, pilus biogenesis by the CU pathway is one of the best-understood bacterial secretion systems.

The P and type 1 pili expressed by uropathogenic *Escherichia coli* have served as models for investigation of the CU pathway. These pili mediate colonization of the urinary tract and are critical virulence factors (8, 9). We describe here methods for purifying the P pilus usher PapC and the type 1 pilus usher FimD from the OM. We also describe methods for purification of periplasmic PapD-PapG and FimC-FimH chaperone-adhesin complexes from P and type 1 pili, respectively. Ushers are purified from OM preparations using a hexahistidine tag (His-tag) and metal affinity chromatography, followed by thrombin cleavage to remove the His tag, and finally gel filtration chromatography to obtain a final homogenous preparation of dimeric usher protein. Chaperone-subunit complexes are purified from periplasm preparations using a His-tagged chaperone, followed by ion exchange chromatography to separate chaperone-subunit complexes from the free chaperone.

2. Materials

2.1. Bacterial Culture

1. LB broth: 25 g/L LB powder in water; sterilize by autoclaving.
2. LB agar plates: Add 15 g/L Bacto Agar to LB broth.
3. Ampicillin stock: 100 mg/mL in water. Sterilize by filtration through a 0.22 μm filter and store at −20 °C.
4. Spectinomycin stock: 100 mg/mL in water. Sterilize by filtration through a 0.22 μm filter and store at −20 °C.
5. Isopropyl-β-d-thiogalactopyranoside (IPTG) stock: 0.1 M in water. Sterilize by filtration through a 0.22 μm filter and store at −20 °C.
6. Arabinose stock: 20 % in water. Sterilize by autoclaving and store at room temperature.
7. 4 L culture flasks with baffles (see Note 1).

8. *E. coli* strain BL21(DE3)omp8 transformed with plasmid pDG2, for expression of the PapC usher containing a C-terminal, thrombin-cleavable His-tag (see Notes 2 and 3). The plasmid confers ampicillin resistance and PapC expression is induced with arabinose.
9. *E. coli* strain BL21(DE3)omp8 transformed with plasmid pNH213, for expression of the FimD usher containing a C-terminal, thrombin-cleavable His-tag (see Notes 2 and 4). The plasmid confers ampicillin resistance and FimD expression is induced with IPTG.
10. *E. coli* strain Tuner transformed with plasmids pHJ9210 and pTN17, for expression of the PapD chaperone containing a C-terminal His-tag and the PapG adhesin (see Notes 5 and 6). The plasmids confer spectinomycin and ampicillin resistance, respectively; PapD is induced with arabinose and PapG with IPTG.
11. *E. coli* strain Tuner transformed with plasmids pETS1000 and pHJ20, for expression of the FimC chaperone containing a C-terminal His-tag and the FimH adhesin (see Notes 5 and 7). The plasmids confer spectinomycin and ampicillin resistance, respectively; FimC is induced with arabinose and FimH with IPTG.
12. Incubator shaker.
13. Spectrophotometer.

2.2. Outer Membrane Isolation and Extraction

1. Centrifuge and associated rotor.
2. Ultracentrifuge and associated rotor.
3. Tube rocker.
4. French press for disruption of bacteria.
5. Protease inhibitor cocktail or tablets.
6. 20 mM Tris–HCl, pH 8.0.
7. French press buffer: 20 mM Tris–HCl, pH 8.0, 1× protease inhibitor cocktail.
8. Sarkosyl stock (sodium lauroylsarcosine): 10 % in water; prepare fresh.
9. Dodecyl-maltopyranoside (DDM) stock: 10 % in water; store at 4 °C.
10. OM extraction buffer: 20 mM Tris–HCl, pH 8.0, 0.3 M NaCl, 1 % DDM.

2.3. Protein Quantification and Thrombin Cleavage

1. Bicinchoninic acid (BCA) protein quantification assay (Pierce) (see Note 8).
2. $CaCl_2$ stock: 2.5 M in water; sterilize by 0.22 μm filtration and store at −20 °C.

3. Novagen restriction grade thrombin (EMD Chemicals).
4. Phenylmethylsulfonyl fluoride (PMSF) stock: 100 mM in Isopropanol; store at room temperature.

2.4. Chromatography for Usher Purification

1. FPLC chromatography system.
2. 150 mL superloop for injection of samples into FPLC.
3. 1 mL HisTrap HP column.
4. 1 mL HiTrap Chelating HP column.
5. 120 mL HiLoad 16/60 Superdex 200 prep grade gel filtration column.
6. Imidazole stock: 1 M in water; store –20 °C.
7. Cobalt chloride stock: 0.2 M in water; store at room temperature.
8. Lauryldimethylamine-oxide (LDAO; Affymetrix) stock: 100 mM in water; store at 4 °C.
9. Usher affinity buffer A: 20 mM Tris–HCl, pH 8.0, 0.3 M NaCl, 10 mM LDAO, 20 mM imidazole. Sterilize by 0.22 μm filtration; store at room temperature.
10. Usher affinity buffer B: 20 mM Tris–HCl, pH 8.0, 0.3 M NaCl, 10 mM LDAO, 0.4 M imidazole. Sterilize by 0.22 μm filtration; store at room temperature.
11. Gel filtration buffer: 20 mM Tris–HCl, pH 8.0, 120 mM NaCl, 5 mM LDAO. Sterilize by 0.22 μm filtration.
12. Centrifugal protein concentrator with a 50,000 molecular weight cutoff (MWCO) membrane.
13. Dialysis membrane with 7,000–10,000 molecular weight cutoff (MWCO).
14. SDS-PAGE system.
15. Coomassie blue stain.

2.5. Periplasm Preparation

1. Centrifuge and associated rotor.
2. 20 mM Tris–HCl, pH 8.0.
3. Spheroplasting buffer: 20 mM Tris–HCl, pH 8.0, 20 % sucrose; sterilize by autoclaving and store at 4 °C.
4. EDTA stock: 0.5 M in water; sterilize by autoclaving and store at room temperature.
5. Lysozyme stock: 30 mg/mL in water; prepare fresh.
6. $MgCl_2$ stock: 2 M in water; sterilize by autoclaving and store at room temperature.
7. DNase I (10,000 U/mL).
8. Periplasm dialysis buffer: 20 mM Tris–HCl, pH 8.0, 0.3 M NaCl.
9. Dialysis membrane with 7,000–10,000 MWCO

2.6. Chromatography for Chaperone-Adhesin Purification

1. FPLC chromatography system.
2. 150 mL superloop for injection of samples into FPLC.
3. 1 mL HisTrap HP column.
4. 1 mL Resource S cation exchange column.
5. Imidazole stock: 1 M in water; store –20 °C.
6. Nickel buffer A: 20 mM Tris–HCl, pH 8.0, 0.3 M NaCl, 20 mM imidazole. Sterilize by 0.22 μm filtration; store at room temperature.
7. Nickel buffer B: 20 mM Tris–HCl, pH 8.0, 0.3 M NaCl, 0.4 M imidazole. Sterilize by 0.22 μm filtration; store at room temperature.
8. MES buffer stock: 1 M in water, pH to 5.8 for PapD-PapG or 5.4 for FimC-FimH (see Note 9).
9. S buffer A: 20 mM MES, pH 5.8 or 5.4.
10. S buffer B: 20 mM MES, pH 5.8 or 5.4, 1 M NaCl.
11. Centrifugal protein concentrator with a 10,000 molecular weight cutoff (MWCO) membrane.
12. SDS-PAGE system.
13. Coomassie blue stain.
14. Dialysis membrane with 7,000–10,000 MWCO.
15. Centrifuge and associated rotor.

3. Methods

Except where noted, all steps following bacterial culture should be carried out on ice or at 4 °C as much as possible. Breakage of bacteria using the French press may be carried out at room temperature, but the French pressure cell should be prechilled at 4 °C and all solutions kept on ice. Similarly, the chromatography steps may be carried out at room temperature, but the collection tubes should be on ice. The chromatography steps are described for an FPLC system and related columns; equivalent chromatography setups may be used instead.

3.1. Purification of the PapC Usher

The PapC protein contains a C-terminal, thrombin-cleavable His-tag and is expressed from an arabinose-inducible promoter. This protocol is for purification from 2 L of total bacterial culture, which yields of ~3 mg of the dimeric PapC usher. Subheadings 3.1.1 and 3.1.2 are described for each 1 L of culture. The resulting output from Subheadings 3.1.1 and 3.1.2 should then be combined for the chromatography procedures beginning at Subheading 3.1.3.

3.1.1. Bacterial Culture

1. For each 1 L of culture, inoculate a 125 mL flask containing 30 mL of LB supplemented with 100 μg/mL ampicillin with bacteria scraped from a frozen stock of BL21(DE3)Omp8/pDG2 (see Note 10).
2. Incubate overnight at 37 °C with shaking at 200 rpm.
3. The next day, pour all of the 30 mL culture (which should be highly turbid due to bacterial growth) into a 4 L flask containing 1 L of LB (pre-warmed to 37 °C, see Note 11) supplemented with 100 μg/mL ampicillin.
4. Incubate at 37 °C with shaking at 200 rpm until the cultures reach an OD_{600} of approximately 0.6 (about 2 h).
5. Induce the cultures for expression of the PapC usher by adding arabinose to 0.1 % final concentration and growing as before for an additional 2 h.

3.1.2. Outer Membrane Isolation and Extraction

1. Transfer the culture to centrifuge bottles and pellet the bacteria by centrifugation at 4,500 × *g*, 15 min, 4 °C. Decant and discard the supernatant (see Note 12).
2. For each 1 L of culture, gently resuspend the bacteria in French press buffer and bring to a total volume of 30 mL with the same buffer.
3. Break the resuspended bacteria by two passes through a French pressure cell at 14,000 psi. The output from the French press should be collected into two 30 mL polypropylene centrifuge tubes on ice (see Note 13).
4. Add 10 mL of 20 mM Tris–HCl, pH 8.0 to each tube and invert several times to mix.
5. Spin out unbroken bacteria by centrifugation at 11,000 × *g*, 20 min, 4 °C. Carefully transfer the supernatants into one 70 mL ultracentrifuge tube containing 2.5 mL of Sarkosyl stock solution (to give a 0.5 % final concentration of Sarkosyl; see Note 14).
6. Rock the tube for 5 min at room temperature.
7. Pellet the OM by ultracentrifugation at 100,000 × *g*, 65 min, 4 °C. Remove and discard the supernatant.
8. Resuspend the OM pellet in 30 mL OM extraction buffer and solubilize the usher from the OM by rocking overnight at 4 °C (see Note 15).
9. Following extraction, spin out the insoluble material by ultracentrifugation at 100,000 × *g*, 45 min, 4 °C; save the supernatant fractions containing the solubilized OM.

3.1.3. Chromatography of the Usher

1. Combine the solubilized OM from 2 L of culture and add imidazole stock solution to a 20 mM final concentration.

2. Capture the His-tagged usher protein by passing over a 1 mL HisTrap HP column (nickel sepharose) using an FPLC apparatus. Inject the sample over the column using a 150 mL superloop. Start with 100 % usher affinity buffer A and use a flow rate of 1 mL/min. Wait for the A_{280} reading to decrease back to baseline, indicating that all of the nonbinding proteins have passed through the column (this is the flow-through fraction). Next, do a three-step elution of 10, 60, and 100 % usher affinity buffer B, waiting at each step for the A_{280} to return to baseline before moving to the next step (see Note 16) (Fig. 1). The usher should elute at 60 % buffer B.
3. Dialyze the peak fractions containing the usher against three changes of gel filtration buffer; let the final dialysis step go overnight (see Note 17).
4. The next day, centrifuge the fractions at 10,000 × *g*, 10 min, 4 °C to remove any protein aggregates; transfer the supernatants to new tubes and store at 4 °C.
5. Combine the dialyzed peak fractions containing the usher together and perform a BCA assay to quantify the protein concentration.
6. To remove the His-tag from the usher, first add $CaCl_2$ stock solution to 2.5 mM final concentration and then add thrombin to a final concentration of 1.0 U thrombin per mg protein (see Note 18). Digest overnight with rocking at room temperature.
7. Add PMSF stock solution to a 100 μM final concentration to inhibit the thrombin, and centrifuge the sample at 10,000 × *g*, 10 min, 4 °C to pellet out any aggregates (see Note 19).
8. Inject the sample over a 1 mL HisTrap HP column again; the starting buffer A in this case is 100 % gel filtration buffer and buffer B is the same as for the first HisTrap column (item 10 Subheading 2.4). Note that for this run, the usher protein with His-tag removed will flow through the column; the flow through fraction is the fraction you want to keep (see Note 20).
9. Analyze a small amount of the flow-through fraction by SDS-PAGE and Coomassie blue staining to check the quality of the protein. Store the fraction at 4 °C.
10. Concentrate the flow-through fraction to 2 mL using a centrifugal protein concentrator with a 50,000 MWCO membrane (see Note 21).
11. Perform a final gel filtration chromatography step using the FPLC by running the sample over a HiLoad 16/60 Superdex 200 prep grade gel filtration column. The running buffer is 100 % gel filtration buffer and flow rate is 1 mL/min.

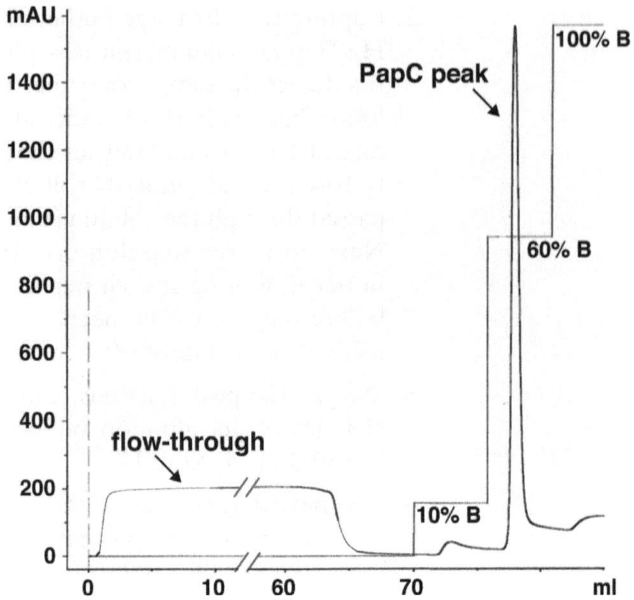

Fig. 1. Nickel affinity purification of the PapC usher. Solubilized OM containing His-tagged PapC was run over a 1 mL HisTrap column. The A_{280} absorbance trace is shown (mAU), along with the elution steps (% usher affinity buffer B). The broad peak at 0 % buffer B is the flow-through fraction. The sharp peak at 60 % buffer B contains the PapC usher protein.

The PapC protein should come off the column in a single major peak at ~58 mL, corresponding to the usher dimer.

12. Store the purified protein at 4 °C or quick-freeze aliquots in liquid nitrogen and store at −80 °C.

3.2. Purification of the FimD Usher

The FimD protein contains a C-terminal, thrombin-cleavable His-tag and is expressed from an IPTG-inducible promoter. We have found that the FimD usher must be expressed at a lower level than PapC, to ensure that FimD is properly inserted and folded in the OM (mislocalization or misfolding of the usher due to overexpression will be evident as usher protein that does not extract from the OM with detergent). This protocol is for purification from 16 L of total bacterial culture, which yields ~1 mg of FimD usher. Subheadings 3.2.1 and 3.2.2 are described for each 1 L of culture. The combined output from 4 L of culture should then be used for the first affinity chromatography step (steps 1–3 of Subheading 3.2.3). Steps 1–3 of Subheading 3.2.3 are repeated three times to process the total 16 L of culture. The combined output from step 3 of Subheading 3.2.3 is then used for steps 4–9 of Subheading 3.2.3.

3.2.1. Bacterial Culture

1. For each 1 L of culture, inoculate a 125 mL flask containing 30 mL of LB supplemented with 100 μg/mL ampicillin with bacteria scraped from a frozen stock of BL21(DE3)Omp8/pNH213 (see Note 10).
2. Incubate overnight at 37 °C with shaking at 200 rpm.
3. The next day, pour all of the 30 mL culture (which should be highly turbid due to bacterial growth) into a 4 L flask containing 1 L of LB (pre-warmed to 37 °C, see Note 11) supplemented with 100 μg/mL ampicillin.
4. Incubate at 37 °C with shaking at 200 rpm for 3 h (see Note 22).
5. Induce the cultures for expression of the FimD usher by adding IPTG to 100 μM final concentration. Grow the induced cultures overnight with shaking at room temperature.

3.2.2. Outer Membrane Isolation and Extraction

1. Transfer the culture to centrifuge bottles and pellet the bacteria by centrifugation at 4,500 × *g*, 15 min, 4 °C. Decant and discard the supernatant (see Note 12).
2. For each 1 L of culture, gently resuspend the bacteria in French press buffer and bring to a total volume of 30 mL with the same buffer.
3. Break the bacteria using a French press and solubilize FimD from the OM as described for PapC (Subheading 3.1.2).

3.2.3. Chromatography of the Usher

1. Combine the solubilized OM from 4 L of culture and add imidazole stock solution to a 20 mM final concentration.
2. Pellet insoluble material by ultracentrifugation at 100,000 × *g* for 45 min at 4 °C.
3. Capture the His-tagged usher protein by passing over a 1 mL HiTrap Chelating HP column freshly charged with 2 mL 0.2 M cobalt chloride (see Note 23). Inject and elute the FimD sample as described for PapC (steps 1–4 Subheading 3.1.3). Repeat this step three times to process the total 16 L of original culture.
4. Combine the dialyzed peak fractions containing the usher together and perform a BCA assay to quantify the protein concentration.
5. Add $CaCl_2$ stock solution to 2.5 mM and digest with thrombin to remove the His-tag as described for PapC (steps 5–6 Subheading 3.1.3).
6. Add PMSF stock solution to 100 μM final concentration to inhibit the thrombin, and centrifuge the sample at 10,000 × *g*, 10 min, 4 °C to pellet out any aggregates (see Note 19).
7. Inject the sample over a 1 mL HiTrap Chelating HP column freshly charged with 2 mL 0.2 M cobalt chloride. Collect the

flow-through fraction as described for PapC (steps 7–9 Subheading 3.1.3).

8. Concentrate the flow-through fraction and run over a gel Filtration column as described for PapC (steps 10–11 Subheading 3.1.3). FimD typically elutes from the gel filtration column in two peaks—one at ~49 mL, which corresponds to the usher dimer (see Note 24) and a second at around ~63 mL, which is the usher monomer.
9. Store the purified protein at 4 °C or quick-freeze aliquots in liquid nitrogen and store at −80 °C.

3.3. Purification of Chaperone-Adhesin Complexes

The PapD and FimC chaperones contain a C-terminal His-tag and are expressed from an arabinose-inducible promoter. The PapG and FimH adhesins are expressed from IPTG-inducible promoters. This protocol is for purification from 4 L of total bacterial culture, which yields ~5 mg of PapDG and ~2 mg of FimCH. Subheadings 3.3.1 and 3.3.2 are described for each 1 L of culture. The resulting output from Subheadings 3.3.1 and 3.3.2 should then be combined for the chromatography procedures beginning at Subheading 3.3.3 (see Note 25).

3.3.1. Bacterial Culture

1. For each 1 L of culture, inoculate a 125 mL flask containing 30 mL of LB supplemented with 100 μg/mL ampicillin and 100 μg/mL spectinomycin with colonies from a LB agar plate of Tuner/pHJ9210 + pTN17 or Tuner/pETS1000 + pHJ20.
2. Incubate overnight at 37 °C with shaking at 200 rpm.
3. The next day, pour all of the 30 mL culture (which should be highly turbid due to bacterial growth) into a 4 L flask containing 1 L of LB (pre-warmed to 37 °C, see Note 11) supplemented with 100 μg/mL ampicillin and 100 μg/mL spectinomycin.
4. Incubate at 37 °C with shaking at 200 rpm until the cultures reach an OD_{600} of approximately 0.6 (about 2 h).
5. Induce the cultures for expression of the chaperone and adhesin by adding arabinose to 0.002 % final concentration and IPTG to 1 mM final concentration and grow as before for before for an additional 2 h.

3.3.2. Periplasm Preparation

1. Transfer the culture to centrifuge bottles and pellet the bacteria by centrifugation at 4,500 × *g*, 15 min, 4 °C. Decant and discard the supernatant (see Note 26).
2. For each 1 L of culture, gently resuspend the bacteria in 20 mM Tris–HCl, pH 8.0 and bring to 300 mL total volume in one centrifuge bottle.
3. Pellet the bacteria as in step 1 above, then resuspend in 20 mL spheroplasting buffer and transfer to a 30 mL polypropylene centrifuge tube.

4. Add 260 μL of 0.5 M EDTA stock solution and 150 μL of lysozyme stock solution, mix, and incubate 40 min on ice (see Note 27).
5. Next, add 260 μL of $MgCl_2$ stock solution and 200 μL of DNase I, and incubate an additional 20 min on ice (see Note 28).
6. Spin out the spheroplasted bacteria by centrifugation at 12,000 × *g*, 30 min, 4 °C. Carefully decant and save the supernatant, which contains the periplasm fraction.
7. Dialyze the periplasm against three changes of periplasm dialysis buffer; let the final dialysis step go overnight (see Note 29).

3.3.3. Chromatography of the Chaperone-Adhesin Complex

1. Combine the dialyzed periplasms from 4 L of culture and filter using a 0.22 μm, 33 mm non-protein binding syringe filter; make note of the total final volume of filtered periplasm (see Note 30).
2. Divide the filtered periplasm into two equal aliquots and store one aliquot at 4 °C (see Note 31).
3. Add imidizole stock solution to 20 mM final concentration to the other aliquot and run over a 1 mL HisTrap HP nickel column to capture the His-tagged chaperone-adhesin complexes (see Note 32), following the protocol described in steps 1–4 Subheading 3.1.3, with the following modifications. Use Nickel buffers A and B. Start with 100 % buffer A and do a four-step elution of 10, 60, and 100 % buffer B. The chaperone-adhesin complexes should be in the 60 % B peak.
4. Analyze the fractions by SDS-PAGE and Coomassie blue staining, using a 15 % gel, to identify fractions containing both the chaperone and adhesion (see Note 33).
5. Combine these fractions together and dilute to 10 mL total volume with nickel buffer A to reduce the possibility of protein aggregation.
6. Dialyze against three changes of S buffer; let the final dialysis step go overnight.
7. The next day, centrifuge the fractions at 10,000 × *g*, 10 min, 4 °C to remove any protein aggregates; transfer the supernatant to a new tube and store at 4 °C. Repeat the nickel column purification with the other aliquot of filtered periplasm and combine the resulting dialyzed samples together.
8. To isolate 1:1 molar ratio chaperone-subunit complexes, run half of the dialyzed samples over a Resource S cation exchange column using the FPLC (see Note 34). Start with 100 % S Buffer A and a flow rate of 1 mL/min, and do a three stage elution as follows: (1) linear gradient from 0–20 % B over 6 mL; (2) linear gradient from 20 to 45 % B over 40 mL (the chaperone-adhesin complexes will be here) (see Note 35); (3) step to 100 % B (Fig. 2). Repeat for the other half of the sample.

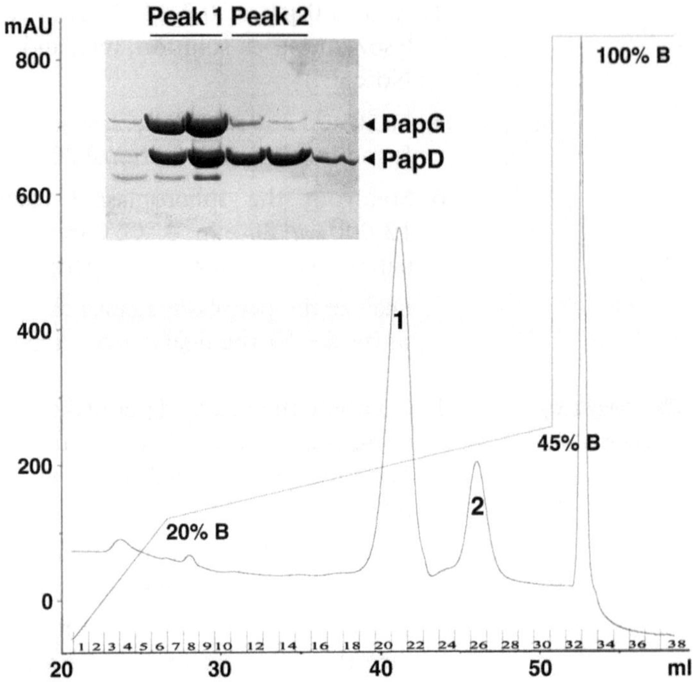

Fig. 2. S column purification of PapDG chaperone-adhesin complex. PapDG complex from the nickel affinity column was run over a 1 mL Resource S cation exchange column. The A_{280} absorbance trace is shown (mAU), along with the elution gradient (% S buffer B). The inset shows a Coomassie-stained SDS-PAGE gel with fractions from the peaks labeled *1* and *2*. Peak 1 contains 1:1 PapD-PapG complex. Peak 2 contains excess free PapD chaperone.

9. Analyze the fractions from the column by SDS-PAGE and Coomassie blue staining to identify fractions containing 1:1 chaperone-adhesion complexes (Fig. 3.2).
10. Concentrate the desired fractions, if necessary, using a centrifugal protein concentrator with a 10,000 MWCO membrane. Store the purified protein at 4 °C.

4. Notes

1. The baffles increase aeration of the culture and allow for increased bacterial growth.
2. BL21(DE3)Omp8 is a multi-porin mutant strain (10), allowing for purification of the usher without contamination from the abundant OM porin proteins. In addition, *E. coli* B strains, including BL21, lack the OM OmpT protease, which can cleave the usher during breakage of the bacteria.

3. Plasmid pDG2 confers ampicillin resistance and encodes *papC* with a C-terminal, thrombin-cleavable His-tag under control of an arabinose-inducible promoter (11).
4. Plasmid pNH213 is derived from plasmid pETS4 (12), confers ampicillin resistance, and encodes *fimD* with a C-terminal thrombin-cleavable His-tag under control of an IPTG-inducible promoter.
5. Tuner is derived from the *E. coli* BL21 strain and has a linear response to IPTG (Novagen).
6. Plasmid pTN17 confers ampicillin resistance and encodes *papGII* under control of an IPTG-inducible promoter (13). The class II version of the PapG adhesin is more stable than class I PapG. Plasmid pHJ9210 confers spectinomycin resistance and encodes *papD* with a C-terminal His-tag under control of an arabinose-inducible promoter (14).
7. Plasmid pETS1000 confers spectinomycin resistance and encodes *fimC* with a C-terminal His-tag under control of an arabinose-inducible promoter (12). Plasmid pHJ20 confers ampicillin resistance and encodes *fimH* under control of an IPTG-inducible promoter (15).
8. The BCA assay is compatible with the detergent LDAO; the Bradford protein assay is not.
9. To prepare the MES buffer, let the MES dissolve a bit, then adjust the pH using a few NaOH pellets, and then adjust to the final desired pH with 1 N or 5 N NaOH solution.
10. Inoculating directly from a frozen stock minimizes chances for the porin-minus BL21(DE3)Omp8 strain to accumulate compensatory mutations.
11. Do not grow more than a 1 L culture in each 4 L flask. A larger culture will result in slower growth. The LB broth should be pre-warmed to 37 °C before inoculation to prevent a lag in growth (the flasks with broth can be placed in a 37 °C incubator the day before). Add the antibiotics just prior to adding the bacteria.
12. The bacterial pellets may be frozen at this step.
13. The French press is highly efficient at disrupting bacteria. Other cell disruption procedures may be used if a French press is not available. For the most efficient breakage of bacteria using the French press, open the outlet valve just enough so that individual drops of liquid can be discerned in the outflow stream.
14. The detergent Sarkosyl selectively solubilizes the cytoplasmic membrane. The OM, with its asymmetric bilayer containing LPS, is not solubilized by Sarkosyl.

15. Resuspend the membrane pellet in 2 mL 20 mM Tris-HCl (pH 8.0), 0.3 M NaCl (no DDM), then make the volume up to a total of 30 mL by adding 24 mL 20 mM Tris -HCl (pH 8.0), 0.3 M NaCl, and 3 mL 10 % DDM stock solution. The OM pellet should appear a translucent white. If the pellet appears yellow, this indicates the presence of cytoplasmic membrane. The waxy OM pellet is difficult to resuspend. Start by carefully breaking up the pellet into solution using a pipet tip or a Pasteur pipette with a sealed tip (seal the Pasteur pipette by melting with a Bunsen burner). Gently pipet the resuspended pieces of the OM pellet up and down, starting with a large bore pipet tip and moving to smaller bore pipet tips, until all the membrane pieces are broken up and evenly suspended.
16. Imidazole absorbs at A_{280}; therefore, the baseline will increase with increasing concentration of imidazole (increasing % B).
17. Run the peak fractions on an SDS-PAGE gel the first time you do the purification to check for the presence of the usher protein (PapC is 89 kDa and FimD is 91 kDa), but generally the whole 60 % B peak should be taken for the subsequent steps. The fractions must be dialyzed the same day as the purification to remove imidazole, which may cause the usher protein to aggregate. After dialysis, the samples may be stored at 4 °C.
18. Novagen thrombin is not always 1 U/μL; the concentration varies with batch.
19. There should not be a large pellet.
20. Running the protein back over the metal affinity column has the added benefit of improving purity by removing contaminants that bind to the column.
21. Use of smaller MWCO will result in excessive concentration of the detergent.
22. OD_{600} should be ~0.8–1.2, but the exact OD is not important.
23. FimD sticks too tightly to a nickel column, and therefore cobalt is used instead.
24. The ~49 mL FimD gel filtration peak elutes earlier than expected for the usher dimer and may represent a mixture of dimers and dimers of dimers.
25. This protocol can be used to purify any chaperone-subunit complex from either the Pap or Fim system, although the non-adhesin subunits do not behave as well as the adhesins for purification.
26. If you wish to freeze the bacterial pellets at this step, first wash the bacteria with 20 mM Tris–HCl, pH 8.0 and then spin down again and freeze the pellets.

27. Treatment of bacteria with lysozyme and ETDA generates spheroplasts. The EDTA weakens the OM barrier by chelating divalent cations that bridge the negatively charged LPS head groups. The weakened OM allows the lysozyme to enter the periplasm and degrade the peptidoglycan.
28. Addition of $MgCl_2$ stabilizes the spheroplasts by stabilizing the cytoplasmic membrane. DNase is added because there is typically some leakage of cytoplasmic constituents into the periplasm, including DNA, which is highly viscous and can interfere with pelleting of the spheroplasts.
29. For the periplasm dialysis, note that the volume will increase due to influx of water. Divide the periplasm into two aliquots and put each in a wide-diameter dialysis tubing such as SnakeSkin (Pierce). Leave room in the dialysis tubing for swelling of the solution (around a quarter to a third of the volume of the periplasm). Remove air from the dialysis tubing and place the dialysis clamps so as to leave free tubing to accommodate swelling.
30. The filters will clog during filtration of the periplasm; roughly one filter will be needed per 10 mL of sample. Some periplasm will be lost during filtration and therefore note the final total volume obtained.
31. After filtering, the periplasm can be stored at 4 °C for at least a week.
32. Only reuse the HisTrap column about three times; discard when looking brownish near top.
33. Because periplasm fractions are being run over the nickel column, the fractions will have a lot of unwanted proteins in them. There will also be excess free chaperone (PapD or FimC). Take only the fractions containing both the chaperone and adhesin. Excess unwanted proteins (including excess free chaperone) can cause aggregation during dialysis. There will be some aggregation, but there should not be a lot. PapD-His runs at 28.5 kDa, PapG at 36 kDa, FimC-His at 23.5 kDa, and FimH at 29 kDa.
34. Running only half the sample over the cation exchange column at once is suggested to prevent overloading of the column.
35. There will be two peaks close together—the first will be the chaperone-adhesin complex, and the second will be free chaperone alone.

Acknowledgments

Work in the Thanassi laboratory is supported by NIH grants GM62987 and AI055621.

References

1. Nuccio SP, Baumler AJ (2007) Evolution of the chaperone/usher assembly pathway: fimbrial classification goes Greek. Microbiol Mol Biol Rev 71:551–575
2. Waksman G, Hultgren SJ (2009) Structural biology of the chaperone-usher pathway of pilus biogenesis. Nat Rev Microbiol 7: 765–774
3. Zav'yalov V, Zavialov A, Zav'yalova G, Korpela T (2010) Adhesive organelles of Gram-negative pathogens assembled with the classical chaperone/usher machinery: structure and function from a clinical standpoint. FEMS Microbiol Rev 34:317–378
4. Sauer FG, Pinkner JS, Waksman G, Hultgren SJ (2002) Chaperone priming of pilus subunits facilitates a topological transition that drives fiber formation. Cell 111:543–551
5. Zavialov AV, Berglund J, Pudney AF, Fooks LJ, Ibrahim TM, MacIntyre S, Knight SD (2003) Structure and biogenesis of the capsular F1 antigen from Yersinia pestis: preserved folding energy drives fiber formation. Cell 113:587–596
6. Phan G, Remaut H, Wang T, Allan WJ, Pirker KF, Lebedev A, Henderson NS, Geibel S, Volkan E, Yan J, Kunze MBA, Pinkner JS, Ford B, Kay CWM, Li H, Hultgren S, Thanassi DG, Waksman G (2011) Crystal structure of the FimD usher bound to its cognate FimC-FimH substrate. Nature 474:49–53
7. Remaut H, Tang C, Henderson NS, Pinkner JS, Wang T, Hultgren SJ, Thanassi DG, Waksman G, Li H (2008) Fiber formation across the bacterial outer membrane by the chaperone/usher pathway. Cell 133:640–652
8. Roberts JA, Marklund B-I, Ilver D, Haslam D, Kaack MB, Baskin G, Louis M, Mollby R, Winberg J, Normark S (1994) The Galα(1–4) Gal-specific tip adhesin of *Escherichia coli* P-fimbriae is needed for pyelonephritis to occur in the normal urinary tract. Proc Natl Acad Sci U S A 91:11889–11893
9. Mulvey MA, Lopez-Boado YS, Wilson CL, Roth R, Parks WC, Heuser J, Hultgren SJ (1998) Induction and evasion of host defenses by type 1-piliated uropathogenic *Escherichia coli*. Science 282:1494–1497
10. Prilipov A, Phale PS, Van Gelder P, Rosenbusch JP, Koebnik R (1998) Coupling site-directed mutagenesis with high-level expression: large scale production of mutant porins from E. coli. FEMS Microbiol Lett 163:65–72
11. Li H, Qian L, Chen Z, Thahbot D, Liu G, Liu T, Thanassi DG (2004) The outer membrane usher forms a twin-pore secretion complex. J Mol Biol 344:1397–1407
12. Saulino ET, Thanassi DG, Pinkner JS, Hultgren SJ (1998) Ramifications of kinetic partitioning on usher-mediated pilus biogenesis. EMBO J 17:2177–2185
13. Li Q, Ng TW, Dodson KW, So SS, Bayle KM, Pinkner JS, Scarlata S, Hultgren SJ, Thanassi DG (2010) The differential affinity of the usher for chaperone-subunit complexes is required for assembly of complete pili. Mol Microbiol 76:159–172
14. Jones CH, Danese PN, Pinkner JS, Silhavy TJ, Hultgren SJ (1997) The chaperone-assisted membrane release and folding pathway is sensed by two signal transduction systems. EMBO J 16:6394–6406
15. Jones CH, Pinkner JS, Roth R, Heuser J, Nicholoes AV, Abraham SN, Hultgren SJ (1995) FimH adhesin of type 1 pili is assembled into a fibrillar tip structure in the *Enterobacteriaceae*. Proc Natl Acad Sci U S A 92:2081–2085

Chapter 4

Experimental Manipulation of the Microbial Functional Amyloid Called Curli

Yizhou Zhou, Daniel R. Smith, David A. Hufnagel, and Matthew R. Chapman

Abstract

Curli are proteinaceous fibrous structures produced on the surface of many gram-negative bacteria. As a major constituent of the extracellular matrix, curli mediate interactions between the bacteria and its environment, and as such, curli play a critical role in biofilm formation. Curli fibers share biophysical properties with a growing number of remarkably stable and ordered protein aggregates called amyloid. Here we describe experimental methods to study the biogenesis and assembly of curli by exploiting their amyloid properties. We also present methods to analyze curli-mediated biofilm formation. These approaches are straightforward and can easily be adapted to study other bacterially produced amyloids.

Key words: Curli, Congo red, Western blot analysis, Plug western blot analysis, Polymerization, Seeding, Pellicle biofilm

1. Introduction

Many enteric bacterial including *Escherichia coli*, *Salmonella* spp., and *Citrobacter* spp. produce a class of proteinaceous amyloid fibers on their cell surfaces called curli (Fig. 1a) (1).

Curli mediate bacterial cell–cell interactions (2), host-pathogen interactions (3–5) and attachment to biotic and abiotic surfaces (6–8). Extracellular matrixes formed by curli and cellulose increase bacterial resistance to environmental stresses and antimicrobial agents (9). Curli also play a critical role in biofilm formation (10–12). Some pathogenic *E. coli* strains and *Salmonella* spp. form a curli-dependent pellicle biofilm at the air–liquid interface (11, 12). Chemicals designed to inhibit curli assembly efficiently abolish the

Anne H. Delcour (ed.), *Bacterial Cell Surfaces: Methods and Protocols*, Methods in Molecular Biology, vol. 966,
DOI 10.1007/978-1-62703-245-2_4, © Springer Science+Business Media New York 2013

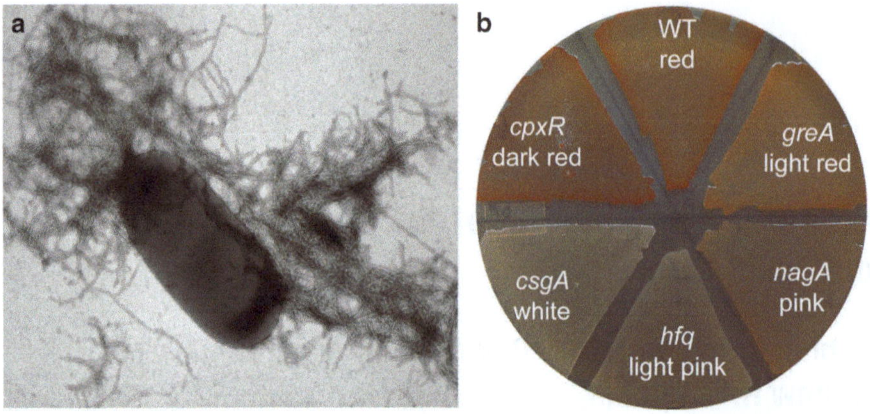

Fig. 1. Curliated *E. coli* stains red on CR agar. (**a**) An EM micrograph of a curliated *E. coli*. (**b**) Use of CR assay to screen for curli mutants. Wild-type BW25113 and mutants as indicated in the figure were grown on YESCA CR agar at 26°C for 48 h.

pellicle formation (12). Therefore, curli provide a new target for the development of antimicrobial and anti-biofilm agents.

The biogenesis of curli is complex and highly regulated (13). At least seven proteins, encoded by divergent operons *csgDEFG* and *csgBAC* (*csg*, curli-specific genes), are involved in fiber assembly (14). CsgD is a transcriptional regulator required for activation of the *csgBAC* operon (14) and the regulation of CsgD is controlled by a complex set of environmental cues (13, 15). CsgA and CsgB are the major and minor subunits of curli fibers, respectively (16, 17). CsgB is associated with the outer-membrane via the C-terminal domain (18). The translocation of CsgA and CsgB to the bacterial surface is mediated by the outer-membrane lipoprotein CsgG (19) and the periplasmic chaperone-like proteins CsgE and CsgF (20, 21). CsgC is involved in the pore activity of CsgG (22).

Once secreted to the cell surface, curli subunits self-assemble into fibers amidst an extracellular environment lacking an energy supply and dynamic physical conditions, yet the assembly process needs to be highly efficient and the resulting fibers need to be resistant to environmental stresses. This makes amyloid the perfect fold for curli. Defined by their shared biophysical characteristics, all amyloids are β-sheet rich, ordered fibers with 6–10 nm widths and extreme resistant to heat, chemical and enzymatic denaturation (23–25). Amyloids are traditionally linked with protein misfolding and neurodegenerative ailments such as Alzheimer's and prion-based encephalopathies (26, 27). Microbial amyloids like curli represent a rapidly growing class of "functional amyloids," and unlike their disease-associated counterparts, are designed to facilitate certain biological tasks (28–31). Functional amyloids have been found in all walks of cellular life including bacteria (curli, TasA, chaplins, FapA, etc.) (32–35), fungi (Sup35, Het-S, etc.) (36, 37), and mammalian cells (Pmel17 and peptide signaling hormones) (38, 39).

Regardless of their origin, all amyloids share distinguishing biophysical properties. Curli, like all amyloids, bind amyloid specific dyes Congo red (CR) and Thioflavin T (ThT), and are extraordinarily stable and resistant to SDS and other chemical denaturation (35). In vitro, CsgA and CsgB polymerize into fibers in a nucleation dependent pattern with a lag phase, an exponential phase and a stationary phase (18, 40). The fibrillation of CsgA is promoted by its own fibers or by CsgB fibers in a process called seeding (18, 40). In vivo, the assembly of CsgA into curli fibers requires the function of CsgB. Without CsgB, CsgA proteins are secreted into the extracellular environment in an SDS-soluble, unpolymerized form (18).

In this chapter, we describe both in vivo and in vitro approaches for analyzing curli assembly and curli-dependent biofilm formation. The CR assay provides a simple method to assess curli production and can be adapted for large-scale screening (14, 35). The western blot assay analyzes the presence and integrity of curli and is able to distinguish between different polymerization states of the CsgA protein (29). We also present approaches to purify CsgA and CsgB from *E. coli* for in vitro characterization of curli subunits (40, 41). The polymerization kinetics of purified proteins and the seeding process can be followed by ThT fluorescence (35, 40). Finally, we describe methods to grow curli-dependent pellicle biofilms and basic approaches for biofilm quantification. These assays are straightforward and can be performed with common lab equipment.

2. Materials

Prepare solutions and liquid media using ultrapure water and store them at room temperature (RT) unless otherwise indicated. Store agar plates at 4°C.

2.1. Congo Red (CR) Assays to Asses Curli Production

1. Luria-Bertani (LB) agar plates: 10 g/L bacto tryptone, 5 g/L yeast extract, 10 g/L sodium chloride, and 17 g/L agar.
2. YESCA agar plates: 10 g/L casamino acids, 1 g/L yeast extract, and 20 g/L agar (see Note 1).
3. CR stock: dissolve 1 g of Congo red in 100 mL of water and filter sterilize. Store at 4°C.
4. Brilliant Blue stock: dissolve 1 g Brilliant Blue G250 dye in 100 mL water and filter sterilize. Store at 4°C.
5. YESCA CR agar plates: Add sterile CR stock and Brilliant Blue stock to YESCA agar after autoclaving to make a final concentration of 50 μg/mL CR and 1 μg/mL Brilliant Blue (see Note 2).

6. KPi buffer: 50 mM potassium phosphate buffer, pH 7.2; make 28.9 mM KH_2PO_4 and 21.1 mM K_2HPO_4 in water.
7. CR solution: 0.5 g/L CR in KPi buffer.

2.2. Western Blot Analysis of Curli Proteins

1. Spectrometer.
2. Acid-resistant centrifugal vacuum concentrator.
3. Microcentrifuge and tubes.
4. Polyvinylidene difluoride (PVDF) and Nitrocellulose (0.2 μM) membranes.
5. Whatman filter papers.
6. Pasteur pipettes with 8 mm diameter.
7. Bath sonicator.
8. Heat blot.
9. Protein electrophoresis apparatus.
10. Semidry transfer apparatus.
11. Wet transfer apparatus.
12. Methanol.
13. Platform shaker.
14. Chemiluminescent substrates.
15. Autoradiography film and cassettes for film exposure.
16. YESCA agar plates: see Subheading 2.1. Prepare thick YESCA agar plates for standard western blot analysis by pouring 25 mL of autoclaved YESCA agar per standard petri dish. For thin YESCA plates used in plug western blots, pour 15 mL of autoclaved YESCA agar per petri dish.
17. KPi buffer: see item 6 Subheading 2.1.
18. Hexafluoroisopropanol (HFIP) or formic acid (FA) (see Note 3).
19. 2× SDS sample buffer: 62.5 mM Tris–HCl, pH 6.8, 5% (v/v) β-mercaptoethanol, 0.01% (m/v) bromophenol blue, 10% (v/v) glycerol, 3% (m/v) SDS.
20. SDS-polyacrylamide gel (SDS-PAGE): 15% separating gel and 3% stacking gel.
21. SDS running buffer: 3.03 g/L Tris base, 14.4 g/L glycine, and 1 g/L SDS.
22. Semidry transfer buffer: 3.03 g/L Tris Base, 0.75 g/L glycine, 38.6 mg/L dithiothreitol (DTT), and 20% (v/v) methanol.
23. Wet transfer buffer: 5.53 g/L CAPS and 10% (v/v) methanol. Use NaOH pellets to adjust the pH to 11.2. Make fresh cold wet transfer buffer before usage.
24. Tris buffered saline–Tween-20 (TBS-T): 8 g/L NaCl, 0.2 g/L KCl, and 3 g/L Tris–HCl. Adjust pH to 7.5 with HCl.

Add Tween-20 to make a final concentration of 0.1% (v/v) Tween-20. Store the TBS-T buffer at 4°C.

25. Blocking solution: 1% (m/v) dry fat-free milk and 1% (m/v) bovine serum albumin (BSA) in TBS-T buffer.
26. Antibodies: CsgA antibody (Proteintech) was raised in rabbits against purified CsgA proteins. CsgB antibody (Proteintech) was raised in rabbits against a peptide fragment of the second repeating unit of CsgB. Store the antibodies at −20°C. Before use, dilute antibodies by 1:10,000 (v/v) for CsgA and 1:7,000 (v/v) for CsgB into blocking solution for primary probing (see Note 4). The diluted antibody in blocking solution can be stored at −20°C and be reused for 3–4 times.
27. Anti-rabbit secondary antibody: peroxidase antibody produced in goat. Store the antibody at −20°C. Dilute the antibody by 1:7,000 (v/v) into blocking solution before use.

2.3. Purification of Curli Subunits

2.3.1. Non-denaturing Purification of CsgA or Truncated CsgB from Bacterial Supernatant

1. Expression strain LSR12/pMC1/pMC3: (see Note 5).
2. LB medium: 10 g/L bacto tryptone, 5 g/L yeast extract and 10 g/L sodium chloride.
3. Antibiotics supplemented LB: LB supplemented with 100 μg/mL Ampicillin and 25 μg/mL chloramphenicol.
4. Isopropyl β-D-1-thiogalactopyranoside (IPTG) stock. Make a 1 M solution in water. Make 1 mL aliquots in 1.5 mL microcentrifuge tubes and store at −20°C.
5. Spectrometer.
6. Centrifuge and bottles.
7. 0.22 μm polyethersulfone bottle-top filters.
8. Kontes 2.5 × 10 cm and 1 × 30 cm glass columns.
9. Nickel-nitrilotriacetic acid agarose (Ni-NTA).
10. KPi buffer: see item 6 in Subheading 2.1. Store at 4°C.
11. Funnel.
12. Elution buffer 1: 100 mM imidazole in KPi buffer. Store at 4°C.
13. Microcentrifuge tubes.
14. Protein concentration determination assay.
15. Sephadex G-25 agarose.
16. 0.02 μm, Anotop filters.

2.3.2. Denaturing Purification of CsgA or CsgB from Bacterial Lysates

1. Expression strain NEB C2566/pNH11: (see Note 6).
2. Expression strain NEB C2566/pAN1: (see Note 6).
3. LB medium: 10 g/L bacto tryptone, 5 g/L yeast extract and 10 g/L sodium chloride.

4. Antibiotics supplemented LB: LB supplemented with 100 μg/mL Ampicillin (Amp) and 25 μg/mL chloramphenicol (Cm).
5. 1 M Isopropyl β-D-1-thiogalactopyranoside (IPTG) stock. Make 1 mL aliquots in 1.5 mL microcentrifuge tubes and store at −20°C.
6. Spectrometer.
7. Centrifuge and bottles.
8. Denaturing solution: 8 M Guanidine hydrochloride (GdnHCl) in KPi buffer.
9. 50 mL conical tubes.
10. Probe sonicator.
11. Nickel-nitrilotriacetic acid agarose (Ni-NTA).
12. Rocker.
13. Disposable polypropylene columns.
14. KPi buffer: see item 6 Subheading 2.1 Store at 4°C.
15. Column wash buffer: 12.5 mM imidazole in KPi buffer. Store at 4°C.
16. Elution buffer 2: 125 mM imidazole in KPi buffer. Store at 4°C.
17. Microcentrifuge tubes.
18. Protein determination assay.
19. 30 kDa centrifugal filter units.
20. Spin desalting columns.

2.4. Polymerization Assay

2.4.1. Polymerization from Monomers

1. Freshly purified CsgA or CsgB proteins.
2. Non-treated, flat-bottom, opaque 96-well microtiter plates.
3. Thioflavin T (ThT) solution: 1 mM of ThT in water. Sterile filter through a 0.22 μm filter and store at 4°C.
4. Transparent stickers for microtiter plates.
5. Fluorescent plate reader.

2.4.2. Seeding assay

1. Freshly formed CsgA fibers or CsgB fibers.
2. Freshly purified monomeric CsgA or CsgB.
3. Probe sonicator.
4. Non-treated, flat-bottom, opaque 96-well microtiter plates.
5. KPi buffer: see item 6 in Subheading 2.1.
6. ThT solution: see item 3 in Subheading 2.4.1.
7. Fluorescent plate reader.

2.5. Curli Dependent Biofilm Formation

2.5.1. Microtiter plate assay for ring-shaped biofilm

1. LB medium: see item 2 Subheading 2.3.1.
2. LB-low salt medium: 10 g/L bacto tryptone, 5 g/L yeast extract.
3. YESCA medium: 10 g/L casamino acid, 1.2 g/L yeast extract (see Note 7).
4. Cell culture treated, flat bottom, transparent 96-well microtiter plates with lids.
5. Crystal violet solution: 0.1% (w/v) in water.
6. 95% (v/v) ethanol.
7. Non-sterile, flat bottom, transparent 96-well microtiter plates.
8. Platform shaker.
9. Plate reader.

2.5.2. Pellicle Biofilm Assay

1. LB medium: see item 2 Subheading 2.3.1.
2. Disposable borosilicate glass tubes with caps.
3. Cell culture treated, flat bottom 24-well microtiter plates with lids.
4. LB-low salt medium: see item 2 in Subheading 2.5.1.
5. YESCA medium: see item 3 in Subheading 2.5.1. (see Note 7).
6. Crystal violet solution: see item 5 in Subheading 2.5.1.
7. Homogenizer.
8. 95% ethanol.
9. Spectrometer.

3. Methods

3.1. Cong Red (CR) Assays to Assess Curli Production

Most *E. coli* lab strains express curli at low temperature and in low salt conditions. Colonies of curliated bacteria stain red when grown on agar containing the diazo dye Cong red (CR). The CR assay provides an ideal assay for large-scale screens and for scoring curli assembly on bacterial surfaces (Fig. 1b). Curli can be detected on the cell surface after growth in the presence of CR or colonies may be post-stained with CR.

3.1.1. CR Assay for Bacteria on YESCA CR Agar

1. Streak out *E. coli* strains from a −80°C glycerol stock onto LB agar plate. Grow the bacteria at 37°C overnight.
2. Pick single colonies and streak out on a YESCA CR agar plate.

3. To induce curli production, grow bacteria on YESCA CR agar at 26°C for 48 h (see Notes 8 and 9).
4. Check the color of the bacterial colonies. Wild-type curli-producing *E. coli* cells stain red on YESCA CR agar, whereas curli defective mutants are usually pink or white. *E. coli* mutants with hyper curli production sometimes stain dark red (see Note 10).

3.1.2. Post-staining CR assay

1. Pick single colonies from an LB agar plate and re-streak out on YESCA agar plate without CR dye.
2. Incubate the bacteria at 26°C for 48 h to induce curli production.
3. After 48 h, gently pour 10 mL of CR solution on the colonies, stain for 5 min without shaking (see Note 11). Pour or vacuum off the CR solution after staining.
4. Gently pour 10 mL of KPi buffer to wash away the remaining CR dye on the agar. Remove the buffer.
5. Check the color of colonies. Curliated bacteria stain red and non-curliated bacteria are white or pink.

3.2. Western Blot Analysis of Curli Proteins

Western blot analysis of whole cell lysate is a standard way to analyze bacteria-associated curli (Fig. 2b). Curli are resistant to SDS treatment and remain as protein polymers unless treated with strong denaturants. Thus curli polymers which contain CsgA and CsgB need to be pretreated with formic acid (FA) or Hexafluoroisopropanol (HFIP) to release SDS-soluble monomers that can be detected by anti-CsgA and anti-CsgB antibodies (see Note 12). However, standard western blot analysis of whole cell lysates cannot distinguish between unpolymerized CsgA proteins that have been secreted into the agar and changes in expression or intracellular stability of CsgA (18, 42). The plug western blot provides an approach to detect CsgA that has been secreted into the agar (Fig. 2a, c).

3.2.1. Western Blot Analysis of Bacterial Lysates

1. Incubate the bacteria on thick YESCA agar plate at 26°C for 48 h to induce curli production.
2. Scrape the bacteria from the agar with a sterile inoculation loop and suspend them in 1 mL of KPi buffer.
3. Measure the optical density of the bacteria suspension in KPi at 600 nm. Normalize the cell number to 1 OD_{600}.
4. Prepare two microcentrifuge tubes for each sample. The sample in one tube will be treated with FA or HFIP. The sample in the other tube is the untreated control.
5. Transfer 150 μL of the normalized bacteria suspension into each tube.

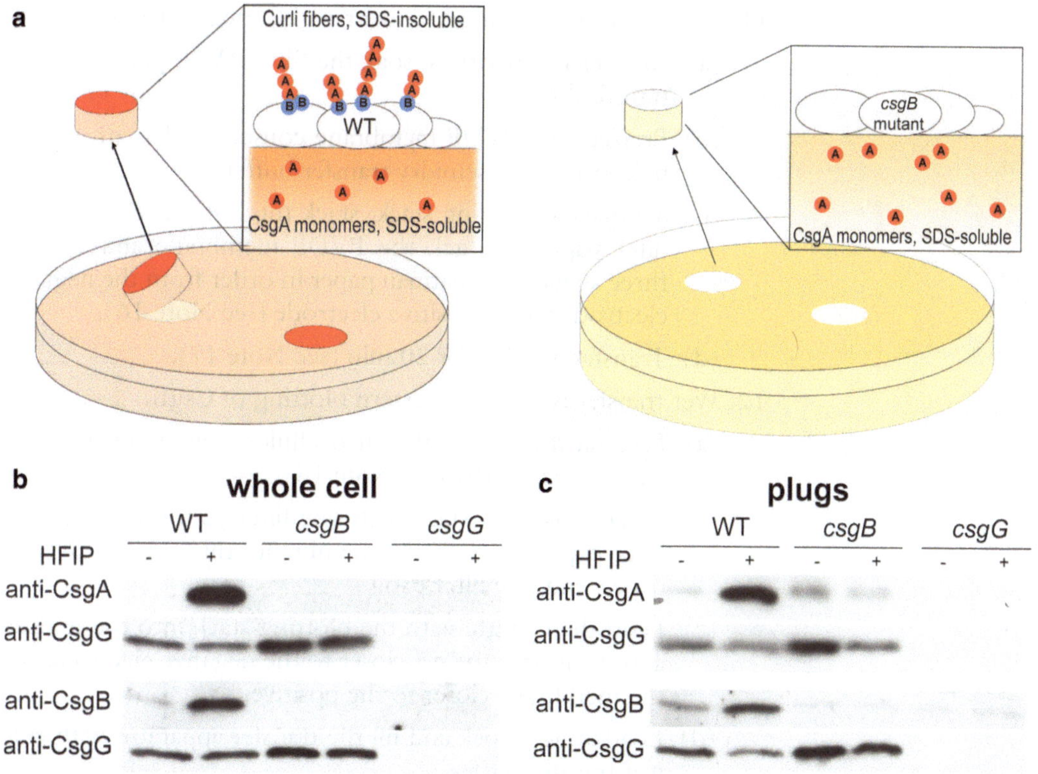

Fig. 2. Western blot analysis of whole cell lysates and plug western analysis. (**a**) Schematic diagrams of the plug assay. An 8-mm circular plug including the bacteria colony and the underlying agar is collected for the western blot analysis. In a wild-type *E. coli* (*left*), CsgA proteins incorporated into curli fibers are SDS-insoluble and cannot migrate into a SDS-PAGE gel unless treated with HFIP or FA. CsgA secreted into the agar is SDS-soluble. In a *csgB* mutant (*right*), CsgA is secreted into the agar in an SDS-soluble, unpolymerized form. (**b**) Western blot analysis of whole cell extracts of wild-type *E. coli*, a *csgB* mutant and a *csgG* mutant. CsgA and CsgB were probed by anti-CsgA antibody and anti-CsgB antibody respectively. Wild-type *E. coli* produces cell-associated, SDS-insoluble CsgA. No bacteria-associated CsgA was detected from the lysates of *csgB* and *csgG*. (**c**) Western blot analysis of plugs containing bacteria and the underlying agar. SDS-soluble CsgA was found in the agar underlying the *csgB* mutant. A *csgG* strain does not secrete any CsgA or CsgB.

6. Pellet down the bacteria by centrifugation at 16,000×*g* for 3 min. Remove the supernatant carefully. Bacteria-associated curli are in the pellet.
7. Resuspend the pellet of one of the duplicates in 70 μL of 100% HFIP or 90% FA briefly (see Note 13).
8. Immediately dry the sample in an acid-resistant centrifugal vacuum concentrator at 45°C for 30 min (see Note 14).
9. Resuspend both the non-treated pellet and the FA or HFIP treated pellet in 150 μL of 2× SDS sample buffer (see Note 15). Boil the sample at 95°C for 10 min.
10. Load 5–7 μL of each sample on a SDS-PAGE gel. Run the samples at 25 mA per gel for 35–40 min.

11. Semidry transfer is used for western blotting of CsgA.
    (a) After electrophoresis, soak the SDS-PAGE gel in semidry transfer buffer.
    (b) Pretreat the PVDF membrane consecutively with methanol, water, and semidry transfer buffer.
    (c) To make a blotting stack, stack three sheets of Whatman filter paper, the gel, the PVDF membrane and another three sheets of Whatman paper in order from the negative electrode to the positive electrode (see Note 16).
    (d) Transfer at 10 V for 20 min (see Note 17).
12. Wet transfer is used for western blotting of CsgB.
    (a) Pretreat the gel and the nitrocellulose membrane with the wet transfer buffer (see Note 18).
    (b) Make a blotting stack with one filter pad, three sheets of filter paper, the gel, the membrane, three sheets of filter paper, and one filter pad.
    (c) Load the cassette with the blotting stack into the transfer apparatus with the gel closer to the negative electrode and the membrane closer to the positive.
    (d) Load an ice block and fill the transfer apparatus with cold wet transfer buffer.
    (e) Transfer at 50 V for 3 h or 12 V for 12 h at 4°C with stirring (see Note 17).
13. After transfer, rinse the membrane with TBS-T buffer. Transfer the membrane into the blocking buffer. Block at RT for at least 1 h or at 4°C overnight.
14. Wash the blot with TBS-T three times, 5 min each time with shaking. Probe with 1:10,000 diluted anti-CsgA or 1:7,000 diluted anti-CsgB antibodies in blocking buffer at RT for 1 h.
15. Wash the blot with TBS-T three times, 5 min each time with shaking. Probe with 1:10,000 diluted anti-rabbit secondary antibody in blocking buffer at RT for 1 h.
16. Wash the blot with TBS-T three times, 5 min each. Develop the blot with chemiluminescent substrates.

3.2.2. Plug Western Blot Analysis

1. Normalize the overnight bacterial culture in LB to 1 OD_{600}.
2. Spot 4 μL of the normalized bacterial culture on a thin YESCA agar plate. Incubate at 26°C for 48 h to induce curli production.
3. Use the wide end of a Pasteur pipette to stab through the agar creating an 8-mm circular plug including the bacteria colony and the underlying agar. For each sample, collect two plugs. One plug will be treated with FA or HFIP, the other is the untreated control. Place the plugs in 1.5 mL microcentrifuge tubes.

4. For each sample, suspend one of the plugs in 100 μL of 90% FA or 100% HFIP and briefly vortex to dissolve the agar. Immediately dry the treated sample in an acid-resistant centrifugal vacuum concentrator at 45°C for 40 min (see Notes 13, 14 and 19).
5. Add 150 μL of 2× SDS sample buffer to both treated and untreated plugs. Sonicate in a bath sonicator for 5 min (see Note 20). Boil samples at 95°C for 10 min.
6. Run the samples on a SDS-PAGE gel, transfer the proteins to a PVDF membrane or nitrocellulose membrane, probe and develop the blots as described in steps 10–16 in Subheading 3.2.1.

3.3. Purification of curli subunits

Both CsgA and CsgB proteins can be purified for in vitro biochemical and structural analysis. Here we describe two approaches (see Note 21). In the first approach (18, 40), a curli-mutant overexpressing the curli secretion apparatus CsgG is used. Overexpressed CsgA or a truncated version of CsgB, which is no longer membrane associated, will be secreted into the liquid medium and can be collected for affinity chromatography (Fig. 3a). Alternatively, CsgA or CsgB without Sec signal sequences are expressed in the cytoplasm and can be purified via a denaturation method (Fig. 3b) (12, 43). For simplicity, we refer to C-terminal 6× his tagged CsgA or CsgB protein as CsgA and CsgB in this section.

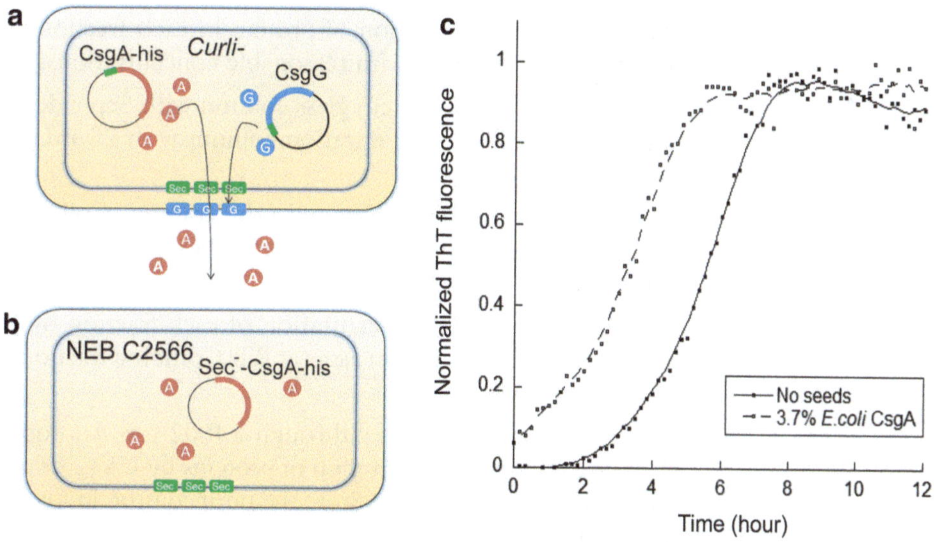

Fig. 3. Purification of curli subunits and the polymerization assay. (**a**) A schematic of CsgA purification from the supernatant. Both C-terminal his$_6$-tagged CsgA and CsgG are over-expressed in a mutant with the whole curli operon deleted. CsgG protein assembles into pore structures on the outer-membrane and secretes CsgA-His$_6$ into the medium. Soluble CsgA in the supernatant can be collected for chromatography. (**b**) A schematic illustrating CsgA purification from the cytoplasm. CsgA-His$_6$ without the Sec signal sequence is over-expressed in NEB C2566. Sec$^-$CsgA-His$_6$ is trapped in the cytoplasm and can be purified from the bacterial lysate. (**c**) Freshly purified CsgA polymerizes with a lag phase, a log phase and a stationary phase (*solid line*). The polymerization is accelerated in the presence of 3.7% preformed, sonicated CsgA fibers indicated by the lack of lag phase (*dashed line*). The polymerization kinetics was followed by ThT fluorescence with excitation at 438 nm and emission at 495 nm. The cutoff filter was set at 475 nm.

3.3.1. Non-denaturing Purification of CsgA or Truncated CsgB from Bacterial Supernatant (see Note 22)

1. For 1 L of desired prep, start an overnight culture of freshly transformed LSR12/pMC1/pMC3 in 25 mL of antibiotics supplemented LB at 37°C.
2. Dilute the overnight culture into 1 L of antibiotics supplemented LB. Incubate with shaking at 37°C to about 1 OD_{600}.
3. Add 250 μL of IPTG stock to induce protein expression for 45 min at 37°C (see Note 23).
4. Pellet down bacteria by centrifugation at 10,000 × *g* for 15 min at 4°C.
5. Collect the supernatant. Filter the supernatant over a 0.22 μm polyethersulfone bottle-top filter (see Note 24).
6. Perform the following purification steps at 4°C unless otherwise stated (see Note 25).
7. Pack a Kontes 2.5 × 10 cm glass column with 4 mL of Ni-NTA. Wash the column with 4 bed volumes (BV) of KPi buffer.
8. Assemble a funnel on top of the glass column (see Note 26). Flow the filtered supernatant over the column and adjust the flow rate to maximum.
9. Wash the column with 10 BV of KPi buffer.
10. Elute with 2 BV of elution buffer 1. Set the flow rate to 20 drops/min. Collect elutes in 1 mL fraction in microcentrifuge tubes. Immediately store the fractions on ice.
11. Measure the concentration of protein in each fraction at RT. Combine the fractions with reasonable concentrations.
12. Prepare a Kontes 1 × 30 cm glass column with Sephadex G-25 agarose. Equilibrate the filtration column with 25 mL of KPi buffer.
13. Load the combined fractions onto the column. Collect elutes in 1 mL fractions in microcentrifuge tubes. Immediately store elutes on ice.
14. Measure the protein concentration of each fraction by UV_{280} or a protein determination assay at RT. Combine fractions with reasonable concentrations.
15. Filter the combined elutes through a 0.02 μm Anotop filter. Measure the final concentration of proteins by UV_{280} or protein determination assay at RT. Store the purified protein for further biochemical assays (see Note 27).

3.3.2. Denaturing Purification of CsgA or CsgB from Bacterial Lysates

1. For 1 L of desired prep, start an overnight culture of NEB C2566/pNH11 or NEB C2566/pAN1 in 20 mL of antibiotics supplemented LB at 37°C.
2. Dilute the culture into 1 L of antibiotics supplemented LB. Incubate at 37°C for around 3 h till the OD_{600} reaches 1.

3. Induce with 500 μL of IPTG stock at 37°C for 1 h.
4. Pellet down the bacteria by centrifugation at 5,000×*g* for 20 min at RT.
5. Discard the supernatant. Store the pellets at –80°C.
6. Use GdnHCl to lyse the cell. For pellets from 500 mL bacterial culture, resuspend the bacteria in 50 mL of denaturing solution with stirring at RT for 1 h or at 4°C overnight.
7. Centrifuge the suspension at 10,000×*g* for 20 min at 4°C. Collect the supernatant in a 50 mL conical tube. Discard the cell debris.
8. Sonicate the supernatant by a probe sonicator at power 5 for three 15-s bursts at RT (see Note 28).
9. Add 1 mL of NiNTA resin to 50 mL of sonicated supernatant. Incubate on a rocker at RT for 1 h.
10. Load the supernatant with NiNTA resin onto a disposable polypropylene column.
11. Wash the resin with 10 BV of KPi buffer followed by 3 BV of column wash buffer to remove nonspecific bound proteins.
12. Elute the proteins from the column with elution buffer 2. Collect elutes in 1 mL fraction in microcentrifuge tubes. Immediately store the fractions on ice.
13. Measure the protein concentration of each fraction. Combine fractions with reasonable concentrations.
14. Run the combined elutes through a 30 kDa centrifugal filter unit at 7,500×*g* for 10 min at 4°C (see Note 29).
15. Use a spin desalting column to remove imidazole. Prewash the spin desalt column with KPi buffer two or three times. Load the resulting filtrate onto the prewashed spin desalting column and centrifuge at 1,000×*g* for 2 min at 4°C. (see Note 30).
16. Measure the final protein concentration. Store the freshly purified protein on ice for future use.

3.4. Polymerization Assay

Freshly purified CsgA or CsgB proteins self-polymerize into amyloid fibers in a nucleation dependent pattern (40). The polymerization can be promoted by the presence of preformed fibers in a process called seeding. Both polymerization kinetics and seeding can be monitored by Thioflavin T (ThT) fluorescence in real time (Fig. 3c). (see Note 31)

3.4.1. Polymerization from monomers

1. Load 100 μL of freshly purified proteins onto a non-treated, flat-bottom, opaque 96-well microtiter plate. Various concentrations of the same protein should be tested as the polymerization kinetics will be concentration dependent (see Note 32). Also, load a well with buffer only as the control.

2. Add 2 μL of ThT solution into each well (see Note 33). Mix ThT with proteins by pipetting.
3. Put a transparent sticker on top of the plate (see Note 34). Wipe the sticker with a Kimwipe.
4. Load the 96-well plate onto a plate reader. Set the excitation wavelength to be 438 nm, the emission wavelength 495 nm and a cut-off filter at 475 nm (see Note 35). Set the plate reader to read the plate every 10 min or 20 min at RT over the course of 24 h. Set the plate reader to shake linearly for 5 s to mix samples prior to each reading.
5. Collect the data and plot the ThT fluorescence arbitrary units against time.

3.4.2. Seeding Assay

1. Prepare seeds. Sonicate 1–3 days old, freshly formed CsgA fibers or CsgB fibers on ice with a probe sonicator on setting two 3× for 15 s each (see Note 36).
2. On the opaque 96-well plate, load 100 μL of freshly purified monomeric protein.
3. Add seeds at 2–10% (w/w) relative to the monomers. Add the same amount of seeds into 100 μL KPi buffer as controls.
4. Add 2 μL of ThT solution to each well, mix by pipetting.
5. Load the microtiter plate on a plate reader and follow the polymerization kinetics as described in steps 4–5 in Subheading 3.4.1.

3.5. Curli Dependent Biofilm Formation

Curli are involved in biofilm formation. In static cultures, *E. coli* MC4100 strains form a ring-shaped biofilm at the air–liquid interface on the inner surface of glass tubes (Fig. 4a) or microtiter plate wells (see Note 37). This type of biofilm can be scored or

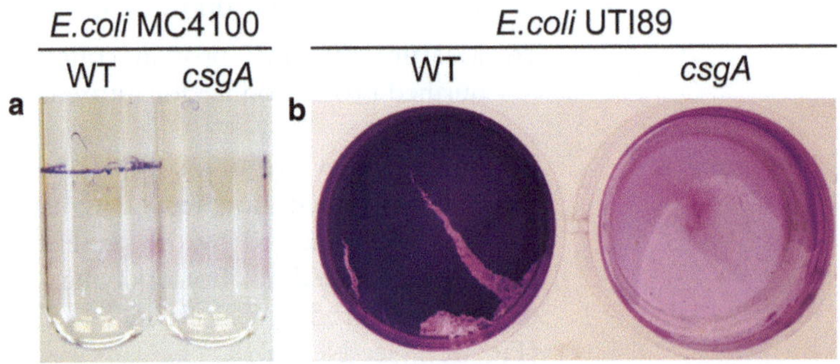

Fig. 4. Curli dependent biofilm formation. (**a**) Wild-type *E. coli* MC4100 in static YESCA medium at 26°C for 4 days generated a ring-shaped biofilm at the air–liquid interface on a glass tube. A MC4100 *csgA* mutant was unable to form such biofilm. (**b**) Wild-type UTI89 grown in static LB-low salt medium at 26°C for 2 days formed a pellicle-like biofilm (*top* view) on a 24-well microtiter plate. The planktonic cells were removed and the pellicle was stained with 0.1% crystal violet. A UTI89 *csgA* mutant cannot form a pellicle under the same condition.

quantified by a high throughput microtiter plate assay adapted from Merritt et al. (44). Some *E. coli* and *Salmonella* spp. lab strains and isolates also form a pellicle-like biofilm floating on the air–liquid interface (Fig. 4b) (11, 12). The pellicle biofilm in these strains is dependent on curli (see Note 38). Curli defective mutants are unable to form a pellicle. The pellicle biofilm can be analyzed by microscopy and can be quantified as described in this section.

3.5.1. Microtiter Plate Assay for Ring-Shaped Biofilm

1. Make an overnight culture of the bacteria of interest in LB at 37°C.
2. Dilute each overnight culture by 1:100 into 200 μL of LB-low salt medium or YESCA medium in wells of a cell culture treated, flat bottom, transparent 96-well microtiter plates with lids. Cover the plate.
3. Incubate the plate at 26°C or 30°C for 3–5 days (see Note 39).
4. Remove the planktonic bacteria from each well by pipetting or by shaking out the liquid.
5. Wash the wells with water to remove the remaining planktonic cells. Submerge the plate in a tray of water and shake out the liquid. One can also add in water and remove the liquid with a multichannel pipet. Repeat this step twice.
6. Add 210 μL of crystal violet solution into each well and stain for 10 min at RT.
7. Discard the crystal violet solution. To remove the nonspecifically attached crystal violet, wash the wells with water three times in a tray or by pipetting.
8. Invert the microtiter plate and vigorously tap the plate on a paper tower to remove the remaining water in the well. Air-dry the plate.
9. Add 200 μL of 95% ethanol into each stained well. Cover the plate and incubate it on a platform shaker at RT for 15 min.
10. Transfer the 100 μL of liquid onto a non-sterile, flat bottom, transparent 96-well microtiter plate. Measure the OD at 600 nm on a plate reader.

3.5.2. Pellicle Biofilm Assay

1. Start an overnight culture of bacteria in LB at 37°C.
2. To grow a pellicle biofilm in a glass tube, add 4 μL of overnight culture into 4 mL of LB-low salt or YESCA medium. To grow a pellicle in wells on a sterile 24-well microtiter plate, add 2 μL of overnight culture into 2 mL of LB-low salt or YESCA medium (see Note 40).
3. Incubate bacteria at 26°C or at 30°C for 2–3 days.
4. Once a pellicle biofilm formed, the morphotype of the pellicle can be visualized directly. The pellicle can also be picked up with tweezers and be analyzed under microscope.

5. The pellicle can be stained with the crystal violet solution for better visualization. Remove planktonic bacteria from the wells and wash the pellicle with water by careful pipetting (see Note 41). Add 2 mL of crystal violet solution into each well. After 5 min staining, remove the crystal violet staining and wash the pellicle with water by pipetting. Add 2 mL of water into each well. The pellicle is stained purple, floating on the water.
6. To quantify the biomass of pellicle biofilms generated in step 3:
   (a) For each strain, prepare at least three duplicates of pellicle on the plate.
   (b) Remove planktonic bacteria. Stain the pellicle with the crystal violet solution for 5 min. Remove the crystal violet and wash the pellicle with 2 mL of water by pipetting twice. Remove the water.
   (c) Add 2 mL of water into each well.
   (d) Prewash the tip of a homogenizer by 95% ethanol followed by water. Dry the tip on paper towel.
   (e) Break the pellicles by homogenizing at power 3 for 20 s (see Note 42). Wash the homogenizer with clean water between each sample.
   (f) Mix 100 μL of homogenized sample with 900 μL water, transfer into a cuvette and measure OD at a wavelength of 600 nm.
   (g) Clean the homogenizer with ethanol and water after quantification.

4. Notes

1. Other than YESCA agar, bacteria also express curli on various types of agar media including LB-low salt agar (45), T-agar (16) and colonization factor antigen (CFA) agar (46). Curli are also detected on the surface of bacteria grown in static liquid YESCA medium or LB-low salt medium at 26°C or 30°C (11, 12).
2. Brilliant Blue increases the color contrast of colonies on the agar.
3. FA and HFIP are corrosive and irritating chemicals. Use them under a chemical fume hood and wear appropriate protective equipment. Store them in a cool, well ventilated place.
4. Both anti-CsgA and anti-CsgB antibody bind nonspecific proteins. The anti-CsgA antibody binds to a nonspecific protein migrating at 40 kDa on SDS-PAGE gels. CsgB antiserum needs to be cleaned using acetone precipitated proteins from a *csgB* mutant before use. 1:10,000 dilution of anti-CsgA or 1:7,000 of anti-CsgB gives a clean, strong signal for western

blotting. Higher concentration of antibody may result in high background. If using antiserum from different resources, a preexperiment is recommended to find the ideal dilution rate of antibody for primary probing.

5. LSR12 is a *csgDEFG csgBA* deletion mutant in *E. coli* C600 background. pMC1 was made by cloning CsgG into the *NcoI* and *BamHI* sites of pTrc99A. pMC3 was made by cloning full length C-terminal His_6-tagged CsgA into the *NdeI* and *EcoRI* sites of pHL3 (40). The strain is resistant to Ampicillin (Amp) and Chloramphenicol (Cm). LSR12/pMC1/pMC3 does not maintain robust expression of CsgA. Thus, we do not recommend long-term glycerol storage of this strain. To get best expression, we normally transform LSR12/pMC1 with pMC3 and keep selection on LB Amp/Cm agar. Fresh transformants are tested for CsgA expression level and the one with best expression is used for large-scale purification. The freshly made strain can be stored as glycerol stock in –80°C freezer for 1 month.
6. NEB C2566 is a T7 expression strain offered by New England Biolabs Inc. The plasmid pNH11 was made by cloning C-terminal His_6-tagged CsgA without the Sec signal sequence into the pET11d vector (12). The plasmid pAN1 was made by cloning C-terminal His_6-tagged CsgB without the Sec signal sequence into the pET11d vector (43). Both NEB C2566/pNH11 and NEB C2566/pAN1 can be stored as a glycerol stock at –80°C.
7. The uropathogenic *E. coli* strain UTI89 forms a robust pellicle biofilm in LB-low salt medium. However, the pellicle formation of this strain in YESCA medium is highly dependent on the brand of yeast extract. We use yeast extract from BD Biosciences.
8. Many *E. coli* and *Salmonella* spp. lab strains express curli at 26 or 30°C. However, some clinical isolates are able to produce curli at 37°C (47).
9. The time of incubation is important. Although curli defective mutants do not stain red after 2 days of incubation on YESCA CR agar, some of them can take up CR dye and eventually turn light pink after more than 2 or 3 days of incubation.
10. The CR phenotypes of most *E. coli* K-12 isolates are completely dependent on curli production. However, many curliated bacteria produce cellulose and other extracellular polysaccharides which also bind CR. For instance, wild-type *S. typhimurium* form red, dry colonies on CR agar, while the colonies of a curli defective mutant is pink and smooth (48). For those strains, multiple approaches such as electron microscopy (EM) and western blotting analysis are recommended.
11. For CR post-staining, we do not shake the agar plates during staining or the washing step. Curli facilitate bacterial adhesive-

ness to biotic and abiotic surfaces. Curli defective mutants are much less adherent on agar surface and can easily be washed away in CR solution or KPi buffer with shaking.

12. Components of the curli system- CsgC, CsgD, CsgF, CsgG, and other regulatory proteins can be analyzed by standard western blotting. No FA or HFIP treatment is needed.
13. FA can acid-hydrolyze proteins including curli subunits. Extended treatment with FA may result in loss of target protein. Drying samples immediately after FA treatment in a centrifugal vacuum concentrator helps to reduce protein loss. HFIP does not cause acid hydrolysis.
14. Use an acid-resistant centrifugal vacuum concentrator since both FA and HFIP are corrosive.
15. Low amounts of FA sometimes remain in bacterial pellets or plugs after speed-drying in a centrifugal vacuum concentrator and will acidify the SDS sample buffer and make it turn yellow. The low pH of samples increases the chance of acid hydrolysis and affects electrophoresis. We recommend adjusting the pH of samples in SDS sample buffer back to neutral by adding several microliters of NaOH before boiling.
16. After the second layer of filter paper, squeeze out any remaining bubbles by rolling a glass pipet over the top, from the center outwards.
17. The semidry transfer apparatus described here is from FisherBiotech and the wet-transfer system is from Bio-Rad. For different models of transfer apparatus, follow the manufacturer's instructions to assemble the blotting stack and to set the voltage and transfer time.
18. CAPS buffer with high pH is used for wet transfer of CsgB. CsgB has a high pI and cannot be transferred efficiently with buffers with lower pH.
19. Make sure that the plug is completely dry after centrifugation. Depending on the agar condition, a longer time may be needed to completely dry plugs and remove FA or HFIP.
20. Bath sonication of samples in 2× SDS sample buffer gives more consistent results for western blot assay, probably because sonication helps break down large chunks of dried agar and release trapped proteins into the sample buffer.
21. The two approaches described here both have advantages and drawbacks. Purification of secreted proteins from the medium avoids denaturation of the target protein. However, the induction and purification typically take 2 days. The supernatant with target proteins cannot be stored for more than 1 day, so fresh supernatants have to be collected each time before purification. Purification from cell lysates is less

time-consuming, and the induced cell pellet can be stored for long-term usage. This method requires denaturation treatment and the yield is low. CsgA purified by the two approaches polymerizes with the same kinetic and forms fibers with the same morphologies. We use CsgA and CsgB as examples to illustrate the purification process. The methods can be easily adapted to purify CsgA and CsgB mutants.

22. We recommend collecting 100 μL of sample from each step of purification including the load, the wash and the elutes. Store the collections on ice for SDS-PAGE electrophoresis. This step analyzes the purity of the target protein.
23. Higher concentration of IPTG or longer induction may result in toxicity and reduce the yield of protein.
24. This step removes any bacteria remaining in the supernatant after centrifugation. It is recommended to use the freshly prepared supernatant for protein purification because CsgA and CsgB are highly aggregative even at low concentrations. The filtered supernatant can be stored at 4°C for at maximum of 1 day.
25. CsgA and CsgB are highly aggregative at RT. Low temperatures reduce the polymerization rate. In order to obtain as much monomeric protein from the column as possible, it is recommended to perform the experiment at 4°C and store the elutes and samples on ice.
26. A funnel on the column greatly accelerates the loading process. To flow 1 L of supernatant through the affinity column by gravity usually takes 10 h. With a funnel the loading time decreases to around 2 h.
27. Always use freshly purified CsgA or CsgB for biochemical assays. CsgA and CsgB cannot be stored at −20 or −80°C. The proteins precipitate out of the solution in the thawing process.
28. The supernatant still contains some bacterial cells and debris including lipids and DNA which will clog the affinity column and reduce flow rate. Sonication helps the supernatant to flow smoothly.
29. The size of monomeric CsgA is around 17 kDa. CsgA starts to aggregate once GdnHCl is removed. A 30 kDa filter helps to remove dimers, oligomers, and other aggregates. It also helps to remove a SlyD protein which binds the nickel affinity column and elutes with the target protein. Follow the product instruction manual to set the speed and time for centrifugation. CsgB aggregates much faster than CsgA. CsgB eluting off the column already form dimers or oligomers and cannot pass through the filter.
30. Different types of spin desalting columns or gel filtration columns may be used. Follow the product instruction manual for this step.

31. The ThT assay described here provides a simple approach to follow the polymerization of curli subunits. Other methods are also described in the literature (49, 50). The aggregation of proteins can be monitored by light scattering (50). The solubility of proteins in SDS also offers an alternative approach. Freshly purified CsgA is SDS soluble and can migrate into a SDS-PAGE gel (51). Overtime, CsgA polymerizes into fibers and are less and less SDS-soluble.
32. Several concentrations of protein, from 10 to 50 μM, may be tested to make sure that the ThT fluorescence does not exceed the reading limit of the plate reader.
33. Depending on the proteins and monomer concentrations, various concentrations of ThT may be tested to make sure that ThT is in excess.
34. Make sure that the edges of the plate are sealed with the sticker to prevent evaporation.
35. Some plate reader models do not have a filter set at 475 nm, instead, the emission bandwidth is 20 nm. These models can also be used for the ThT assay.
36. The amount of time for sonication affects the seeding ability of seeds. Longer sonication generates more free ends of fibers to interact with monomers and increase the seeding ability.
37. The ring shaped biofilm formed by *E. coli* MC4100 is dependent on curli. A *csgA* mutant or *csgB* mutant is unable to produce a biofilm. In many other *E. coli* and *Salmonella* strains, type I-pili and other extracellular structures instead of curli play a major role in assembly of such kind of biofilm.
38. Curli play a major role in pellicle biofilm formation. Other extracellular structures including flagella and cellulose are also important for robust pellicle biofilm formation.
39. The incubation time varies in different media and at different temperature. Bacteria grow faster at 30°C. *E. coli* MC4100 forms a ring-shaped biofilm in LB-low salt in 3 days, while in YESCA medium it takes 4–5 days.
40. The morphology of pellicle biofilms varies in different media. Uropathogenic *E. coli* isolate UTI89 forms wrinkled pellicles in YESCA, but a smooth biofilm in LB-low salt.
41. Wild-type UTI89 form a robust pellicle biofilm. However, the pellicles formed by some of the mutants are fragile and may be damaged by vigorous pipetting.
42. Depending on the model of homogenizer, different power levels and times may be tested to make sure that the pellicle is fully broken apart.

Acknowledgments

We thank members of the Chapman laboratory for helpful discussions and review of this manuscript. This work was supported by the National Institutes of Health Grant AI073847.

References

1. Zogaj X, Bokranz W, Nimtz M, Romling U (2003) Production of cellulose and curli fimbriae by members of the family Enterobacteriaceae isolated from the human gastrointestinal tract. Infect Immun 71(7):4151–4158
2. Prigent-Combaret C, Prensier G, Le Thi TT, Vidal O, Lejeune P, Dorel C (2000) Developmental pathway for biofilm formation in curli-producing Escherichia coli strains: role of flagella, curli and colanic acid. Environ Microbiol 2(4):450–464
3. Tükel CRM, Humphries AD, Wilson RP, Andrews-Polymenis HL, Gull T, Figueiredo JF, Wong MH, Michelsen KS, Akçelik M, Adams LG, Bäumler AJ (2005) CsgA is a pathogen-associated molecular pattern of Salmonella enterica serotype Typhimurium that is recognized by Toll-like receptor 2. Mol Microbiol 58(1):289–304
4. Tukel C, Wilson RP, Nishimori JH, Pezeshki M, Chromy BA, Baumler AJ (2009) Responses to amyloids of microbial and host origin are mediated through Toll-Like receptor 2. Cell Host Microbe 6(1):45–53
5. Gophna U, Barlev M, Seijffers R, Oelschlager TA, Hacker J, Ron EZ (2001) Curli fibers mediate internalization of Escherichia coli by eukaryotic cells. Infect Immun 69(4):2659–2665
6. Johansson C, Nilsson T, Olsen A, Wick MJ (2001) The influence of curli, a MHC-I-binding bacterial surface structure, on macrophage-T cell interactions. FEMS Immunol Med Microbiol 30(1):21–29
7. Olsen A, Jonsson A, Normark S (1989) Fibronectin binding mediated by a novel class of surface organelles on Escherichia coli. Nature 338(6217):652–655
8. Kikuchi T, Mizunoe Y, Takade A, Naito S, Yoshida S (2005) Curli fibers are required for development of biofilm architecture in Escherichia coli K-12 and enhance bacterial adherence to human uroepithelial cells. Microbiol Immunol 49(9):875–884
9. Uhlich GA, Cooke PH, Solomon EB (2006) Analyses of the red-dry-rough phenotype of an Escherichia coli O157: H7 strain and its role in biofilm formation and resistance to antibacterial agents. Appl Environ Microbiol 72(4):2564–2572
10. Austin JW, Sanders G, Kay WW, Collinson SK (1998) Thin aggregative fimbriae enhance Salmonella enteritidis biofilm formation. FEMS Microbiol Lett 162(2):295–301
11. Weiss-Muszkat M, Shakh D, Zhou Y, Pinto R, Belausov E, Chapman MR, Sela S (2010) Biofilm formation by and multicellular behavior of Escherichia coli O55:H7, an atypical enteropathogenic strain. Appl Environ Microbiol 76(5):1545–1554
12. Cegelski L, Pinkner JS, Hammer ND, Cusumano CK, Hung CS, Chorell E, Aberg V, Walker JN, Seed PC, Almqvist F, Chapman MR, Hultgren SJ (2009) Small-molecule inhibitors target Escherichia coli amyloid biogenesis and biofilm formation. Nat Chem Biol 5(12):913–919
13. Barnhart MM, Chapman MR (2006) Curli biogenesis and function. Annu Rev Microbiol 60:131–147
14. Hammar M, Arnqvist A, Bian Z, Olsen A, Normark S (1995) Expression of two csg operons is required for production of fibronectin- and Congo red-binding curli polymers in Escherichia coli K-12. Mol Microbiol 18(4):661–670
15. Gerstel U, Romling U (2003) The csgD promoter, a control unit for biofilm formation in Salmonella typhimurium. Res Microbiol 154(10):659–667
16. Collinson SK, Emody L, Muller KH, Trust TJ, Kay WW (1991) Purification and characterization of thin, aggregative fimbriae from Salmonella enteritidis. J Bacteriol 173(15): 4773–4781
17. Hammar MBZ, Normark S (1996) Nucleator-dependent intercellular assembly of adhesive curli organelles in Escherichia coli. Proc Natl Acad Sci U S A 93:6562–6566

18. Hammer ND, Schmidt JC, Chapman MR (2007) The curli nucleator protein, CsgB, contains an amyloidogenic domain that directs CsgA polymerization. Proc Natl Acad Sci U S A 104(30):12494–12499
19. Robinson LSAE, Hultgren SJ, Chapman MR (2006) Secretion of curli fibre subunits is mediated by the outer membrane-localized CsgG protein. Mol Microbiol 59(3):870–881
20. Nenninger AA, Robinson LS, Hammer ND, Epstein EA, Badtke MP, Hultgren SJ, Chapman MR (2011) CsgE is a curli secretion specificity factor that prevents amyloid fibre aggregation. Mol Microbiol 81(2):486–499
21. Nenninger AA, Robinson LS, Hultgren SJ (2009) Localized and efficient curli nucleation requires the chaperone-like amyloid assembly protein CsgF. Proc Natl Acad Sci U S A 106(3):900–905
22. Taylor JD, Zhou YZ, Salgado PS, Patwardhan A, McGuffie M, Pape T, Grabe G, Ashman E, Constable SC, Simpson PJ, Lee WC, Cota E, Chapman MR, Matthews SJ (2011) Atomic resolution insights into curli fiber biogenesis. Structure 19(9):1307–1316
23. Sunde M, Blake C (1997) The structure of amyloid fibrils by electron microscopy and X-ray diffraction. Adv Protein Chem 50:123–159
24. Sunde M, Serpell LC, Bartlam M, Fraser PE, Pepys MB, Blake CCF (1997) Common core structure of amyloid fibrils by synchrotron X-ray diffraction. J Mol Biol 273(3):729–739
25. Nordstedt C, Naslund J, Tjernberg LO, Karlstrom AR, Thyberg J, Terenius L (1994) The alzheimer a-beta-peptide develops protease resistance in association with its polymerization into fibrils. J Biol Chem 269(49):30773–30776
26. Glenner GG, Wong CW (1984) Alzheimer's disease: initial report of the purification and characterization of a novel cerebrovascular amyloid protein. Biochem Biophys Res Commun 120(3):885–890
27. Prusiner SB (1996) Molecular biology and pathogenesis of prion diseases. Trends Biochem Sci 21(12):482–487
28. Blanco LP, Evans ML, Smith DR, Badtke MP, Chapman MR (2012) Diversity, biogenesis and function of microbial amyloids. Trends Microbiol 20:66–73
29. Wang X, Chapman MR (2008) Curli provide the template for understanding controlled amyloid propagation. Prion 2(2):57–60
30. Badtke MP, Hammer ND, Chapman MR (2009) Functional amyloids signal their arrival. Sci Signal 2(80):pe43
31. Fowler DM, Koulov AV, Balch WE, Kelly JW (2007) Functional amyloid—from bacteria to humans. Trends Biochem Sci 32(5): 217–224
32. Elliot MA, Karoonuthaisiri N, Huang J, Bibb MJ, Cohen SN, Kao CM, Buttner MJ (2003) The chaplins: a family of hydrophobic cell-surface proteins involved in aerial mycelium formation in Streptomyces coelicolor. Genes Dev 17(14):1727–1740
33. Dueholm MS, Petersen SV, Sønderkær M, Larsen P, Christiansen G, Hein KL, Enghild JJ, Nielsen JL, Nielsen KL, Nielsen PH, Otzen DE. Functional amyloid in Pseudomonas. Mol Microbiol. 2010 Jun 21. [Epub ahead of print]. PMID: 20572935
34. Romero D, Aguilar C, Losick R, Kolter R (2010) Amyloid fibers provide structural integrity to Bacillus subtilis biofilms. Proc Natl Acad Sci U S A 107(5):2230–2234
35. Chapman MR, Robinson LS, Pinkner JS, Roth R, Heuser J, Hammar M, Normark S, Hultgren SJ (2002) Role of Escherichia coli curli operons in directing amyloid fiber formation. Science 295(5556):851–855
36. King CYTP, Gross H, Gebert R, Aebi M, Wüthrich K (1997) Prion-inducing domain 2–114 of yeast Sup35 protein transforms in vitro into amyloid-like filaments. Proc Natl Acad Sci U S A 94(13):6618–6622
37. Dos Reis SC-SB, Forge V, Lascu I, Bégueret J, Saupe SJ (2002) The HET-s prion protein of the filamentous fungus Podospora anserina aggregates in vitro into amyloid-like fibrils. J Biol Chem 227(8):5703–5706
38. Fowler DM, Koulov AV, Alory-Jost C, Marks MS, Balch WE, Kelly JW (2006) Functional amyloid formation within mammalian tissue. PLoS Biol 4(1):100–107
39. Maji SKPM, Perrin MH, Sawaya MR, Jessberger S, Vadodaria K, Rissman RA, Singru PS, Nilsson KP, Simon R, Schubert D, Eisenberg D, Rivier J, Sawchenko P, Vale W, Riek R (2009) Functional amyloids as natural storage of peptide hormones in pituitary secretory granules. Science 325(5938):328–332
40. Wang X, Smith DR, Jones JW, Chapman MR (2007) In vitro polymerization of a functional Escherichia coli amyloid protein. J Biol Chem 282(6):3713–3719
41. Wang X, Zhou Y, Ren JJ, Hammer ND, Chapman MR (2010) Gatekeeper residues in the major curlin subunit modulate bacterial amyloid fiber biogenesis. Proc Natl Acad Sci U S A 107(1):163–168

42. Wang X, Chapman MR (2008) Sequence determinants of bacterial amyloid formation. J Mol Biol 380(3):570–580
43. Hammer ND, McGuffie BA, Zhou Y, Badtke MP, Reinke AA, Brännström K, Gestwicki JE, Olofsson A, Almqvist F, Chapman MR. The C-Terminal Repeating Units of CsgB Direct Bacterial Functional Amyloid Nucleation. J Mol Biol. 2012 Sep 21;422(3):376-89. Epub 2012 Jun 7. PMID: 22684146
44. Merritt JH, Kadouri DE, O'Toole GA. Growing and analyzing static biofilms. Curr Protoc Microbiol. 2005 Jul;Chapter 1:Unit 1B.1. PMID: 18770545
45. Werner Bokranz XW, Tschäpe H, Römling U (2005) Expression of cellulose and curli fimbriae by Escherichia coli isolated from the gastrointestinal tract. J Med Microbiol 54:1171–1182
46. Collinson SK, Doig PC, Doran JL, Clouthier S, Trust TJ, Kay WW (1993) Thin, aggregative fimbriae mediate binding of Salmonella enteritidis to fibronectin. J Bacteriol 175(1):12–18
47. Bian Z, Brauner A, Li Y, Normark S (2000) Expression of and cytokine activation by Escherichia coli curli fibers in human sepsis. J Infect Dis 181(2):602–612
48. Römling U (2005) Characterization of the rdar morphotype, a multicellular behaviour in Enterobacteriaceae. Cell Mol Life Sci 62(11):1234–1246
49. O'Nuallain B, Williams AD, Westermark P, Wetzel R (2004) Seeding specificity in amyloid growth induced by heterologous fibrils. J Biol Chem 279(17):17490–17499
50. Wright CF, Teichmann SA, Clarke J, Dobson CM (2005) The importance of sequence diversity in the aggregation and evolution of proteins. Nature 438(7069):878–881
51. Evans ML, Schmidt JC, Ilbert M, Doyle SM, Quan S, Bardwell JC, Jakob U, Wickner S, Chapman MR (2011) E. coli chaperones DnaK, Hsp33 and Spy inhibit bacterial functional amyloid assembly. Prion 5(4): 323–34

Chapter 5

Visualization of Gram-positive Bacterial Pili

Chungyu Chang, I-Hsiu Huang, Antoni P.A. Hendrickx, and Hung Ton-That

Abstract

Pili or fimbriae are recognized as essential virulence determinants assembled on the bacterial surface. Gram-positive bacteria produce covalently linked pilus structures that are distinct from gram-negative counterparts. In this chapter, we describe three commonly used techniques to extract, detect, and visualize pili from gram-positive bacteria: (1) Western blot analysis, (2) Immuno-Electron Microscopy, and (3) Atomic Force Microscopy.

Key words: Gram-positive bacteria, Sortase, Pilus detection, Fimbriae, Western blots, Immuno-electron microscopy, Atomic force microscopy

1. Introduction

First identified by Duguid et al. in 1955 (1), pili were shown to be hair-like structures protruding out from the surface of gram-negative bacteria. Pili were later detected on the surface of gram-positive bacterium *Corynebacterium renale* by electron microscopy in 1968 by Yanagawa et al. (2). Similar "fibril-like" appendages on the surface of another gram-positive bacterium *Actinomyces naeslundii* were also observed by Girard and Jacius in 1974 (3). However, the mechanism of pilus assembly in gram-positive bacteria was not known until 2003, when it was shown in *Corynebacterium diphtheriae* by the study of SpaA pili (4).

Encoded by the *spaA-srtA-spaB-spaC* locus, the SpaA pilus is composed of SpaA forming the pilus shaft, SpaC at the tip, and SpaB located at the base and along the pilus structure (4). In addition to the cell wall sorting signal (CWS) with a conserved LPXTG motif (5) present in all Spa pilins, SpaA contains a pilin motif *YPK*

Anne H. Delcour (ed.), *Bacterial Cell Surfaces: Methods and Protocols*, Methods in Molecular Biology, vol. 966, DOI 10.1007/978-1-62703-245-2_5, © Springer Science+Business Media New York 2013

with the conserved lysine residue required for pilus assembly (4). According to the current biphasic model of pilus assembly (6), pilin-specific sortase SrtA catalyzes transpeptidation reactions utilizing the LPXTG motif and the lysine residue of the YPK motif, first linking SpaA to SpaC, and then SpaA to SpaA. Pilus polymerization is terminated when SpaB enters the pilus base by a similar transpeptidation reaction requiring a lysine residue of SpaB (7). SpaB is then anchored to the cell wall peptidoglycan by the house keeping sortase SrtF (8).

Sortase-mediated pilus assembly has been subsequently described in many other gram-positive bacteria, including *Actinomyces oris*, *Enterococcus faecalis*, *Bacillus cereus*, *Enterococcus faecium* and several species of streptococci such as *Streptococcus agalactiae* (GBS), *Streptococcus pneumoniae*, *Streptococcus pyrogenes* (GAS), *Streptococcus parasanguis*, *Streptococcus salivarius*, and *S. sanguis* (4, 9–18). Furthermore, recent studies of *B. cereus*, *S. agalactiae*, and *S. pyogenes* pili (12, 19, 20) lend support to the aforementioned biphasic model as described in *C. diphtheriae*.

Like gram-negative pili, the gram-positive counterparts are virulent determinants. It has been shown that gram-positive pili are involved in many facets of bacterial pathogenesis, including adhesion, bacterial co-aggregation, biofilm formation, and modulation of immunity (6, 21–24) (20, 25). More recently, pili have been employed as vaccine candidates against streptococcal pathogens (23, 26–28). As the genomes of gram-positive pathogens continue to be released, pilus gene clusters are subsequently identified by BLAST and bioinformatics analysis. It is thus important to establish a standard methodology to detect and visualize these covalently linked pilus structures.

1.1. Western Blotting

A combination of SDS-PAGE and electrophoretic transfer to porous membrane has become the standard method for analyzing proteins since the method was first described in 1979 by Towbin et al. (29). In 1997, using various cell wall lysing enzymes such as lysostaphin, amidase, and muramidase, Ton-That et al. successfully purified anchor peptides of staphylococcal surface proteins Seb and confirmed that sortase anchored proteins are linked to the cell wall peptidoglycan (30). Since then, enzymatic solubilization of surface proteins from the cell wall peptidoglycan followed by SDS-PAGE and Western blot analysis using anti-serum raised against pilin subunits has become a valuable and essential tool to study pilus assembly in many gram-positive bacteria (4, 10, 31, 32). Here, we described cellular fractionation of *C. diphtheriae* cells in which cell wall proteins including pilus polymers are released by treatment with muramidase. Trichloroacetic acid and acetone precipitated *C. diphtheriae* pilus polymers can be separated by 3–12% gradient SDS-PAGE and immunoblotted with anti-Spa antiserum. The oligomeric pilus structure can be visualized as a high molecular weight (HMW) peptide ladder.

1.2. Immunogold Transmission Electron Microscopy

Microscopes have been essential for the study of microorganisms and microbial surfaces as they not only revealed bacterial cell morphology, but also the structures of peptidoglycan, crystalline surface layers, surface proteins, and pili/fimbriae. Transmission electron microscopy is a special microscopic technique in which highly detailed images of a specimen are obtained when a beam of electrons generated typically by a tungsten filament is allowed to transmit through (33). Images are magnified and detected on a fluorescent screen, photographic film, or camera sensors. Different imaging methods have been developed to collect information from electron waves exiting from the sample to form an image, in which the most common method for visualizing bacterial cells is through bright field imaging of stained samples. Whole bacterial cells or ultrathin sections prepared by plastic/gold embedding can be placed upon small grids composed of meshed copper or nickel. Additional support of the grid can be accomplished by the inclusion of Formvar and/or carbon coated film. Cells expressing pilus polymers are labeled with specific anti-sera raised against pilus components and secondary antibodies conjugated with gold particles, followed by staining with uranyl acetate to enhance image contrast. Different components of the pilus polymers can be distinguished by labeling with different sizes of gold particles.

1.3. Atomic Force Microscopy

Another unique form of microscopy is atomic force microscopy (AFM), which was invented in 1986 (34). During the past two decades, AFM has become a very powerful tool and contributed, together with X-ray crystallography and electron microscopy, to the rise of the nanomicrobiology field (34–36). Conventional AFM can visualize the three-dimensional organization and distribution of single molecules on live bacterial surfaces or single proteins with nanometer (nm^3) to atomic resolution (37).

The basic principle of AFM is the use of a very sharp tip (cantilever) which is scanned over the surface of a sample in both x and y directions, thereby allowing to monitor the force between the tip and the sample with piconewton sensitivity (35, 38). The sample is mounted on a piezoelectric scanner that can move in three dimensions with very high accuracy. During scanning of a sample, a laser beam is emitted by a laser diode and focussed onto the free end of the cantilever. Differences in deflection of the reflected laser beam are recorded with a position-sensitive photodiode. Acquisition of an image typically takes minutes, depending on the size of the scan area, scan rate and resolution. Different AFM operating modes exist and can be used for imaging, however, the most commonly used imaging modes for microbiological samples are the direct contact mode and dynamic (tapping) mode. These two modes differ in the way the cantilever scans over the sample surface.

Some limitations of AFM include the relatively slow scanning time and inability to incorporate immune-gold labeling. A major advantage of AFM over electron microscopy however, which typically requires high vacuum conditions, is that it requires no fixing, staining, metal coating, or immunogold labeling of samples and it can be used in different experimental set-ups, such as in physiological buffer solutions or in air. Taken together, this makes AFM a useful technique for the nanoscale analysis of cell surface-exposed pili, which has recently been employed for both gram-negative and gram-positive bacteria (39–43).

2. Materials

Water used here for preparing solutions is ultrapure (Milli-Q water) unless mentioned differently.

2.1. Cell Fractionation

1. Media: heart infusion broth (HIB) and agar (HIA).
2. Antibiotics, as needed.
3. Orbital shaker.
4. Spectrophotometer.
5. Microcentrifuge and associated tubes.
6. Rotisserie rotator (with temperature control, or able to fit inside an incubator).
7. SMM buffer: 0.5 M Sucrose, 10 mM $MgCl_2$, 10 mM Maleate, pH 6.8. Filter-sterilize.
8. Muramidase buffer: 2 mM acetic acid, 48 mM sodium acetate. Mix 100 μL of 0.2 N acetic acid with 2.4 mL of 0.2 M sodium acetate and 7.5 mL of H_2O. Filter to sterilize.
9. Muramidase stock: Dissolve 10,000 Units of lyophilized mutanolysin from *Streptomyces globisporus* ATCC 21553 with 2 mL of muramidase buffer to give a concentration of 5,000 Units/mL (see Note 1).
10. 0.2 M phenylmethylsulfonyl fluoride (PMSF) in isopropanol.
11. Trichloroacetic acid (TCA). 100% stock in water.
12. Acetone (undiluted).
13. Dry ice.

14. Water bath.
15. Tabletop ultracentrifuge associated rotor.

2.2. Sodium Dodecyl Sulfate (SDS) Polyacrylamide Gel Electrophoresis and Western Blot

1. Water bath.
2. Neutralization buffer: 1 M Tris–HCl, pH 9.0.
3. 3–12% gradient SDS-PAGE: acryamide 40% solution, 1.875 M Tris–HCl, pH 8.75, 3 M NaCl, 0.2 M EDTA, pH 8.0, 10% ammonium persulfate (APS), *N*, *N*, *N*, *N'*-tetramethyl-ethylenediamine (TEMED).
4. Sample loading buffer: 0.625 mL of 1 M Tris–HCl, pH 6.8, 2 mL of 10% SDS, 2 mL of glycerol, 0.4 mL of β-mercaptoethanol, 5 mL of 6 M urea (see Note 2).
5. Pre-stained protein markers with high molecular weight standards.
6. 10× Running buffer: 30 g Tris base, 140 g glycine, 10 g SDS in 1 L of H_2O. For 1× solution, dilute with H_2O before use.
7. 10× Transfer buffer: 30 g Tris base, 145 g glycine in 1 L of water. For 10 L of 1× buffer, combine 1 L of 10× transfer buffer with 2 L of methanol and 7 L of water.
8. 10× Phosphate-buffered saline (PBS): 1.37 M NaCl, 27 mM KCl, 100 mM Na_2HPO_4, 20 mM KH_2PO_4, pH 7.4. Autoclave to sterilze.
9. 1× PBS-T: 5 mL Tween 20, 100 mL 10× PBS, 895 mL H_2O.
10. Blocking solution: 5 g nonfat powdered milk in 100 mL 1× PBS-T.
11. Pilin-specific primary antibody raised from rabbits and HRP-conjugated anti-rabbit antibody (see Note 3).
12. Luminolstock:0.22g5-Amino-2,3-dihydro-1,4-phthalazinedione in 5 mL of DMSO.
13. *p*-Coumaric stock: 0.038 g of *p*-Coumaric acid in 2.5 mL of DMSO.
14. 1 M Tris–HCl, pH 8.5: 121 g of Tris base in 700 mL of H_2O and adjust the pH with 6 N HCl to pH 8.5 before adding additional H_2O to 1 L.
15. 3% Hydrogen peroxide (H_2O_2).
16. ECL developing Solution A: 50 mL of 1 M Tris–HCl, pH 8.5, 443 mL of H_2O, 5 mL of Luminol stock, 2.2 mL of *p*-Coumaric stock. Keep solution A in a dark or foil-covered bottle at 4°C.
17. ECL Solution: Mix 20 mL of ECL developing solution A and 60 μL of 3% H_2O_2 immediately before use.

18. Immobilon-P PVDF transfer membrane.
19. Complete gel electrophoresis and transferring systems.
20. Peristaltic pump for making gradient SDS-PAGE gels.
21. Mini-Protean 3 Multi-Casting Chamber (Bio-Rad, Hercules, CA, USA) and Gradient maker (Hoefer, Holliston, MA, USA).
22. Platform shaker.
23. 4″ ¾″. × 3″ ¼″ containers, such as small tip boxes for holding PVDF blots.
24. Image processing equipment.

2.3. Immunogold Electron Microscopy

1. Media: Heart infusion broth (HIB) or agar (HIA), Blood agar.
2. Eppendorf tubes.
3. Tabletop microcentrifuge.
4. Formvar carbon-coated Nickel grids, 200 mesh (Electron Microscopy Science (EMS) Hatfield, PA, USA) (see Note 4).
5. 0.1 M NaCl: 0.58 g of NaCl in 100 mL H_2O and filter-sterilize. Store at room temperature.
6. 10× PBS buffer: 1.37 M NaCl, 27 mM KCl, 100 mM Na_2HPO_4, 20 mM KH_2PO_4, pH 7.4. Autoclave to sterilize. For 1× working solution, dilute with H_2O.
7. Wash buffer: 1% Bovine Serum Albumin in 1× PBS. Filter-sterilize the solution before use.
8. Blocking buffer: 0.1% Gelatin in 1× PBS. Filter-sterilize the solution before use.
9. Primary antiserum: polyclonal (unpurified or purified) or monoclonal antibody made against the target of interest (see Note 5).
10. Gold-labeled secondary antiserum: if the primary antibody was made from rabbit, the secondary antibody will be the gold-labeled anti-rabbit IgG with gold particle size ranging from 6 to 18 nm (see Note 6).
11. Staining solution: 1% uranyl acetate, dihydrate in H_2O. Filter-sterilize the solution before use (see Note 7).
12. Inoculating loops.
13. Parafilm.
14. Petri dishes.
15. Dumont N4AC tweezers and Dumont #5 tweezers (EMS, Hatfield, PA, USA).
16. Grid storage box (EMS, Hatfield, PA, USA).
17. Transmission Electron Microscope (see Note 8).
18. Digital camera with installed software.

2.4. Atomic Force Microscopy

1. Spectrophotometer.
2. Phosphate buffer saline (PBS): see item 6 in Subheading 2.2.
3. 2% glutaraldehyde in PBS.
4. 4% paraformaldehyde in PBS.
5. Microcentrifuge.
6. Carbon mica disks, fixed on a metal disk (Ted Pella, Inc, Redding, CA, USA).
7. Aero-Duster.
8. Mica disk tweezers (Ted Pella, Inc, Redding, CA, USA).
9. Asylum Research MFP-3D-BIO atomic force microscope (AFM) on an inverted optical microscope placed in an acoustic and vibration isolation enclosure (Asylum Research, Goleta CA, USA).
10. AC240TS cantilevers (Asylum Research, Goleta CA, USA).
11. Asylum MFP-3D software (Asylum MFP-3D software).

3. Methods

3.1. Cell Fractionation

1. Inoculate culture into fresh HIB with appropriate antibiotics if needed. Incubate overnight at 37°C in an orbital shaker.
2. Next day, dilute overnight culture into fresh 10 mL HIB. Incubate at 37°C in orbital shaker.
3. Monitor growth until desired OD_{600} has reached. Collect and normalize OD_{600} of all cultures to 1.0 (see Note 9).
4. Pellet cells in microcentrifuge for 5 min at 8,000 RPM (6,010 x *g*). Transfer the medium fraction carefully into new 1.5 mL tube. Store pellets on ice while spinning down the medium fraction for 5 additional minutes at 15,000 RPM (21,130 x *g*) to ensure any remaining cells are pelleted down. Transfer the medium fraction to new 1.5 mL tube.
5. Wash the cell pellet with 1 mL SMM buffer and then spin down for 5 min at 8,000 RPM (6,010 x *g*). Discard the buffer and then resuspend the pellet into fresh 1 mL SMM buffer.
6. Add 60 μL of muramidase stock and 5 μL of 0.2 M PMSF solution per tube and mix briefly. Incubate in 37°C with constant rotation on rotisserie rotator for 4 h or until protoplasts are observed (see Note 10).
7. TCA precipitate proteins from the medium fraction (obtained in step 4 above) by adding and appropriate volume of 100% TCA to obtain a final concentration of 7.5%. Mix briefly by hand and place the tubes on ice for at least 30 min. Spin down for 20 min

Table 1
3–12% Gradient SDS-PAGE gels

Ingredient	3% PAGE components (mL)	12% PAGE components (mL)
40% Acrylamide	3	20
1.875 M Tris–HCl, pH 8.75	8	8
3 M NaCl	0.4	0.4
0.2 M EDTA, pH 8.0	0.4	0.4
10% APS	0.2	0.2
TEMED	0.028	0.028
H_2O to 40 mL	27.972	10.972

at 15,000 RPM (21,130 x *g*) at 4°C and remove TCA carefully. Add 500 μL of acetone and subsequently concentrate the pellet by spinning down for 10 min at 15,000 RPM (21,130 x *g*). Air-dry the pellet and store at 4°C (see Note 11).

8. After cell wall lysis, spin down the protoplast sample (obtained in step 6 above) at 15,000 RPM in the microcentrifuge for 5 min to separate the cell wall fraction (supernatant) and the protoplast fraction (pellet).
9. Remove the supernatant carefully (corresponding to the cell wall fraction) and transfer to a new 1.5 mL tube. Proceed with TCA precipitation and acetone wash as indicated in step 7.
10. To separate cell membrane fraction from the cytoplasmic fraction, resuspend the pellet obtained in step 8 above (corresponding to the protoplast fraction) in 200 μL SMM buffer, and subject the suspension to three freeze–thaw cycles (dry ice and 37°C water bath). Spin down the suspension at 100,000 × *g* in a tabletop ultracentrifuge for 30 min at 4°C to separate the cell membrane fraction (pellet) from the cytoplasmic fraction (supernatant). TCA precipitate the cytoplasmic fraction and wash the pellet with acetone. Air-dry both the membrane and cytoplasmic pellets. Store at 4°C (see Note 12).

3.2. Sodium Dodecyl Sulfate (SDS) Polyacrylamide Gel Electrophoresis and Western Blot

1. Prepare SDS gradient gels. The quantity of each ingredient shown in Table 1 is designed for making 12 gradient gels using Mini-Protean 3 Multi-Casting Chamber and Gradient maker. A peristaltic pump is used to facilitate the flow of the gel-making solution. Each gel uses 0.75 mm Spacer plate and 15-well 0.75 mm comb (see Note 13). A 3–12% gradient gel is used to

separate the HMW pili present in the medium and cell wall fractions (see Note 14).

2. Add 50 μL of Sample loading buffer to each pellet and allow the buffer to dissolve the pellet for about 10–15 min without pipette mixing (see Note 15).
3. Prepare a boiling water bath, and place the tightly capped tubes into the water bath for 10 min. Centrifuge with a desktop microcentrifuge to collect all samples and then vortex briefly to ensure all samples have been properly mixed (see Note 16).
4. Load 5 μL of each sample per well. Add protein standards for each gel and electrophorese at 200 V until dye front has reached the bottom of the gel (see Note 17).
5. Following electrophoresis, open gel plate with a plastic spatula. Transfer gel to a container with transfer buffer. Soak for at least 10 min (see Note 18).
6. Assemble the transfer apparatus and initiate electrophoretic transfer at 100 V or 0.6 Amp for 1 h.
7. After transfer, the PVDF membrane can be stored dry or proceed to blocking in a small 4″ ¾″.×3″ ¼″ container with 20 mL of blocking solution. Shake for 1 h on a platform shaker.
8. Incubate with primary antibody for 1 h, and then wash three times with 1× TBST for about 3 min each time.
9. Incubate with 1 μL of secondary antibody (HRP-conjugated) and wash three times with 1× TBST with for 3 min each time.
10. Prepare the ECL solution (see item 17 Subheading 2.2) and pour into the container. Incubate on platform shaker for 2 min.
11. Develop the blot immediately using an image processing equipment (Fig. 1).

3.3. Immunogold Electron Microscopy

1. The entire procedures for this protocol are performed at room temperature and all the washing and incubation solutions are placed on the surface of parafilm unless mentioned differently. A pair of Dumont #5 tweezers is used for transferring the grids between different steps.
2. A loop-size of the fresh cell culture from the HIA plate, which was incubated overnight at 37°C, is gently resuspended in 1 mL of 0.1 M NaCl in a 1.5 mL eppendorf tube (see Note 19). The suspended cells are spun down at 8,000 RPM (6,010 x *g*) for 4 min. The supernatant is aspirated away after the spin.
3. If the fresh culture is made from HIB medium, concentrate 1 mL of culture into a pellet by centrifuging at 8,000 RPM (6,010 x *g*) for 4 min before washing with 0.1 M NaCl.

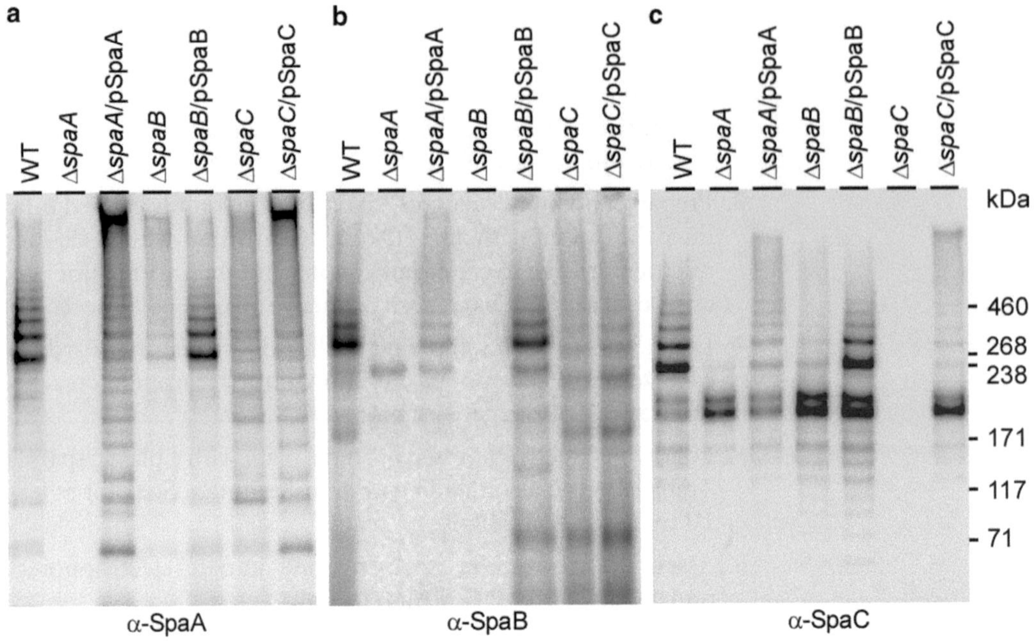

Fig. 1. Western blot analysis of pilus polymers. Pilus polymers isolated from the bacterial cell wall of various strains by muramidase were separated by SDS-PAGE and probed with antibodies specific for (**a**) the pilus shaft SpaA (α-SpaA), (**b**) the minor pilin SpaB (α-SpaB), and (**c**) the tip pilin SpaC (α-SpaC). The strains shown are the wild-type strain (WT), deletion strains denoted with Δ, and deletion strains complemented by a specific plasmid.

4. Add 1 mL of 1× PBS to the cell pellet from step 2 or 3 and resuspend it very gently (no vortexing). The washed cells are spun at 8,000 RPM (6,010 x *g*) for 4 min and the supernatant is aspirated away after the spin.
5. Gently resuspend the cell pellet with 1 mL of 1× PBS again.
6. Hold the edge of a formvar carbon-coated grid with a pair of Dumont N4AC tweezers. Pipette 7 μL of cells from step 5 onto the coated side of the grid (in this case, the shiny side) and wait for at least 1 min (see Note 20).
7. The grid needs to be washed three times in 500 μL drops of wash buffer after step 6. Prepare three drops of wash buffer on a sheet of parafilm (see Note 21). At the end of step 6, turn the grid sample side down onto the first wash buffer droplet and let the grid float on the surface. Gently wash the grid by pipetting the wash buffer up and down underneath the grid for at least 60 times before transferring the grid (see Note 22) to the surface of the second drop by using a pair of Dumont #5 tweezers. Repeat washing on the surface of the next two droplets. It is extremely important not to let the grid sink to the bottom of the droplet (see Note 23).
8. Prepare a drop of blocking buffer onto another sheet of parafilm that has been placed inside a Petri dish. Place the grid from

step 7 (with sample side face down) to the top of the blocking buffer droplet. Close the lid of the Petri dish and incubate for 1 h (see Note 24).

9. Prepare 100 μL of the primary antibody solution by diluting the antibody appropriately (1:100 for *C. diphtheriae* α-SpaA) with wash buffer and placed the drop on top of another piece of parafilm inside a Petri dish (see Note 25).
10. After blocking (step 8 above), rinse the grid in a fresh drop of wash buffer (see step 7 above) once before transferring it into the primary antibody solution, prepared in step 9 above.
11. Cover the Petri dish with a lid and incubate for 1 h.
12. Subsequently, the grid is washed three times in wash buffer (see step 7 above).
13. After washing, the grid is transferred onto the surface of a drop of blocking buffer and incubated for an hour (see step 8 above).
14. Prepare the secondary gold-labeled antibody. The secondary antibody is diluted 1:20 in the wash buffer before use. Transfer and float the grid from step 13 onto the surface of the secondary antibody solution for 1 h (see Note 26).
15. If the sample requires only single labeling, at the end of the 1 h incubation (step 14 above), the grid is rinsed five times with sterile water (see step 7 above for instruction). After thorough washing by water, the grid is ready for negative staining using uranyl acetate as described in step 17.
16. If the sample requires double labeling, the grid from step 14 will be washed three times with sterile water drops and then rinsed three times in wash buffer before repeating steps 8–15, using another primary antiserum against a second target of interest and a gold-labeled secondary antibody with a different size gold particle from the first labeling (see Note 27).
17. For uranyl acetate staining, pick up the grid from the surface of the water droplet by holding the edge of the grid with the sample side up. Drop 7 μL of 1% uranyl acetate directly on top of the grid and wait for 1 min; then immediately suck up the solution with a Kimwipe and air-dry the grid on the Kimwipe (see Note 28).
18. Examine the grids with a transmission electron microscope set to bright field mode with appropriate magnifications (Fig. 2) (see Note 29).

3.4. Atomic force microscopy

1. Prepare bacterial cells by streaking cells from an overnight culture onto a plate, and add antibiotics when required for plasmid maintenance. Incubate the plate overnight at 30°C or 37°C.

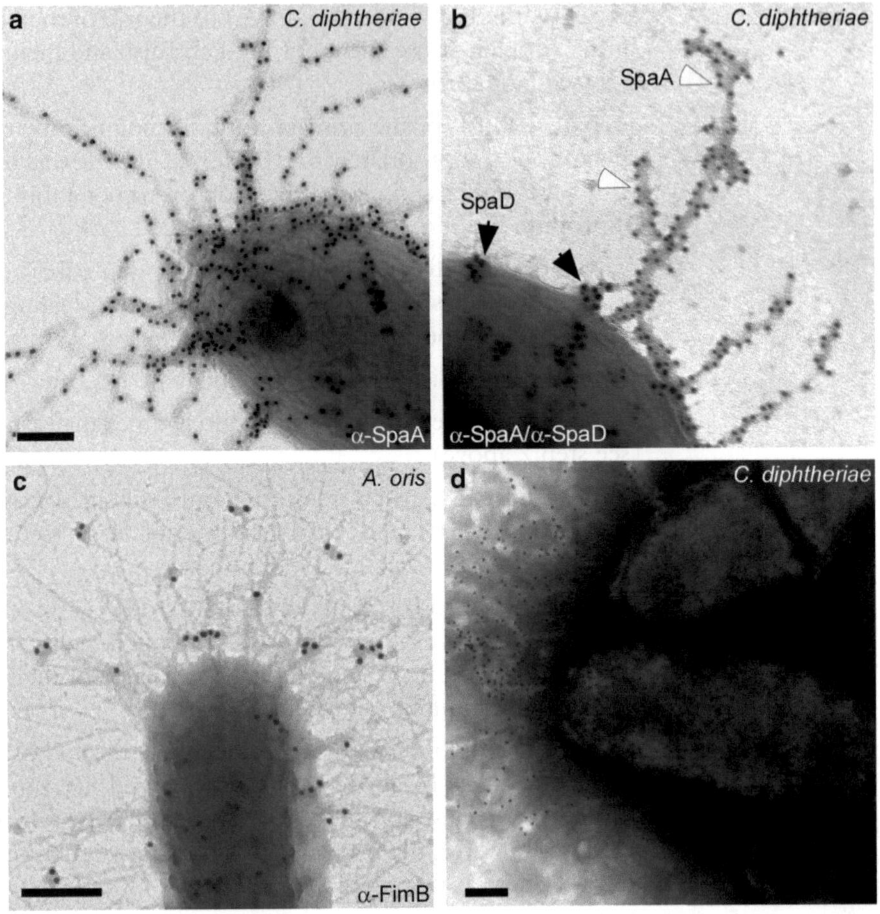

Fig. 2. Detection of ***C. diphtheriae*** and ***A. oris*** pili by immuno-electron microscopy. Bacterial cells were immobilized on Formvar/Carbon nickel grids, probed with pilin-specific antibodies, followed by goat anti-rabbit IgG conjugated to gold particles. Samples were then stained with 1% uranyl acetate and viewed by transmission electron microscopy. (**a**) Shown are ***C. diphtheriae*** cells over-expressing SpaA pili. (**b**) Double-labeling of ***C. diphtheriae*** cells was performed, whereby over-expressed SpaA pili were labeled with α-SpaA/IgG-conjugated 12 nm gold particles and SpaD pili labeled with α-SpaD/IgG-conjugated 18 nm gold particles. Reproduced from Gaspar and Ton-That. Copyright © American Society for Microbiology, (45) (**c**) ***A. oris*** MG1 cells were labeled with antibodies against the tip pilin FimB (α-FimB). (**d**) ***C. diphtheriae*** cells were over-stained with uranyl acetate, resulting in darkening of cells and the surrounding area. Scale bars of 0.1 μm.

2. Scrape the cells from the plate and resuspend them into 1 mL of sterilized H_2O and measure the optical density at 660 nm. The OD_{600} should be at least ~1.0 or approximately 1×10^9 CFU/mL.
3. Take 100 μL of the bacterial suspension in water and spin down at 4,000 RPM to remove the supernatant. Resuspend the pellet in 100 μL of 2% glutaraldehyde or 4% paraformaldehyde and incubate for 1 h at room temperature.
4. Spin the cells at 3,500 RPM for 5 min using a microcentrifuge.

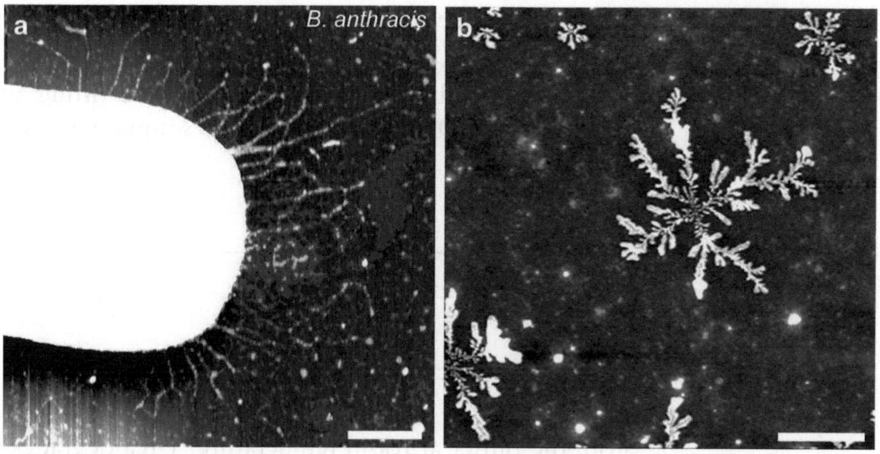

Fig. 3. Atomic force microscopy (AFM) scans of pili. (**a**) AFM image of a Sterne cell overproducing variant ATCC14579 pili. Scale bar is 0.5 μm. (**b**) An AFM scan of an artifact is shown. Note that if impure samples are analyzed, deposition of salt and/or debris precipitates occurs on the mica disk. Scale bar is 5 μm.

5. Remove excess fixative by discarding the supernatant and wash three times with ultrapure H_2O (see Note 30).
6. Spot 5 μL of fixed bacteria in H_2O onto a freshly cleaved mica disk, without covering (see Note 31).
7. Incubate at room temperature until the sample is completely dry. This typically takes 15 min (see Note 32).
8. If desired, the drying process can be shortened by blowing air onto the mica disk with an Aero-Duster.
9. Set up the settings of the Asylum Research MFP-3D-BIO atomic force microscope (AFM) using the Asylum MFP-3D software.
10. Carefully place the Olympus AC240TS cantilever (~70 kHz resonant frequency, 10 nm nominal tip radius, and 2 N/m spring constant) in the cantilever holder and assemble the cantilever holder onto the AFM head.
11. Select the scan area using the light microscope and perform an initial scan in air of 20 × 20 μm. For an initial scan, use a resolution of 256 × 256 pixels (see Note 33).
12. Select the area of interest using the Asylum MFP-3D software (see Note 34).
13. Collect tapping mode images with a resolution of at least 512 × 512 pixels with a line scan rate of 0.5 Hz, a set point between 650 and 950 mV and an integral gain between 4 and

8. Faster scan rates (1 Hz) will break the cantilever when whole cells are imaged (see Note 35).
14. Process images by flattening and/or plane fitting with the Asylum MFP-3D software to remove scanning artifacts when required (Fig. 3) (see Note 36).

4. Notes

1. The muramidase stock should be aliquoted and stored at −20°C.
2. Store the buffer at room temperature. Urea degrades rapidly, so it is best not to use sample buffer older than 1 week. This particular buffer is recommended only for visualizing high-molecular-weight products such as pili but urea is not required for detection of sortases.
3. The antibody does not have to be raised from rabbit. Any animal source can be used as long as the secondary antibody is specific to the animal.
4. It is recommended to use Formvar/Carbon-coated nickel grids as it provides the strongest support for bacteria.
5. In general, higher amounts of antibody are required for immunogold electron microscopy (IEM) as compared to Western blot analysis. For example, a 1:20,000 diluted SpaA antibody was used for Western blot analysis, while the same antibody was diluted at 1:100 for IEM.
6. Larger size gold particles are recommended as they are easier to see under IEM. If double labeling is performed, 12 and 18-nm gold particles are two good options.
7. Store the solution at −20°C when not in use. Thaw completely and vortex vigorously before use as uranyl acetate tends to precipitate in solution. Before proceeding with uranyl acetate staining, the grids have to be thoroughly washed with water.
8. A JEOL JEM-1400 Transmission Electron Microscope was used in Fig. 2.
9. A plate culture can also be used instead of a liquid culture if the collection of secreted pili is not required for the experiment.
10. Do not allow incubation to occur overnight as this could result in lysing of the protoplast too. Lysing of cell wall can be monitored over time by taking a few microliters of the suspension and view it under a light microscope. For rod-shaped bacterial cells, protoplasts would appear circular under the microscope. Muramidase alone has been shown to be effective in lysing the cell wall in *C. diphtheriae, A. oris,* and Streptococcal species,

while a mixture of both lysozyme and muramidase might be required when working with other gram-positive bacteria. In the case of *Staphylococcus aureus*, it has been reported that incubating 1 mL of exponentially growing staphylococci with 50 μg/mL of lysostaphin for 10 min is sufficient (44).

11. It is important to remove as much TCA as possible before acetone addition. Similarly, removal of acetone from the sample is also critical as any residual liquid left could affect the drying of the pellet. If SDS-PAGE of the extracted protein sample is not performed on the same day, simply store the dried pellet at −20 or 4°C.

12. Cell membrane and cytoplasmic fractions are generally not used to visualize pilus polymers but are often collected to visualize other proteins of interest. For example, membrane-bound sortases can be detected using these fractions.

13. After the gels are casted, wrap individual gels with wet paper towels and store at 4°C in a sealed bag. Be sure to check the integrity of each well before using.

14. 3–12% gradient gels have been used successfully to visualize both pilus polymers and monomers in the range of 40 to >460 kDa. For detection of sortases using cell membrane and cytoplasmic fractions, a fixed percentage gel with higher acrylamide content is suggested.

15. As protein samples tend to get stuck in the tip, even when using a low-binding tip, it is recommended to not disturb the pellet after sample buffer addition. Allow the buffer to solubilize the pellet gradually at room temperature before boiling the samples. If the color of the buffer changes to yellow or orange after it comes into contact with the pellet, it means there are still traces of TCA left in the sample, and this will prevent the sample from being dissolved completely in the sample buffer. When this occurs, add 1 or 2 μL of Neutralization buffer.

16. After boiling, quick-spin the sample using a desktop microcentrifuge to make sure all the liquid is collected at the bottom of the tube. Vortex briefly and check to see no pellet remains in the bottom. Dissolving the pellet completely is critical as the sample might be trapped in the well during electrophoresis.

17. If using a 3–12% gradient gel running at 200 V, it takes approximately 45 min for the dye front to reach the bottom of the gel. Sometimes users might wish to allow the gel to run longer to obtain a better resolution of the high-molecular weight pilus polymers. Be sure to use pre-stained protein standards to monitor progress.

18. Soak gradient gels in transfer buffer for at least 10 min is important because it prevents the gel from shrinkage/distortion during transfer.
19. To resuspend the cell pellet as gently as possible, cut the tip off a 1-mL pipette tip to make a wider opening in order to reduce shearing.
20. Hold the grid only at the edge, if the tweezers hold on too much of the grid it could cause cracking of the support film. Remove bacterial culture off the grid by touching the edge of the grid to a Kimwipe.
21. Be sure to fix the parafilm to the bench or petri dish surface securely. One way to do so is by adding a few drop of water onto the work bench and placing the parafilm tightly on top of it.
22. Pipetting the wash buffer up and down underneath the grid will easily lead to bubbles, which should be avoided.
23. If the grid sinks to the bottom of the droplet, it is not recommended to continue, since later washing would be difficult and the coating might come off along with the sample in later steps.
24. Covering the lid helps to prevent the block buffer from drying out.
25. Mix the primary antibody solution before placing the grid on top of the droplet. Be sure not to make any bubbles. Dilution of the primary antibody will depend on the titer.
26. In order to use the secondary antibody sparingly, 20–30 μL of working solution is sufficient for one grid. Cover to avoid evaporation.
27. When possible, the second primary antibody should be obtained from a different animal than the first primary antibody, as this would minimize cross-labeling. If both primary antibodies are obtained from same type of animal, it is still possible to obtain specific binding of the gold-labeled secondary antibody by extensive washing and blocking.
28. Turn the grid sample side down on a piece of Kimwipe to absorb the remaining solution on the grid surface, and then turn the grid sample side up and let it air-dry. The grid now is ready for TEM examination.
29. When inserting a stained grid into the sample holder for TEM, it is not critical which side of the grid faces up. Start viewing the grid with a lower magnification and zoom in until gold particles and pili are clearly visible.
30. Washing the sample to remove residual salts or other chemicals will prevent the formation of salt crystals and/or artifacts on the mica disk.

31. Always use freshly cleaved mica to reduce the presence of small particles on the disk. Aim to use a low concentration of bacterial cells; this will also reduce the background.
32. Longer drying time (and low concentration of cells or pili) typically yields better results as this prevents attractive or repulsive forces on the cantilever during scanning.
33. Search for single cells using the light microscope; avoid scanning cell aggregates as this will easily damage the cantilever tip. In case a pilus extract is analyzed, perform multiple low resolution scans to find an area of interest.
34. If an area of interest has been found, use the "nice zoom zoom" function of the software to perform a scan of a smaller scan area with a higher resolution.
35. For whole cells, use a low scan rate (0.5 Hz) as faster scan rates (1 Hz) will break the cantilever during imaging.
36. Visualization of bacterial cells with pili is challenging since the height difference between cells and pili is very large. Performing first or second order flattening and adjustment of the height scale may help.

Acknowledgments

This work was supported by grants from National Institute of Allergy and Infectious Diseases (AI061381) and National Institute of Dental and Craniofacial Research (DE017382) to HTT.

References

1. Duguid JP, Smith IW, Dempster G, Edmunds PN (1955) Non-flagellar filamentous appendages (fimbriae) and haemagglutinating activity in Bacterium coli. J Pathol Bacteriol 70: 335–348
2. Yanagawa R, Otsuki K, Tokui T (1968) Electron microscopy of fine structure of *Corynebacterium renale* with special reference to pili. Jpn J Vet Res 16:31–37
3. Girard AE, Jacius BH (1974) Ultrastructure of *Actinomyces viscosus* and *Actinomyces naeslundii*. Arch Oral Biol 19:71–79
4. Ton-That H, Schneewind O (2003) Assembly of pili on the surface of *Corynebacterium diphtheriae*. Mol Microbiol 50:1429–1438
5. Ton-That H, Schneewind O (2004) Assembly of pili in Gram-positive bacteria. Trends Microbiol 12:228–234
6. Mandlik A, Swierczynski A, Das A, Ton-That H (2008) Pili in Gram-positive bacteria: assembly, involvement in colonization and biofilm development. Trends Microbiol 16:33–40
7. Mandlik A, Das A, Ton-That H (2008) The molecular switch that activates the cell wall anchoring step of pilus assembly in gram-positive bacteria. Proc Natl Acad Sci USA 105:14147–14152
8. Swaminathan A, Mandlik A, Swierczynski A, Gaspar A, Das A, Ton-That H (2007) Housekeeping sortase facilitates the cell wall anchoring of pilus polymers in *Corynebacterium diphtheriae*. Mol Microbiol 66:961–974
9. Cisar JO, Vatter AE, Clark WB, Curl SH, Hurst-Calderone S, Sandberg AL (1988) Mutants of *Actinomyces viscosus* T14V lacking type 1, type 2, or both types of fimbriae. Infect Immun 56:2984–2989
10. Nallapareddy SR, Singh KV, Sillanpaa J, Garsin DA, Hook M, Erlandsen SL, Murray BE (2006) Endocarditis and biofilm-associated

pili of *Enterococcus faecalis*. J Clin Invest 116: 2799–2807

11. Hendrickx AP, Bonten MJ, van Luit-Asbroek M, Schapendonk CM, Kragten AH, Willems RJ (2008) Expression of two distinct types of pili by a hospital-acquired *Enterococcus faecium* isolate. Microbiology 154:3212–3223
12. Budzik JM, Marraffini LA, Schneewind O (2007) Assembly of pili on the surface of *Bacillus cereus* vegetative cells. Mol Microbiol 66:495–510
13. Barocchi MA, Ries J, Zogaj X, Hemsley C, Albiger B, Kanth A, Dahlberg S, Fernebro J, Moschioni M, Masignani V, Hultenby K, Taddei AR, Beiter K, Wartha F, von Euler A, Covacci A, Holden DW, Normark S, Rappuoli R, Henriques-Normark B (2006) A pneumococcal pilus influences virulence and host inflammatory responses. Proc Natl Acad Sci USA 103:2857–2862
14. Wu H, Mintz KP, Ladha M, Fives-Taylor PM (1998) Isolation and characterization of Fap1, a fimbriae-associated adhesin of *Streptococcus parasanguis* FW213. Mol Microbiol 28: 487–500
15. Levesque C, Vadeboncoeur C, Chandad F, Frenette M (2001) *Streptococcus salivarius* fimbriae are composed of a glycoprotein containing a repeated motif assembled into a filamentous nondissociable structure. J Bacteriol 183:2724–2732
16. Fives-Taylor PM, Thompson DW (1985) Surface properties of *Streptococcus sanguis* FW213 mutants nonadherent to saliva-coated hydroxyapatite. Infect Immun 47:752–759
17. Mora M, Bensi G, Capo S, Falugi F, Zingaretti C, Manetti AG, Maggi T, Taddei AR, Grandi G, Telford JL (2005) Group A *Streptococcus* produce pilus-like structures containing protective antigens and Lancefield T antigens. Proc Natl Acad Sci USA 102:15641–15646
18. Lauer P, Rinaudo CD, Soriani M, Margarit I, Maione D, Rosini R, Taddei AR, Mora M, Rappuoli R, Grandi G, Telford JL (2005) Genome analysis reveals pili in Group B *Streptococcus*. Science 309:105
19. Nobbs AH, Rosini R, Rinaudo CD, Maione D, Grandi G, Telford JL (2008) Sortase A utilizes an ancillary protein anchor for efficient cell wall anchoring of pili in *Streptococcus agalactiae*. Infect Immun 76:3550–3560
20. Smith WD, Pointon JA, Abbot E, Kang HJ, Baker EN, Hirst BH, Wilson JA, Banfield MJ, Kehoe MA (2010) Roles of minor pilin subunits Spy0125 and Spy0130 in the serotype M1 *Streptococcus pyogenes* strain SF370. J Bacteriol 192:4651–4659
21. Proft T, Baker EN (2009) Pili in Gram-negative and Gram-positive bacteria - structure, assembly and their role in disease. Cell Mol Life Sci 66:613–635
22. Rogers EA, Das A, Ton-That H (2011) Adhesion by pathogenic corynebacteria. Adv Exp Med Biol 715:91–103
23. Soriani M, Telford JL (2010) Relevance of pili in pathogenic streptococci pathogenesis and vaccine development. Future Microbiol 5:735–747
24. Mandlik A, Swierczynski A, Das A, Ton-That H (2007) *Corynebacterium diphtheriae* employs specific minor pilins to target human pharyngeal epithelial cells. Mol Microbiol 64:111–124
25. Maisey HC, Hensler M, Nizet V, Doran KS (2007) Group B streptococcal pilus proteins contribute to adherence to and invasion of brain microvascular endothelial cells. J Bacteriol 189:1464–1467
26. Maione D, Margarit I, Rinaudo CD, Masignani V, Mora M, Scarselli M, Tettelin H, Brettoni C, Iacobini ET, Rosini R, D'Agostino N, Miorin L, Buccato S, Mariani M, Galli G, Nogarotto R, Nardi Dei V, Vegni F, Fraser C, Mancuso G, Teti G, Madoff LC, Paoletti LC, Rappuoli R, Kasper DL, Telford JL, Grandi G (2005) Identification of a universal Group B streptococcus vaccine by multiple genome screen. Science 309:148–150
27. Margarit I, Rinaudo CD, Galeotti CL, Maione D, Ghezzo C, Buttazzoni E, Rosini R, Runci Y, Mora M, Buccato S, Pagani M, Tresoldi E, Berardi A, Creti R, Baker CJ, Telford JL, Grandi G (2009) Preventing bacterial infections with pilus-based vaccines: the group B streptococcus paradigm. J Infect Dis 199:108–115
28. Nuccitelli A, Cozzi R, Gourlay LJ, Donnarumma D, Necchi F, Norais N, Telford JL, Rappuoli R, Bolognesi M, Maione D, Grandi G, Rinaudo CD (2011) Structure-based approach to rationally design a chimeric protein for an effective vaccine against Group B Streptococcus infections. Proc Natl Acad Sci USA 108:10278–10283
29. Towbin H, Staehelin T, Gordon J (1979) Electrophoretic transfer of proteins from polyacrylamide gels to nitrocellulose sheets: procedure and some applications. Proc Natl Acad Sci USA 76:4350–4354
30. Ton-That H, Faull KF, Schneewind O (1997) Anchor structure of staphylococcal surface proteins. I. A branched peptide that links the carboxyl terminus of proteins to the cell wall. J Biol Chem 272:22285–22292
31. Mishra A, Devarajan B, Reardon ME, Dwivedi P, Krishnan V, Cisar JO, Das A, Narayana SV,

Ton-That H (2011) Two autonomous structural modules in the fimbrial shaft adhesin FimA mediate Actinomyces interactions with streptococci and host cells during oral biofilm development. Mol Microbiol 81:1205–1220

32. Dramsi S, Caliot E, Bonne I, Guadagnini S, Prevost MC, Kojadinovic M, Lalioui L, Poyart C, Trieu-Cuot P (2006) Assembly and role of pili in group B streptococci. Mol Microbiol 60:1401–1413
33. Egerton RF (2005) Physical principles of electron microscopy: an introduction to TEM, SEM, and AEM. Springer Science+Business Media, New York
34. Binnig G, Quate CF, Gerber C (1986) Atomic force microscope. Phys Rev Lett 56: 930–933
35. Dufrene YF (2008) Towards nanomicrobiology using atomic force microscopy. Nat Rev Microbiol 6:674–680
36. Muller DJ, Dufrene YF (2008) Atomic force microscopy as a multifunctional molecular toolbox in nanobiotechnology. Nat Nano technol 3:261–269
37. Sugimoto Y, Pou P, Abe M, Jelinek P, Perez R, Morita S, Custance O (2007) Chemical identification of individual surface atoms by atomic force microscopy. Nature 446:64–67
38. Liu S, Wang Y (2010) Application of AFM in microbiology: a review. Scanning 32:61–73
39. Arce FT, Carlson R, Monds J, Veeh R, Hu FZ, Stewart PS, Lal R, Ehrlich GD, Avci R (2009) Nanoscale structural and mechanical properties of nontypeable Haemophilus influenzae biofilms. J Bacteriol 191:2512–2520
40. Biais N, Higashi DL, Brujic J, So M, Sheetz MP (2010) Force-dependent polymorphism in type IV pili reveals hidden epitopes. Proc Natl Acad Sci USA 107:11358–11363
41. Tripathi P, Dupres V, Beaussart A, Lebeer S, Claes IJ, Vanderleyden J, Dufrene YF (2012) Deciphering the Nanometer-Scale Organization and Assembly of Lactobacillus rhamnosus GG pili using atomic force microscopy. Langmuir 28:2211–2216
42. Foroni E, Serafini F, Amidani D, Turroni F, He F, Bottacini F, O'Connell Motherway M, Viappiani A, Zhang Z, Rivetti C, van Sinderen D, Ventura M (2011) Genetic analysis and morphological identification of pilus-like structures in members of the genus *Bifidobacterium*. Microb Cell Fact 10(Suppl 1):S16
43. Alegre-Cebollada J, Perez-Jimenez R, Kosuri P, Fernandez JM (2010) Single-molecule force spectroscopy approach to enzyme catalysis. J Biol Chem 285:18961–18966
44. Bae T, Banger AK, Wallace A, Glass EM, Aslund F, Schneewind O, Missiakas DM (2004) *Staphylococcus aureus* virulence genes identified by bursa aurealis mutagenesis and nematode killing. Proc Natl Acad Sci USA 101:12312–12317
45. Gaspar AH, Ton-That H (2006) Assembly of distinct pilus structures on the surface of *Corynebacterium diphtheriae*. J Bacteriol 188: 1526–1533

Chapter 6

Single Cell Microfluidic Studies of Bacterial Motility

Adrien Ducret, Olivier Théodoly, and Tâm Mignot

Abstract

A large number of bacterial species move smoothly on solid surfaces in the absence of extracellular organelles. In the deltaproteobacterium *Myxococcus xanthus*, this surface motion, termed gliding motility, involves a novel macromolecular machinery Agl-Glt. During the motility process, the Agl-Glt system, an integral envelope protein complex, is assembled on the ventral side of the cell. Doing so, the complex couples surface adhesion to the activity of the Agl motility motor. On the cytosolic side, the Agl-Glt system is linked to the bacterial actin cytoskeleton MreB. It is proposed that motility is produced when surface immobilized Agl-Glt complexes produce traction on a rigid track, possibly the MreB cables. Testing this hypothesis directly requires both microfluidic techniques to perturb the motility process with inhibitors (i.e., A22, CCCP) and state-of-the-art microscopy techniques (i.e., TIRF and AFM). These approaches require a microscopy chamber where the cells glide in liquid on a non-agar substrate. Here, we describe a straightforward coating procedure to construct a chitosan-functionalized microfluidic chamber that fulfills these requirements. This set up circumvents all the disadvantages of traditional agar-based assays, providing new grounds for high-resolution experiments. We also describe simple image processing to maximize the quality of data representation. In theory, our procedure could be used for any bacterial system that adheres to chitosan.

Key words: Coating procedures, Surface motility, *Myxococcus xanthus*, Microfluidics, Light microscopy, Adhesion

1. Introduction

Bacterial motility is a widespread phenomenon allowing bacteria to colonize new environmental niches, infect hosts, or develop multicellular communities, biofilms, or fruiting bodies (1, 2). While the mechanism by which bacteria swim in liquid using flagella has been studied in great details for several decades, how bacteria crawl on solid surfaces is much less understood. In bacteria, surface motility

Anne H. Delcour (ed.), *Bacterial Cell Surfaces: Methods and Protocols*, Methods in Molecular Biology, vol. 966,
DOI 10.1007/978-1-62703-245-2_6, © Springer Science+Business Media New York 2013

is not operated by a universal system but can be powered by distinct macromolecular systems (3). For example, flagella themselves can promote a form of surface motility on hard-wet surfaces, a process known as swarming motility (4). In twitching motility, another type of surface motion, a Type-IV pilus forms retractile fibers at the bacterial pole, which pull cells forward following their attachment to the substrate (5, 6). Finally, gliding rod-shaped bacteria move smoothly along their long axis, in the absence of any visible organelles. The molecular basis of this process has long been mysterious and is only starting to emerge in the Bacteroidetes and the deltaproteobacteria.

In the Bacteroidetes, studies in *Flavobacterium jonhsonniae* suggest that motility is powered by a protein treadmilling system along the cell surface (7). Motility genes have been identified but the identity of the molecular machinery remains unclear. In the deltaproteobacterium *Myxococcus xanthus*, the Agl-Glt system has recently been characterized unambiguously as the motility machinery (8–10). This macromolecular complex involves at least 15 proteins and consists of the modular assemblage of a three-protein motor embedded in the inner membrane and a larger transducer complex that spans all three layers of the bacterial envelope. The motility motor (Agl), a member of the Tol/Exb/Mot family of proton channels, uses the proton motive force rather than ATP to energize the motility process (10, 11). During the last 5 years, work from several laboratories has led to a model of Agl-Glt function during motility. It is proposed that following direct interaction with the MreB-actin cytoskeleton (12), Agl-Glt complexes assemble dynamically on the ventral side of the cells and contact the substrate (8) (Fig. 1). More specifically, the Agl-Glt complexes may traffic along MreB cables, aggregating on the ventral side of the cell as they latch onto the substrate (Fig. 1). Agl motor traction against the immobilized Glt transducer would thus power motility. Proving this model requires new experimental systems, for example experimental state of the art light microscopy approaches (see below).

Studies of the gliding motility mechanism have long been restricted to genetics and electron microscopy. These approaches have failed to reveal the mechanism because in the absence of obvious motility organelle (i.e., a Type-IV pilus), it was difficult to determine if the effects of mutations are direct. Electron microscopy proved relatively unhelpful because EM images chemically fixed cells and it is often difficult to relate structures to specific molecular complexes. Recently, a major advance in the study of the motility mechanism has been the use of fluorescence microscopy techniques to track the localization of motility proteins in live motile cells, which suggested the existence of the Agl-Glt complex and led to its characterization (8, 9, 13). Until now, these time-lapse experiments have used thin agar pads as a substrate, a standard

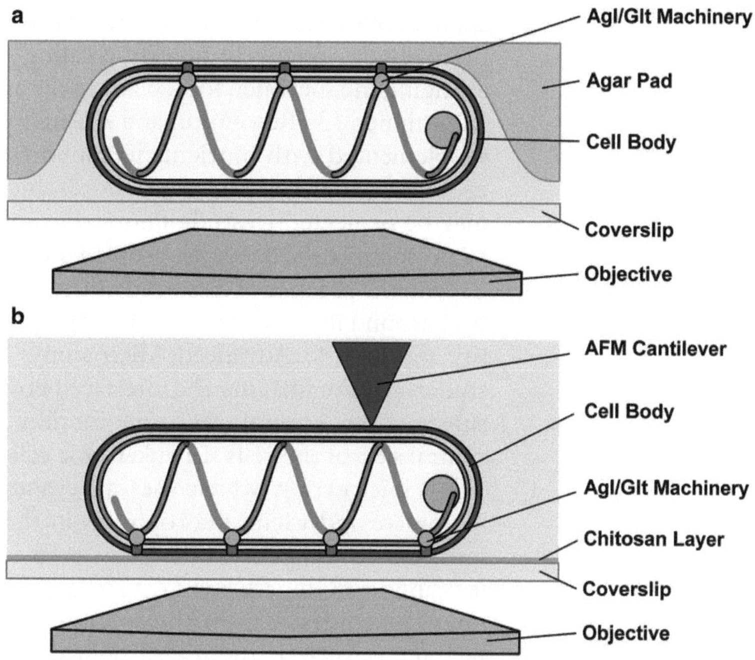

Fig. 1. Schematic representations of gliding *Myxococcus* cells on a standard agar pad system (**a**) vs. the Chitosan-chamber system (**b**). (**a**) In the agar pad system, the substrate and the microscope objective are on opposite faces of the cell body. Agl-Glt complexes are represented against the substrate. The *big circle* at the pole represents Agl-Glt assembly sites (8, 13). (**b**) By contrast, in the Chitosan chamber, the cell establishes contacts directly on the coverslip. The contact zone is thus closer to the objective allowing advanced microscopy like TIRF. Note that the cell is directly accessible to an Atomic Force Microscopy cantilever or optical tweezer.

convenient method with broad applications in bacterial cell biology. Unfortunately, the agar pad technique has several major limitations: (1) Changing the medium rapidly, for example to assess the effects of a drug, is difficult. In the past, we have developed a system where a microfluidic chamber is mounted on top of an agar pad (14). In this system, the fluids are not in direct contact with bacteria but diffuse passively through the thin agar layer. The system gave satisfactory results but was only applicable to small molecules with little unspecific interactions with the agar layer. Also, due to the diffusion constraints, the kinetics of drug effect was slow. Moreover, it was not always easy to flush the drugs efficiently from the agar gel and the effects could not always be reversed. Lastly, on occasions, the shear stress applied by the flow detached the agar pad, frustrating many experiments. (2) Desiccation. Motility studies, for example, the development of multicellular patterns over the course of several hours or studies of the cell cycle, require stable wet media. Agar pads are not ideal for these applications because in the absence of an external moisture source, the pads dry relatively quickly creating major cellular stress and changing the specimen focal plane as the agar layer becomes thinner and thinner. (3) Resolution and sensitivity. Agar pads are generally supplemented with nutrients to support motility, creating a background when observed by epi-fluorescence. While this background may be neglected to study fluorescent specimens with high signal to noise ratios, it becomes problematic to analyze weaker signals. Even in the absence of dissolved nutrient, agar pads show significant background fluorescence, hampering rigorous studies of low intensity signals. (4) Advanced Microscopy. High-resolution motility studies require imaging the interface between the cell body and the substrate, for example to define motility protein dynamics on the ventral side of the cells. In eukaryotic cells, this is typically done by TIRF microscopy, which uses an evanescent wave to selectively illuminate and excite fluorophores in the region of the specimen immediately adjacent to the glass-water interface (15). Such studies cannot be done when an agar layer of variable thickness separates the coverslip from the specimen. Finally, other microscopy techniques such as the use of optical tweezers and Atomic Force Microscopy require the cells to be directly accessible and thus immerged in liquid on agar-free substrates.

The solution to all these limitations is a small volume microfluidic chamber where the cells can move directly on glass while immersed in liquid (Fig. 1) (16). However, since most bacteria, especially *Myxococcus xanthus*, do not naturally adhere to glass, adhesion and therefore motility requires functionalization of the glass coverslip. A major challenge is to design a functionalization procedure that allows both binding and motility. For example, highly adhesive molecules like poly-L-lysine adsorb the cells to the glass surface so effectively that they remain stuck. On the other hand, on other coatings (for example cellulose coatings, unpublished) the cells quickly resuspend in the liquid phase. We were able to overcome these difficulties using chitosan, a linear polysaccharide composed of randomly distributed β-(1–4)-linked D-glucosamine (deacetylated unit) and *N*-acetyl-D-glucosamine (acetylated unit). Here, we describe a straightforward procedure to make chitosan-coated PDMS microfluidic chambers allowing the study of *Myxococcus* gliding motility under liquids and potentially any other bacterial cellular process, provided that the model bacterium binds the chitosan substrate. We show how this system can be used to test the reversible effect of a proton gradient uncoupler (CCCP, carbonyl cyanide *m*-Chlorophenylhydrazone) on *Myxococcus* motility (Fig. 2). Finally, we describe basic straightforward methods for rigorous image analysis of the experiments (Fig. 3).

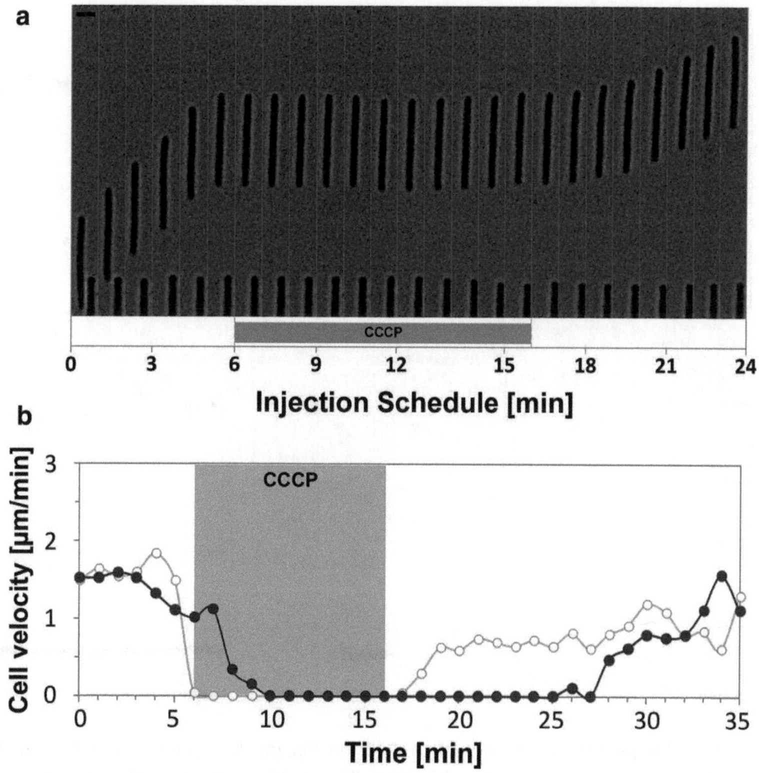

Fig. 2. Drug perturbation of motility in the chitosan chamber (**a**) Cell movements before, during and after CCCP injection are shown by phase contrast. The cell was straightened as described in Subheading 3. Note that cell movement stops very rapidly and reversibly upon addition of CCCP. Scale bar = 1 μm. (**b**) Computed cell velocity as a function of time (min) during the experiment. *Open Circle*: chitosan chamber. *Closed Circle:* standard agar-pad system (10). Note that in the chitosan system, the cells stop and resume movement within 1 min after CCCP addition and washing against several minutes in the agar system.

2. Materials

2.1. Mold Preparation for the Microfluidic Chamber

1. Clewin software (from WieWeb Software).
2. Film Photomask: mask of resolution >20 μm.
3. Standard 76 × 26 mm microscope slides.
4. Dry film photoresist: thickness 100 μm (WBR™ 2000 Series).
5. Thermal film laminator.
6. Photolithography masks aligner.
7. Precision hot plate.
8. 1 % (w/w) sodium carbonate (Na_2CO_3, purity >99 %) in deionized water.

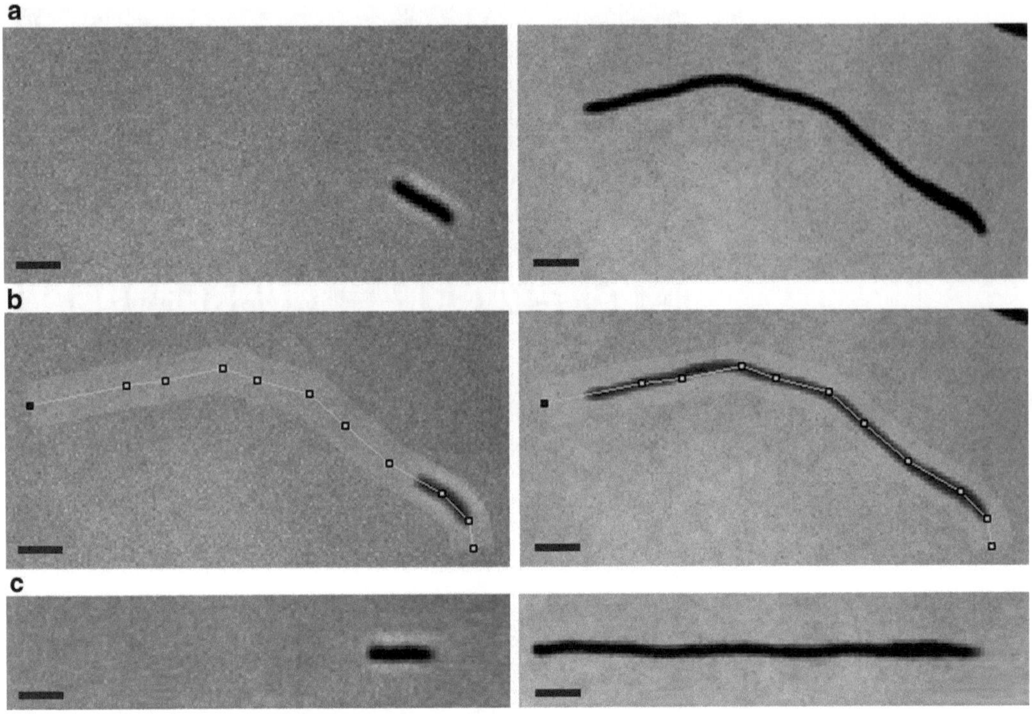

Fig. 3. Computational straightening of the gliding cells. (**a**) Moving Cell (*left panel*) and resulting curved trajectory (*right panel*) are shown by phase contrast images. (**b**) Characteristic segmented lines recapitulating the cell's trajectory are shown overlaid on the phase contrast images shown in (**a**). *White square* correspond to nodes. (**c**) Resulting straightened images of the cell (*left panel*) and its trajectory (*right panel*). Scale bar = 1 μm.

9. 99 % ethanol.
10. Filtered (0.2 μm) deionized water.

2.2. Microfluidic Chamber Preparation

1. Polydimethylsiloxane (PDMS): Sylgard 184 Silicone Elastomer Kit (Base and curing agent).
2. Incubator oven.
3. 170-mm-thick standard glass coverslips.
4. 5 % DECON 90 (w/w) diluted in deionized water.
5. 99 % Ethanol.
6. Filtered (0.2 μm) deionized water.
7. Micro-Punch, Tip Diameter 1.0 mm.
8. Plasma cleaner.
9. Polyethylene tubing: inner diameter 0.015 in. × outer diameter 0.04 in.
10. 1 mL Luer syringes.
11. Luer needles: 27 G × 3/4 in.

2.3. Bacterial Cells Preparation

1. *Myxococcus xanthus* strains: DZ2 (TM108), DZ2 ΔpilA (TM478).
2. 1 M MOPS buffer (3-morpholinopropanesulfonic acid), pH 7.6.
3. CYE medium: 10 g/L Bacto Casitone peptone, 5 g/L yeast extract, 1 g/L $MgCl_2$, 10 mM MOPS, adjusted to pH 7.4 with 5 M NaOH, and sterilized (20 min at 121 °C).
4. Centrifuge.
5. TPM-buffer: 10 mM Tris-HCl, 8 mM $MgSO_4$, 100 mM KH_2PO_4. Adjust the pH to 7.2 with 1 M KOH. Filter or autoclave. Store at room temperature.
6. 100 mM $CaCl_2$: filter or autoclave. Store at room temperature.
7. TPM-Ca-Buffer: 1 mM $CaCl_2$ in TPM-buffer. Store at room temperature.

2.4. Flow Chamber Functionalization and Cells Visualization

1. 1 mL syringes.
2. 2 M Acetic acid. Freshly prepared.
3. Chitosan 100×: prepare a 15 mg/mL stock solution in 2 M acetic acid (see Note 1). Prepare freshly. The stock solution is stable for less than 1 h.
4. Chitosan Coating Solution: dilute Chitosan 100× 100-fold with deionized water (see Note 2). Prepare freshly. Use the solution immediately after preparation.
5. TPM-Ca-Buffer: see item 7 in Subheading 2.3.
6. Automated inverted epifluorescence microscope.

2.5. CCCP Injection

1. CCCP (Carbonyl cyanide 3-chlorophenylhydrazone) stock: 20 mg/mL in DMSO. Store at −20 °C in aliquots.
2. TPM-Ca-Buffer: see item 7 in Subheading 2.3.
3. 10 μM CCCP: Prepare freshly in TPM-Ca-Buffer.

2.6. Image Analysis

1. Metamorph (Molecular devices).
2. Excel 2000 (Microsoft).
3. ImageJ 1.43 software.

3. Methods

3.1. Mold Preparation for the Microfluidic Chamber

1. Design a chamber of dimensions 500 μm×4 mm using the Clewin software.
2. Send the Clewin designs to a specialized film photomasks manufacturer for lithography.

3. Wash a glass microscope slide first with 99 % Ethanol and then with deionized water (previously filtered at 0.2 μm) in a clean room.
4. Sequentially deposit four 100 μm thick layers of dry film photoresist on the glass slide using the Laminator at 80 °C.
5. Place the sample on a hot plate at 110 °C for 30 s.
6. Insulate the sample through the photomask set at 35 W/cm^2 ($\lambda = 405$ nm) for 65 s using the mask aligner (16).
7. Reveal the sample with a 1 % solution of Na_2CO_3 for 20 min.
8. Rinse with deionized water.

The final dimensions of the channels of the positive molds are 500 μm wide, 4 mm long and 400 μm high.

3.2. Microfluidic Chamber Preparation

1. Prepare a 1/10 (w/w) mixture of curing agent/base PDMS.
2. Stir thoroughly then place under vacuum for 1 h to remove bubbles.
3. Pour the curing agent/base PDMS mixture on a positive mold of channels, prepared in Subheading 3.1.
4. Place under vacuum for 1 h to remove bubbles.
5. Place in an incubator oven at 65 °C for at least 3 h.
6. Unmold the PDMS replica and punch 1 mm holes with the Micro-Punch tool to plug inlet and outlet tubing.
7. Clean the PDMS replica and a glass coverslip with 5 % DECON 90, then 99 % Ethanol and finally filtered (0.2 μm) deionized water. Add: Dry with a flow of nitrogen gas.
8. Activate the glass coverslip and the PDMS replica with the plasma cleaner for 5 min and 1.5 min, respectively, in air at a pressure of 1,000 mtorr.
9. Immediately seal the glass coverslip and PDMS replica together, by placing the replica on the coverslip with the microchannels facing the coverslip. Intimate contact should spread spontaneously without pressing on the replica.
10. Place in an incubator oven at 65 °C for at least 3 h.
11. Insert the PE tubings directly in the holes prepared in the PDMS as inlet and outlet ports.
12. Insert needles in the PE tubing to connect syringes.

3.3. Bacterial Cells Preparation

1. Inoculate CYE medium with the chosen bacterial strain and appropriate antibiotic(s). Grow with aeration at 32 °C overnight (see Note 3).
2. Centrifuge 1 mL of cells suspension for 3 min at 5,000 × *g* at room temperature; discard the supernatant.

3. Thoroughly resuspend the cells in 1 mL of TPM-Ca-buffer.
4. Repeat the washing procedure once.

3.4. Flow Chamber Functionalization and Cells Visualization

1. Using a 1-mL syringe, inject 1 mL of distilled water at 0.5 mL/min into the flow chamber (see Note 4).
2. Inject 1 mL of Chitosan coating solution.
3. Incubate 30 min at room temperature.
4. Rinse the Chitosan coating solution using 1 mL of deionized water.
5. Inject 1 mL of TPM-Ca-Buffer.
6. Inject 1 mL of cell suspension prepared in Subheading 3.3.
7. Place the flow chamber on the microscope stage, using an appropriate custom-made holder.
8. Allow the cells to attach to the substrate (for ~20 min), then rinse unbound cells using 1 mL of TPM-Ca-Buffer (see Note 5).
9. Chose image acquisition parameters (see Note 6).

3.5. CCCP Injection

1. Inject 1 mL of 10 μM CCCP.
2. Incubate 10 min.
3. Rinse the cells using 1 mL of TPM-Ca-Buffer.

3.6. Image Analysis

1. The multidimensional data sets obtained on the automated microscope are saved into 16-bits TIFF stack format for Further processing using the ImageJ software.
2. To correct for background fluctuations over time, the stack images are normalized via the "Process>Enhance Contrast>Normalize All Slices" function of ImageJ using appropriate "saturated pixels" value (i.e., 0.4 %).
3. To correct for image defects caused by uneven illumination in fluorescent or phase contrast microscopy, the background intensity is corrected via the "Process>Substract Background" function of ImageJ using appropriate "Rolling Ball Radius" value (i.e., 10 pixels). For phase contrast images, check the "light background" radio button.
4. To minimize drift in each plane, the stack images are aligned via the "stackreg" plug-in (http://bigwww.epfl.ch/thevenaz/stackreg/). In the stackreg plug-in, the "Translation method" is preferred to correct the microscope stage drift.
5. Cell tracking is done either manually or automatically using the Metamorph track object function or a custom-made cell tracking program (14). The data sets obtained are subsequently plotted using Excel (Fig. 2).

3.7. Micrographs Representations

Gliding *Myxococcus* cells are often curved precluding the localization of fluorescent proteins along the cell body over time. Computational projections allow straightening of the cells for better spatiotemporal resolution (Fig. 3).

1. Draw a straight line or segmented line if the trajectory is curved recapitulating a cell's trajectory (see Note 7).
2. Use the "Edit>Selection>Straighten" function to straighten the trajectory. For stack images, check the "Process Entire Stack" radio button. Choose appropriate "Line Width" (i.e., 20 pixels) and generate the straightened image.
3. To create images sequence, straightened stack images can be assembled via the "Image>Stacks>Make Montage ..." function. Choose appropriate number of columns (i.e., 25), number of rows (i.e., 1), as well as scale factor (i.e., 1) to generate the images sequence.

4. Notes

1. Chitosan dissolves slowly in acetic acid. Mix thoroughly before use and then centrifuge 3 min at 5,000 × *g* to remove bubbles. Failure to completely dissolve chitosan results in heterogeneous surface functionalization. To mix the Chitosan stock solution, the use of a mixer mill (i.e., Retsch) that shakes the sample radially is preferred (30 Hz for 5 min).
2. The Chitosan Stock solution is very viscous. Pipette slowly to avoid bubble formation.
3. Cells should be in mid- to late-log phase of growth (less than an A_{600nm} of 1).
4. When solutions are injected manually, use 1 mL Luer syringes to limit the flow rate in the flow chamber.
5. Inject slowly (i.e., 0.5 mL/min) to limit shear stress and cell detaching.
6. Setting up the acquisition conditions properly is of highest importance to obtain images that can be quantified. To maintain the point of interest within a specimen in sharp focus at all times, the microscope is equipped with "The Perfect Focus System" (PFS). Fields were manually or automatically defined with a motorized stage and stored (*X*, *Y*, *Z*, PFS-offset) in a custom-made automation system designed for time-lapse experiments (14). The images were recorded with a CoolSNAP HQ 2 (Roper Scientific) and 100×/1.4 DLL objective. The resulting images measured 0.064 μm/pixel. Excitation light was emitted by a 120 W metal halide light and the signals were

monitored using appropriate filters (480ex/535em). All fluorescence images were acquired with a minimum exposure time to minimize bleaching and phototoxicity effects.

7. The positions and the number of nodes in the segmented line can create artifacts on the resulting straightened image. If needed, the positions and number of nodes can be recursively optimized back on the original image for correction.

Acknowledgments

We thank Yong Zhang for suggesting Chitosan as a potential motility substrate. AD was funded by an HFSP young investigator grant (HFSP—RGY0075/2008) to TM. This work was supported by a C-Nano PACA grant (AAP09) to OT and TM.

References

1. Shrout JD, Tolker-Nielsen T, Givskov M, Parsek MR (2011) The contribution of cell-cell signaling and motility to bacterial biofilm formation. MRS Bull 36:367–373
2. Curtis PD, Taylor RG, Welch RD, Shimkets LJ (2007) Spatial organization of Myxococcus xanthus during fruiting body formation. J Bacteriol 189:9126–9130
3. McBride MJ (2001) Bacterial gliding motility: multiple mechanisms for cell movement over surfaces. Annu Rev Microbiol 55:49–75
4. Kearns DB (2010) A field guide to bacterial swarming motility. Nat Rev Microbiol 8: 634–644
5. Merz AJ, So M, Sheetz MP (2000) Pilus retraction powers bacterial twitching motility. Nature 407:98–102
6. Skerker JM, Berg HC (2001) Direct observation of extension and retraction of type IV pili. Proc Natl Acad Sci U S A 98:6901–6904
7. McBride MJ (2004) Cytophaga-flavobacterium gliding motility. J Mol Microbiol Biotechnol 7:63–71
8. Luciano J et al (2011) Emergence and modular evolution of a novel motility machinery in bacteria. PLoS Genet 7:e1002268
9. Nan B, Mauriello EMF, Sun I-H, Wong A, Zusman DR (2010) A multi-protein complex from Myxococcus xanthus required for Bacterial gliding motility. Mol Microbiol 76:1539–1554
10. Sun M, Wartel M, Cascales E, Shaevitz JW, Mignot T (2011) Motor-driven intracellular transport powers bacterial gliding motility. Proc Natl Acad Sci 108:7559–7564
11. Nan B et al (2011) Myxobacteria gliding motility requires cytoskeleton rotation powered by proton motive force. Proc Natl Acad Sci U S A 108:2498–2503
12. Mauriello EMF et al (2010) Bacterial motility complexes require the actin-like protein, MreB and the Ras homologue, MglA. EMBO J 29:315–326
13. Mignot T, Shaevitz JW, Hartzell PL, Zusman DR (2007) Evidence that focal adhesion complexes power bacterial gliding motility. Science 315:853–856
14. Ducret A et al (2009) A microscope automated fluidic system to study bacterial processes in real time. PLoS One 4:e7282
15. Berginski ME, Vitriol EA, Hahn KM, Gomez SM (2011) High-resolution quantification of focal adhesion spatiotemporal dynamics in Living cells. PLoS One 6:e22025
16. Ng JMK, Gitlin I, Stroock AD, Whitesides GM (2002) Components for integrated poly(dimethylsiloxane) microfluidic systems. Electrophoresis 23: 3461–3473.

Chapter 7

Chromatographic Analysis of the *Escherichia coli* Polysialic Acid Capsule

Susan M. Steenbergen and Eric R. Vimr

Abstract

Polysialic acid capsules are the major virulence factors in *Escherichia coli* K1, K92, and groups B and C meningococci. The sialic acid monomers (2-keto-3-deoxy-5-acetamido-7,8,9-D-*glycero*-D-*galacto*-nonulosonic acids) comprising these homopolymeric polysaccharide chains can be selectively modified with 1,2-diamino-4,5-methylenedioxy-benzene to produce highly fluorescent quinoxalinone derivatives distinguished by their elution times during reverse phase chromatography. Here, we describe methods to release the constituent capsular polysialic acid monomers, and to detect and quantify them by sensitive fluorometry. There are relatively few 2-keto acids in bacteria, making it possible to rapidly analyze samples even without prior purification of capsular polysaccharides.

Key words: *Escherichia coli*, Polysialic acid capsules, Virulence factors, Release of sialic acid monomers, Reverse phase chromatography, Fluorometry

1. Introduction

The physiochemical, immunological, and host-interactive properties of bacterial cell surfaces are dominated by polysaccharides composed of one or more types of carbohydrate monomers. Capsular polysaccharides represent one class of cell surface structure that receives intensive research scrutiny due to its value as target for vaccine and new drug development (1, 2). Capsules are usually linked to the bacterial cell surface and are most often composed of one or two different carbohydrates. The *Escherichia coli* capsular polysialic acid chains are homopolymers composed of about 200 sialic acid (Neu5Ac) residues connected by alpha-2,8-glycoketosidic linkages. The chains are attached to the surface by an acid labile phosphodiester bond such that, depending on growth conditions, the capsule is continually sloughed into the medium. Growth in

Anne H. Delcour (ed.), *Bacterial Cell Surfaces: Methods and Protocols*, Methods in Molecular Biology, vol. 966,
DOI 10.1007/978-1-62703-245-2_7,

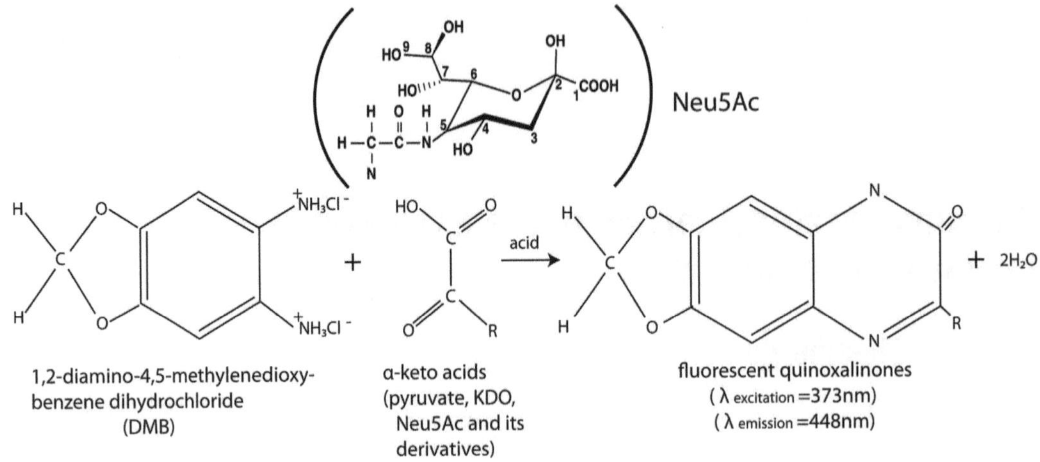

Fig. 1. Fluorescent labeling of alpha-keto acids. DMB reacts with 2-keto acids under acidic conditions to produce fluorescent quinoxalinones that are detected after excitation at 373 nm and emission at 448 nm. The solution structure of Neu5Ac is shown in parentheses with carbons-1–9 numbered by the *subscripts*. 3-deoxy-D-*manno*-octulosonic acid (KDO) is an alpha-keto sugar that is a usual component of lipopolysaccharide; depending on the preparative conditions it can be a contaminant in the analysis of polysialic acids, as can pyruvate that accumulates in some bacterial strains (*see* ref. (4)).

highly buffered defined chemical medium maximizes the amount of capsule retained at the cell surface (3). We have adapted a variety of preparative and release procedures that allow chemical analyses of virtually any sample regardless of its purity (4). These methods are described in his article.

Most clinical *E. coli* K1 isolates are variably modified with O-acetyl groups (CH_3–COO–) attached to carbon-7 or -9 of the Neu5Ac monomer (Fig. 1). These groups, but not the acetamido at carbon-5, are base labile and can be lost during analytical or preparative procedures. Therefore, attention must be given to how the bacteria are grown and handled during preparation for subsequent analysis by labeling with 1,2-diamino-4,5-methylenedioxybenzene (DMB) (Fig. 1). The variable O-acetylation at carbons-7 or -9 is controlled by a random translational switch and expression of a specific polysialic acid O-acetyltransferase encoded by *neuO* (4–9). The enzyme uses acetyl-coenzyme A as donor to transfer acetyl groups to Neu5Ac residues during polymerization of the nascent polysialic acid chains inside the cell. Modified capsule chains are ultimately transported to the outer surface of the outer bacterial membrane where the constituent sialic acids, either with or without prior purification, can be analyzed after enzymatic or chemical hydrolysis of the polysaccharides. The degree of O-acetylation is quantified by reacting the resulting alpha-keto acids with DMB, giving rise to fluorescent quinoxalinones that are separated by reverse phase chromatography. Hara and associates first described the DMB labeling method to measure the degree of O-acetylated sialic acids in serum glycoproteins (10). Our modified methods describe the use of the DMB labeling procedure for analyzing bacterial polysialic acids. To designate acetylation of Neu5Ac, the abbreviations Neu5,7Ac_2, Neu5,8Ac_2, Neu5,9Ac_2, or Neu4,5Ac_2 identify O-acetyl groups at the indicated carbon positions, respectively. The other major Neu5Ac derivative has an *N*-glycolyl group instead of acetamido at carbon-5 and is abbreviated Neu5Gc; this derivative elutes earlier than Neu5Ac during reverse phase chromatography but is not known to be a component of any bacterial surface polysaccharide.

2. Materials

Clinical *E. coli* K1 isolates are class 2 pathogens that must be grown and harvested using biosafety level 2 procedures. Organic effluent and wastes should be collected and disposed of according to local chemical hazard requirements. Neither spent growth media nor live bacteria should be disposed of in sinks without prior sterilization.

2.1. Analysis of Standards

1. 2-Keto acids: Dissolve each of 1 mg Neu5Ac, or any of its commercially available derivatives (see Note 1), 1 mg KDO, or 1 mg pyruvate in 0.1–1 mL water. These standards should be stored at -20°C.
2. Bovine submaxillary gland mucin (SGM): Dissolve 1 mg Type I-S SGM in 0.1–1 mL water (see Note 2). Store at -20°C.
3. 4 M acetic acid: Take 230 mL of glacial acetic acid and bring it to 1 L with water.
4. Incubator: 80 and 50°C incubator. Samples at 50°C should be covered with aluminum foil to block light.
5. Centrifuge: 3 min at 16,000 × *g*
6. DMB labeling reagent: Dissolve 0.63 mg DMB and 1.2 mg sodium hydrosulfite in 200 μL of 1.4 M acetic acid; add 22 μL beta-mercaptoethanol and vortex the mixture for a few seconds to dissolve the components. Store protected from light for up to 1 week at 4°C (see Note 3).
7. 0.22 μm nylon Spin-X LC microfuge tubes.
8. HPLC Solvent: Mix 45 mL HPLC-grade acetonitrile, 35 mL methanol, and 420 mL ultrapure water, mix with stir bar and add another 500 mL of water. Pass the solvent through a 47 mm diameter, 0.2-μm porosity nylon filter into a side-arm Erlenmeyer flask containing magnetic stir bar; mix under vacuum to degas the solvent (see Note 4).
9. Reverse phase column.

10. HPLC system with fluorescence detector and associated software for data management (see Note 5).
11. 1 M NaOH.
12. 2 M acetic acid: Take 115 mL of glacial acetic acid and bring it to 1 L with water.
13. 50% methanol: Mix 500 mL of methanol with 500 mL ultra-pure water. Pass the solvent through a 47 mm diameter, 0.2-μm porosity nylon filter into a side-arm Erlenmeyer flask containing magnetic stir bar; mix under vacuum to degas the solvent (see Note 4).

2.2. Growing and Harvesting Bacterial Cells

1. LB medium: Weigh 10 g tryptone, 5 g NaCl, and 5 g yeast extract. Dissolve the ingredients in 600 mL water in a 2 L Erlenmeyer flask then bring to 1 L in a graduated cylinder, transfer back to the flask or aliquot into smaller containers and sterilize (see Note 6). Store at room temperature.
2. Defined minimal medium: Weigh 10 g casamino acids, 5 g glucose, 12.5 g $Na_2PO_4{\cdot}2H_2O$, 0.9 g KCl, and 0.6 g $MgSO_4{\cdot}7H_2O$. Dissolve ingredients in 700 mL water and pH to 7.8 with concentrated NaOH. Bring to 1 L with water. Filter sterilize (see ref. (3) and Note 7).
3. Spectrophotometer.
4. Water Bath Shaker.
5. Centrifuge and associated rotor.

2.3. Analysis of Polysialic Acids from Intact Bacteria

1. 4 M acetic acid: see item 3, Subheading 2.1.
2. 1.4 M acetic acid: Take 80.5 mL of glacial acetic acid and bring it to 1 L with water.
3. Centrifugal vacuum concentrator and/or lyophilizer.
4. Centrifuge and associated rotor.
5. DMB labeling reagent: see item 6, Subheading 2.1.
6. HPLC solvent and system: see items 8–10, Subheading 2.1.

2.4. Semi-Purification and Analysis of Polysialic Acids

1. Modified phosphate buffered saline (PBS): Dissolve 8 g NaCl, 0.2 g KCl, 1.15 g Na_2HPO_4, and 0.2 g KH_2PO_4 in 800 mL of water. Adjust pH to 7.4 with 6 M HCl and bring to 1 L final volume. Sterilize by autoclave or filtration.
2. Pyridine buffer (100 mM): Add 79 mL pyridine to 500 mL water and bring pH to 5.0 with glacial acetic acid. Bring solution to 1 L and store away from light in the refrigerator.
3. Water Bath Shaker.
4. Centrifuge and associated rotor.
5. Lyophilizer.

6. Dialysis tubing: 12,000 MW cutoff.
7. Vacuum centrifuge.
8. DMB labeling reagent: see item 6 Subheading 2.1.
9. HPLC solvent and system: see items 8–10 Subheading 2.1.

3. Methods

Carry out all procedures except bacterial growth at room temperature unless indicated otherwise. Bacteria should be grown at 37°C with vigorous (200–250 rpm) shaking.

3.1. Analysis of Standards

1. Mix equal volumes of SGM and 4 M acetic acid to give 2 M final concentration. Incubate at 80°C for 2 h to hydrolyze sialic acid residues while retaining O-acetyl groups. After hydrolysis, centrifuge at 16,000 × *g* and discard any pellet.
2. Mix 20 μL of SGM supernatant, the Neu5Ac standards, or other standards with 20 μL of 2× DMB labeling reagent. Incubate in the dark at 50°C for 2–2.5 h. Prepare SGM-derived sialic acids or other standards for chromatography by centrifuging them through 0.22 μm nylon Spin-X LC Corning-Costar microfuge tubes.
3. Inject 6–10 μL sample and perform reverse phase chromatography in an ascending manner. The column will run at about 2,000 psi with a flow rate of 1 mL/min. Each run takes 30–35 min. For SGM, expect an elution profile similar to that shown in Fig. 2a (see Note 2).
4. To confirm the presence of O-acetyl groups in samples or standards, mix 20 μL of solution with 4 μL of 1 M NaOH. Incubate for 30 min at 37°C and neutralize with 2 μL of 2 M acetic acid or simply proceed to the DMB labeling reaction. Figure 2b shows an example of sialic acids derived from SGM that was treated with base prior to DMB labeling and chromatography.

3.2. Growing and Harvesting Bacterial Cells

1. Inoculate LB with a single bacterial colony and grow to saturation. Dilute culture 1:200 into the desired volume of medium and grow to an $A_{600} = 0.6$ with vigorous shaking (see Note 8). It is important to grow bacteria in a sterile container with at least 2× the volume of the medium to attain proper aeration and reproducible growth rates. Cultures can be grown to saturation, but capsule production is maximal during the exponential phase of growth.
2. Collect bacteria by centrifugation at 5,000 × *g* for 7 min and discard supernatant. It is generally not necessary to wash pellets derived from 5 to 100 mL cultures. Simply invert containers

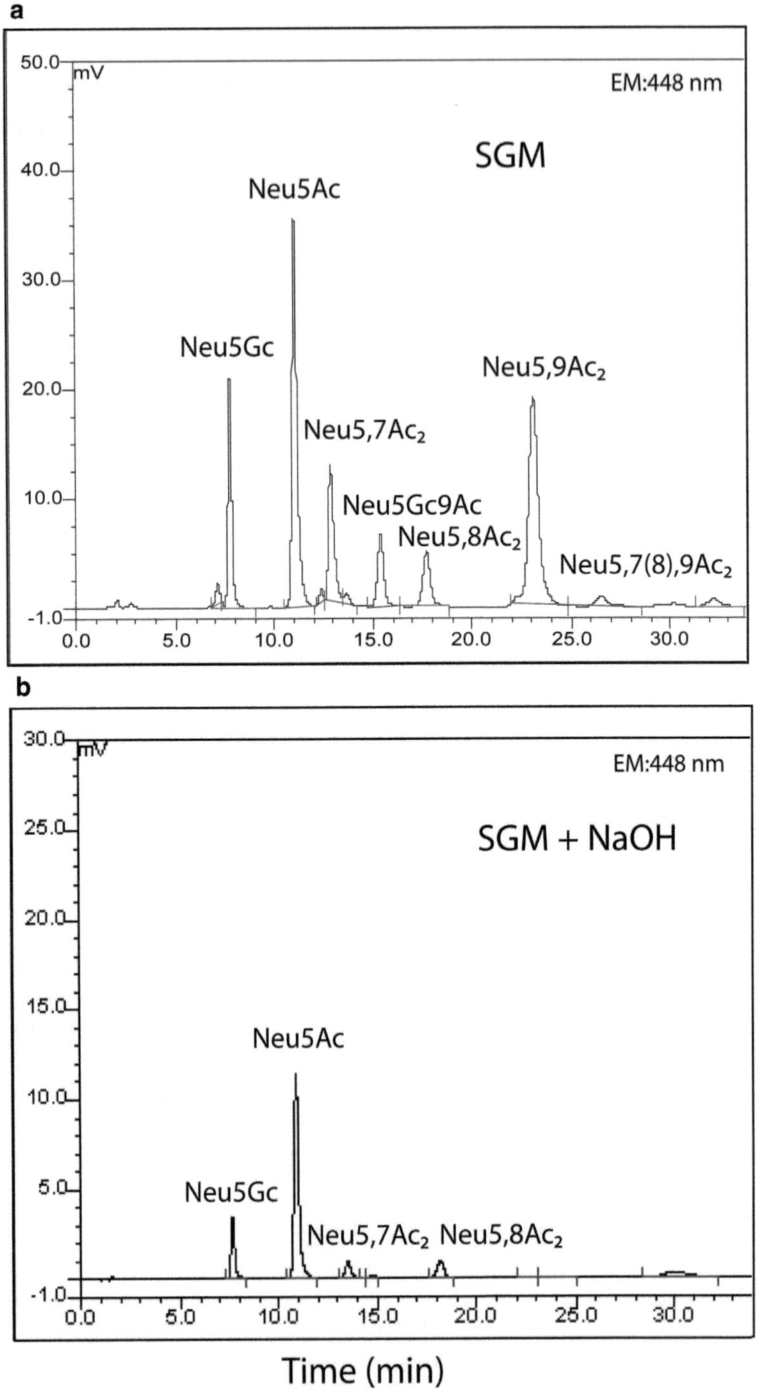

Fig. 2. Chromatographic analysis of SGM-derived standards. (**a**) SGM is a glycoprotein with about 15% sialic acids as terminal residues on the constituent sugar chains. Peaks represent the biologically most common sialic acids as indicated. (**b**) Base hydrolysis collapses acetylated sialic acids to their Neu5Ac or Neu5Gc forms.

over paper towels for a few minutes to allow residual liquid to drain, and then wipe sides with Kimwipes folded on forceps of suitable size. Alternatively, wash pellets one or more times with water, PBS, or desired buffer and resuspend to 1:10 or less the original volume of the bacterial culture. Store samples in an ice bucket or freeze at −20°C for later use.

3.3. Analysis of Polysialic Acids from Intact Bacteria

1. Mix the desired volume of bacterial sample with an equal volume of 4 M acetic acid and hydrolyze as described in step 1 Subheading 3.1. If preferred, monomers may be obtained by enzymatic hydrolysis using commercially available sialidase(s), but due to the simplicity of acid hydrolysis only this method is described. Depending on the sample there will be some amount of precipitate that should be removed by centrifugation prior to proceeding to the next step.
2. It is usually useful to concentrate the samples prior to DMB labeling. Concentration of small volumes (up to 2 mL) is conveniently done in a vacuum centrifuge under medium heat. After concentrating, centrifuge at 16,000 × *g* for 5 min and discard any precipitate. Lyophilization is recommended for larger sample volumes. Dried samples should be dissolved in the desired volume of 1.4 M acetic acid; remove any precipitate by centrifugation.
3. Perform DMB labeling (see step 2 Subheading 3.1) and chromatography (see step 3 Subheading 3.1). Figure 3 shows examples of results for a strain lacking modification (EV36), and SP#4, which is a naturally occurring clinical K1 isolate that, because of the particular nature of its translational switch, expresses *neuO* constitutively. Because there is much more lipopolysaccharide than polysialic acid in the bacteria, KDO is an inevitable contaminant in the analysis, but is readily distinguished from Neu5Ac or its derivatives by the lower retention time (see Note 9).

3.4. Semi-Purification and Analysis of Polysialic Acids

1. Grow bacteria in defined minimal medium and collect by centrifugation. Wash 2× with PBS, discarding supernatant after each wash.
2. Resuspend bacterial pellet in 0.1 volume of cold 100 mM pyridine buffer; incubate with shaking for 2 h at 37°C. Remove intact bacteria by centrifugation at 5,000 × *g* for 7 min, and lyophilize supernatant to dryness.
3. Dissolve the dried material in the desired volume of pyridine buffer and remove any precipitate by centrifugation. Dialyze against three changes of 50 mM pyridine buffer or water in a cold room (see Note 10).

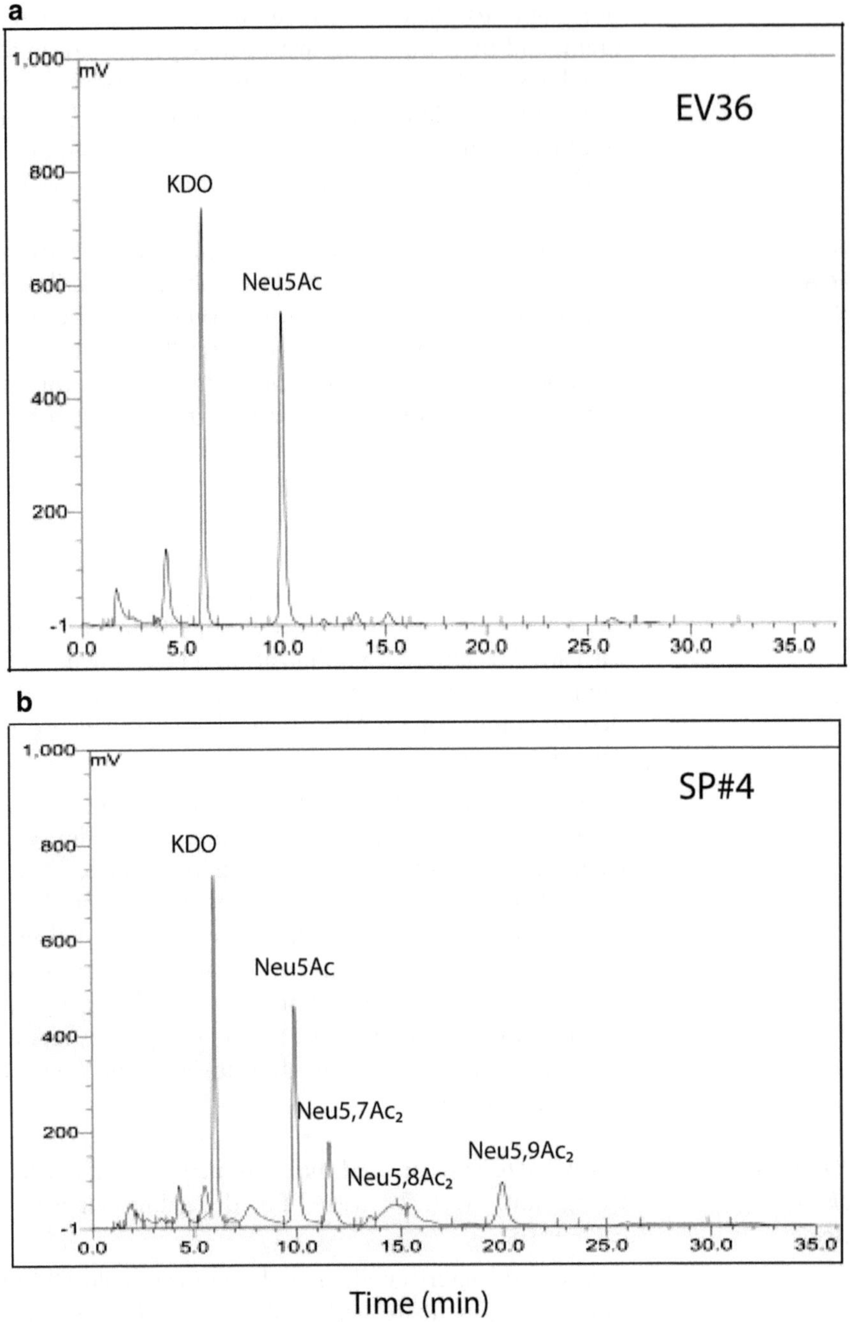

Fig. 3. Analysis of capsule using intact bacteria. (**a**) Strain EV36, lacking *neuO*, produces little or no modified sialic acid. (**b**) Strain SP#4 constitutively expresses *neuO* (*see* ref. (4)).

4. Concentrate by vacuum centrifugation and carry out DMB labeling and chromatography. Expect 13–15 mg of polysialic acid isolated from 200 mL of bacteria grown to saturation in this medium. Figure 4a shows the results with a negative control

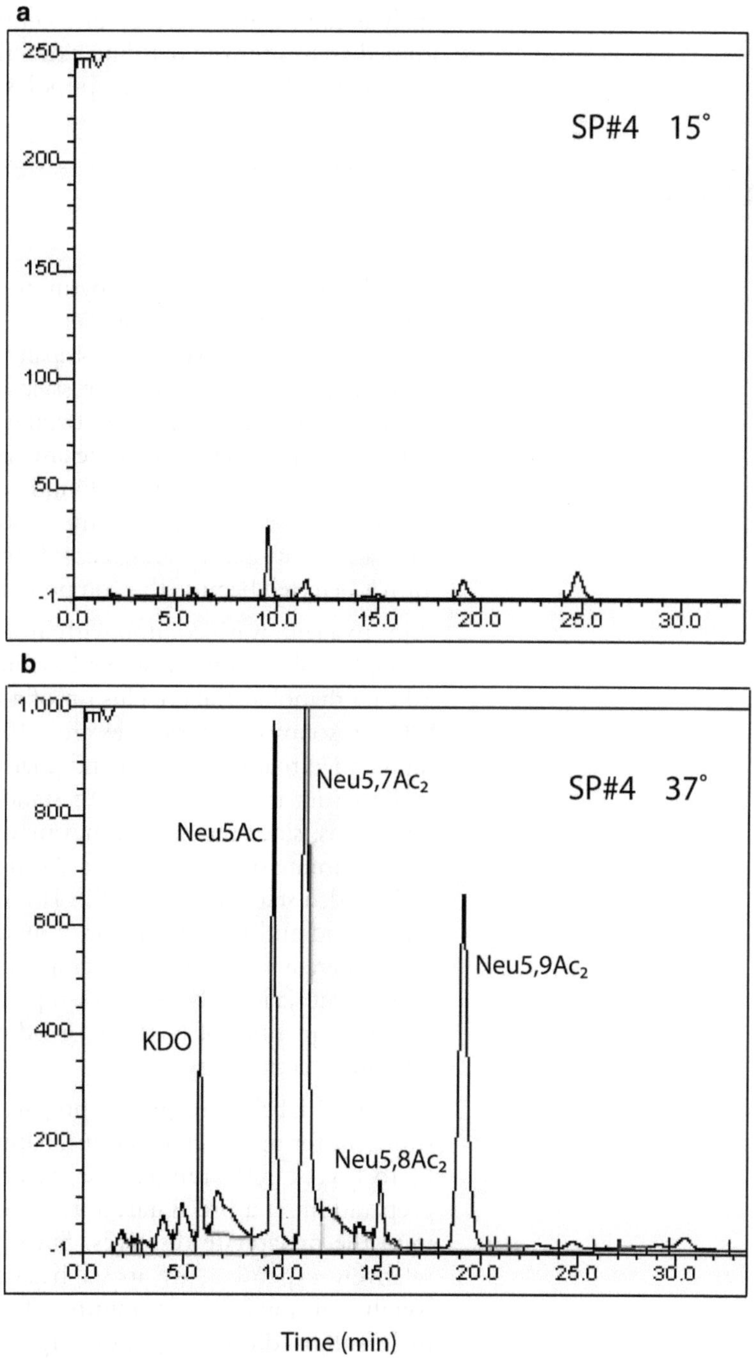

Fig. 4. Semi-purified polysialic acid from strain SP#4 grown at permissive and nonpermissive temperatures. As described in the text, bacteria grown at 15°C synthesize little or no capsular polysaccharide (**a**) relative to the normal growth temperature of 37°C (**b**).

sample of strain SP#4 grown at 15°C, a growth condition that inhibits polysialic acid synthesis. By contrast, SP#4 grown at 37°C shows the expected presence of O-acetylated sialic acids (Fig. 4b). If desired, the relatively low amount of KDO in the analysis can be prevented by ultracentrifugation at $100,000 \times g$ immediately after the pyridine extraction step to remove small membrane blebs containing lipopolysaccharide.

4. Notes

1. Neu5Ac, Neu5Gc, KDO, and pyruvate are purchased from Sigma. The only commercial supplier of Neu5,9Ac$_2$ is Applied Biotechnology, Austria. This company also markets Neu4,5Ac$_2$, which is not a component of polysialic acids but is found as a constituent of some mammalian glycoconjugates. Any sialic acid with a potential reducing end should be amenable to DMB labeling, but with the exception of those found on SGM, we have tested Neu4,5Ac$_2$ (see ref. (9)), and the 2-deoxy-anhydro derivative of sialic acid from Sigma. As expected, no labeling of the anhydro form is detected because this compound lacks a free reducing end.
2. Prior to a run, wash column with 50% methanol, then chromatography solvent until a stable baseline is achieved. Wash with 50% methanol at the conclusion of a day's set of experiments. When resolution deteriorates after 200 or more runs, it is recommended that the column and guard be replaced. It is important to run at least a Neu5Ac standard prior to each set of unknowns, since retention times may change from day to day. The minimum detection limit is about 2 pmol/DMB derivative. If desired, a standard curve of 2–10 or 20 pmol Neu5Ac can be established and used to quantify unknown sample quantities. It is generally most useful to simply compare relative amounts of different sialic acids within a sample, which can be spiked with a known quantity of Neu5Ac or other standard(s) depending on sample complexity.
3. The DMB labeling reagent is prepared as a 2× solution that is sufficient for up to ten reactions. Although it can be stored for up to 1 week in the refrigerator, we prefer to use it shortly after preparation. If a precipitate forms, heat the solution at 50°C until the precipitate dissolves. Since the DMB component is relatively expensive, we use a balance with five decimals to weigh sub-milligram amounts of reagents. Store sodium hydrosulfite (dithionite) under argon gas, and beta-mercaptoethanol in a fume hood. Store DMB powder at −20°C.
4. The relative proportions of chromatography solvent ingredients may need adjustment depending on the separation desired. The original procedure (see ref. (9)) uses 9:7:84 (v:v:v) acetonitrile–methanol–water. We modified the ratio to achieve good separation of Neu5Gc, Neu5Ac, and Neu5,9Ac$_2$ standards with our column and HPLC equipment.

5. As long as a fluorescence detector and data management system are available, any HPLC system or reverse phase column should be adequate. We use the Dionex system because of its versatility for different kinds of carbohydrate analyses.
6. The ingredients can be either purchased separately or as a preweighed mix. Sterilize in autoclave for 20 min.
7. Dissolve medium components in order in water with stir bar spinning in a 1–2 L beaker. Bring to the desired volume and filter sterilize. Store at 4°C. Luxurious growth is obtained by including glucose and casamino acids as carbon sources in a highly phosphate-buffered salts solution. The buffering prevents any significant drop in pH such that the labile phosphodiester linkage between the terminal nonreducing sialic acid residue and phospholipid membrane anchor is stabilized during growth. Casamino acid is a casein digest; we use NZ Amine from Sigma. Glycerol can be substituted for glucose in the event that catabolite repression is not desired.
8. Bacterial strains are stored at −80°C in 15–20% glycerol. We have stored strains this way with no apparent phenotypic change after at least 30 years. A small amount of frozen cells is inoculated onto an LB petri dish solidified with 1.5% agar. The dish is streaked for single colony isolation and incubated at the desired temperature, usually 37°C overnight. Dishes containing grown bacteria may be wrapped with Parafilm and stored for up to 2 weeks at 4°C.
9. Because K1 polysialic acid is a repeating homopolymer with connecting alpha-2,8-linkages, Neu5,8Ac_2 is likely produced after hydrolysis by a spontaneous transesterification, or branch migration of acetyl groups from carbon positions 7–9. Alternatively, it is conceivable that Neu5,8Ac_2, synthesized by an O-acetyltransferase other than NeuO prior to polymerization is used to terminate chain growth (*see* ref. (4)). Thus, the methods described here can be used to study sialic acid metabolism in general, and not just the analysis of capsular polysaccharide chains described here.
10. Pyridine also will extract low molecular weight intracellular compounds, and if monomeric sialic acids or pyruvate are overproduced these will be detected in the analysis of polysialic acid. To distinguish between sialic acids derived from polymer or the intracellular environment, dialyze using standard 12000-MWCO membrane tubing.

Acknowledgments

This work was supported by NIH grant AI042015. We thank Bahaa Fadl-Alla for expert technical assistance.

References

1. Vimr ER, Kalivoda KA, Deszo EL, Steenbergen SM (2004) Diversity of microbial sialic acid metabolism. Microbiol Mol Biol Rev 68:132–153
2. Gasparini R, Panatto D (2011) Meningococcal glycoconjugates vaccines. Hum Vaccin 7: 170–182
3. Pelknonen S, Häyrinen J, Finne J (1988) Polyacrylamide gel electrophoresis of the capsular polysaccharides of *Escherichia coli* K1 and other bacteria. J Bacteriol 170:2646–2653
4. Lee YC, Vann WF, Vionnet J, Wright LF, Vimr ER (2006) Separate pathways for O-acetylation of polymeric and monomeric sialic acids and identification of sialyl O-acetyl esterase in *Escherichia coli* K1. J Bacteriol 188: 6195–6206
5. Deszo EL, Steenbergen SM, Freedberg DI, Vimr ER (2005) *Escherichia coli* K1 polysialic acid O-acetyltransferase gene, *neuO*, and the mechanism of capsule form variation involving a mobile contingency locus. Proc Natl Acad Sci U S A 102:5564–5569
6. Steenbergen SM, Vimr ER (2008) Biosynthesis of the *Escherichia coli* K1 group 2 polysialic acid capsule occurs within a protected cytoplasmic compartment. Mol Microbiol 68:1252–1267
7. King MR, Vimr RP, Steenbergen SM, Spanjaard L, Plunkett G 3rd, Blattner FR, Vimr ER (2007) *Escherichia coli* K1-specific bacteriophage CUS-3 distribution and function in phase-variable capsular polysialic acid O-acetylation. J Bacteriol 189:6447–6456
8. King MR, Steenbergen SM, Vimr ER (2007) Going for baroque at the *Escherichia coli* K1 cell surface. Trends Microbiol 15:196–202
9. Steenbergen SM, Jirik JL, Vimr ER (2009) *YjhS* (*NanS*) is required for *Escherichia coli* to grow on 9-O-acetylated N-acetylneuraminic acid. J Bacteriol 191:7134–7139
10. Hara S, Yamaguchi M, Takemori Y, Furuhata K, Ogura H, Nakamura M (1989) Determination of mono-O-acetylated N-acetylneuraminic acids in human and rat sera by fluorometric high-performance liquid chromatography. Anal Biochem 179:162–166

Chapter 8

Analysis of Exopolysaccharides in *Myxococcus xanthus* Using Confocal Laser Scanning Microscopy

Wei Hu, Renate Lux, and Wenyuan Shi

Abstract

Confocal laser scanning microscopy (CLSM) is a useful research tool to explore the exopolysaccharides (EPS) in bacterial biofilms. Here, we describe the analysis of different biofilms of *Myxococcus xanthus* formed in a modified chamber slide system with CLSM. In conjunction with several specific fluorescent probes, the EPS within *M. xanthus* biofilms can be visualized and analyzed in situ.

Key words: Bacterial biofilms, Exopolysaccharides, Confocal laser scanning microscopy, Fluorescent staining, Myxococcus xanthus

1. Introduction

Microbial biofilms grown on solid surfaces are highly structured and often embedded within extracellular matrix (1, 2). The extracellular matrix of biofilms is a mixture of different biochemical polymers including exopolysaccharides (EPS), proteins, nucleic acids, lipids, and other components released from cells that act as a physical substratum to maintain the biofilm structure and functions (3, 4). Among them, EPS has been considered as the key component for the formation of complex three-dimensional structures of many bacterial biofilms (5).

The ability to visually detect EPS and its distribution within microbial biofilms would be a key step to study the role of EPS for biofilm formation. Among various imaging tools, confocal laser scanning microscopy (CLSM) has been a common choice since it allows in situ analysis of the bacterial biofilms (6) and avoids the collapse of extracellular matrix that frequently occurs during the

Anne H. Delcour (ed.), *Bacterial Cell Surfaces: Methods and Protocols*, Methods in Molecular Biology, vol. 966,
DOI 10.1007/978-1-62703-245-2_8, © Springer Science+Business Media New York 2013

fixation and dehydration procedures for electron microscopic observation (7). In order to effectively explore EPS with CLSM, fluorescent probes that could specifically bind EPS are needed. Since bacterial EPS are a complex group of polymers containing a variety of monosaccharides with acyl and other substituents (4, 5), the choices of EPS-specific probes include fluorescently labeled lectins (8), antibodies (9), and other polymers (10). After recording the images in digital format by CLSM, quantitative three-dimensional image analysis are often performed (6), using several software programs such as COMSTAT (11), DAIME (12), PHLIP (http://sourceforge.net/projects/phlip/), and CONAN (http://www.biocom-online.de/index.html).

Here, we describe the application of CLSM, in conjunction with various fluorescent staining methods to examine the three-dimensional distribution of EPS in the native *Myxococcus xanthus* biofilm, which is one of the most complex, well-organized microbial biofilms (13, 14). On a solid surface, when starved, *M. xanthus* cells undergo a developmental process to form a specific multicellular structure called fruiting body (15). In submerged cultures, by controlling the concentration of Ca^{2+}, highly organized developmental biofilms (submerged fruiting bodies) and simple non-developmental biofilms of *M. xanthus* can be cultivated, respectively (16). Our approach is to form these *M. xanthus* biofilms in a modified chamber slide system, which allows us to directly observe these biofilms with CLSM. At the same time, based on the compositional analysis of *M. xanthus* EPS (17), fluorescent dye conjugated lectins and antibodies are used to specifically stain the EPS in *M. xanthus* biofilms. The technique described in this paper allows the direct visualization and quantification of *M. xanthus* EPS in situ to reveal the characteristics of EPS localization and distribution within its biofilms.

2. Materials

2.1. Cell Cultures

1. Bacterial strain: *M. xanthus* DK1622 (18).
2. CYE medium: 10 g Casitone, 5 g yeast extract, 2.09 g 4-Morpholinopropanesulphonic acid (MOPS), and 0.48 g $MgSO_4$ dissolved in 1 L distilled water. Adjust pH to 7.6. Sterilize by autoclaving and store at room temperature.
3. Aerobic shaker for broth cultures.
4. Aerobic incubator for biofilm growth.
5. VIS-spectrophotometer with a 1-cm path length cuvette.
6. Centrifuge and associated rotors.

7. Sodium pyruvate stock solution: 1 g sodium pyruvate dissolved in 10 mL distilled water. Sterilize by filtration and store at 4 °C.
8. MMC buffer: 209 mg MOPS, 48 mg $MgSO_4$, and 22 mg $CaCl_2$ dissolved in 100 mL of distilled water. Adjust pH to 7.6. Sterilize by filtration and store at 4 °C.
9. MMC-agar medium: Dissolve 209 mg of MOPS, 48 mg of $MgSO_4$, and 22 mg of $CaCl_2$ in 80 mL distilled water, add 1.5 g of granular agar, adjust volume to 99 mL with additional distilled water, and adjust pH to 7.6. Sterilize by autoclaving and cool to 60 °C. Add 1 mL of sodium pyruvate stock solution for a final volume of 100 mL, and store at room temperature.
10. Modified 8-well Lab-Tek®II Chamber Slide® System (Thermo Nunc, NY) with bottom slides replaced with No. 1.5 thickness (0.17 mm) glass coverslips (see Note 1). Sterilize the modified chamber slide system with UV radiation before use.
11. Humid chamber, which consists of a sealed plastic container with a moist paper towel at the bottom.
12. MOPS buffer: 209 mg MOPS and 96 mg $MgSO_4$ dissolved in 100 mL of distilled water. Adjust the pH to 7.6. Sterilize by filtration and store at 4 °C.
13. Glass beads (3 mm of diameter) sterilized by autoclaving.

2.2. Staining of Cells and EPS

1. Phosphate-buffered saline (PBS): 0.8 g NaCl, 0.02 g KCl, 0.144 g Na_2HPO_4, and 0.024 g KH_2PO_4 in 100 mL of distilled water. Adjust pH to 7.4, sterilize by autoclaving and store at room temperature.
2. SYTO9 stock solution: 5 mM SYTO® 9 dye in dimethyl sulfoxide (DMSO), commercially available (Molecular Probes). Divide the 100-μL solution into 10 aliquots, and store at −20 °C protected from light (see Note 2).
3. Alexa633-WGA stock solution: 1 mg of lyophilized Alexa Fluor® 633 conjugated wheat germ agglutinin (Molecular Probes) in 1 mL of PBS. Divide the solution into 10 aliquots, store at −20 °C protected from light (see Note 3).
4. MOPS buffer: see item 12 Subheading 2.1.
5. Microcentrifuge.
6. SYTO9 staining solution: 1 μL of SYTO9 stock solution in 1 mL MOPS (5 μM final concentration of SYTO9). Store at 4 °C and protect it from light. Make fresh.
7. SYTO9/Alexa633-WGA staining solution: Thaw the Alexa633-WGA stock solution, and spin 5 min at maximum speed in a microcentrifuge and retain the supernatant (see Note 4).

Add 5 μL of Alexa633-WGA stock solution supernatant and 1 μL of SYTO9 stock solution in MOPS buffer to a final Volume of 1 mL. Store at 4 °C and protect from light. Make fresh.

8. GlnNAc-Ab antibody solution: 1 μL of Anti-*O*-linked *N*-Acetylglucosamine monoclonal antibody in 200 μL MOPS buffer. Make fresh.
9. Alexa546-2nd Ab solution: 1 μL of Alexa Fluor® 546 conjugated goat anti-mouse secondary antibody (Molecular Probes) in 200 μL MOPS buffer. Before use, spin 5 min at maximum speed in a microcentrifuge and retain the supernatant (see Note 4). Store at 4 °C and protect it from light. Make fresh.

2.3. CLSM Instrument

1. Zeiss LSM 5 PASCAL confocal laser scanning microscope. The scanning module of the system is mounted onto an inverted microscope Axiovert 200 M. An argon laser source and a helium–neon laser source are provided with the CLSM system.
2. 63× water-immersion objective (Achroplan/NA 0.95), provided with the CLSM system (see Note 5).
3. 63× oil-immersion objective (Plan-Apochromat/NA1.4), provided with the CLSM system (see Note 5).
4. Immersol® immersion oil.

2.4. Software

1. LSM 5 Pascal software 3.2 (Carl Zeiss), for processing the images acquired by CLSM (see Note 6)
2. COMSTAT Program (see Note 7), to determine the amount of stained EPS (11).

3. Methods

3.1. Cell Cultures

3.1.1. Growth of Fruiting Body on Agar Layer

1. Grow *M. xanthus* cells in CYE medium in an aerobic shaker (250 rpm) at 32 °C to log phase (OD_{600nm} = 0.5–1.0) (see Note 8).
2. Harvest the cells by centrifugation at 8,000 × *g* for 5 min; wash the cell pellet three times in MMC buffer, and resuspend to 5×10^9 cells/mL (OD_{600nm} = 10) in MMC buffer.
3. Pour 80 μL of melted MMC-agar in a well of the modified 8-well chamber slide system, and prepare agar layers about 1 mm thickness (see Note 9). After the agar is solidified, the chamber slide system can be stored in a humid chamber at 4 °C for up to 7 days (see Note 10).

4. Spot 10 μL *M. xanthus* cell suspension on the agar layer of each well.
5. Put the chamber slide system in a humid chamber and incubate at 32 °C for 3–5 days. The developmental process and fruiting body formation of *M. xanthus* can be observed (see Note 11).

3.1.2. Growth of Submerged Biofilms on Coverslips

1. Grow *M. xanthus* cells in CYE medium (see step 1 in Subheading 3.1.1).
2. Harvest cells by centrifugation at 8,000 × *g* for 5 min.
3. Resuspend the cell pellet in MOPS buffer.
4. Vortex the suspension for 3 min in the presence of 3 mm glass beads to disrupt cell clumps (see Note 12).
5. Remove the glass beads and wash the cell pellet three times in MOPS buffer, and concentrate to 5×10^8 cells/mL ($OD_{600nm} = 1$) in MOPS buffer (see Note 13).
6. To cultivate submerged biofilms under starvation condition (non-developmental biofilm), add 200 μL cell suspension into a well of the modified 8-well chamber slide system and incubate in a humid chamber at 32 °C. After 24 h, a thin layer of non-developmental biofilm structure is formed on the bottom of the chamber.
7. To cultivate submerged fruiting bodies (developmental biofilms):
   (a) Add 10 μL cell suspension and 190 μL CYE medium (use pipet tip to mix) to a well of the chamber slide system
   (b) Incubate in a humid chamber at 32 °C for 24 h. Cells can form a semi-adherent layer on the bottom of the well.
   (c) Gently and completely remove the CYE medium by aspiration and replenish with the same volume of MMC buffer (see Note 14).
   (d) Incubate the cells in a humid chamber at 32 °C for 24 h, after which the initial submerged fruiting body structures can be observed.

3.2. Staining of Cells and EPS

3.2.1. WGA and SYTO9 Staining of Cells and EPS in Biofilms

1. For the submerged biofilms, carefully aspirate the culture medium from the well before staining. For biofilms (fruiting bodies) on agar, stain them directly.
2. Add 150 μL of SYTO9/Alexa633-WGA staining solution into a well of the 8-well chamber slide system (see Note 15).
3. Incubate at room temperature for 30 min in dark (see Note 16).
4. View the specimens using CLSM without washing (see Note 17) (Fig. 1). All nucleic acid-containing cellular structures are revealed by the green-fluorescent stain SYTO9 (see Note 18), and carbohydrates present in the EPS portion of the

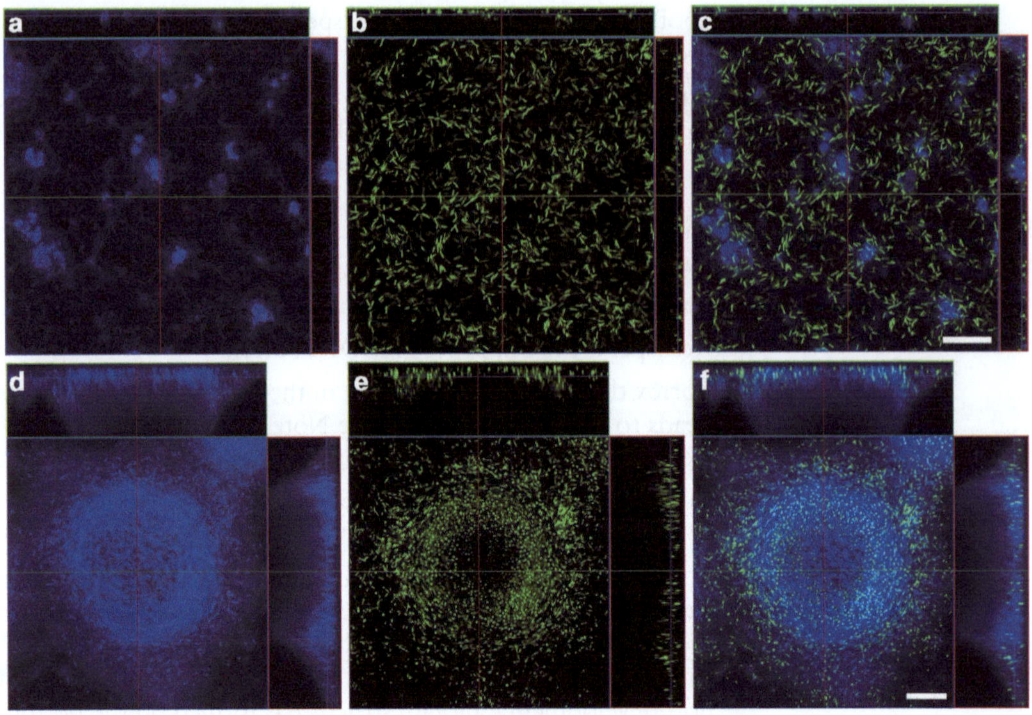

Fig. 1. Exopolysaccharide (EPS) distribution in *M. xanthus* strain DK1622 biofilms. Each picture shows the orthogonal sections of one layer of the *z*-stack collected for a representative structure with a 63× oil-immersion objective using CLSM as described. Submerged non-developmental biofilms are shown in panels (**a–c**), and submerged fruiting bodies are shown in panels (**d–f**) (staining of DK1622 fruiting bodies formed on agar surface exhibits similar structure (9)). The EPS pattern revealed in biofilms using Alexa 633 conjugated wheat germ agglutinin (Alexa633-WGA) is shown in panels (**a**) and (**d**). The distribution of *M. xanthus* cells labeled with SYTO9 is shown in panels (**b**) and (**e**). Scale bars, 20 μm. Panels (**c**) and (**f**) are overlaid images.

extracellular matrix are labeled with the lectin WGA conjugated with blue-fluorophore (Alexa633) (see Note 19).

3.2.2. Antibody and SYTO9 Staining of Cells and EPS in Biofilms

This is an alternative assay. The EPS in *M. xanthus* biofilms is stained with a GlnNAc-Ab antibody. This type of staining is performed only in submerged culture because the repeated washing steps required during this procedure were found to disrupt the biofilm structures formed on a dry agar surface (9).

1. Carefully aspirate the culture medium from the well containing the submerged biofilms.
2. Add 100 μL GlnNAc-Ab solution into a well of the 8-well chamber slide system and incubate at room temperature for 1 h.
3. Wash off the unbound antibody by carefully adding and removing 500 μL MOPS buffer twice.

4. Add 100 μL Alexa546-2nd Ab solution into the well and incubate in dark at room temperature for 1 h.
5. Wash off the unbound antibody by carefully adding and removing 500 μL MOPS buffer twice.
6. Add 100 μL SYTO9 solution in the well.
7. Incubate in the dark at room temperature for 30 min.
8. View the specimen using CLSM. All nucleic acid-containing cellular structures are revealed by the green-fluorescent stain SYTO9, and carbohydrates present in the EPS portion of the extracellular matrix are labeled with red-fluorophore (Alexa546) by the antibodies (see Note 20).

3.3. CLSM Observation

1. Examine biofilms on agar surface with a 63× water-immersion objective, and all the submerged biofilms with a 63× oil-immersion objective.
2. Use an argon laser with an excitation wavelength at 488 nm and a 505–530 nm band-pass emission filter to reveal SYTO9 fluorescence (see Note 21).
3. Use a helium–neon laser with an excitation wavelength at 543 nm and a 560 nm long-pass emission filter to reveal Alexa546 fluorescence (see Note 21).
4. Use a helium–neon laser with an excitation at 633 nm and a 650 nm long-pass emission filter to reveal Alexa633 fluorescence (see Note 21).
5. Acquire individual images from random positions or specific structures of biofilms.
6. To perform subsequent quantitative analysis, acquire the *z*-stack images at 1.0 μm intervals through the biofilm; the number of images in each stack will vary according to the thickness of the biofilm.

3.4. Image Processing and Data Analysis

1. Use the three-dimensional measurement option integrated in the LSM 5 Pascal software to determine the size of biofilm structures.
2. Convert the CLSM files into TIF files for further calculation by LSM 5 Pascal software or ImageJ software.
3. Convert the TIF images from gray-scale images into black-and-white images for analysis by the COMSTAT program. This conversion requires setting a threshold by the operator (see Note 22).
4. Quantify the EPS stained with WGA or antibody in *M. xanthus* biofilms using COMSTAT, to obtain parameters such as biovolume and average thickness (11).

4. Notes

1. This modified chamber slide system is suitable for high-resolution microscopy using oil or water immersion objectives on typical inverted CLSM. Attach new coverslides every time before the cultivation of biofilms. Different Lab-Tek®II Chamber Slide® Systems, such as 1, 2 or 4-well configurations, can be employed according to the purposes of the experiments.
2. Since SYTO9 dye binds to nucleic acids and DMSO is known to facilitate the entry of organic molecules into tissue, the SYTO9 DMSO stock solution should be handled with extreme caution. Wear gloves when handling the solution, and pour all the solutions containing SYTO9 dye through activated charcoal before disposal.
3. Do not add sodium azide to the stock solution. Alexa633-WGA stock solution can be stored at −20 °C for a month, and repeated freezing and thawing should be avoided.
4. This treatment eliminates any protein aggregates that may have formed in the stock solution, which can reduce the nonspecific background binding and staining.
5. Since the thickness of agar layer is about 1 mm, the biofilms on agar surface can only be viewed through the water-immersion objective, which has a 2 mm working distance. The submerged biofilms can be examined through the oil-immersion objective, which has a better numeric aperture versus the water-immersion objective.
6. An alternative is the ImageJ software, which is available at http://rsbweb.nih.gov/ij/.
7. COMSTAT is a script in Matlab 7.0 (The MathWorks) and is available at http://www.im.dtu.dk/comstat.
8. Assays for *M. xanthus* biofilm formation generally require the cells to be harvested from vegetatively grown broth cultures at log phase, otherwise, the cells may be defective in fruiting body formation or in attachment onto the surface under submerged conditions.
9. The thickness of agar layer must be less than 1.5 mm, so the specimen can be viewed through the water-immersion objective on CLSM.
10. The agar layer should be pre-warmed by incubating at 32 °C for 30 min just before use.
11. During the extended incubation, supply water to the paper towel in the humid chamber to prevent desiccation of the relatively thin agar layer.

12. Cells must be fully resuspended and no clumps of cells should be visible. However, the time for vortex treatment with glass beads should be controlled, since drastic treatment may reduce the cell viability.
13. The submerged fruiting bodies of *M. xanthus* can be grown in the presence of Ca^{2+}, while submerged non-developmental biofilms is only formed without Ca^{2+} in the medium (16). This washing step is important to remove the residual nutrients and Ca^{2+} in the cell pellets.
14. Avoid disrupting the attached cells by slowly aspirating or adding the buffer along the side of the well.
15. This procedure should be performed very slowly and carefully, since the early stage biofilms can be easily destroyed.
16. The dyes are sensitive to light, and therefore the incubation should be performed in the dark. Short-term exposure to dim light will not cause any damage to the dyes.
17. For SYTO9 staining, the dye displays a ~40-fold increased fluorescence intensity upon nucleic acids binding, thus no washing step is needed. The binding of Alexa633-WGA to EPS results a strong fluorescence while the unbound dye exhibits marginal background fluorescence (9). Therefore, unwashed samples can be directly viewed under CLSM to minimize disturbance of the biofilm structures.
18. SYTO9 is a cell-membrane-permeable green-fluorescent dye that shows much enhanced fluorescence upon binding nucleic acids, which has been used to stain RNA and DNA in both live and dead bacterial cells. If needed, the cell-membrane-impermeable DNA-binding dye, such as red-fluorescent dye propidium iodide, can be added to distinguish live (with intact cell membranes) and dead (with compromised cell membranes) cells.
19. The main specificities of WGA are for *N*-acetylglucosamine (GlnNac) and *N*-acetylneuraminic acid (sialic acid) residues (19, 20), which has been reported to be present in the EPS of *M. xanthus* (17). Therefore, Alex633 conjugated WGA can be used to label the EPS within the biofilms. Alternatively, fluorophore conjugated derivative of the concanavalin A (ConA) can be used to stain EPS, which selectively binds to α-d-glucose and α-d-mannose (two sugars known to be present in the *M. xanthus* EPS) (9, 17, 20).
20. This primary antibody recognizes β-1,3 linked *O*-linked GlcNAc residues of carbohydrate, which are present in *M. xanthus* EPS (17). This staining reveals virtually the same type of EPS structure as WGA staining (9) (Fig. 1).

21. The combination of dyes should be carefully determined, and different dyes should be distinguished and visualized in separated channels of CLSM.
22. This includes a manual adjustment of the threshold value by comparing the original gray-scale image with the black-and-white image, and the best value is chosen to give the most accurate conversion. This threshold value is then fixed and used for all image stacks.

Acknowledgment

This work was supported by National Institutes of Health Grant GM54666 (to W.S.), Grant 2012TS007 from SDU, Grants 2009BSB01405 and 2R2012CM003 from SD-NSF, and Grant 30870020 from Natural Science Foundation of China (to W.H.).

References

1. Dickschat JS (2010) Quorum sensing and bacterial biofilms. Nat Prod Rep 27:343–369
2. Lopez D, Vlamakis H, Kolter R (2010) Biofilms. Cold Spring Harb Perspect Biol 2:a000398
3. Stoodley P, Sauer K, Davies DG, Costerton JW (2002) Biofilms as complex differentiated communities. Annu Rev Microbiol 56: 187–209
4. Flemming HC, Wingender J (2010) The biofilm matrix. Nat Rev Microbiol 8:623–633
5. Sutherland I (2001) Biofilm exopolysaccharides: a strong and sticky framework. Microbiology 147:3–9
6. Neu TR, Manz B, Volke F, Dynes JJ, Hitchcock AP, Lawrence JR (2010) Advanced imaging techniques for assessment of structure, composition and function in biofilm systems. FEMS Microbiol Ecol 72:1–21
7. Merroun ML, Ben Chekroun K, Arias JM, Gonzalez-Munoz MT (2003) Lanthanum fixation by *Myxococcus xanthus*: cellular location and extracellular polysaccharide observation. Chemosphere 52:113–120
8. Neu T, Swerhone GD, Lawrence JR (2001) Assessment of lectin-binding analysis for *in situ* detection of glycoconjugates in biofilm systems. Microbiology 147:299–313
9. Lux R, Li Y, Lu A, Shi W (2004) Detailed three-dimensional analysis of structural features of *Myxococcus xanthus* fruiting bodies using confocal laser scanning microscopy. Biofilms 1:293–303
10. Lawrence JR, Wolfaardt GM, Korber DR (1994) Determination of diffusion coefficients in biofilms by confocal laser microscopy. Appl Environ Microbiol 60:1166–1173
11. Heydorn A, Nielsen AT, Hentzer M, Sternberg C, Givskov M, Ersboll BK et al (2000) Quantification of biofilm structures by the novel computer program COMSTAT. Microbiology 146:2395–2407
12. Daims H, Lucker S, Wagner M (2006) DAIME, a novel image analysis program for microbial ecology and biofilm research. Environ Microbiol 8:200–213
13. O'Toole G, Kaplan HB, Kolter R (2000) Biofilm formation as microbial development. Annu Rev Microbiol 54:49–79
14. Kolenbrander PE, Jakubovis NS, Chalmer NI (2008) Multispecies interactions and biofilm community development. In: Whitworth DE (ed) Myxobacteria: multicellularity and differentiation. ASM Press, Washington, DC, p 453
15. Diodati ME, Gill RE, Plamann L, Singer M (2008) Initiation and early developmental events. In: Whitworth DE (ed) Myxobacteria: multicellularity and differentiation. ASM Press, Washington, DC, p 520
16. Kuner JM, Kaiser D (1982) Fruiting body morphogenesis in submerged cultures of *Myxococcus xanthus*. J Bacteriol 151: 458–461
17. Behmlander RM, Dworkin M (1994) Biochemical and structural analyses of the

extracellular matrix fibrils of *Myxococcus xanthus*. J Bacteriol 176:6295–6303

18. Kaiser D (1979) Social gliding is correlated with the presence of pili in *Myxococcus xanthus*. Proc Natl Acad Sci U S A 76:5952–5956
19. Wright CS (1984) Structural comparison of the two distinct sugar binding sites in wheat germ agglutinin isolectin II. J Mol Biol 178:91–104
20. Goldstein IJ, Hayes CE (1978) The lectins: carbohydrate-binding proteins of plants and animals. Adv Carbohydr Chem Biochem 35:127–340

Chapter 9

Assessment of Multidrug Efflux Assemblies by Surface Plasmon Resonance

Elena B. Tikhonova and Helen I. Zgurskaya

Abstract

Surface plasmon resonance (SPR) is a powerful tool for kinetic analyses of protein–protein interactions. Here we describe the application of this method to study interactions of membrane proteins involved in multidrug efflux in *Escherichia coli*. These so-called multidrug efflux pumps comprise an inner membrane transporter, a periplasmic membrane fusion protein and an outer membrane channel. The three components are assembled into a protein conduit that enables bacteria to expel multiple drugs directly into the external medium bypassing the periplasm. The "across-two-membranes" transport mechanism makes tripartite transporters poorly amenable to biochemical analyses. Two out of three components are located in different membranes and must be studied in detergents or reconstituted into lipid bilayers to retain their functionality. Furthermore, cytoplasmic and exoplasmic domains of all three components perform different functions and must be oriented in a specific way to assemble functional complexes. The SPR approach provides means to overcome some of these problems in studies of protein assemblies that function in the context of two membranes.

Key words: Surface plasmon resonance, Membrane protein interactions, Binding kinetics

1. Introduction

1.1. Multicomponent Drug Transporters

Gram-negative bacteria are notoriously more resistant to antibiotics than gram-positive bacteria. For a long time, the outer membrane was believed to be the major reason why Gram-negative pathogens are recalcitrant to antibiotic treatment (1). The seminal finding that *Escherichia coli* cells contain multidrug efflux transporters changed dramatically our view on how these bacteria resist antibiotics (2, 3). Current models postulate that intrinsic multidrug resistance of gram-negative bacteria is the result of synergy between reduced uptake of drugs across the outer membrane and active drug efflux from the inner membrane (4–6). This synergy is

Anne H. Delcour (ed.), *Bacterial Cell Surfaces: Methods and Protocols*, Methods in Molecular Biology, vol. 966,
DOI 10.1007/978-1-62703-245-2_9, © Springer Science+Business Media New York 2013

possible because multidrug efflux transporters of gram-negative bacteria function together with proteins belonging to the Membrane Fusion Protein (MFP) family (7). MFPs are located in the periplasm and act on both membranes to enable drug efflux across the whole cell envelope directly into the medium. On the inner membrane (IM), MFPs associate with drug efflux transporters and stimulate their activities (8–10). On the outer membrane (OM) they recruit OM channels and possibly enable expulsion of drugs into the medium (11, 12). The three components form large multi-protein assemblies that traverse both the inner and outer membranes of gram-negative bacteria. Working together as a well-coordinated team, they achieve the direct extrusion of substrates from the cytoplasm and/or the periplasm into the medium.

Drug transporters that associate with MFPs and OM channels can belong to any of the three major superfamilies of proteins: RND (Resistance-Nodulation-Cell Division), ABC (ATP-Binding Cassette) and MF (Major Facilitator) Superfamilies (13–16). These transporters are structurally and mechanistically very diverse. ABC transporters are driven by ATP hydrolysis, whereas drug efflux by RND and MF pumps is coupled to transport of protons. MF transporters are thought to function as monomers, whereas ABC and RND transporters are dimers and trimers, respectively. Surprisingly, MFPs that associate with these transporters are structurally very similar. Furthermore, in *E. coli* and other enterobacteria multidrug transporters share the same OM channel. How transporters of different architecture and molecular mechanism assemble into tripartite complexes with MFPs and OM channels and how transporter activities are coupled to efflux across the OM are the main questions of our research program.

Our studies encompass representatives of all three superfamilies and include the best studied RND transporter AcrB functioning with MFP AcrA and the OM channel TolC, MF transporter EmrB in the complex with MFP EmrA and TolC and ABC transporter MacB with MFP MacA and TolC (9, 17–20). Co-purification and genetic experiments agreed that inner membrane complexes between a transporter and a MFP pre-assemble without TolC. However, Surface Plasmon Resonance (SPR) studies provided the major insight into the mechanism of assembly and dynamic interactions between components of multidrug efflux complexes (18, 19). The technique allows determination of the dissociation rate constants for pairwise interactions within complexes as well as manipulation of the assembly of the complexes by performing sequential binding reactions in the presence and absence of cofactors. Using this technique we investigated kinetics of TolC interactions with three MFPs AcrA, EmrA, and MacA and measured affinity between these proteins (18). We also succeeded in reconstitution of AcrAB–TolC interactions in real-time and established critical parameters in the assembly of this complex (19). The SPR continues to be our major approach in reconstitution of MacAB–TolC and EmrAB–TolC complexes.

1.2. Surface Plasmon Resonance

SPR is a powerful tool for the analysis of biomolecular interactions and its applications are expanding rapidly. The main advantage of SPR in comparison to other techniques is that no particular molecular property (such as fluorescence or radioactive label) in either of the interacting molecules is required. Furthermore, SPR monitors the binding of molecules in real-time and therefore is more potent in analyses of reaction kinetics. In addition, the technique does not require large amounts or high concentrations of ligands and accommodates broad range of physicochemical conditions.

Commercially available and lab assembled instrumental setups to detect surface plasmon resonance have been reviewed recently (21–24). Among the most popular optical biosensors is the Biacore instrument, which provides a very sensitive and user-friendly platform for targeted and high-throughput studies. In a typical Biacore experiment, a ligand molecule is immobilized onto an optical sensor surface and its interacting partner, an analyte, is injected into a flow cell over the surface. Changes in the refractive index, close to the sensor surface, are monitored versus time to record the association phase (Fig. 1). In the dissociation phase the analyte is washed out by buffer and the loss of bound analyte from the sensor is recorded as a function of time. These data are used to calculate kinetic constants that describe the interaction between the analyte and the ligand. Such experiments are carried out in four basic steps: (1) preparation of the appropriate sensor chip, (2) determination

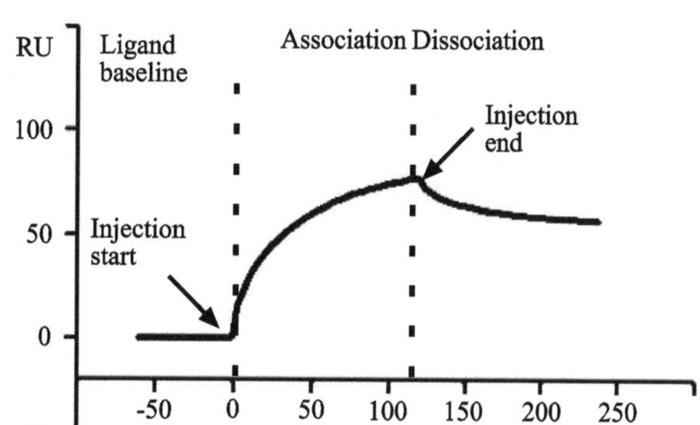

Fig. 1. A typical sensorgram of protein–protein interactions. The sensorgram is a plot of changes in the resonance signal as a function of time. The ligand baseline is a continuous flow of a buffer over an immobilized ligand. The association step is an injection and capture of an analyte on the ligand surface. The dissociation step is an injection of a buffer and disassembly of the complex. The start and end of the analyte injection are indicated by *arrows*. SPR signal is expressed in response units (RU) and represents the mass of the analyte molecule accumulated on the chip surface. Here the binding response is normalized by subtraction of nonspecific binding of the analyte on a control surface and represents only ligand–analyte interactions. All interactions are shown in real time.

of regeneration conditions, (3) optimization of conditions for the interaction between the ligand and analyte, (4) actual experiments and data analyses. In this Chapter, we discuss these steps in the design of SPR experiments as applied to protein–protein and drug–protein interactions in multidrug efflux transporters. We will emphasize specific requirements and limitations of membrane proteins.

1.3. Immobilization Approach and the Chemistry Behind It

The choice of chemistry for ligand immobilization is the first and critical step in the design of SPR experiments. Several criteria should be considered when designing a ligand surface. (1) Physicochemical properties of a ligand such as pI, solubility, and stability define whether the chosen immobilization chemistry will be effective for this ligand and whether or not the ligand will retain its activity after immobilization. (2) Depending on the purpose of SPR experiments, i.e., qualitative analysis of interactions or fine kinetics, orientation of the ligand on the surface should be carefully considered. Random orientation of ligand's binding sites on the surface could create heterogeneity in affinities and as result complicate or make impossible kinetic studies (25). (3) Nonspecific binding of analytes to the sensor surface could affect the quality of data and their subsequent analyses. Thus, properties of analytes and expected conditions for interactions should also be considered during selection of sensor surfaces and chemistry for immobilization.

We found two approaches that are useful in immobilization of membrane proteins by modification of the surface: (1) through amine and thiol coupling of ligand without protein labeling (direct coupling) or (2) through the high affinity interactions of a labeled ligand with the chip surface (capturing technique).

For coupling through ligand's primary amino and thiol groups, we used the CM5 chip (Biacore), which contains carboxymethylated dextran chains linked to the gold surface. For amine coupling, carboxyl residues are activated by EDC (1-ethyl-3-(3-dimethylaminopropyl)-carbodiimide) and converted into reactive esters by NHS (*N*-hydroxysuccinimide). The activated groups then interact with primary amino groups of a ligand (see Note 1). Ligands immobilized by this technique are randomly oriented on the sensor surface unless the number of accessible amino groups on the ligand is restricted to specific domains (cytoplasmic loops in membrane proteins) or positions. A thiol immobilization strategy includes additional steps. After activation of the CM5 chip surface with EDC/NHS, reactive esters are modified with PDEA (2-(2-pyridinyldithio)ethaneamine) to introduce active disulfide groups on the surface. A reduced ligand with available -SH groups is then injected over the activated surface to form S-S bonds between the ligand and the chip surface. The number of thiol groups in membrane proteins is usually limited and they are frequently disulfide bonded or located in transmembrane domains where they are inaccessible for chemical modifications unless a protein is denatured. Therefore, a site-specific mutagenesis approach is used to introduce a strategically positioned cysteine residue that can be used for an oriented immobilization of a functionally competent ligand.

An alternative to the covalent immobilization is capturing a ligand on the chip coated with a high affinity target such as ligand-specific antibodies or streptavidin. In our experiments, we used a streptavidin coated (SA) chip. No activation of the SA chip surface is needed to immobilize membrane proteins and this immobilization is the least intrusive for a protein structure. However, a ligand should be biotinylated prior capturing. The biotinylation could be carried out through primary amines or thiols native to the ligand or engineered by site-directed mutagenesis. High affinity biotin–streptavidin interactions provide a stable linkage between the ligand protein and the chip surface, which is comparable in its strength to a covalent bond.

Procedures and rationales for selection of amino acid residues for mutagenesis are specific for each protein and will not be discussed here. We also omitted protein purification protocols because details vary depending on a protein. The methods described below are applicable to a wide range of soluble and membrane proteins. We will describe procedures for the preparation and immobilization of ligands with the example of the MFP AcrA and the RND transporter AcrB. Experimental protocols for kinetics and complex assembly analyses are also included.

2. Materials

2.1. Immobilization of AcrA to the CM5 Chip Surface by Amine Coupling

1. Buffer A: 20 mM HEPES-KOH, pH 7.4, 150 mM NaCl, 1 mM phenylmethylsulfonyl fluoride (PMSF) in MilliQ or ultrapure water.
2. Amicon Ultra-0.5 and Amicon Ultra-4 YM30 centrifugal devices with molecular weight cut off 30 kDa.
3. Syringes and syringe acetate 0.22 μm filters.
4. Bovine Serum Albumin (BSA).
5. Protein assay kit.
6. Biacore 3000 system (GE HealthCare).
7. CM5 sensor chip (GE HealthCare).
8. Running Buffer A: 20 mM MES-KOH pH 5.9, 150 mM NaCl, 0.05% *n*-Dodecyl-β-d-maltoside (DDM) in MilliQ water.
9. Amine coupling immobilization buffer: 10 mM sodium acetate in MilliQ water, pH 4.8

10. CHAPS buffer: 20 mM 3-[(cholamidopropyl) dimethylammonio]-1-propanesulfonate (CHAPS) in Running Buffer A (see item 8 of this section).
11. 0.4 M EDC (1-ethyl-3-(3-dimethylaminopropyl)-carbodiimide hydrochloride) in water.
12. 0.1 M NHS (*N*-hydroxysuccinimide) in MilliQ water.
13. EDC/NHS mix: 1:1 mixture of 0.4 M EDC and 0.1 M NHS, prepared right before injection into the Biacore.
14. 0.5 M Ethanolamine in MilliQ water.

2.2. Thiol Coupling of AcrAS362C to the CM5 Surface

1. TCEP gel slurry: 50% (Tris(2-carboxylethyl)phosphine hydrochloride)—agarose (TCEP-agarose) in water (disulfide reducing gel).
2. Microbiospin column.
3. Equilibration buffer: 20 mM HEPES-KOH, pH 7.0, 500 mM NaCl, 5 mM Na_2EDTA.
4. Bovine Serum Albumin (BSA).
5. Protein assay kit.
6. Biacore 3000 system (GE HealthCare).
7. CM5 sensor chip (GE HealthCare).
8. Running Buffer A: see item 8 Subheading 2.1.
9. CHAPS buffer: see item 10 Subheading 2.1.
10. EDC/NHS mix: see item 13 Subheading 2.1.
11. PDEA solution: 80 mM PDEA (2-(2-pyridinyldithio)ethaneamine) in 0.1 M sodium borate, pH 8.5. Freshly prepared.
12. Thiol-coupling immobilization buffer: 10 mM sodium acetate in distilled water, pH 4.8.
13. Cysteine/NaCl solution: 50 mM cysteine, 1 M NaCl in 0.1 M sodium acetate, pH 4.8.

2.3. Capture of Biotinylated AcrAS362C on a Streptavidin Coated (SA) Chip

1. Phosphate-buffered saline (PBS): 137 mM NaCl, 2.7 mM KCl, 10 mM Na_2HPO_4, 2 mM KH_2PO_4.
2. MP-biotin solution: 20 mM Maleimide-PEG$_2$-Biotin (MP-biotin) in PBS buffer.
3. NAP™-5 Column Sephadex™ G-25 DNA grade (GE HealthCare).
4. NAP-5 column equilibration buffer: 20 mM Tris–HCl, pH 7.0, 500 mM NaCl, 5 mM Na_2EDTA in MilliQ water.
5. Spectra/Por Dialysis membrane with molecular weight cut off 8,000 Da.
6. 0.22 μm syringe filter.

7. HABA solution: dissolve 24.2 mg of HABA (4′-hydroxyazobenzene-2-carboxylic acid) in 9.9 ml of ultrapure water. Add 0.1 ml of 1 M NaOH. Filter to remove any particulates. Keep the solution at 4°C.
8. HABA/avidin solution: mix 10 mg of avidin, 600 μl of HABA solution and 19.4 ml of PBS. This solution can be stored at 4°C for up to 2 weeks.
9. UV-visible spectrophotometer.
10. SDS-PAGE electrophoresis system.
11. Electrode buffer: 3.0 g Tris–HCl, 14.42 g Glycine, 1 g SDS in 1 l of MilliQ water.
12. PVDF (polyvinylidene fluoride) transfer membrane.
13. Transfer buffer: 25 mM Tris–HCl, 192 mM glycine, 10% methanol, in MilliQ water.
14. Washing buffer: 20 mM Tris–HCl, pH 8.0, 500 mM NaCl, 0.05% Tween 20, in MilliQ water.
15. Blocking solution: 5% nonfat dry milk in washing buffer.
16. S-AP solution: 0.001 mg/ml solution of Streptavidin-conjugated Alkaline Phosphatase (S-AP) prepared in the washing buffer supplemented with 1% nonfat dry milk.
17. AP Buffer: 100 mM Tris–HCl, pH 9.5, 100 mM NaCl, 5 mM $MgCl_2$, 0.05% Tween 20, in MilliQ water.
18. BCIP (5-bromo-4-chloro-3-indolyl phosphate) stock solution: 15 mg/ml in 100% *N,N*-dimethylformamide (DMF).
19. NBT (nitro blue tetrazolium) stock solution: 30 mg/ml in 70% DMF.
20. Staining solution: 0.15 mg/ml BCIP, 0.3 mg/ml NBT in AP buffer.
21. SA sensor chip (GE HealthCare).
22. Biacore 3000 system (GE HealthCare).
23. Running Buffer B: 20 mM Tris–HCl, pH 7.5, 150 mM NaCl, 0.05% DDM buffer, in MilliQ water.
24. NaOH/NaCl solution: 50 mM NaOH, 1 M NaCl in distilled water.
25. CHAPS buffer: see item 10 Subheading 2.1.

2.4. Data Collection and Normalization

1. Running Buffer A or Running Buffer B: see item 8 of Subheading 2.1 or item 23 of Subheading 2.3, respectively.
2. Biacore 3000 system (GE HealthCare).
3. Biacore plastic vials with caps.
4. CHAPS buffer.
5. BIAevaluation 3.2 RC1 software (GE HealthCare).

3. Methods

3.1. Immobilization of AcrA to the CM5 Chip Surface by Amine Coupling

1. Transfer purified AcrA protein into buffer A using a YM30 centrifugal device. Centrifuge the sample at 14,000×*g* for 10–30 min if operating Amicon Ultra-0.5 or at 4,000×*g* when using Amicon Ultra-4. Filter the ligand using 0.22 μm syringe filter and determine the protein concentration using BSA as a standard.
2. Prime the Biacore system and equilibrate a CM5 chip with the Running Buffer A (see Note 2).
3. Dilute AcrA to a final concentration of 0.006 mg/ml in amine coupling immobilization buffer (see Note 3).
4. Wash the protein-free control flow cell (flow cell 1) and the ligand surfaces flow cell (flow cell 2) twice for 1 min at the flow rate of 30 μl/min with CHAPS buffer.
5. Activate both surfaces by injecting the EDC/NHS mix prepared right before the injection. Inject the mixture for 2 min at the 5 μl/min flow rate (Fig. 2a, *step a*).
6. Inject the ligand sample (Fig. 2a, *step b*) for 3.5 min at the 5 μl/min rate into flow cell 2. Treat the control surface (flow cell 1) with protein-free amine coupling immobilization buffer (Fig. 2a, *step c*).
7. Block the remaining active groups on both the control and AcrA surfaces by a 5 min injection of 0.5 M ethanolamine at a 10 μl/min flow rate.
8. To estimate the amount of AcrA coupled to the chip surface first calculate the difference in response units (RU) before and after immobilization (338 RU for amine coupled AcrA in Fig. 2a). To convert RU into a molar concentration of the protein on the chip surface use a conversion factor (see Note 4). For preparation of a higher density surface increase the concentration of the ligand and/or activation and injection times.
9. Store the chip in the Running Buffer A at 4°C.

Fig. 2. (continued) same as in (**a**); step (b) is a 4 min injection of PDEA at 5 μl/min to introduce thiol groups onto the surface; steps (c) and (d) are the same as steps (b) and (c) in (**a**) but the Cys-containing AcrAS362C mutant protein was used instead of AcrA; step (e) is a 4 min injection of cysteine/NaCl at a 5 μl/min flow rate to block unreacted groups. The amount of the immobilized AcrA shown here is 217 RU. (**c**) Immobilization of the biotinylated AcrAS362C ligand on a SA chip. Step (a) is the cleaning of the surface by repeated 1 min injections of NaOH/NaCl at a flow rate 40 μl/min; step (b) is the washing the surface with a 20 mM CHAPS injection for 1 min at 40 μl/min; step (c) is an injection of 0.134 μg/ml of biotinylated AcrAS362C for 4 min at 10 μl/min. The amount of the immobilized AcrA shown here is 194 RU.

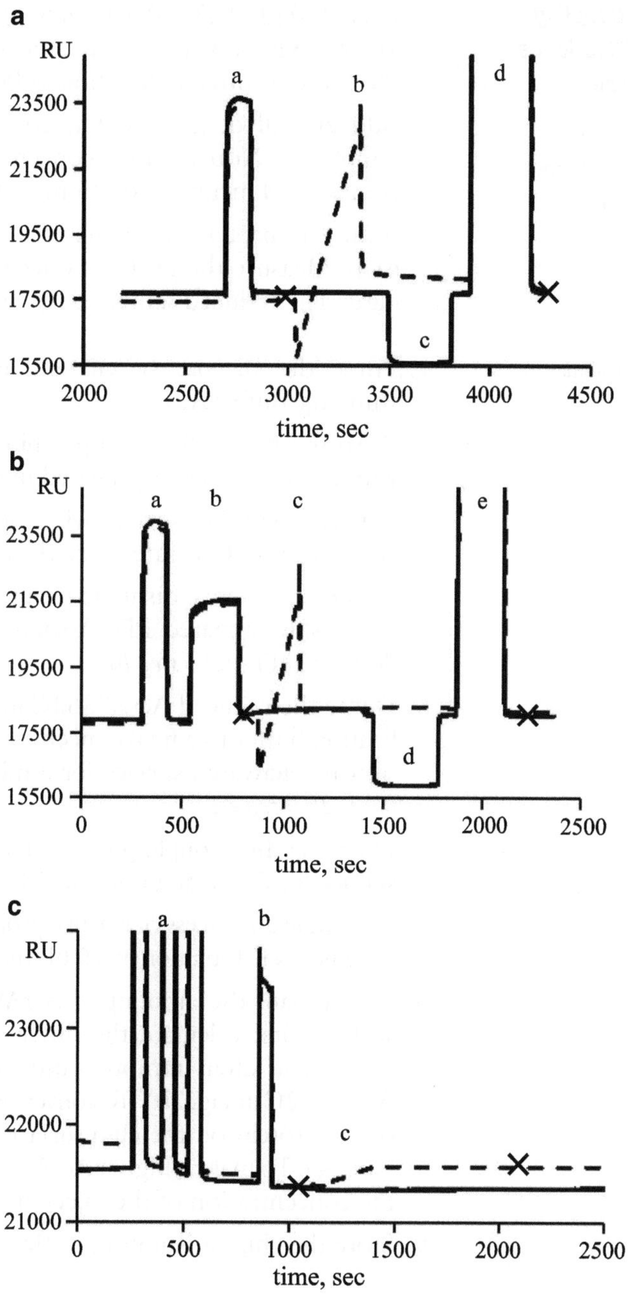

Fig. 2. Immobilization of AcrA using different techniques. (**a**) Amine coupling on a CM5 chip. SPR signals of sample and control flow cells are shown as *dashed* and *solid lines*, respectively. Step (a) is an activation of the surface by a 2 min injection of EDC/NHS mixture at 5 μl/min; steps (b) and (c) are injections of 0.006 mg/ml of AcrA into the sample flow cell and the acetate buffer into both the sample and control flow cells, respectively, for 5 min at 5 μl/min; step (d) is the blocking of unreacted groups by a 5 min injection of ethanolamine at 10 μl/min. The difference in RU before and after immobilization (marked by X) corresponds to the amount of the ligand immobilized on the chip. The amount of the immobilized AcrA shown here is 338 RU. (**b**) Thiol coupling on a CM5 chip. Step (a) is the

3.2. Thiol Coupling of AcrAS362C to the CM5 Surface

3.2.1. Reduction of AcrAS362C Mutant Using TCEP Gel

1. Load 400 μl of TCEP gel slurry (use two volumes of slurry per a protein sample volume) into a microbiospin column. Equilibrate the column with equilibration buffer (see Note 5).
2. Add 200 μl of purified Cys-containing AcrAS362C protein sample (see Note 6) at the concentration of ~1.0 mg/ml to the column and incubate for 45 min at the room temperature.
3. Elute the protein from the column in four fractions 200 μl each. Measure the protein concentration using BSA as a standard. The reduced sample is ready for immobilization.

3.2.2. Immobilization Procedure

1. Prime the Biacore system and equilibrate a CM5 chip in Running Buffer A.
2. Wash the flow cells at 30 μl/min flow rate twice 1 min each with 20 mM CHAPS prepared in Running Buffer A.
3. Activate the CM5 surface with the EDC/NHS mix for 2 min at a 5 μl/min flow rate (Fig. 2b, *step a*).
4. Introduce reactive disulfide groups on the surface by injecting the freshly prepared PDEA solution for 4 min at a 5 μl/min flow rate (Fig. 2b, *step b*).
5. Dilute the reduced AcrAS362C in the thiol coupling immobilization buffer to a final concentration 0.006 mg/ml and inject over the activated surface for 5 min at a flow rate of 5 μl/min (Fig. 2b, *step c*).
6. Inject the thiol coupling immobilization buffer over the control surface for 5 min at a flow rate of 5 μl/min (Fig. 2b, *step d*).
7. Deactivate any excess of reactive groups by a 4 min injection of the cysteine/NaCl solution at a 5 μl/min flow rate (Fig. 2b, *step e*).
8. To estimate the amount of AcrAS362C coupled to the chip surface, first calculate the difference in response units (RU) before and after immobilization (217 RU for thiol coupled AcrAS362C in Fig. 2b). To convert RU into a molar concentration of the protein on the chip surface use a conversion factor (see Note 4). To make a higher AcrAS362C density surface, increase the concentration of the injected protein to 20–50 μg/ml.
9. Store the chip in Running Buffer A at 4°C.

3.3. Capture of Biotinylated AcrAS362C on a Streptavidin Coated (SA) Chip

3.3.1. Preparation and Biotinylation of Reduced AcrAS362C with Maleimide-PEG$_2$-Biotin

1. Reduce the cysteine residue of purified AcrAS362C as described in Subheading 3.2.1 (see Notes 5 and 6).
2. Prepare the MP-biotin solution immediately prior biotinylation (see Note 7).
3. Mix the reduced protein and the biotin solution at a molar ratio 20:1 of MP-biotin per protein. For example, for 300 μl of the reduced AcrAS362C protein in the concentration of 0.45 mg/ml add 3.4 μl of the MP-biotin solution. Incubate the mixture on ice for 2 h.

3.3.2. Removal of Free Biotin from the Biotinylated Protein by Gel Filtration and Dialysis

1. Load the biotin–protein mixture onto a NAP-5 column equilibrated in NAP-5 column equilibration buffer. Elute the protein with the same buffer and collect 0.5 ml fractions. The protein will be present in the first fraction while free biotin will be eluted with 2–2.5 ml.
2. To ensure removal of free biotin from the biotinylated protein, perform an extensive overnight dialysis with two changes of the buffer used in the gel filtration. Carry out dialysis at 4°C using Spectra/Por dialysis membranes with an 8,000 Da molecular weight cut off.
3. After dialysis, filter the protein sample through a 0.22 μm syringe filter to remove any particulate. Measure the concentration of the biotinylated protein. The protein sample is suitable for immobilization onto a SA chip for at least 2 weeks.

3.3.3. Quantification of Biotin–Protein Molar Ratio by HABA Assay

1. Pipette 900 μl of HABA/avidin solution into a cuvette and measure the absorbance at 500 nm using a UV-visible spectrophotometer (value A1). A1 should be in the 0.9–1.3 range.
2. Add 100 μl of a biotinylated protein sample to the same cuvette, mix and measure the absorbance again (value A2). If A2 is less than 0.3, dilute the sample and repeat the measurement (see Note 8).
3. Calculate the difference in absorbance before and after addition of the sample (value $A3 = 0.9 \times A1 - A2$).
4. Calculate the concentration of the biotinylated protein in mmol/ml (value P= (protein concentration, mg/ml protein)/ (molecular weight of protein, mg/mmol)).
5. Calculate the concentration of biotin in mmol/ml of the reaction mixture (value $B = A3/(34{,}000 \times l)$, where 34,000 $M^{-1}\,cm^{-1}$ is the extinction coefficient for HABA/avidin mixture and l is the path length of the cuvette in cm).
6. Calculate a Biotin–Protein molar ratio in the biotinylated sample as $B \times 10/P$, where 10 is a dilution factor of the protein sample taken into a cuvette.

3.3.4. Analysis of Protein Biotinylation by Blotting with Streptavidin-Conjugated Alkaline Phosphatase

The HABA assay described in Subheading 3.3.3 is designed to measure total amounts of biotin present in a labeled protein solution. Blotting of biotinylated samples with streptavidin conjugated to alkaline phosphatase (S-AP) should be used to confirm that PEG_2-Biotin is attached to a Cys-containing protein (Fig. 3).

1. Load 1 μg of a biotinylated sample per well and separate by sodium dodecyl sulfate polyacrylamide gel electrophoresis (SDS-PAGE) in the electrode buffer.
2. Transfer proteins from the gel onto a PVDF membrane at 70 V for 1 h in transfer buffer.

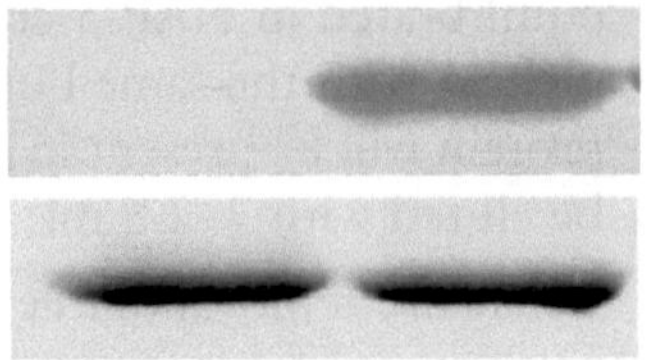

Fig. 3. Anti-biotin blotting of purified AcrAS362C. The biotinylated (+) and unlabeled (–) AcrAS362C samples (1 μg of each) were separated on a 10% SDS-PA gel and either Coomassie Brilliant Blue stained (*lower panel*) or transferred onto a PVDF membrane and probed with the streptavidin-conjugated alkaline phosphatase. Visualization of blotting was done by incubation with NBT/BCIP substrates (*upper panel*).

3. Block the membrane overnight at 4°C using the blocking solution.
4. Wash the membrane with 10 ml of the washing buffer three times for 10 min and then incubate for 1 h with 10 ml of S-AP solution. Remove the excess of S-AP by washing the membrane three times with 10 ml of the washing buffer for 10 min.
5. Visualize biotinylated proteins by incubating the membrane with 5 ml of. Stop the alkaline phosphatase reaction by addition of 0.5 ml of 0.5 M Na_2EDTA, pH 8.0.

3.3.5. Immobilization Procedure

1. Prime the Biacore system and a SA chip in Running Buffer B (see Note 9).
2. Clean the chip surface by injecting NaOH/NaCl solution for 1 min at 40 μl/min flow rate three times (Fig. 2c, *step a*).
3. Clean the chip surface by injecting 20 mM CHAPS buffer for 1 min at a 40 μl/min flow rate (Fig. 2c, *step b*).
4. Inject a 0.134 μg/ml solution of biotinylated AcrAS362C for 4 min at the flow rate of 10 μl/min (Fig. 2c, *step c*) into a flow cell. A protein-free flow cell prewashed with NaOH/NaCl solution and CHAPS buffer should be used as a control.
5. Estimate the amount of AcrAS362C captured on the chip surface by calculating the difference in the baseline before and after immobilization (194 RU for AcrA S362C, Fig. 2c). See Note 4 for a conversion of RU into protein concentration units. To prepare a high density surface, use a higher concentration of the biotinylated ligand during immobilization. Store the chip at 4°C in the running buffer.

3.4. Data Collection and Normalization

1. Prepare analyte samples in the appropriate running buffer. To transfer a protein from a storage buffer into the running buffer perform extensive dialysis using Spectra/Por dialysis membrane with the molecular weight cut off 8,000 Da against the

running buffer or use centrifugal devices YM30 to change the buffer and concentrate the sample (see Subheading 3.1).

2. Filter the sample and re-measure the protein concentration as needed.
3. Make a set of 5–6 twofold serial dilutions of analytes. For protein analytes, use concentrations ranging from nM to low μM. For low molecular weight analytes (substrates, cofactors), use high μM to high mM concentrations. Transfer samples into Biacore plastic vials with caps. Place the vials into a thermo regulated rack inside the Biacore instrument. It is important to equilibrate and maintain the temperature of the sample to minimize the effect of temperature on the refractive index, mass transfer, and binding kinetics.
4. Inject one concentration of the sample over the ligand and control surfaces simultaneously. The choice of the injection mode depends on the experiment (see Note 10). To avoid rebinding of analytes and mass transport effects use high flow rates (30–50 μl/min). The contact time should be chosen based on whether or not the binding response reaches a steady state.
5. Repeat injections of the same concentration of the analyte at least two to three times. Regenerate the surface between each injection by injecting CHAPS buffer (see Note 11). To account for instrumental noise and discrepancy in buffer compositions, inject the analyte buffer and proceed with double referencing (see Note 12).
6. To acquire a response signal specific to the analyte–ligand interaction, open a sensorgram file using BIAevaluation software (Biacore). Align the injection start time for the original sensorgrams and subtract the signal on the control surface from the total response on the ligand surface (see Note 12). Perform normalization for all injections.

3.5. Optimization of Binding Response (see Note 13)

1. Interactions between membrane proteins are strongly affected by detergents. A preliminary screening of detergents to identify those compatible with an assay could be done by diluting an analyte in buffers containing different detergents and injecting it over the ligand and control surfaces. Protein-free buffers should be injected as well for double referencing (see Note 12). After normalization of binding responses, one can determine how different detergents affect a SPR signal and whether or not association and dissociation rates depend on a detergent. The effect of a detergent could be further tested by running a sensorgram with a continuous flow of the running buffer containing the selected detergent. For further experiments choose the detergent in which an analyte generates a

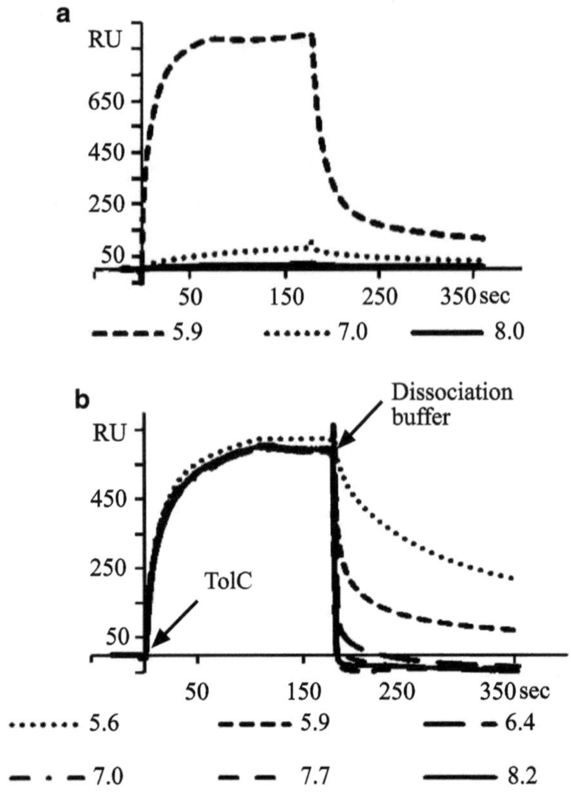

Fig. 4. Application of different injection modes in studies of AcrAB-TolC. (**a**) Effect of pH on AcrA–TolC interactions (adapted from (18)). AcrA was immobilized on a CM5 chip at a density 2,225 RU. The analyte, TolC, at a concentration of 0.5 μM was transferred into running buffers at pH 5.9 (*dashed lines*), 7 (*dotted lines*) and 8 (*solid lines*) and injected over the AcrA surface equilibrated in the respective running buffers. The buffers used are 20 mM MES-KOH, pH 5.9, or 20 mM Tris–HCl, pH 7.0 and pH 8.0 supplemented with 150 mM NaCl and 0.05% DDM. The data shown are normalized by subtracting the SPR signal of the control protein-free surface from that of the ligand surface. (**b**) Effect of pH on the stability of AcrA–TolC complex (modified from (18)). The Co-Inject mode was used to study TolC dissociation in buffers of different pH. During the association phase, 0.25 μM TolC (*arrow*) was injected over a AcrA 2,225 RU surface in running buffer at pH 5.9 for 3 min at a 30 μl/min flow rate. During the dissociation phase, buffers with different pH were injected at the same flow rate (shown by *arrow*).

reproducible and reliable binding response and which does not affect the activity and stability of the immobilized ligand (see Note 14).

2. To suppress nonspecific electrostatic interactions, apply the same approach as described above but use buffers with pH ranging from 5 to 9 and vary the salt concentration in the 150–500 mM range (see Fig. 4a, b and Note 10). At least 150 mM NaCl or KCl should be included into all running buffers. Salt concentrations above 500 mM however, could destabilize protein interactions (see Note 15).

3. To study the role of cofactors ($MgCl_2$, ATP, substrates, inhibitors and accessory proteins), prepare serial dilutions of cofactors and analytes in the running buffer and inject over the ligand and control surfaces. Then, mix a cofactor with an analyte at varying molar ratios and inject the mixture again. Normalize the results by subtracting a binding response obtained from the separate injections of an analyte or a cofactor. If a cofactor is expected to act only at the dissociation phase, use the Co-Inject mode (see Note 10 and Fig. 4b) to deliver the cofactor only during the complex disassembly. If a cofactor is needed for the active conformation of the immobilized ligand, include it into the running buffer.

3.6. Optimization of Surface Regeneration

Quantitative analyses of interactions require that after each injection of an analyte the chip surface must be regenerated to its original pre-injection state without any changes in the activity of the immobilized ligand (see Note 11).

1. Inject an analyte over the ligand and control surfaces. Normalize sensorgrams and determine the specific binding response (RU1).
2. Make a brief 0.5–1 min injection of a regeneration solution at a flow rate of 30–50 μl/min. Do not expose the surface to regeneration agents for a longer time. Start with the mildest possible solution that target specific, e.g., electrostatic, interactions. If regeneration is successful the baseline will return to its pre-injection level. If the analyte remains bound to the ligand, increase the harshness of the regeneration solution and repeat the regeneration step until all of the bound analyte is removed from the surface.
3. Repeat the injection of the analyte as in step 1 and determine the specific binding response for the second injection (RU2). If RU1 is equal to RU2, the surface remains active and the regeneration solution is applicable for further analyses of the particular analyte and ligand pair. If RU2 is low or no binding is detected, the injection of the regeneration solution resulted in the loss of the ligand activity. In this case, preparation of a new immobilized ligand and further analysis of regeneration conditions are needed.

3.7. Data Analysis

Evaluation of stoichiometries (S) of analyte–ligand complexes and kinetics of binding reactions obtained on different surfaces could help determine whether the interactions under study are sensitive to immobilization conditions.

1. Normalize the data and determine the response (RU_{an}) value at the steady state for each concentration of the analyte. The RU_{an} will be the response value taken a few seconds before the end of the association phase. To calculate the stoichiometry

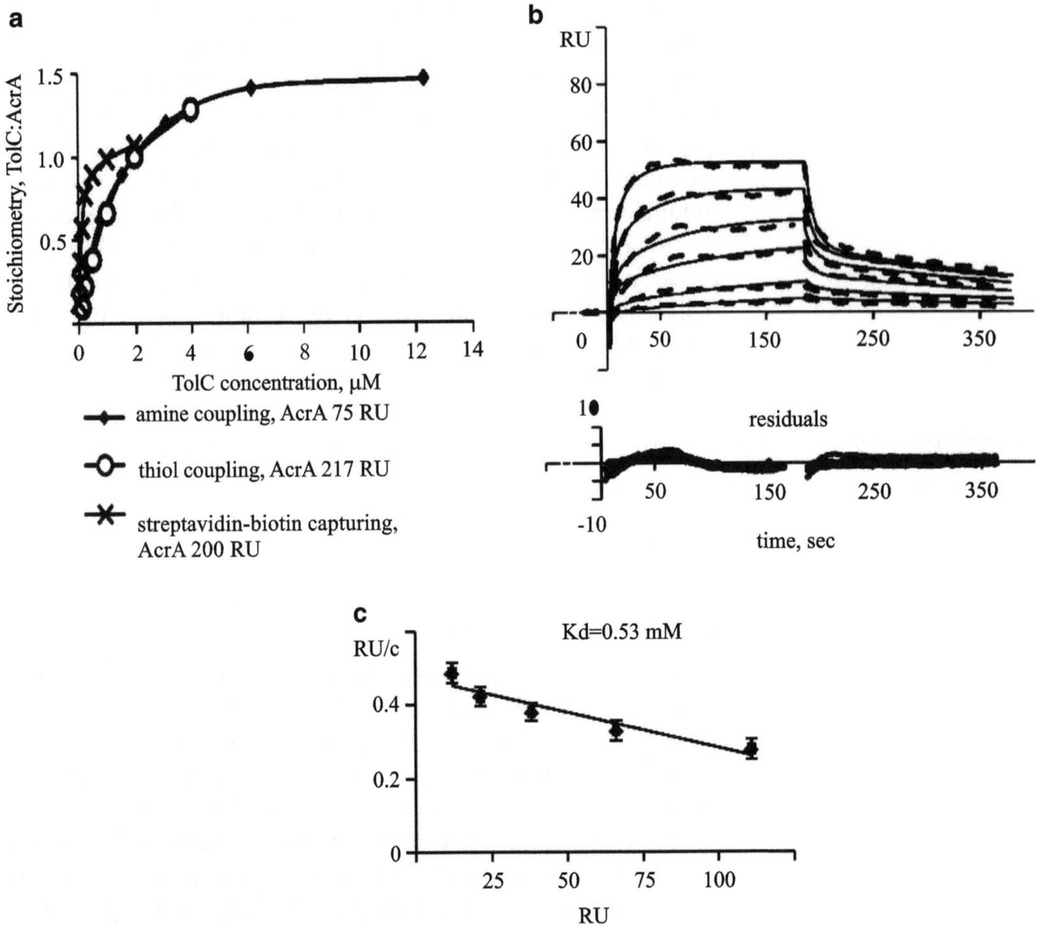

Fig. 5. Examples of various approaches to data analysis. (**a**) Stoichiometry of AcrA–TolC complex from SPR data obtained on different surfaces. Steady-state SPR signals (RU) obtained from injections of doubling TolC concentrations over the indicated AcrA surfaces were taken a few seconds before the end of each association phase. The stoichiometry of AcrA-TolC was calculated as described in the Methods. For all three AcrA surfaces, the stoichiometry of TolC–AcrA complex was found to be 1.5:1 at saturation. (**b**) Kinetic analysis of AcrA–TolC interactions (adapted from (18)). AcrA was immobilized on a CM5 surface at a density of 75 RU. TolC was injected in 20 mM Mes-KOH, pH 5.8, 150 mM NaCl, 0.03% DDM running buffer at doubling concentrations, increasing from 0.0625 to 2.0 μM. In the *upper panel*, the *dashed lines* are the experimental normalized sensorgrams, and the *solid grey lines* are the global fit to the Heterogeneous Ligand model. The *lower panel* shows the residuals calculated for the fit. (**c**) Equilibrium analysis of novobiocin binding to immobilized AcrB (adapted from (19)). The AcrBS1043C mutant (2,040 RU) was immobilized on a SA chip. Novobiocin was injected in 20 mM Tris–HCl, pH 7.5, 150 mM NaCl, 0.03% DDM at doubling concentrations, increasing from 25 to 200 μM. Equilibrium binding responses for each novobiocin concentration were collected a few seconds prior to dissociation. A Scatchard plot of three independent experiments is shown. The equilibrium dissociation constant Kd was calculated as the negative reciprocal of the slope. Error bars are standard deviations.

use the equation $S = (RU_{an} / RU_{lig}) \times (MW_{lig} / MW_{an})$, where RU_{an} is a steady-state binding response of the analyte, RU_{lig} is the amount of the immobilized ligand, MW_{lig} and MW_{an} are molecular weights of the ligand and the analyte, respectively. An example of a stoichiometry graph for AcrA–TolC interaction collected on different chip surfaces is shown in Fig. 5a.

Here, despite different immobilization techniques, the binding capacity of an AcrA surface and the stoichiometry of TolC–AcrA at the surface saturation are very similar.

2. To study kinetics of binding reactions use a low ligand density surface and conditions that give a high specific signal/noise ratio. To determine kinetic parameters (rate constants k_{on} and k_{off}), the BIAevaluation software (Biacore) can be used to fit data into one of the several common bimolecular interaction models. For this purpose, first align binding curves collected at increasing concentrations of an analyte to the start of each injection and normalize the data to exclude a nonspecific binding. Delete any parts of the curve which might affect the fitting procedure. Start with a global fitting using a simplest 1:1 (Langmuir) binding model. Examine the obtained fit and its residual plot. The latter calculates the difference between experimental and theoretical data. The χ^2 value is a standard statistical value characterizing the closeness of the fit and experimental data. Typically a fit is considered to be good if χ^2 is lower than 2 (BIAevaluation Software Handbook, Biacore). If the experimental and fitted curves are not closely aligned, test other models describing more complex binding interactions (see Note 16).
3. If the kinetic analysis cannot be achieved due to the fast association and dissociation rates, perform an equilibrium analysis of binding responses. Inject serial dilutions of an analyte for a time sufficient for each binding response reach a steady state. Acquire equilibrium binding response (RU). Plot RU (*X* axis) against the analyte concentration (*Y* axis). Calculate the equilibrium dissociation constant by fitting data into a kinetic model or if applicable, by linearization and analysis of a Scatchard plot (see Note 17).

4. Notes

1. The major disadvantage of the amino coupling approach is the random orientation of a ligand on the chip surface. All exposed primary amines of the protein will react during immobilization creating a heterogeneous population of the ligand on the chip surface. Random orientation of the immobilized ligand creates heterogeneity of the surface sites, which makes kinetic studies ambiguous (25). Before applying this immobilization method to study MFP–TolC complex (18), we analyzed the sequence and structure of AcrA. We found that the majority of lysine residues in AcrA (77%) are located in the β-barrel and lipoil-binding domains. Thus, during the amino-coupling procedure,

AcrA is predominantly cross-linked to the surface via these domains, with the TolC-binding α-hairpins extended away from the surface. In agreement with this analysis, AcrA mutant lacking the α-hairpin did not bind TolC (18).

2. To prevent contamination of the surface and tubing with aggregates, all buffers, reagents, samples should be filtered to remove possible particulates before using in an SPR experiment.

3. For optimal immobilization of a ligand, low ionic strength buffers should be used during coupling. The pH of an immobilization buffer should be below pI of the protein to be coupled on the surface. For example, the estimated pI of MFP proteins MacA, AcrA, and EmrA are in the range of 5.4–6.8. We used 10 mM sodium acetate buffer pH 4.8 for amine coupling of these proteins onto a CM5 chip (18). Buffers containing amines should be excluded.

4. A change in the response signal by 1,000 RU is equivalent to the change in a surface concentration of 1 ng protein/mm^2 or approximately 6–10 mg/ml in a bulk protein concentration (BIAtechnology Handbook, Biacore). Therefore, 338 RU of AcrA immobilized on a CM5 chip by amine coupling (Fig. 2a) corresponds to the concentration of ~80 μM, 217 RU of AcrA immobilized by thiol coupling (Fig. 2b) is 51 μM and 194 RU of AcrA immobilized on a SA chip (Fig. 2c) is 46 μM. The amount of a ligand for immobilization depends on a specific experiment. Low density surfaces with the lowest amount of ligand showing a reproducible binding response are used for kinetic studies of interactions. Kinetic characterization of MFP–TolC interactions was done on surfaces containing as low as 75 RU for AcrA and 108 RU for EmrA, whereas specific MacA-TolC binding response was not detectable for MacA surface below 200 RU ((18) and unpublished data). To detect binding of low molecular weight molecules (e.g., substrates and inhibitors of MDR pumps), higher immobilization levels of a ligand on a chip surface might be needed. The lowest level of immobilized AcrB S1043C on the SA chip, which gave us a detectable binding response with efflux inhibitors, was 450 RU ((19) and unpublished data).

5. TCEP reagent covalently linked to 4% cross-linked beaded agarose eliminates the need to separate a protein sample from the reducing reagent. The reduced protein should be used for immobilization or biotinylation within 2 h. Use 20 mM Hepes-KOH, pH 7.0, 500 mM NaCl, 5 mM EDTA buffer when preparing samples for immobilization on a CM5 chip. For biotinylation and subsequent immobilization onto a SA chip, a Tris–HCl based buffer can be used as well.

6. Site-specific coupling of a ligand using strategically positioned Cys residues can be used to reduce heterogeneity of the immobilized ligand. The choice of a Cys position depends on the protein. In the case of AcrA, unique Cys residues were introduced into a Cys-less mutant protein close to either the N- or C-terminus (AcrACys) so that after immobilization the α-helical hairpin, the site of TolC binding, will be fully exposed into solution (18). Both the N- and C-terminal Cys mutants of AcrA immobilized onto CM5 chips showed comparable TolC binding responses and kinetic properties (data not shown). An additional consideration is that the expression, folding and function of Cys mutants should be similar to the wild-type protein. To create Cys mutants we designed primers and followed the protocol of the QuickChange XL Site-directed mutagenesis kit (Stratagene). In vivo complementation assays showed that AcrACys mutants are fully functional. Mutant proteins were purified following the same procedure as for the wild-type protein (18).
7. For modification of the SH-group in AcrAS362C use sulfhydryl-reactive biotin with a polyethylene glycol (PEG) spacer arm. Maleimide groups react specifically with reduced sulfhydryls to form stable thioether bonds. The PEG spacer arm helps with the stability and solubility of a biotinylated sample.
8. Biotin–avidin interactions are strong non-covalent interactions with the affinity constant ~10^{15} M^{-1}. Because of the high affinity to avidin, biotin will displace HABA from the complex with avidin thereby decreasing the HABA absorbance. The difference in absorbance before and after addition of a biotinylated sample will reflect the amount of biotin in the solution. For example, a concentration of the biotinylated AcrAS362C after dialysis was found to be 0.143 mg/ml. The A1 and A2 values were measured to be 1.30 and 1.16, respectively. Given the molecular weight of AcrAS362C is 40,574 Da, the A3, B and P values were calculated to be 0.01406, 0.414×10^{-6} and 3.537×10^{-6}, respectively. Finally, the calculated Biotin:Protein ratio was found to be 1.17. We concluded that contamination with free biotin was insignificant and the biotinylated sample can be used for immobilization onto a SA chip.
9. Another advantage of using a SA chip is the neutral pH of the immobilization buffer. Both the amine and thiol coupling procedures require low pH (pH 4.8) during immobilization to promote the electrostatic pre-concentration of positively charged ligands on the chip surface. Some proteins can be unstable under such conditions and might be aggregated or inactivated on the surface. For membrane proteins, a detergent is added to the immobilization buffer. The immobilization buffer for AcrBS1043C contained 0.05% dodecyl maltoside (DDM).

10. Biacore 3000 instrument offers several modes of analyte injections (Biacore 3000 Instrument Handbook, BIAcore). We used a QuickInject mode for the analyte delivery during optimization of binding responses or running conditions. QuickInject consumes smaller amounts of analytes when compared to other modes of injection and requires less time for the injection preparation. Sensorgrams presented in Fig. 4a were obtained using this mode of injection. The Co-Injection mode can be used in experiments that require consecutive injections of two different analytes. In this mode, the injection of a second analyte will start immediately after the injection of the first analyte. We applied this method to study for example, the effect of pH on the stability of MFP–TolC complexes (Fig. 4b, (18)) (18). For kinetic studies we used the KinInject mode, which provides less dispersion of a sample, more accurate response measurements and kinetic analyses (Fig. 5b, (18)).
11. Regeneration of the ligand surface (removal of a tightly bound analyte) is an important requirement for acquiring reproducible and good quality experimental data. The incomplete regeneration will result in a baseline drifting and a partial loss of the ligand activity. Preliminary experiments devoted to selection of the best regeneration solution should be carried out for each immobilized ligand, analytes and chip surface. The ligand should retain its biological activity after each regeneration step. Depending on the nature of the ligand–analyte interactions under study, one should use reagents which will disrupt either electrostatic or hydrophobic interactions or both. The extent of electrostatic interactions can be manipulated with pH (an acidic regeneration using for example 10 mM glycine, pH 2.0 or an alkaline one with 10 mM HEPES-NaOH), or ionic strength (1 M NaCl or up to 4 M $MgCl_2$) of the regeneration buffer. Hydrophobic interactions could be targeted with 50% ethylene glycol or detergents. Optimization of the surface regeneration could be done with different solutions included in the regeneration kit available from Biacore (GE HealthCare). In our studies, we used 20 mM CHAPS prepared in running buffers with pH in the range of 6–8. This detergent was chosen because both immobilized AcrA and AcrB proteins retained their activity and the analyte binding was highly reproducible. Injections of ethylene glycol or SDS, as well as high pH, inactivated AcrB surfaces. In addition, optimization of conditions for the ligand regeneration might provide invaluable insights into the complex assembly and stability.
12. Double referencing to normalize SPR sensorgrams helps to remove a nonspecific signal originated from a sample buffer, a baseline drift or an instrument noise (26). For this purpose, a protein buffer is diluted the same way as the analyte sample and

injected over both the ligand and the control surfaces. The normalized responses are obtained by subtracting SPR responses generated by buffer injections from the analyte responses already corrected onto the binding to a no-ligand control surface.

13. Optimization of experimental conditions helps to decrease nonspecific binding and perhaps improve regeneration of the surface. The effect of detergents, salts, pH of running buffers, the presence of cofactors and accessory proteins should be carefully examined.

14. Detergents are added to running buffers in concentrations above their critical micelle concentrations (CMC) to keep purified membrane proteins in solution. In addition, detergents can be involved in and stabilize the network of bonds formed between protein partners. The choice of a detergent for SPR experiments could be based on previous data (if any available) on the effect of detergents on the activity and stability of the membrane proteins under study. In our optimization experiments, we used 0.2% Triton X-100, 0.03–0.05% DDM (dodecyl maltoside) and 1% OG (octyl glucoside). AcrA, AcrB and TolC were most stable in DDM-containing buffers and generated highly reproducible SPR signals. Therefore, DDM became our choice of detergent for studies of the assembly of drug efflux complexes (19).

15. We routinely use 150 mM NaCl or KCl in all running buffers. This salt concentration was found to be optimal for keeping proteins in active conformations and for monitoring bi- and tripartite interactions in AcrAB–TolC complex. In our experiments, raising NaCl concentration to 250 mM reduced specific binding responses.

16. Figure 5b shows an example of experimental data for AcrA–TolC interactions fitted globally into a "Heterogeneous ligand-parallel reaction" model. This model assumes the presence of two populations of a ligand on the surface with fast and slow dissociation rates. The choice of this model was supported by studies of kinetics of AcrA–TolC interactions on different chip surfaces and by finding that at high concentrations (as those on the surface) AcrA forms oligomers. In most cases, in addition to data quality controls described here and in other publications, the biological significance and reliability of a chosen kinetic model should also be verified by alternative biochemical methods and functional assays. For a more detailed analysis of surface heterogeneity in SPR experiments readers are referred to a recent method article by Schuck and Zhao (25).

17. In our studies, binding kinetics of small molecules such as the AcrAB-TolC substrate novobiocin and its inhibitor MC-207, 110 to immobilized AcrB was too fast for reliable calculations

of k_{on} and k_{off} from the sensorgrams (19). For these analytes, Scatchard plots were generated by plotting equilibrium binding responses (RU, *X* axis) for doubling concentrations of analytes as a function of a response over an analyte concentration (RU/c, *Y* axis). Equilibrium constants were calculated from the slopes of these plots (Fig. 5c).

References

1. Nikaido H, Vaara M (1985) Molecular basis of bacterial outer membrane permeability. Microbiol Rev 49:1–32
2. Lomovskaya O, Lewis K (1992) Emr, an Escherichia coli locus for multidrug resistance. Proc Natl Acad Sci U S A 89:8938–8942
3. Ma D, Cook DN, Alberti M, Pon NG, Nikaido H, Hearst JE (1993) Molecular cloning and characterization of acrA and acrE genes of Escherichia coli. J Bacteriol 175:6299–6313
4. Lomovskaya O, Zgurskaya HI, Totrov M, Watkins WJ (2007) Waltzing transporters and 'the dance macabre' between humans and bacteria. Nat Rev Drug Discov 6:56–65
5. Nikaido H (2001) Preventing drug access to targets: cell surface permeability barriers and active efflux in bacteria. Semin Cell Dev Biol 12:215–223
6. Zgurskaya HI, Krishnamoorthy G, Tikhonova EB, Lau SY, Stratton KL (2003) Mechanism of antibiotic efflux in Gram-negative bacteria. Front Biosci 8:s862–s873
7. Dinh T, Paulsen IT, Saier MH Jr (1994) A family of extracytoplasmic proteins that allow transport of large molecules across the outer membranes of gram-negative bacteria. J Bacteriol 176:3825–3831
8. Aires JR, Nikaido H (2005) Aminoglycosides are captured from both periplasm and cytoplasm by the AcrD multidrug efflux transporter of Escherichia coli. J Bacteriol 187:1923–1929
9. Tikhonova EB, Devroy VK, Lau SY, Zgurskaya HI (2007) Reconstitution of the Escherichia coli macrolide transporter: the periplasmic membrane fusion protein MacA stimulates the ATPase activity of MacB. Mol Microbiol 63:895–910
10. Zgurskaya HI, Nikaido H (1999) Bypassing the periplasm: reconstitution of the AcrAB multidrug efflux pump of Escherichia coli. Proc Natl Acad Sci U S A 96:7190–7195
11. Andersen C, Hughes C, Koronakis V (2001) Protein export and drug efflux through bacterial channel-tunnels. Curr Opin Cell Biol 13:412–416
12. Thanabalu T, Koronakis E, Hughes C, Koronakis V (1998) Substrate-induced assembly of a contiguous channel for protein export from E. coli: reversible bridging of an inner-membrane translocase to an outer membrane exit pore. EMBO J 17:6487–6496
13. Paulsen IT, Chen J, Nelson KE, Saier MH Jr (2001) Comparative genomics of microbial drug efflux systems. J Mol Microbiol Biotechnol 3:145–150
14. Saier MH Jr, Beatty JT, Goffeau A, Harley KT, Heijne WH, Huang SC, Jack DL, Jahn PS, Lew K, Liu J, Pao SS, Paulsen IT, Tseng TT, Virk PS (1999) The major facilitator superfamily. J Mol Microbiol Biotechnol 1:257–279
15. Tseng TT, Gratwick KS, Kollman J, Park D, Nies DH, Goffeau A, Saier MH Jr (1999) The RND permease superfamily: an ancient, ubiquitous and diverse family that includes human disease and development proteins. J Mol Microbiol Biotechnol 1:107–125
16. Saier MH Jr, Paulsen IT, Sliwinski MK, Pao SS, Skurray RA, Nikaido H (1998) Evolutionary origins of multidrug and drug-specific efflux pumps in bacteria. FASEB J 12:265–274
17. Modali SD, Zgurskaya HI (2011) The periplasmic membrane proximal domain of MacA acts as a switch in stimulation of ATP hydrolysis by MacB transporter. Mol Microbiol 81:937–951
18. Tikhonova EB, Dastidar V, Rybenkov VV, Zgurskaya HI (2009) Kinetic control of TolC recruitment by multidrug efflux complexes. Proc Natl Acad Sci U S A 106:16416–16421
19. Tikhonova EB, Yamada Y, Zgurskaya HI (2011) Sequential mechanism of assembly of multidrug efflux pump AcrAB-TolC. Chem Biol 18:454–463
20. Tikhonova EB, Zgurskaya HI (2004) AcrA, AcrB, and TolC of Escherichia coli form a stable intermembrane multidrug efflux complex. J Biol Chem 279:32116–32124

21. Hearty S, Conroy PJ, Ayyar BV, Byrne B, O'Kennedy R (2010) Surface plasmon resonance for vaccine design and efficacy studies: recent applications and future trends. Expert Rev Vaccines 9:645–664
22. Jason-Moller L, Murphy M, Bruno J (2001) Overview of biacore systems and their applications. Curr Protoc Protein Sci. 19: Unit 19.3
23. Maynard JA, Lindquist NC, Sutherland JN, Lesuffleur A, Warrington AE, Rodriguez M, Oh S-H (2009) Surface plasmon resonance for high-throughput ligand screening of membrane-bound proteins. Biotechnol J 4:1542–1558
24. Piliarik M, Vaisocherova H, Homola J (2009) Surface plasmon resonance biosensing. Methods Mol Biol 503:65–88
25. Schuck P, Zhao H (2010) The role of mass transport limitation and surface heterogeneity in the biophysical characterization of macromolecular binding processes by SPR biosensing. Methods Mol Biol 627:15–54
26. Myszka DG (1999) Improving biosensor analysis. J Mol Recognit 12:279–284

Chapter 10

Fluorescence Microscopy and Proteomics to Investigate Subcellular Localization, Assembly, and Function of the Type II Secretion System

Tanya L. Johnson, Aleksandra E. Sikora, Ryszard A. Zielke, and Maria Sandkvist

Abstract

Investigation of secretion systems is often critical to understanding the virulence mechanisms of bacterial pathogens. With estimates as high as 30–40% of proteins secreted or localized to the cell envelope, information about the subcellular localization and organization of secretion complexes and identification and functional characterization of their substrates are key steps toward understanding these intricate systems. Here we describe a protocol using fluorescent live-cell imaging of fusion proteins that can provide a powerful tool to potentially examine the localization, assembly, and role of each component in the secretion complex. In addition, we describe protocols for the identification of secreted substrates using 1D SDS-PAGE coupled with nano-liquid chromatography (LC) and tandem mass spectrometry (MS/MS), and isobaric tagging for absolute quantification (iTRAQ) coupled with two-dimensional LC and MS/MS. Both experimental approaches are applicable to any similar study of membrane transport systems.

Key words: iTRAQ, Proteomics, Secretome, Fluorescence microscopy, Subcellular localization

1. Introduction

Fluorescence microscopy imaging techniques can be applied to single bacterial cells to investigate transcription activity, protein localization, protein stability, protein–protein interactions, and protein mobility in vivo. The list of examples of specific subcellular protein localization in bacteria is growing, and recent studies are beginning to reveal general mechanisms behind the localization (for review, see ref. 1). Defining the localization of a bacterial

Tanya L. Johnson and Aleksandra E. Sikora have contributed equally to this work.

Anne H. Delcour (ed.), *Bacterial Cell Surfaces: Methods and Protocols*, Methods in Molecular Biology, vol. 966, DOI 10.1007/978-1-62703-245-2_10, © Springer Science+Business Media New York 2013

protein and following its movement over time has resulted in important insights into the molecular mechanisms of complex bacterial systems. Following a particular protein's change in localization upon deletion of potential interacting partners, for example, is an excellent method for visualizing protein–protein interactions between components of complexes in bacterial cells that may be difficult to detect by other methods (2). Although the small size of bacteria complicates experiments, in practice, the limit of possible experiments is primarily determined by the intensity of the fluorescent signal that can be measured and the resolution of the imaging system.

We prepare samples for live cell imaging by embedding bacteria on a flat agarose pad, supplemented with culture medium. This technique permits the immediate capture of images with the option of incubating the microscope slides to allow the cells to grow and divide on a plane surface. A time course of images lets us follow changes in a protein's fluorescence intensity and distribution during the life cycle of the bacterial cell. The first protocol below describes fluorescent imaging of chromosomally expressed GFPmut2 fused to an inner membrane protein of the type II secretion (T2S) system of *Vibrio cholerae*.

Fast development of proteomic techniques helps to broaden our knowledge and understanding of the constituents of the bacterial cell envelope, surface exposed proteins, and secreted proteins (secretome). Particularly, studies aimed at defining these proteins in pathogenic bacteria are important for biomarker discovery, vaccine development, and identifying potential novel drug targets. Proteomic analysis of secreted and membrane proteins has many experimental challenges because of the nature of these proteins. Secreted proteins are often low in abundance, while the cell envelope proteins are difficult to solubilize and separate due to their hydrophobicity, basic charge, and often high molecular weight. To meet these challenges, several experimental steps need to be considered and include, but are not limited to (1) efficient fractionation of bacterial cells, (2) protein enrichment (finding a suitable method for protein precipitation), (3) protein separation (in gel or gel-free and liquid chromatography), and (4) protein identification and quantification.

In this chapter, we present two independent proteomic approaches that we have successfully applied to profile the T2S dependent secretome of *V. cholerae*: (1) one-dimensional (1D) SDS-PAGE coupled with nano-liquid chromatography (LC) and tandem mass spectrometry (MS/MS), and (2) isobaric tagging for absolute quantification (iTRAQ) coupled with two-dimensional liquid chromatography (2D-LC) and MS/MS. Using these approaches allowed us to identify 16 novel proteins specifically secreted by the *V. cholerae* T2S system (3).

The coupling of 1D SDS-PAGE with nano-LC and MS/MS analysis described in the first proteomic approach has led to the identification of 187 proteins in culture supernatants of *V. cholerae* (based on at least two peptides with 1% False Discovery Rate)(3). A similar experimental methodology has identified 60 unique membrane or membrane-associated proteins in *Helicobacter pylori* (4). The non-gel based iTRAQ coupled 2D-LC/MS/MS strategy was originally described in 2004 (5), and since then has been applied in many quantitative global proteomic studies. This proteomic method allows for concurrent analyses of protein contents in four, six or eight separately labeled biological samples in a single experiment by the utilization of up to eight different isobaric tags (6, 7).

2. Materials

2.1. Fluorescence Microscopy

2.1.1. Culture Growth

1. Luria–Bertani (LB) agar: commercially available.
2. M9 minimal medium: commercially available.
3. Incubator-shaker.
4. Spectrophotometer.
5. Antibiotics.
6. Inducer.

2.1.2. Slide Preparation

1. Microwave oven.
2. 1.5% low-melting agarose solution: 150 mg of SeaPlaque low melt agarose in 10 ml of fresh M9 minimal medium.
3. Water bath.
4. Precleaned superfrost 25 × 75 × 1 mm microscope slides.
5. Microscope cover glass 24 × 60 mm.
6. Slide warmer.

2.1.3. Visualization of the Bacteria

1. Fluorescence microscope equipped with an oil immersion objective (1.4 numerical aperture).
2. Digital camera.
3. GFP HC HiSN zero shift filter cube, with a 450- to 490-nm excitation filter and a 500- to 550-nm barrier filter, for visualizing GFP fluorescence.
4. Imaging software.
5. Photo processing software.

2.2. Secreted Protein Identification

2.2.1. 1D SDS-PAGE and 1D LC/MS/MS

1. LB medium: commercially available.
2. LB agar plates.
3. Spectrophotometer.
4. Microcentrifuge.
5. 1-ml syringe.
6. 0.2 μm nylon filter.
7. Ultracentrifuge.
8. Ultracentrifuge tubes.
9. Pyrogallol red-molybdate-methanol (PRMM) : 0.05 mM pyrogallol red, 0.16 mM sodium molybdate, 1.0 mM sodium oxalate, 50 mM succinic acid, 20% (vol/vol) methanol in H_2O, adjusted to pH 2.0 with HCl (see Note 1).
10. pH paper indicators.
11. Acetone.
12. SDS-PAGE loading buffer.
13. 4–12% Bis-Tris Gel.
14. MES SDS Running Buffer.
15. Colloidal Coomassie: 17% ammonium sulfate, 34% methanol, 3% phosphoric acid, 0.1% Coomassie brilliant blue G-250. Add 420 ml of ddH_2O, 35.5 ml of 85% phosphoric acid, and 170 g of ammonium sulfate. Stir until completely dissolved. Add 340 ml of methanol (a white precipitates will form) and continue to stir (solution A). Prepare solution B by adding 1 g of Coomassie brilliant blue G-250 in 100 ml of *hot* ddH_2O and heat the solution until boiling. A microwave oven can be used. Pour solution B into solution A, add H_2O to 1,000 ml, and continue to stir for a couple of minutes. (see Note 2).
16. 10% glacial acetic acid.
17. Tissue paper.
18. HPLC grade water.
19. Razor blades.
20. Light box for visualization of proteins following SDS-PAGE and Coomassie staining.
21. 1.5-ml microcentrifuge tube.
22. 50 mM ammonium bicarbonate.
23. Acetonitrile (ACN).
24. Reducing solution: 10 mM dithiothreitol in 50 mM ammonium bicarbonate (HPLC-grade H_2O).
25. Alkylation solution: 55 mM iodoacetamide in 50 mM ammonium bicarbonate (HPLC-grade H_2O).

26. Trypsin solution: 10 μg/ml trypsin in 50 mM ammonium bicarbonate (HPLC grade H_2O).
27. Extraction solution 1: 1% formic acid and 2% ACN in HPLC grade H_2O.
28. Extraction solution 2: 60% ACN and 0.1% trifluoroacetic acid (TFA) in HPLC grade H_2O.
29. Reconstitution buffer: 10% ACN and 0.1% TFA in HPLC grade H_2O.
30. Vacuum concentrator.

2.2.2. iTRAQ Labeling

1. LB medium: commercially available.
2. LB agar plates.
3. 4-l culture flask.
4. Spectrophotometer.
5. Ultracentrifuge.
6. 1-l bottles with 0.2 μm filter.
7. Complete Mini Protease Inhibitor Cocktail Tablets.
8. Reagents for protein concentration determination (assays: Bradford, Lowry, or BCA).
9. Sodium deoxycholate.
10. 100% trichloroacetic acid.
11. Acetone.
12. Phosphate buffered saline (PBS): 3.2 mM Na_2HPO_4, 0.5 mM KH_2PO_4, 1.3 mM KCl, 135 mM NaCl, pH 7.4.
13. iTRAQ reagent (see Note 3).
14. 0.1% TFA in HPLC grade H_2O.
15. Vacuum concentrator.

2.2.3. 2D LC/MS/MS

1. Strong cation-exchange (SCx) buffer A: 5 mM KH_2PO_4 pH 2.7, 30% (v/v) ACN in HPLC grade H_2O.
2. SCx buffer B: SCx buffer A with 500 mM KCl in HPLC grade H_2O.
3. PolySULFOETHYL Aspartamide™ strong cation-exchange column 200×2.1 mm; 5 μm; 200 Å
4. HPLC system.
5. Vacuum concentrator.
6. Solvent A: 0.1% TFA in HPLC grade H_2O.
7. Solvent B: 90% ACN, 0.1% TFA in HPLC grade H_2O.
8. 1100 Series nano HPLC equipped Micro-FC fraction collector/spotter (Agilent Technologies, Santa Clara, CA, USA).

9. Reversed-phase cartridge (Zorbax C18, 5 mm by 0.3 mm i.d., 5 μm particles).
10. Reverse phase column (Zorbax 300 SB C18 column, 75 μm × 150 mm, 3.5 μm particles).
11. Micro Tee adaptor.
12. Matrix: 2 mg/ml α-cyano 4-hydroxy cinnamic acid in 50:49:1 ratio of HPLC grade water–isopropanol–acetic acid and containing 10 mM ammonium phosphate.
13. Infusion pump.
14. Stainless steel MALDI target plate LC separation.
15. Applied Biosystems 4800 Proteomics Analyzer (TOF/TOF) with 200 Hz Nd:YAG laser (Life Technologies).

2.2.4. Data Analysis

ProteinPilot software (version 3.0) (Life Technologies).

3. Methods

3.1. Fluorescence Microscopy

In choosing a fluorophore reporter and tagging of the protein of interest, there are multiple fluorescent proteins, such as GFP or mCherry, which allow easy visualization in bacterial cells. Using gene fusion technology, fluorescent proteins can be fused to either the amino- or carboxy-terminal end of the candidate proteins, keeping in mind that some fluorescent proteins have specific folding requirements (see Notes 4 and 5). If possible, use chromosomally encoded fusion proteins for all experiments (see Notes 6 and 7). A strain expressing non-fused fluorescent protein from the *lac* locus can also be constructed to serve as a cytoplasmic localization marker and control for experimental samples.

3.1.1. Culture Growth

1. Inoculate bacterial cells from −80°C stocks onto agar plates supplemented with antibiotics if necessary. Incubate plates until single colonies are visible.
2. Use a single colony from the plate and inoculate 2–5 ml of liquid M9 minimal medium supplemented with antibiotics if necessary. Grow the cells in liquid medium overnight with shaking (see Notes 8 and 9).
3. The following morning, the overnight culture of cells can be imaged (proceed to Subheading 3.1.2 Slide Preparation) or diluted in fresh growth medium. To image mid- to late-log cultures, dilute the cultures 1:50–100 in growth medium supplemented with antibiotics or inducers (e.g., IPTG or arabinose) if necessary (see Note 8).
4. Grow the cells to the desired growth phase/time of induction. Monitor cell growth by OD_{600} readings (see Note 8).

5. If using a growth medium that has a low level of intrinsic auto-fluorescence like M9 minimal medium, the cells can be removed from the culture flask or tube and imaged directly (see Note 10). It is also possible to treat the cell culture with stains such as FM 4–64 or DAPI to label membranes and DNA respectively, at this time.

3.1.2. Slide Preparation

1. Use a microwave oven to prepare the 1.5% low-melting agarose solution. After ensuring the agarose is completely dissolved, place the solution in a 55°C water bath to cool (see Note 11).
2. After cooling, supplement the agarose with inducer (e.g., IPTG or arabinose) or other compounds, such as antibiotics, at this time if required.
3. Place microscope slides and coverslips on a slide warmer set to 55°C (see Note 12).
4. Pipet 800–1,000 μl of agarose onto the slide, ensuring the entire area is fully covered (see Note 13).
5. Flick the excess agarose into a trash container while holding the sides of the slide (see Note 14).
6. Place the slide with the agarose pad on a flat surface and allow to cool at room temperature for 1–1.5 min.
7. Pipet between 10 and 20 μl of cell culture onto the center of the agarose pad (without touching the agarose with the pipet tip) and allow to dry/cool at room temperature for 1–1.5 min (see Note 15).
8. Carefully place the coverslip over the agarose pad. Ideally, the cells will be uniformly distributed across the slide and the number of air bubbles trapped between the coverslip and the agarose will be minimal.
9. Allow the agarose to fully solidify at room temperature for 5–10 min (see Note 16).

3.1.3. Visualization of the Bacteria

1. Place a drop of immersion oil in the center of the top surface of the coverslip. Place the slide in the sample holder of the fluorescence microscope equipped with a 100× oil immersion objective (1.4 numerical aperture), making optical contact with the objective lens.
2. Switch on the phase contrast or differential interference contrast (DIC) illumination and bring the cells into focus, selecting a suitable group of cells to be imaged on the basis of their being trapped near each other with their long axis parallel to the coverslip surface.
3. Finely adjust to ensure the majority of bacteria are in focus and image the bacteria (~50 ms exposure). Start camera acquisition and open the laser shutter to excite GFP (450- to 490-nm

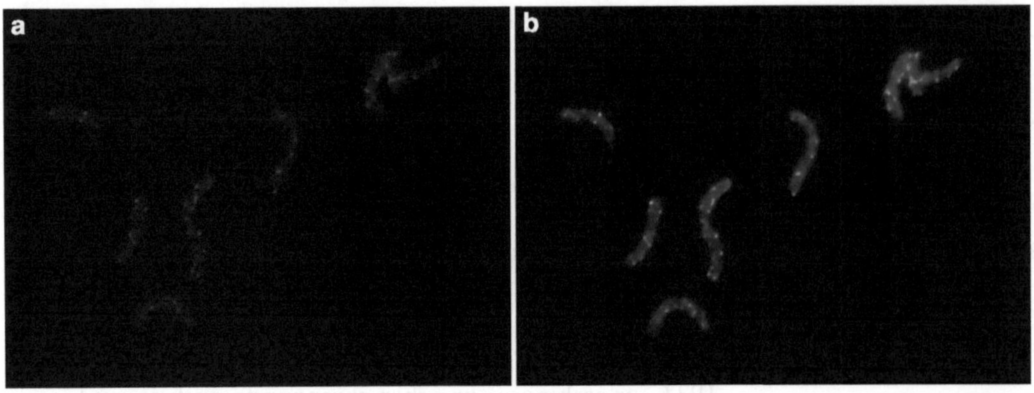

Fig. 1. Fluorescent microscopy image before and after intensity adjustment in Adobe Photoshop CS2. (**a**) A .jpg image of chromosomally expressed GFP-EpsC in *V. cholerae* cells captured with a 2 s exposure with all pixels visible (unadjusted). (**b**) The same image after adjusting the input levels of pixels using Adobe Photoshop CS2. GFP-EpsC is localized to fluorescent foci along the lengths of the cells.

excitation filter and a 500- to 550-nm barrier filter) and capture a fluorescent image (2–3 s exposure). The parameters for the time of acquisition of the fluorescent image will need to be optimized for the particular biological system under investigation (see Notes 17–19).

4. Save and analyze the captured images with the microscope's imaging software. For quantitation of fluorescent results, aim to image and count ~150–200 cells from three separate experiments. For presentation, image input levels and image file types can be adjusted with a photo processing software (see Note 20 and Fig. 1).

3.2. Secreted Protein Identification

3.2.1. 1D SDS-PAGE and 1D LC/MS/MS

1. Streak out bacterial strains (e.g., wild type and isogenic mutant) on LB agar plates with appropriate supplements to obtain single colonies. Inoculate a single bacterial colony of each strain into 5 ml LB. Culture bacteria at 37°C with aeration to the late stationary phase of growth (16 h) (see Note 8). Measure the optical density of the culture (A_{600}).
2. Prepare culture supernatant. This is a two-step process in which bacterial cells and outer membrane vesicles are separated from the soluble, secreted proteins that are present in the culture supernatant.
   (a) Centrifuge the desired amount of bacterial culture (e.g., 1.5 ml) in a microcentrifuge for 10 min, 4,000 × *g*. Transfer 1 ml of the culture supernatant to a 1 ml syringe and passage it through a 0.2 μm nylon filter to remove any residual bacterial cells.
   (b) Transfer the filtered supernatant to an ultracentrifuge tube and centrifuge at 170,000 × *g*, 4°C, for 3 h, to remove membrane vesicles.

3. Transfer 950 μl of the supernatant to a fresh microcentrifuge tube and add an equal amount of PRMM. Immediately adjust the pH of the mixture to 2.8–3 for optimal precipitation. Use paper indicators to measure pH.
4. Incubate the sample for 1–2 h at room temperature, and subsequently overnight at 4°C (see Note 21).
5. Harvest the precipitated proteins by centrifugation at 10,000×g, 4°C, 1 h.
6. Carefully decant the supernatant and wash the precipitated material twice with 500 μl ice-cold acetone (see Note 22). Each time repeat the centrifugation as in step 5.
7. Remove the remaining acetone (see Note 23), and air-dry the sample by placing the open microcentrifuge tube in a rack at 37°C. The pellet is ink-blue when dry and easily falls from the walls of the tube.
8. Reconstitute the proteins in SDS-PAGE loading buffer. The sample needs to be incubated at 90°C for 15–30 min and mixed by frequent vortexing in order to solubilize all precipitated proteins.
9. Prepare for SDS-PAGE gel electrophoresis (see Note 24). Match the samples by equivalent OD_{600} and load onto the gel (MES buffer 4–12%) and resolve for a short length, ~2 cm at 200 V constant (see Note 25).
10. Place the gel in a clean staining container and stain the gel with the Colloidal Coomassie solution, gently agitating at room temperature overnight (see Note 26).
11. Destain the gel in 10% glacial acetic acid; placing tissue paper inside the staining container to absorb the stain. Alternatively, rinse the gel several times with ultrapure water in the presence of tissue paper (see Note 27).
12. Cut out the entire fragment of the gel containing proteins bands using a clean razor blade. Placing the gel on a transparent plastic sheet on a light box facilitates the visualization of the stained proteins.
13. For each sample, cut the gel fragment into smaller pieces and place them together into a sterile 1.5 ml microcentrifuge tube (see Note 28).
14. Add 50 μl of 100 mM ammonium bicarbonate and 50 μl of acetonitrile (ACN) and incubate at 37°C for 10 min. Aspirate mixture and repeat the procedure once or until gel plugs are lightly blue.
15. Dehydrate gel plugs by adding 100 μl of ACN and incubate at 37°C for 10 min. Aspirate the solution and incubate for additional 10 min at 37°C.

16. Reduce proteins by addition of 100 μl of the reducing solution and incubate at 37°C for 20 min.
17. Block the cysteines by addition of 100 μl of the alkylation solution and incubate at 37°C for 20 min to prevent disulfide bond formation.
18. Rinse gel plugs with 100 μl of water (HPLC grade) once.
19. Dispense 100 μL of ACN into each tube and incubate at 37°C for 5 min. Aspirate the solution. Repeat once.
20. Remove any residual ACN by evaporation (place the samples at 37°C for additional 5 min).
21. Digest the proteins overnight at 37°C in 40 μl of trypsin solution.
22. Extract the tryptic fragments by addition of 55 μl of extraction solution 1 and incubate at room temperature for 30 min. Remove supernatant into fresh microcentrifuge tube.
23. Re-extract gel plugs by adding 50 μl of extraction solution 2 and incubate at room temperature for 30 min.
24. Combine the extract from step 23 with the extract from step 22.
25. Use a vacuum concentrator to dry the extracts without heating.
26. Add 50 μl of reconstitution buffer to reconstitute the peptides and subject to reverse phase chromatography followed by MS/MS analysis as described below (Subheading 3.2.3, step 7).

3.2.2. iTRAQ Labeling

1. Streak out bacterial strains (e.g., wild type and isogenic mutant) on LB agar plates with appropriate supplements to obtain single colonies. Inoculate a single bacterial colony of each strain into 5 ml LB. Culture bacteria at 37°C with aeration to the stationary phase of growth, and back dilute the cultures 1:100 in 1 l of prewarmed LB in a 4-l flask.
2. Grow bacterial cultures with agitation to the late stationary phase of growth (16 h, OD_{600} about 4, see Note 8).
3. Separate culture supernatants from the cells by centrifugation (4,000 × *g*, 30 min) followed by filtration through 0.2 μm filters into 1-l bottles.
4. Add ½ tablet of protease inhibitor cocktail to the filtered supernatant.
5. Measure the total protein concentration in filtered supernatant (see Note 29).
6. Add sodium deoxycholate to each culture supernatant containing 300 μg protein to a final concentration of 0.015%. Place a stirring bar in each bottle.

7. Incubate the mixture with stirring for 10 min at 37°C (see Note 30).
8. Precipitate the proteins by addition of 100% trichloroacetic acid (final concentration 10%) and incubate the mixture with stirring for 1 h at 4°C.
9. Centrifuge the samples, 8,000 × *g*, 20 min, to pellet the protein precipitate.
10. Wash the pellet twice with ice-cold acetone and reconstitute the proteins in 0.05% SDS in phosphate buffered saline.
11. Measure the total protein concentration in the reconstituted samples.
12. Precipitate a total of 100 μg proteins from each sample overnight in 80% acetone.
13. Centrifuge the samples at 10,000 × *g* at 4°C and wash the proteins pellets in 100% acetone. Repeat twice.
14. Reconstitute protein pellets and proceed with cysteine blocking, trypsin digestion and iTRAQ labeling to chemically tag the free amines of all peptides according to the manufacturer's instructions. A different isobaric tag will be used for each sample (see Note 31).
15. Incubate the mixture at room temperature for 1 h and then stop the reaction by adding 250 μl of 0.1% TFA.
16. Combine the reaction products from every sample (technical/biological replicas), dry in vacuum concentrator and store at −20°C before 2D-LC.

3.2.3. 2D LC/MS/MS

1. Load samples for the first dimension liquid chromatography samples reconstituted in SCx buffer A on the PolySULFOETHYL Aspartamide™ strong cation-exchange column (Poly LC, see Note 32).
2. Fractionate the iTRAQ-labeled peptides using a HPLC system.
3. Elute peptides using buffer SCx A and buffer SCx B with a flow rate of 200 μl/min using the following gradient: 2% B from 0 to 5 min, 8% B to 5.1 min, 18% B to 20 min, 34% B to 32 min, 60% B to 40 min, 98% B to 40.1 min to 50 min, 2% B to 50.1 min and held to the end of the run at 60 min.
4. Collect 2-min fractions throughout and combine to 12 fractions for subsequent reverse phase (RP-HPLC-MS) analysis.
5. Dry the collected fractions in a vacuum concentrator.
6. Reconstitute each fraction with 45 μl of 0.1% TFA.

7. Separate peptides using an 1100 Series nano HPLC equipped Micro-FC fraction collector/spotter (see Note 33). First, use a reversed-phase cartridge to desalt and concentrate peptides. Elute peptides on a reverse phase column and fractionate with the following gradient of solvent B: 0 min, 6.5% B; 12 min, 18% B; 80 min, 70% B; 80.1 min, 100% B; 82 min, 100% B; 82.1 min, 6.5% B; 96 min, 6.5% B at a flow rate of 0.4 μl/min throughout the gradient.
8. Mix the column effluent (using a micro Tee) with matrix delivered with an infusion pump at 1.3 μl/min.
9. Spot fractions at 20 s intervals onto a stainless steel MALDI target plate.
10. Acquire the MS and MS/MS spectra on an Applied Biosystems 4800 Proteomics Analyzer (TOF/TOF) in positive ion reflection mode with a 200 Hz Nd:YAG laser operating at 355 nm, and accelerating voltage 20 kV with 400 ns delay.
11. For MS/MS spectra, use the collision energy at 2 keV and air as the collision gas. Both MS and MS/MS data will be then acquired in the sample wells using the instrument default calibration.
12. Obtain the MS spectra with the minimum possible laser energy in order to maintain the best resolution.
13. Collect single-stage MS spectra for the entire samples first, and in each sample well MS/MS spectra may be acquired from the 12 most intense peaks above the signal to noise the ratio threshold of 30.

3.2.4. Data Analysis

1. Analyze MS/MS spectra using the Paradigm search algorithm, which is a part of the ProteinPilot software (version 3.0) (see Note 34) against the SwissProt target-decoy database for the analyzed organism (e.g., *V. cholerae*) with the following settings:

   Sample type: Identification (for 1D LC MS/MS analysis) or iTRAQ 4plex (for iTRAQ 2D LC MS/MS analysis).

   Cysteine alkylation: Iodoacetamide (for 1D LC MS/MS analysis) or MMTS (for iTRAQ 2D LC MS/MS analysis).

   Digestion: Trypsin.

   ID Focus: Biological Modifications.

2. Use the ProGroup Algorithm built within ProteinPilot software to perform the statistical analysis on the identified peptides to determine the minimal set of identifications. Use thorough ID search and False Discovery Rate calculations (e.g., 1%). Only proteins with p values <0.05 associated with a ratio will be included for further analysis.

4. Notes

1. The pyrogallol-red-molybdate-methanol (PRMM) method has proven useful for concentrating low abundant proteins in bacterial culture supernatants with much better efficiency than other more conventional protocols based on trichloroacetic acid and acetone (8–10). PRMM should be protected from light and stored in a brown color glass bottle at room temperature.
2. The stain will not dissolve well unless the volume and composition are accurate. A good staining solution should have a dark blue color.
3. Depending on the number of samples, four-plex or eight-plex iTRAQ reagents might be used and simultaneously analyzed.
4. The inability of GFP to fold properly in the periplasm if exported by the Sec machinery (11) can be circumvented by use of a GFP variant that is targeted through the Tat system, which exports folded proteins, or by the use of a newly developed superfolder GFP (12, 13). The long half-life of stable GFP, which is a potential problem for gene expression studies, can be reduced by the use of destabilized variants.
5. Some proteins may not allow for tagging at one end due to disruption of protein function. In order to verify that the tagged protein is functional, perform complementation assays. It is also advisable to perform immunoblot analysis to determine if the fusion protein is subject to degradation.
6. Producing chromosomally encoded fluorescent-tagged protein, as opposed to using the more common but less ideal plasmid expression, results in protein expression levels that are closer to those present in unmodified strains. Because of this, the biological systems under investigation are visualized significantly closer to the native conditions, and therefore the observations are likely more relevant to physiological processes.
7. An alternative to fluorescent proteins is the use of small-molecule dyes (e.g., fluorescein), which are brighter than fluorescent proteins and can be targeted to proteins by means of antibody conjugates or protein/peptide tags. Antibody labeling is relatively specific but normally requires cell fixation, limiting its use studying dynamic processes and possibly introducing morphological changes to the cells and cell envelope complexes (14). High-affinity tags such as FlAsH (15) allow labeling of proteins in living cells, but their application is hindered by high background fluorescence and the necessity for specific staining

procedures. As a result, the use of fluorescent proteins is favored in most studies of protein localization in bacteria.

8. The specific growth conditions (medium, temperature and time of growth) may vary from species to species. We use M9 minimal medium whenever possible for microscopy due to the low auto-fluorescence of this medium.
9. Alternatively, it is possible to resuspend a bacterial colony from a plate in a small volume of liquid M9 minimal medium and proceed with imaging if desired (if doing so, proceed to step 5).
10. If the cells were grown in a medium that is auto-fluorescent (e.g., LB), the cells should be centrifuged briefly and resuspended in buffer such as phosphate buffered saline and imaged. This will not, however, allow for time course experiments.
11. Microwaving in a 50 ml conical tube works well for small volumes (i.e., 15 ml or less) of agarose. Do not allow the agarose to boil over.
12. A slide warmer is not absolutely necessary, but it does allow you to construct multiple slides before the agarose solidifies.
13. Ensure that the entire slide is coated evenly with agarose by using the pipet tip to spread the melted agarose from side to side and around the edges of the slide.
14. A significant portion of the agarose will come off the slide; however, the entire slide should remain coated with a thin layer of agarose.
15. When the cell culture is pipetted onto the agarose pad, the culture should remain in place and "bead up" on the surface of the agarose. If the culture instead dissipates into the agarose, the agarose was not cool enough and the slide preparation should be started again at step 4.
16. If the coverslip is easily dislodged from the agarose pad after cooling, the cooling steps 6 and 7 were too long, and the agarose solidified prior to the coverslip placement.
17. Plasmid expression of fluorescent proteins can usually be imaged with short acquisition times, while chromosomal expression of fluorescent proteins generally requires longer (i.e., seconds) acquisition parameters unless the fusion protein is expressed from a strong promoter.
18. For live-cell imaging, it is critical to avoid unnecessary exposure of the sample which could lead to bleaching and phototoxicity. For that reason, initial focusing should always be done with phase contrast or DIC illumination. To obtain a good fluorescence image, it may be necessary to adjust the exciting light intensity, exposure time and camera gain.

19. When using any imaging system it is important to keep settings constant so that the fluorescence intensities are comparable between samples.
20. In Adobe Photoshop CS2, for example, you can choose Image>Adjust>Levels from the toolbar to see how many pixels there are of each level of brightness. Drag the black and white markers inward to the edges of the charted pixels to change the visible pixels. Check the preview and click *OK* once the cells are visible (Fig. 1). It is important to perform the same image adjustments for all samples in order to ensure that results can be compared.
21. We and others (8) found that the prolonged time of incubation increases the efficacy of protein precipitation.
22. Keep acetone at –20°C.
23. You can carefully decant the supernatant or remove it with 1,000 μl pipette tips.
24. Always handle the gel and samples with gloves to avoid contamination with other proteins (e.g., albumins, keratin).
25. The pyrogallol red will dissociate from the proteins and the red dye will migrate within the gel. This step allows for the removal of the pyrogallol red from the proteins.
26. Usually 30 min of staining allows for visualization of the most abundant proteins.
27. Keep the gel in 1% acetic acid at 4°C until further use.
28. Cut the fragment of the gel from each lane (corresponding to each sample) into smaller pieces and each sample's pieces together in a 1.5 ml microcentrifuge tube.
29. A total of 200 μg protein is recommended to analyze each sample in two technical replicas. As there may be some protein loss in subsequent steps it is recommended to proceed with a minimum of 300 μg protein.
30. The centrifuge bottles can be placed on a rotatory shaker as well.
31. Depending on the number of analyzed samples use four plex, six plex, or eight plex iTRAQ reagents. For example, if analyzing technical replicas for both wild type and mutant samples, four plex iTRAQ reagents are required.
32. You can use SCx spin columns (Nest Group, Southborough, MA, USA) instead and then elute peptides using a stepwise gradient.
33. You can use an equivalent brand of nanoflow high performance liquid chromatograph in an on-line or off-line system.
34. The newest version of the ProteinPilot software should always be applied for the analyses.

References

1. Shapiro L, McAdams HH, Losick R (2009) Why and how bacteria localize proteins. Science 326:1225–1228
2. Lybarger SR, Johnson TL, Gray MD, Sikora AE, Sandkvist M (2009) Docking and assembly of the type II secretion complex of *Vibrio cholerae*. J Bacteriol 191:3149–3161
3. Sikora AE, Zielke RA, Lawrence DA, Andrews PC, Sandkvist M (2011) Proteomic analysis of the *Vibrio cholerae* type II secretome reveals new proteins, including three related serine proteases. J Biol Chem 286:16555–16566
4. Carlsohn E, Nystrom J, Karlsson H, Svennerholm AM, Nilsson CL (2006) Characterization of the outer membrane protein profile from disease-related *Helicobacter pylori* isolates by subcellular fractionation and nano-LC FT-ICR MS analysis. J Proteome Res 5:3197–3204
5. Ross PL, Huang YN, Marchese JN, Williamson B, Parker K, Hattan S, Khainovski N, Pillai S, Dey S, Daniels S et al (2004) Multiplexed protein quantitation in *Saccharomyces cerevisiae* using amine-reactive isobaric tagging reagents. Mol Cell Proteomics 3:1154–1169
6. Ow SY, Salim M, Noirel J, Evans C, Rehman I, Wright PC (2009) iTRAQ underestimation in simple and complex mixtures: "the good, the bad and the ugly". J Proteome Res 8:5347–5355
7. Noirel J, Ow SY, Sanguinetti G, Jaramillo A, Wright PC (2008) Automated extraction of meaningful pathways from quantitative proteomics data. Brief Funct Genomic Proteomic 7:136–146
8. Caldwell RB, Lattemann CT (2004) Simple and reliable method to precipitate proteins from bacterial culture supernatant. Appl Environ Microbiol 70:610–612
9. de Barsy M, Jamet A, Filopon D, Nicolas C, Laloux G, Rual JF, Muller A, Twizere JC, Nkengfac B, Vandenhaute J et al (2011) Identification of a *Brucella* spp. secreted effector specifically interacting with human small GTPase Rab2. Cell Microbiol 13:1044–1058
10. Le Blastier S, Hamels A, Cabeen M, Schille L, Tilquin F, Dieu M, Raes M, Matroule JY (2010) Phosphate starvation triggers production and secretion of an extracellular lipoprotein in *Caulobacter crescentus*. PLoS One 5:e14198
11. Feilmeier BJ, Iseminger G, Schroeder D, Webber H, Phillips GJ (2000) Green fluorescent protein functions as a reporter for protein localization in *Escherichia coli*. J Bacteriol 182:4068–4076
12. Aronson DE, Costantini LM, Snapp EL (2011) Superfolder GFP is fluorescent in oxidizing environments when targeted via the Sec translocon. Traffic 12:543–548
13. Dinh T, Bernhardt TG (2011) Using superfolder green fluorescent protein for periplasmic protein localization studies. J Bacteriol 193:4984–4987
14. Schnell U, Dijk F, Sjollema KA, Giepmans BN (2012) Immunolabeling artifacts and the need for live-cell imaging. Nat Methods 9:152–158
15. Griffin BA, Adams SR, Tsien RY (1998) Specific covalent labeling of recombinant protein molecules inside live cells. Science 281:269–272

Chapter 11

Pore Formation by T3SS Translocators: Liposome Leakage Assay

Eric Faudry, Caroline Perdu, and Ina Attrée

Abstract

Gram-negative bacteria utilize a dedicated membrane-embedded apparatus, the type III secretion system (T3SS), to inject proteins into host cells. The passage of the proteins across the target membrane is accomplished by a proteinaceous pore—the translocon—formed within the host-cell cytoplasmic membrane. Translocators bound to their chaperones can be expressed in *Escherichia coli* and subsequently dissociated from the chaperone by guanidine treatment. The pore formation properties of the translocators can then be studied by an in-vitro liposome leakage assay. Sulforhodamine-B is encapsulated within lipid vesicles during liposome preparation. At high concentration, this fluorochrome exhibits self-quenching limiting fluorescence emission. Upon pore formation, liposome leakage leads to the dilution of Sulforhodamine-B in the medium and fluorescence emission increases. Alternatively, fluorochromes coupled to large dextran molecules can be encapsulated in order to estimate pore dimensions. Here we describe protein expression and purification, dye-liposome preparation, and leakage assay conditions.

Key words: Protein expression and purification, Pore, Liposome, Lipid vesicle, LUV, Sulforhodamine-B, Fluorescence

1. Introduction

Pore-forming toxins are a widespread feature of bacterial poisoning, exhibiting a variety of modes of action with some common characteristics enabling their classification in different families (1, 2). Among them, the translocators of the Type III Secretion System (T3SS) display some unique traits: they are bound to a chaperone in the bacterial cytoplasm, from which they are separated for their export through the secretion needle, and then they form a pore in the host-cell membrane (3). In addition, this pore is thought to remain linked to the secretion needle and allows the passage of effector proteins.

Anne H. Delcour (ed.), *Bacterial Cell Surfaces: Methods and Protocols*, Methods in Molecular Biology, vol. 966, DOI 10.1007/978-1-62703-245-2_11,

Pseudomonas aeruginosa translocators PopB and PopD are present in the bacterial cytosol in complex with their chaperone PcrH, and PopB/PcrH and PopD/PcrH complexes can be expressed in a heterologous system and purified. Upon secretion by the T3SS, the chaperone–translocators complexes are dissociated and the chaperone remains in the bacterial cytosol (4). This dissociation of the complexes can be promoted in vitro with the recombinant proteins by lowering the pH or by addition of chaotropic salts. This last procedure is preferable because it prevents spontaneous aggregation which is a propensity of these intrinsically partially folded proteins (5–9). We describe here the conditions for heterologous expression as well as the purification and dissociation of the complex. An expression vector was engineered in order to add a hexa-His tag to the chaperone, allowing purification of the complex under native conditions and the separation from the untagged translocators under denaturing conditions (Fig. 1). The prepared translocators then display reproducible and consistent pore-forming properties.

In the study of protein activity directed towards membranes, liposomes (reconstituted lipid vesicles) are an invaluable tool for the deciphering of membrane binding, oligomerization within the membrane and pore formation (10). Liposomes can be composed of multiple concentric lipid bilayers (Multilamellar Lipid Vesicle or MLV) or a single membrane (unilamellar vesicles). Such vesicles can be classified according to their dimensions into Giant, Large or Small Unilamellar Vesicles (GUV, LUV, and SUV; respectively) (11). The size of the vesicles determines membrane curvature but also relates to vesicle stability, with LUVs of diameters ranging from 80 to 200 nm displaying optimum stability.

Upon dispersion in an aqueous solution, phospholipids spontaneously assemble into vesicles. Different techniques may be employed to achieve vesicle formation in buffer: sonication, detergent removal, hydration at appropriate temperatures combined or not with freeze–thaw cycles, and reverse phase evaporation. Sonication directly leads to SUVs formation whereas the other techniques generally give MLVs that should be subsequently transformed into LUVs by extrusion through filters of controlled porosity. On the other hand, GUVs are generally obtained by electroformation or blister formation (12, 13).

Here we describe a technique for obtaining LUVs by reverse phase evaporation and their utilization to evaluate pore forming activity of T3SS translocators. During vesicle formation, the fluorochrome Sulforhodamine-B is entrapped at high concentration. At this concentration, fluorescence emission is limited by self quenching. When a pore is formed in the vesicle membrane, dye release and dilution in the external buffer leads to an increase in fluorescence that can be monitored by fluorescence spectroscopy.

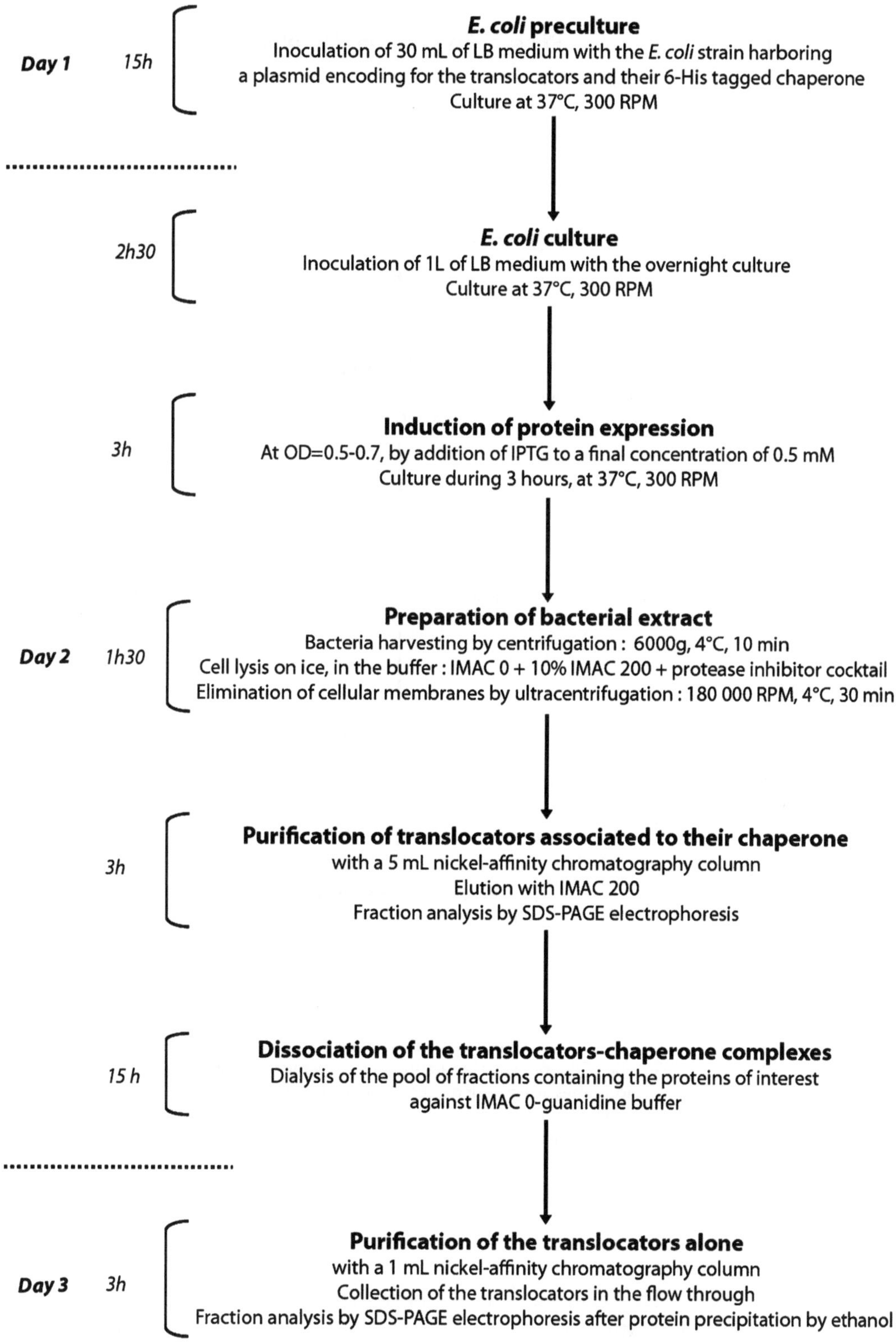

Fig. 1. Diagram of the expression and purification of translocators.

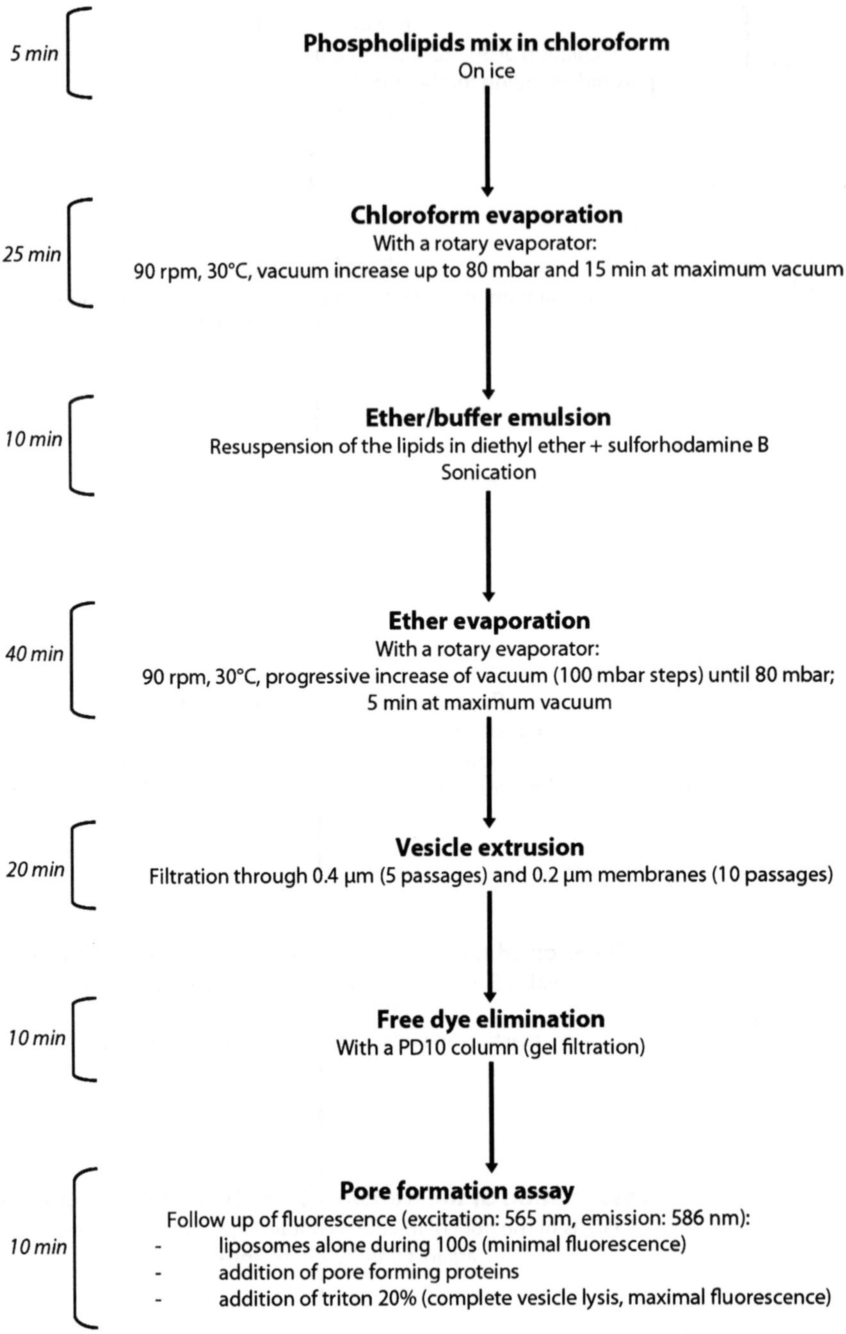

Fig. 2. Diagram of the vesicle preparation and pore formation assay.

The main steps are: evaporation of the chloroform from the lipid mixture; formation of an emulsion with diethyl ether, lipids and buffer; elimination of diethyl ether by controlled evaporation; vesicle extrusion; free dye elimination and dye release from the vesicles by fluorescence monitoring (Fig. 2).

Different fluorochromes can be encapsulated within the LUVs. Sulforhodamine-B displays a high quantum yield allowing high sensitivity and fluorescence properties fairly insensitive to pH and buffer composition. Thus, this assay can be successfully used for the study of pore forming proteins in general and T3SS in particular (6, 14, 15).

2. Materials

2.1. Translocator Expression and Purification

1. *E. coli* expression strain harboring a plasmid designed for the co-expression of a T3SS translocator and its chaperone. The chaperone coding sequence must be fused to a hexa-His tag (see Note 1).
2. LB medium: 10 g/l Tryptone, 5 g/l Yeast Extract, 10 g/l NaCl. For culture, use 30 ml in a 100-ml Erlenmeyer flask, or 1 l in a 3-l Erlenmeyer flask, or 2 × 500 ml in two 2-l Erlenmeyer flasks.
3. Orbital shaker.
4. 100 mM IPTG (isopropyl-beta-d-thiogalactopyranoside): add 119 mg to 5 ml of deionized water. Sterilize by filtration through a 0.22 μm membrane.
5. Kanamycin.
6. Spectrophotometer to measure optical density at 600 nm.
7. Ultracentrifuge and associated rotors.
8. IMAC-0 buffer: 25 mM HEPES, pH 8.0, 500 mM NaCl. Make 250 ml and filter through a 0.22 μm membrane.
9. IMAC-200 buffer: IMAC-0 supplemented with 200 mM Imidazole. Make 200 ml and filter through a 0.22 μm membrane.
10. Protease inhibitor cocktail without chelating agent.
11. 50 mM $NiSO_4$ solution. Make 70 ml.
12. IMAC-0-guanidine buffer: 25 mM HEPES, pH 8.0, 6 M guanidine hydrochloride. Make 250 ml and filter through a 0.22 μm membrane.
13. IMAC-200-guanidine buffer: IMAC-0-guanidine supplemented with 200 mM Imidazole. Make 200 ml and filter through a 0.22 μm membrane.
14. 5 and 1-ml metal affinity chromatography columns.
15. Low pressure liquid chromatography apparatus.
16. Dialysis bag with 10–12 kDa cut-off.
17. French Press, sonicator, or another cell-disrupting apparatus.
18. Ethanol.

2.2. Vesicle Preparation and Extrusion

The volumes and quantities are given to produce 3 ml of vesicles at a global lipid concentration of 20 mM.

1. Standard rotary evaporator with vacuum and bath temperature control.
2. 50 ml spherical evaporation flask.
3. Phospholipids in chloroform (10–40 mg/ml), stored under argon or nitrogen at −20°C. The average molecular weights of natural lipids are indicated by the manufacturer.
4. Chloroform (analytical grade).
5. Diethyl ether (analytical grade).
6. Tris solution: 25 mM Tris–HCl pH 8.0. Make 50 ml.
7. Sulforhodamine-B (see Note 2) solution: 25 mM Tris–HCl pH 8.0, 50 mM Sulforhodamine-B. Make 3 ml.
8. Sonicator.
9. 5-ml hemolysis tube or equivalent.
10. 5-ml standard plastic syringe.
11. Steel Syringe Type Holder.
12. 0.4 and 0.2 μm Nucleopore Polycarbonate membranes (13 mm diameter) (see Note 3).
13. Desalting gravity column.

2.3. Dye Leakage Fluorescence Assay

1. Fluorimeter.
2. Quartz cell for fluorescence measurement (1-cm path length) with 3–4 mm magnetic bar for stirring.
3. 20% Triton-X 100 solution in water. Make 1 ml.

3. Methods

3.1. Translocator Expression and Purification

The conditions for expression are described for *P. aeruginosa* T3SS translocators using the BL21(DE3)Star *E. coli* strain and expression plasmids based on pET-30b (Kan^r). Conditions need to be adapted to each specific translocator (see Note 1).

1. At the end of the afternoon of Day 1, inoculate 30 ml of LB medium supplied with kanamycin (50 μg/l) with one colony of *E. coli* expression strain harboring the bi-cistronic plasmid. Grow overnight at 37°C under orbital shaking (300 rpm).
2. Early on Day 2, inoculate 1 l of LB medium supplied with kanamycin (50 μg/l) with the overnight culture (use the whole 30 ml of the overnight culture with an OD usually above 2.5). Incubate at 37°C under orbital shaking (300 rpm).

3. When the OD measured at 600 nm is between 0.5 and 0.7, add IPTG to the culture at a final concentration of 0.5 mM and further incubate for 3 h.
4. Harvest the cells by centrifugation at 6,000 × *g*, 4°C, for 10 min and resuspend them on ice in 30 ml of IMAC-0 supplemented with 10% of IMAC-200 (final concentration of imidazole: 20 mM) and the protease inhibitor cocktail at the concentration recommended by the manufacturer.
5. Disrupt the cells using a French Press at 13,000 psi, three passages (see Note 4). The suspension has to be kept as cold as possible and completion of disruption can be assessed by analyzing the OD at 600 nm that should decrease and reach a plateau.
6. Clear the suspension by ultracentrifugation at 180,000 × *g*, 4°C, for 30 min.
7. During ultracentrifugation, equilibrate a 5-ml metal affinity chromatography column by successively applying 50 ml of H_2O, 50 ml of 50 mM $NiSO_4$, and 50 ml of IMAC-0 supplemented with 10% of IMAC-200. Use a flow rate of 5 ml/min.
8. Load the clarified-supernatant on the equilibrated column at 4 ml/min. Load again the flow-thru solution to maximize binding of the proteins of interest.
9. Wash the column with 10 ml of IMAC-0 supplemented with 25% of IMAC-200 to elute nonspecifically bound proteins.
10. Elute the translocators complexed to their 6His-tagged chaperone with 40 ml of IMAC-200. Collect 2-ml fractions.
11. Analyze the fractions by standard SDS-PAGE and pool the fractions with the highest concentration and purity. Make pools of 4–6 ml.
12. Dialyze the pooled-fractions overnight at 4°C against IMAC-0-guanidine buffer. The volume of dialysis should be nine times the total volume of the fractions in order to reach a final concentration at the end of dialysis of 25 mM HEPES, 50 mM NaCl, 5.4 M Guanidine, 20 mM Imidazole (see Note 5).
13. On Day 3 equilibrate the 1-ml column as in step 7, except that IMAC-guanidine buffers should be used instead of IMAC buffers with volumes of 10 ml instead of 50 ml and a flow rate of 1 ml/min.
14. Subsequently load the dialyzed fractions at 0.5 ml/min in a closed-loop configuration, by connecting the column outlet to the sample-containing tube.
15. Disconnect the closed loop and collect the flow-thru by pumping IMAC-0-guanidine supplemented with 10% of IMAC-200-guanidine. The flow-thru sample should now contain the free translocator, while the 6His-tagged chaperone remains trapped in the column.

16. Fractions can be analyzed by SDS-PAGE after ethanol-precipitation (see Note 6).
17. Determine protein concentration by absorbance measurement at 280 nm using calculated extinction coefficients.
18. Samples should be stored at 4°C and used within 1 month. A 50-fold dilution is sufficient to provide spontaneous refolding of the T3SS translocators and membrane permeabilizing activity.

3.2. Vesicle Preparation and Extrusion

The liposome preparation is described for vesicles made of egg-phosphatidylcholine and soy-phosphatidylserine (8/2; mol/mol), but other lipid composition can be adapted with various lipids (phospholipids, cholesterol, sphingomyelin…) and molar ratios.

1. In a 50 ml evaporation flask kept on ice, introduce 3.70 ml of 10 mg/ml phosphatidylcholine in chloroform and 0.96 ml of 10 mg/ml phosphatidylserine in chloroform.
2. Set the water bath temperature of the rotary evaporator to 30°C.
3. Connect the evaporation flask to the rotary evaporator, and start rotation at 90 rpm. Gradually increase the vacuum to maximal setting, avoiding solvent boiling.
4. Maintain evaporation for 15 min. After evaporation, a white film of lipids should be present on the flask wall and no chloroform odor should be noticed.
5. With the flask on ice, introduce 1.5 ml of diethyl ether in the evaporation flask to dissolve the lipids.
6. Subsequently introduce 1.5 ml of the Sulforhodamine-B solution (see Note 7). Vortex the flask energetically for 10 s.
7. Using a 3-mm probe, sonicate the biphasique mixture inside the evaporation flask in order to produce a homogeneous emulsion (see Note 8). The power and duration of sonication may vary according to the apparatus (see Note 9).
8. Immediately connect the evaporation flask to the rotary evaporator with the water bath temperature set at 30°C.
9. Start the rotation at 90 rpm and subsequently set the vacuum to 900 mbar. Gradually and slowly increase the vacuum in a step-wise fashion. Typically, the pressure is reduced by 100 mbar every 3 min until it reaches the minimum of 80 mbar (see Note 10).
10. Maintain maximal vacuum for 5 min. A dark (due to the Sulforhodamine-B color), fairly viscous and homogeneous solution should be obtained.
11. Transfer the vesicle solution to a 5-ml plastic hemolysis tube and wash the wall of the evaporation flask with 1.5 ml of

Sulforhodamine-B solution. Mix the two solutions in the hemolysis tube.

12. Insert a 0.4 μm Nucleopore Polycarbonate membrane in a steel Syringe Type Holder with the shiny side of the membrane oriented towards the inlet of the holder. Tightly close the assembly.
13. Push the vesicle suspension through the membrane with a 5-ml plastic syringe and subsequently push air to collect the solution remaining inside the filter holder. Fill the syringe with the filtered suspension and repeat this step for a total of five passages (see Note 11).
14. Repeat steps 12 and 13 two times with new 0.2 μm filters. Passages through the filter should become easier. The extrusion with a total of 15 passages through the filters results in monodispersed Large Unilamellar Vesicles (LUVs) of the expected size.
15. Equilibrate the desalting column (see Note 12) with 25 ml of Tris solution, according to manufacturer's instructions.
16. Load 1 ml of LUV suspension on the column and elute with Tris solution (see Note 13).
17. The final LUV suspension should be stored at 4°C, protected from light, and used within 5 days. We typically obtain LUVs with a radius of 90 ± 2 nm and a polydispersity of 7%, as determined by Dynamic Light Scattering.

3.3. Dye Leakage Fluorescence Assay

1. The fluorimeter should be used in kinetics mode, with excitation at 565 nm and emission at 586 nm. Homogeneous magnetic stirring should be provided.
2. Introduce in the quartz cell the buffer of choice (see Note 14) and the vesicle preparation at a final concentration of lipids of 10 μM in 2 ml.
3. Start fluorescence recording for baseline measurement.
4. After 100 s introduce the purified pore-forming proteins (see Note 15) without stopping the fluorescence recording. The proteins could be introduced either as refolded proteins by prior dilution in a buffer devoid of guanidine, or as unfolded proteins in guanidine directly introduced as a small volume in the vesicle solution, provided that the final concentration of guanidine falls below 100 mM.
5. At the end of the recording, add 40 μl of the 20% Triton-X 100 solution in order to measure the fluorescence maximum, due to full leakage of the encapsulated dye. This value reflects the exact quantity of vesicle used in the assay. A typical recording is shown in Fig. 3.

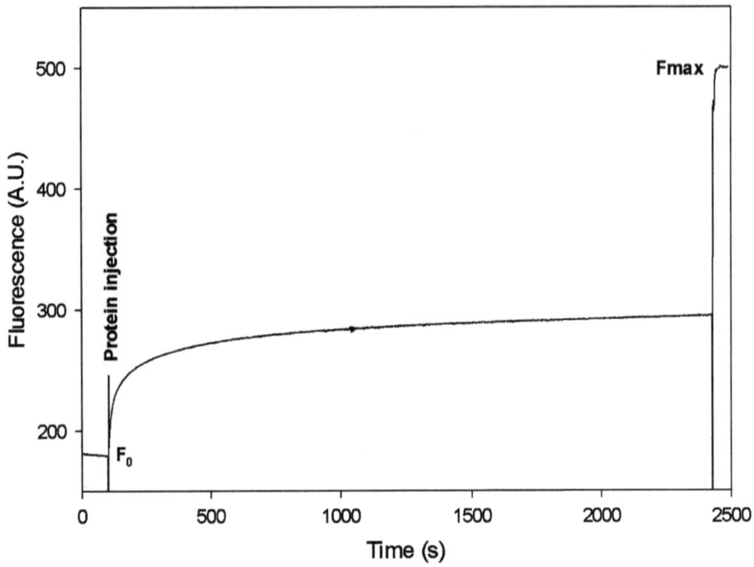

Fig. 3. Typical recording from a pore formation assay. Dye efflux was monitored by fluorescence measurement after the addition (at 100 s) of a 10 nM equimolar mixture of PopBD to a 2 ml suspension of 10 μM LUV. At the end of the kinetics recording, Triton-X 100 was introduced in the solution at a 0.4% final concentration to induce maximum dye release. The fluorimeter settings were: excitation at 565 nm, emission at 586 nm with slits of 3 nm and 5 nm, respectively.

6. The baseline and maximum fluorescence measurements allow fluorescence normalization using the following equation:

$$F(t)_{\text{norm}} = \left(F(t) - F_0\right) / \left(F_{\text{max}} - F_0\right).$$

   Where F_0 is the baseline fluorescence level before protein addition and F_{max} is the level after addition of 20% Triton-X 100 at the end of each assay.

7. For quantitative comparisons of kinetics measurements, end-plateau levels, initial rates or other kinetics parameters can be considered depending on the properties of the protein of interest.

4. Notes

1. *E. coli* strain, plasmid construction and expression conditions should be tested and optimized for each translocator–chaperone pair. The protocol gives general guidelines based on our experience with *P. aeruginosa* T3SS translocators (5, 6, 8, 9) and pET-30b.
2. For estimation of pore size, dextran-FITC of different hydrodynamic radii can be entrapped within the vesicles (6). In this case, the dye concentration should be adapted.

3. Membranes with other pore sizes can be used, depending on the desired diameter of the LUVs to be prepared.
4. A sonicator or another cell-disrupting device can also be used. Conditions used to break the bacteria depend on each device.
5. For *P. aeruginosa* translocators, a final concentration after dialysis of 5.4 M Guanidine was shown to be sufficient to dissociate the translocator–chaperone complex. The final 20 mM Imidazole prevents nonspecific retention of the translocator on the chromatography column, while enabling the retention of the 6His-tagged chaperone.
6. Precipitation of concentrated proteins can be carried out as follows: mix 50 μl of sample with 450 μl of cold absolute ethanol. Incubate overnight (or at least 1 h) at −20°C and centrifuge for 20 min at 14,000 × *g*, 4°C. Wash the pellet with 500 μl Ethanol 90% and centrifuge again in the same conditions. Air-dry the pellets and dissolve the proteins in 50 μl Laemmli sample buffer 1×.
7. The dark coloration due to Sulforhodamine-B precludes the direct observation of the solution turbidity and homogeneity which is an indication of the presence of lipid suspension and vesicle formation. Thus, one could omit Sulforhodamine-B in a first try of the technique in order to facilitate the monitoring of each step.
8. During sonication, the flask must be kept on ice and the probe must not touch the glass wall to avoid flask rupture or uneven sonication.
9. For example, we use a 550 W sonicator XL (Misonix) set at 30% power in three pulses of 15 s separated by 5 s on ice.
10. This is the critical point of the procedure: vacuum and temperature should be set in order to avoid any boiling of the solution in the evaporation flask. Boiling would prevent homogeneous evaporation and vesicle formation. The first steps of ether evaporation (i.e., until the pressure decreases to 600 mbar) can be performed without immersion of the flask in the water bath. The evaporation flask can be put at 30°C at the end of a step of increased vacuum. The viscosity of the solution first increases and then decreases. Every step of increased vacuum should be maintained until no modification of the aspect of the solution is observed. A very viscous solution with apparent heterogeneity can be observed at one step with some lipid mixture but should revert to a homogeneous turbid solution at higher vacuum settings.
11. The first passages thru the filter usually require a strong pressure on the syringe. Be careful to firmly tighten the filter holder and the syringe. The filter holder could be set on top of the hemolysis tube to push vertically on the syringe. Alternatively, a

dedicated extrusion device from Avanti Polar could be used. The filter can get plugged after the first passage: in this case, use a new filter. Moreover, gloves are recommended for this filtration step as small amounts of colored solution could leak between the filter holder and the syringe.

12. When entrapping dextran-FITC of large size (superior to 20 kDa), a gel filtration column with higher exclusion limit should be used. Otherwise, free dextran-FITC would elute in the void volume along with the vesicles.
13. Because of their large size, LUVs elute prior to free Sulforhodamine-B. Only the concentrated (darker) fractions are collected in a 2-ml microfuge tube while the lighter fractions at the beginning and the end of the peak are discarded. Dilution of the vesicles during this step should be taken into consideration in order to estimate the final concentration. Alternatively, lipid concentration can be determined with a phosphorus assay.
14. Buffer composition can be adjusted to test the effects of pH, salts or other additives. Sulforhodamine-B fluorescence is usually not influenced by buffer composition, in particular by changes in pH.
15. Protein concentrations may vary from 1 to 100 nM according to the protein binding affinity for vesicle and pore activity. Lipid–protein ratio should be high enough to ensure that most of the protein binds to the vesicles. Otherwise, liposome leakage would reflect pore activity as well as partition between buffer and vesicle. Binding affinity should be estimated by titration using ultracentrifugation or other binding assay.

Acknowledgments

The work in our lab is supported in part by CEA-DSV programs and the French cystic fibrosis association "Vaincre la Mucoviscidose." The authors are grateful to Michel Ragno for optimization of expression and purification procedures, and to Vincent Forge and Grégory Vernier for helpful discussions and technical help to set up the vesicle protocol.

References

1. Parker MW, Feil SC (2005) Pore-forming protein toxins: from structure to function. Prog Biophys Mol Biol 88:91–142
2. Tilley SJ, Saibil HR (2006) The mechanism of pore formation by bacterial toxins. Curr Opin Struct Biol 16:230–236
3. Mueller CA, Broz P, Cornelis GR (2008) The type III secretion system tip complex and translocon. Mol Microbiol 68:1085–1095
4. Cornelis GR (2006) The type III secretion injectisome. Nat Rev Microbiol 4: 811–825

5. Faudry E, Job V, Dessen A et al (2007) Type III secretion system translocator has a molten globule conformation both in its free and chaperone-bound forms. FEBS J 274:3601–3610
6. Faudry E, Vernier G, Neumann E et al (2006) Synergistic pore formation by type III toxin translocators of Pseudomonas aeruginosa. Biochemistry 45:8117–8123
7. Romano FB, Rossi KC, Savva CG et al (2011) Efficient isolation of Pseudomonas aeruginosa type III secretion translocators and assembly of heteromeric transmembrane pores in model membranes. Biochemistry 50:7117–7131
8. Schoehn G, Di Guilmi AM, Lemaire D et al (2003) Oligomerization of type III secretion proteins PopB and PopD precedes pore formation in Pseudomonas. EMBO J 22:4957–4967
9. Wager B, Faudry E, Attree I, et al. (2012) Pore properties of the type III secretion system translocon of Pseudomonas aeruginosa, Submitted.
10. Dalla Serra M, Menestrina G (2003) Liposomes in the study of pore-forming toxins. Methods Enzymol 372:99–124
11. Lasic DD (1988) The mechanism of vesicle formation. Biochem J 256:1–11
12. Delcour AH, Martinac B, Adler J et al (1989) Modified reconstitution method used in patch-clamp studies of Escherichia coli ion channels. Biophys J 56:631–636
13. Meleard P, Bagatolli LA, Pott T (2009) Giant unilamellar vesicle electroformation from lipid mixtures to native membranes under physiological conditions. Methods Enzymol 465:161–176
14. Agafonov A, Gritsenko E, Belosludtsev K et al (2003) A permeability transition in liposomes induced by the formation of Ca2+/palmitic acid complexes. Biochim Biophys Acta 1609:153–160
15. Galloux M, Vitrac H, Montagner C et al (2008) Membrane Interaction of botulinum neurotoxin A translocation (T) domain. The belt region is a regulatory loop for membrane interaction. J Biol Chem 283:27668–27676

Chapter 12

Isolation of Bacterial Type IV Machine Subassemblies

Mayukh K. Sarkar, Seyyed I. Husnain, Simon J. Jakubowski, and Peter J. Christie

Abstract

The bacterial type IV secretion systems (T4SSs) deliver DNA and protein substrates to bacterial and eukaryotic target cells generally by a mechanism requiring direct contact between donor and target cells. Recent advances in defining the architectures of T4SSs have been made through isolation of machine subassemblies for further biochemical and ultrastructural analysis. Here, we describe a protocol for isolation and characterization of VirB protein complexes from the paradigmatic VirB/VirD4 T4SS of *Agrobacterium tumefaciens*. This protocol can be adapted for isolation of T4SS subassemblies from other gram-negative bacteria as well as gram-positive bacteria. The biological importance of isolated T4SS subcomplexes can be assessed by assaying for copurification of trapped or cross-linked substrates. This can be achieved with a modified form of the chromatin immunoprecipitation (ChIP) assay termed transfer DNA immunoprecipitation (TrIP). Here, a TrIP protocol is described for recovery of formaldehyde-cross-linked DNA substrate–channel subunit complexes from cells employing T4SSs for conjugative DNA transfer.

Key words: Type IV secretion, Core complex, Affinity chromatography, DNA conjugation, Protein translocation, Transport, VirB/VirD4, TrIP, ChIP

1. Introduction

The Type IV Secretion Systems (T4SSs) are found in many species of gram-negative and -positive bacteria, as well as some species of Archaea (1). The largest and most widely distributed T4SS subfamily are the conjugation machines; these machines mediate the formation of direct cell-to-cell contact and then transfer of mobile DNA elements from bacterial donor to recipient cells. A second T4SS subfamily, termed the effector translocators, deliver protein effectors or other macromolecules from bacterial pathogens to eukaryotic cells during infection. A third, small subfamily of

Mayukh K. Sarkar and Seyyed I. Husnain contributed equally to this work.

Anne H. Delcour (ed.), *Bacterial Cell Surfaces: Methods and Protocols*, Methods in Molecular Biology, vol. 966, DOI 10.1007/978-1-62703-245-2_12, © Springer Science+Business Media New York 2013

T4SSs promote transfer of DNA across the gram-negative cell envelope independently of any target cell contact; specifically, a *Neisseria gonorrheae* T4SS delivers fragments of chromosomal DNA to the extracellular milieu and a *Helicobacter pylori* T4SS coordinates with competence proteins to import DNA from the milieu.

Although the bacterial conjugation systems have been subjects of intensive study for many years, isolation of the translocation channels for biochemical and structural analysis has proven challenging. Recent advances in biochemical fractionation have enabled progress in development of structural models of T4SS channels along with correlative functional analyses of structural motifs. Biochemical fractionation and mutational studies have identified T4SS subunits or domains required for elaboration of stable machine subassemblies or extracellular pili (2–13). T4SS machine subunits or soluble domains also have been purified and crystallized. Using a nomenclature devised for the paradigmatic *Agrobacterium tumefaciens* VirB/VirD4 T4SS, X-ray structures have been solved for homologs of two hexameric ATPases termed VirD4 and VirB11 (14–16). Other solved structures exist for a pilus-associated subunit VirB5 (17), a complex of two outer membrane-associated subunits, lipoprotein VirB7 and VirB9 (18) and periplasmic domains of VirB8 and VirB10 (19, 20). Recently, a large subassembly termed a core complex consisting of homologs of VirB7, VirB9, and VirB10 subunits was isolated and structurally analyzed by cryoelectron microscopy and X-ray crystallography (21, 22).

In parallel with biochemical enrichment of T4SS machine subassemblies, an in vivo formaldehyde (FA) cross-linking procedure adapted from the chromatin immunoprecipitation (ChIP) assay was developed for identifying close interactions between translocating substrate DNA and subunits of the T4SS channel (23). Termed transfer DNA immunoprecipitation or TrIP, this assay supplied evidence that the DNA forms close interactions with the *A. tumefaciens* VirD4 and VirB11 ATPases located predominantly at the cytoplasmic entrance to the secretion channel, polytopic VirB6 and bitopic VirB8 at the inner membrane, and VirB2 pilin and VirB9 in the periplasm and outer membrane (23–25). Together with the recent structural data, results of the TrIP studies present a view of how DNA substrates pass through the cell-envelope-spanning T4SS channel (26).

Central to the developing understanding of T4SS architecture and function is the isolation of machine subassemblies without and with trapped substrate. The following protocols are described for (1) use of affinity tags for isolation and enrichment of VirD4/VirB T4SS subassemblies from *A. tumefaciens*, and (2) FA-cross-linking of DNA substrates for recovery of DNA–channel subunit complexes with the TrIP assay.

2. Materials

Prepare all solutions using nanopure water (which is done by purifying deionized water to attain a sensitivity of 18 MΩ cm at 25°C). Prepare and store all reagents at room temperature (unless otherwise indicated). Waste disposal regulations should be carefully followed when disposing of waste materials.

2.1. Induction of A. tumefaciens vir Genes

1. Incubator shaker.
2. Spectrophotometer.
3. MG/L agar: Agar (15 g) in 1 l of MG/L medium. Autoclave and store at room temperature.
4. MG/L medium: Dissolve 30 g Tryptone, 30 g NaCl, 15 g Yeast extract, 30 g Mannitol, 7 g Sodium glutamate, 1.5 g K_2HPO_4, 0.6 g $MgSO_4{\cdot}7H_2O$, and 12 μg Biotin in 6 l of water. Adjust the pH to 7.2 with 1 M NaOH. Autoclave and store at room temperature.
5. *Agrobacterium* AB buffer: Dissolve 4 g MES (2-(*N*-morpholino) ethanesulfonic acid) and 10 g Glucose 10 g in 900 ml of water, adjust the pH to 5.5 with 12 N HCl, and bring to 1 l. Autoclave and store at room temperature.
6. Phosphate buffer: Dissolve 60 g K_2HPO_4 and 20 g Na_2HPO_4 in 900 ml of water. Adjust the pH to 7.0 and bring to 1 l. Autoclave and store at room temperature.
7. 20× AB salts: Dissolve 20 g NH_4Cl, 6 g $MgSO_4{\cdot}7H_2O$, 3 g KCl, 0.2 g $CaCl_2$, and 50 mg $FeSO_4{\cdot}7H_2O$ in 1 l of water. Autoclave and store at room temperature.
8. 2% Yeast extract: Dissolve 2 g of yeast extract in 100 ml of water and autoclave.
9. 100 mM Acetosyringone (4′-Hydroxy-3′,5′-dimethoxyacetophenone): Dissolve 0.196 g of acetosyringone in 10 ml of Dimethyl sulfoxide (DMSO). Store in the dark at −20°C.
10. ABIM (AB induction media): Add 5 ml 20× AB salts, 200 μl Phosphate buffer, 1 ml of 2% yeast extract, and 200 μl of 100 mM acetosyringone per 100 ml of *Agrobacterium* AB buffer.
11. Antibiotics for plasmid maintenance.

2.2. Cell Lysis and Membrane Solubilization

1. Centrifuge and associated rotor.
2. Ultracentrifuge and associated rotor.
3. Cell Lysis Buffer (500 ml): 50 mM Tris–HCl pH 7.4, 150 mM NaCl, 1% Triton-X.

4. Lysate Buffer: 10 mM Tris–Cl (pH 8.0), 0.5 M NaCl, 5 mM EDTA, 0.5% LDAO.
5. Lysozyme solution: Add 300 mg of lysozyme to 10 ml of water and filter-sterilize.
6. DNase I solution: Add 100 mg DNase I to 10 ml of water and filter-sterilize.
7. 5 M NaCl: make 500 ml. Autoclave.
8. 1 M $MgCl_2$: make 500 ml. Autoclave.
9. EDTA-free protease inhibitor tablet.
10. 10× Cell Lytic B detergent mix.
11. 10% DDM (*n*-Dodecyl-β-D-Maltopyranoside) in water.
12. 10% LDAO (*n*-Dodecyl-*N*,*N*-Dimethylamine-*N*-Oxide) in water.

2.3. Isolation of Subcomplexes by Affinity Pull-Down

1. Tube rotator.
2. Eppendorf tubes.
3. Strep-Tactin Sepharose: immobilized streptavidin resin used for purification of Strep-tag II fusion proteins. Store at 4°C.
4. Anti-FLAG M2 Agarose beads: Anti-FLAG resin used for purification of FLAG tagged fusion proteins. Store at –20°C.
5. D-Desthiobiotin: 50 mM stock solution in NH_4OH. Store at 4°C. This is used for elution of Strep-tag II fusion proteins.
6. Lysate buffer: see item 4 in Subheading 2.2.
7. Elution buffer 1: 2.5 mM desthiobiotin in lysate buffer.
8. 2× SDS Sample loading buffer (Laemmli's buffer): dissolve 20 mg bromophenol blue in 32.5 ml of water, 20 ml of glycerol, 5 ml of β-mercaptoethanol, 30 ml of 10% SDS, and 12.5 ml of 4× Resolving Gel buffer, and bring the volume to 100 ml with water. The sample loading buffer is aliquoted in eppendorf tubes and stored at –20°C.
9. Tris buffered saline (TBS; 10×): 1.5 M NaCl, 0.1 M Tris–HCl, pH 7.4.
10. 3× FLAG peptide: 5 μg/μl in TBS. This is used for elution of FLAG tagged fusion proteins.
11. Elution buffer 2: dilute threefold the 3× FLAG Peptide into lysate buffer.
12. Microcentrifuge.
13. Syringe and needle.

2.4. Further Enrichment by CsCl Density Gradient Centrifugation

1. 1 CsCl: 76% (w/v) dissolved in water.
2. Lysate buffer: see item 4 in Subheading 2.2.
3. Beckman ultracentrifuge.
4. Beckman SW55 Ti. Swinging bucket rotor.

2.5. SDS-PAGE and Staining

2.5.1. SDS-PAGE

1. Acrylamide/bisacrylamide (30%/0.8%): 30% (w/v) acrylamide, 0.8% (w/v) bis(*N*,*N'*-methylene) acrylamide. Store at 4°C (see Note 1).
2. SDS: 10% SDS (w/v).
3. Resolving Gel Buffer (4×): 1.5 M Tris–HCl, pH 8.8, filter and store at room temperature.
4. Stacking Gel Buffer (4×): 0.5 M Tris–HCl, pH 6.8, filter and store at room temperature.
5. Ammonium persulfate (AP): Prepare a 10% (w/v) solution in water before use.
6. *N*,*N*,*N'*,*N'*-tetramethyl-ethylenediamine (TEMED). Store at 4°C.
7. 10× Tris–Glycine: 30 g Tris base, 144 g Glycine in 1 l of water.
8. SDS-PAGE running buffer: 200 ml of 10× Tris–glycine, 20 ml of 10% SDS, 1.78 l water. Store at room temperature.
9. Separating gel: prepare 40 ml (for four gels) of 12.5% polyacrylamide separating gel by mixing 12.7 ml of water, 10.0 ml of 4× resolving gel buffer, 16.7 ml of acrylamide/bisacrylamide (30%/0.8%), 0.6 ml of 10% AP, and 20 μl of TEMED.
10. Stacking gel: prepare 12.75 ml (for four gels) of stacking gel by mixing 2.5 ml of water, 6.35 ml of water, 2.50 ml of stacking gel buffer, 1.25 ml of acrylamide/bisacrylamide (30/0.8%), 150 μl of 10% AP, and 10 μl of TEMED.
11. Prestained molecular weight marker.
12. SDS-PAGE apparatus.

2.5.2. Coomassie Staining

1. Coomassie stain solution: dissolve 100 mg of Coomassie Brilliant Blue R250 in 50 ml methanol, 10 ml glacial acetic acid, and water to a final volume of 100 ml.
2. Destaining solution: 45 ml ethanol, 10 ml glacial acetic acid, 45 ml of water.
3. Platform shaker.

2.5.3. Western Blotting and Immunostaining

1. Nitrocellulose membranes.
2. Western blot transfer buffer: 100 ml of 10× Tris–Glycine (see item 7 Subheading 2.5.1), 200 ml of methanol, 700 ml of water.
3. Methanol.
4. Aqueous transfer apparatus.
5. Tris buffered saline (TBS; 10×): 1.5 M NaCl, 0.1 M Tris–HCl, pH 7.4.
6. TTBS: 0.05% Tween-20 in 1× TBS.

7. Blocking solution: 5% Nonfat milk in 1× TBS. Store at 4°C.
8. Diluent solution: 5% Nonfat milk in TBST. Store at 4°C.
9. Whatman blotting paper.
10. Carbonate buffer: 8.40 g $NaHCO_3$ and 0.203 g $MgCl_2$ dissolved in 900 ml of water. Adjust the pH to 9.81 with 1 M NaOH and bring the volume to 1 l with water.
11. NBT solution: Dissolve 30 mg of Nitro-blue tetrazolium (NBT) in 1 ml of 70% dimethylformamide (DMF).
12. BCIP solution: Dissolve 15 mg of 5-bromo-4-chloro-3-indolyl phosphate (BCIP) in 1 ml of 100% DMF.
13. Alkaline phosphatase staining solution: Add 100 μl of NBT solution and 100 μl of BCIP solution to 10 ml of carbonate buffer, just prior to adding to membrane.
14. Platform shaker.
15. Primary antibody.
16. Secondary antibody: conjugated to alkaline phosphatase.

2.6. Transfer Immunoprecipitation (TrIP) Assay

2.6.1. In Vivo Formaldehyde Cross-linking

1. MG/L medium: see item 4 in Subheading 2.1.
2. ABIM: see item 10 in Subheading 2.1.
3. Centrifuge and associated rotor.
4. Spectrophotometer.
5. 20 mM sodium phosphate buffer, pH 6.8.
6. Cross-linking buffer: 0.1% (v/v) formaldehyde (FA) in 20 mM sodium phosphate buffer (pH 6.8). Prepare fresh before each use.
7. Formaldehyde: 37% w/v (see Note 2).
8. Buffer A: 50 mM Tris–HCl (pH 6.8), 2 mM EDTA, 1% β-mercaptoethanol, 1% SDS.
9. Buffer B: 150 mM Tris–HCl (pH 8.0), 0.5 M sucrose, 10 mM EDTA.
10. Lysozyme solution: 1 mg/ml of lysozyme in Buffer B. Prepare fresh before each use.
11. Triton X-100.
12. 1 M Glycine: Dissolve 18.8 g glycine in water (may require gentle heating) and bring up to 250 ml with water.
13. Protease inhibitors cocktail: 5× solution, EDTA-free, in 25 mM $MgCl_2$.
14. Rotating wheel.

2.6.2. Immunoprecipitation

1. Protein A-Sepharose CL4B.
2. Anti-VirB polyclonal antibodies.
3. Centrifuge and associated rotor.

4. Buffer B: see item 9 in Subheading 2.6.1.
5. Buffer B supplemented with 1% Triton X-100.
6. Buffer B supplemented with 0.1% Triton X-100.
7. 10 mM Tris–Cl, pH 6.8.
8. 30% (w/v) Trichloroacetic acid (TCA): store at 4°C.
9. Acetone: Store at −20°C.
10. Speed Vac.
11. 4× Resolving Gel buffer: see item 3 Subheading 2.5.1.
12. 2× SDS Sample loading buffer: see item 8 Subheading 2.3.

2.6.3. PCR Amplification

1. Primers for PCR amplification of the desired transferred DNA (T-DNA) or chromosomal DNA fragments.
2. TAQ Polymerase.
3. dNTPs: 2.5 mM stock.
4. Thermocycler.
5. Agarose.
6. Ethidium bromide.
7. Agarose gel electrophoresis apparatus.

3. Methods

The following protocols have been developed for (1) isolation of subassemblies of the *A. tumefaciens* VirB/VirD4 T4SS and (2) recovery of formaldehyde-cross-linked complexes of translocating T-DNA substrate and channel subunits. The fractionation procedure is adapted from protocols described for isolation of type III secretion needle complexes from *Salmonella typhimurium* by the Galan laboratory (27, 28) and the T4SS "core" complex encoded by the *E. coli* pKM101 conjugation system by the Waksman laboratory (21, 22). By use of this fractionation procedure, we enrich VirB subcomplexes composed minimally of the VirB4, VirB7, VirB9, and VirB10 subunits that assemble as ring-shaped structures when analyzed by negative-stain electron microscopy. These structures closely resemble ring-shaped "core" complexes of the *E. coli* pKM101 conjugation machine; these are composed of TraN, TraO, and TraF, which are homologs of the *A. tumefaciens* VirB7, VirB9, and VirB10 subunits, respectively (21, 22). By use of the FA-cross-linking protocol termed TrIP (*Tr*ansfer DNA *I*mmuno*P*recipitation) (23), we identify close contacts between the DNA substrate and six of the *A. tumefaciens* T4SS subunits - VirD4, VirB11, VirB6, VirB8, VirB2, and VirB9. Our current model is that these subunits assemble as the translocation channel within the ring-shaped VirB7/VirB9/VirB10 "core" complex (26).

3.1. Induction of A. tumefaciens vir Genes

1. Streak *A. tumefaciens* strains from –80°C stocks onto MG/L agar containing appropriate antibiotics for plasmid maintenance. Incubate for 1–2 days at 18°C. Inoculate *A. tumefaciens* strains producing Strept-tag II- or FLAG-tagged VirB proteins of interest from MG/L agar plates into 5 ml of MG/L medium containing appropriate antibiotics.
2. Grow culture overnight with shaking at 28°C, then inoculate culture into 400 ml of MG/L medium and continue incubating with shaking to an OD_{600} of 0.5–0.8.
3. Harvest cells by centrifugation and resuspend in 2 l ABIM to an OD_{600} of 0.1–0.2.
4. Incubate with shaking for 12–14 h at 20°C (see Notes 3 and 4).

3.2. Cell Lysis and Membrane Solubilization

1. Spin cells down and resuspend in 12.5 ml Cell Lysis buffer (see Note 5).
2. Stir slowly using magnetic stirrer at room temperature, and add one complete EDTA-free protease inhibitor tablet, drop by drop of 250 μl of lysozyme solution, and drop by drop of 250 μl of DNAse I solution in that order.
3. Place in an ice water bath and incubate slowly with stirring for 1 h.
4. Add 650 μl of 10× Cell Lytic B detergent mix, 500 μl of 10% LDAO, and 500 μl of 10% DDM and incubate at 37°C for 30 min.
5. Add 150 μl of 1 M $MgCl_2$ slowly with stirring, incubate with stirring for 15 min.
6. Add 1.6 ml of 5 M NaCl slowly under stirring, incubate with stirring for 1 h at 37°C (see Note 6).
7. Pellet debris by centrifugation at 12,000 × *g* at 4°C for 30 min. Repeat if lysate appears cloudy (see Note 7).
8. Recover high molecular weight complexes from clarified lysates by ultracentrifugation at 181,000 × *g* for 2 h at 4°C. (see Note 8).
9. Resuspend pelleted material in 1 ml of Lysate Buffer. Suspension of all material can take several hours. Pool all suspended material (see Note 9).

3.3. Isolation of Subcomplexes by Affinity Pull-Down

We use the FLAG tag (29) or Strep-tag II (engineered from a short sequence of eight amino acids that bind to the biotin binding pocket of Streptavidin) (30) for isolation of VirB subassemblies by affinity chromatography. The tagged proteins are initially assayed for effects on T4SS machine function, as monitored by virulence and pilus production. We use only functional tagged proteins for enrichments of the VirB subassemblies.

1. Perform an affinity pulldown of the tagged subunits/complexes in the following way.
   (a) For affinity pulldown of Strep-tag II-tagged subunits/complexes: Mix 100 μl of Strep-tactin sepharose (50% suspension) with 1 ml of suspended material from step 9 Subheading 3.2 in an eppendorf tube. Incubate overnight at 4°C with gentle tube inversion/mixing. Elute Strep-tag II tagged subunit/complexes from resin by adding 0.8 ml of Elution Buffer 1. Mix by inversion at room temperature for 1 h.
   (b) For affinity pulldown of FLAG-tagged subunits/complexes: Mix 100 μl anti-FLAG M2 agarose beads with 1 ml of the sample from **step 9** Subheading 3.2 in an eppendorf tube. Incubate overnight at 4°C with gentle tube inversion/mixing. Elute the FLAG-tagged fusion proteins from anti-FLAG M2 agarose beads by adding 500 μl of elution buffer 2. Mix by inversion at room temperature for 1 h.
2. Remove Strep-tag II- and FLAG-beads by centrifugation at 16,000 × *g* for 1 min in a microcentrifuge to pellet resin. Collect resin-free supernatant using a syringe needle.
3. At this point, eluted proteins should be analyzed by SDS-PAGE and Coomassie staining, and development of western blots with anti-VirB antibodies (*see* Subheading 3.5).

3.4. Further Enrichment by CsCl Density Gradient Centrifugation

Complexes recovered by affinity chromatography are next subjected to further enrichment by one or two rounds of CsCl density centrifugation. Additional enrichment steps can include centrifugation through sucrose density gradients or gel filtration chromatography; development of further purification steps is empirical and not described in detail here (see Note 10).

1. Material recovered in the eluates from the FLAG or Streptactin resins is added to ultracentrifuge tubes containing 1.63 ml of 76% CsCl. Make sample up to 2.85 ml with Lysate Buffer if there is a shortage of sample volume. Mix by inversion and then centrifuge in a Beckman SW55 Ti swing-bucket rotor at 340,000 × *g* for 12 h at 4°C.
2. Fractionate tubes into ~0.25 ml fractions and analyze by SDS-PAGE and immunostaining (*see* Subheading 3.5 below).
3. To remove the CsCl and concentrate the sample, pool the desired fractions, dilute eight to tenfold with Lysate Buffer, ultracentrifuge at 181,000 × *g* and resuspend in ~200 μl of Lysate buffer.

3.5. SDS-PAGE and Staining

For 12.5% SDS-PAGE gel electrophoresis, carry out the following procedures at room temperature, unless otherwise specified. Five SDS-polyacrylamide gels are prepared by this method. The gels can be stored for several days wrapped in cellophane at 4°C.

3.5.1. SDS-PAGE

1. Cast a separating acrylamide gel within an 8 cm × 10 cm × 1.5 mm gel cassette as specified in item 9 of Subheading 2.5.1. Allow space for the stacking gel and overlay with ethanol or water (see Note 11).
2. Remove the ethanol or water used for overlaying the resolving gel, wick dry with a piece of Whatman paper, being careful not to touch the gel, and then add the stacking gel. Insert a 10-well gel comb immediately without introducing air bubbles (see Note 12).
3. Place the gel cassette in a gel tank and fill it with SDS-PAGE running buffer. Heat the protein samples for 5 min in boiling water. Load about 20 μl of protein sample onto each well. Also load a protein standard in one of the wells (10 μl/well; 2 μg/marker/lane). Electrophorese at 15 mA when the sample is entering the stacking gel, and then continue at 25 mA when the sample enters the resolving gel and then continue at that current until the dye front has reached the bottom of the gel.
4. After electrophoresis, pry the gel plates open with the use of a blunt razor or spatula. The gel remains on one of the glass plates. Then proceed to Subheading 3.5.2 for Coomassie staining or Subheading 3.5.3 for Western blotting and immunostaining.

3.5.2. Coomassie Staining

1. Transfer the gel carefully to a container containing Coomassie stain and place it on a shaker at RT for 1 h.
2. Carefully discard the Coomassie stain and wash the gel once with water. Add destaining solution to the container, so that the gel is submerged and incubate for 15 min. Repeat this step three times.
3. You should be able to see distinct protein bands of your samples.

3.5.3. Western Blotting and Immunostaining

1. Transfer the gel to a container with Western blot transfer buffer.
2. Cut a nitrocellulose membrane to the size of the gel and immerse in methanol. Rinse once in distilled water and once in Western blot transfer buffer. Hydrate two Whatman blotting papers cut to the size of the gel in the transfer buffer.
3. To perform the protein transfer with an aqueous transfer apparatus, lay the nitrocellulose membrane over one Whatman paper. Place the gel on top of the membrane, and lay another

Whatman paper on top of the gel. Roll over the "sandwiched" gel-nitrocellulose membrane using a pipette to make sure that there are no air bubbles trapped.

4. Place the sandwiched package within the transfer cassette and put it vertically in a holder within the transfer tank. The nitrocellulose should face the positive (+ve) side in the transfer apparatus.
5. Perform the transfer for 2 h at 200 mA or overnight at 100–120 mA. Turn on the stir bar and water with a hose linked from the apparatus into the sink.
6. After transfer, take apart the apparatus. Discard the gel and the Whatman papers.
7. Place the nitrocellulose membrane in blocking solution. Incubate at room temperature with gentle shaking for 45 min.
8. Wash the membrane 3× with TTBS for 5 min each.
9. Add primary antibody at 1:1,000-fold dilution in diluent solution and incubate 2 h to overnight at room temperature with gentle shaking.
10. Wash the membrane as in step 8.
11. Add secondary antibody at 1:20,000- to 1:500,000- fold dilution in diluent solution and incubate for 1–2 h at room temperature with gentle shaking.
12. Wash the membrane as in step 8.
13. Wash the membrane 1× with TBS for 5 min.
14. Prepare the alkaline phosphatase staining solution as described in item 13 in Subheading 2.5.3, mix properly and then pour over the nitrocellulose membrane.
15. When the protein bands develop, discard the developing solution and wash the blot in water. Dry on paper towel.

    Using the above fractionation protocol, we detect appreciable amounts of the VirB4, VirB7, VirB9, and VirB10 subunits in immunoblots developed with antisera raised against each of these subunits (see Fig. 1). Surprisingly, we do not detect other VirB subunits even though the *A. tumefaciens* strains subjected to analysis produce all of the T4SS subunits and are phenotypically wild type with respect to virulence, e.g., T-DNA transfer, and pilus biogenesis. When the isolated material is subjected to 1% uranyl acetate negative staining on a carbon-coated grid followed by transmission electron microscopy using a JEOL 1400 microscope, we detect ring-shaped complexes of a uniform size (see Fig. 2). These complexes resemble those of the TraN/O/F "core" complex purified from *E. coli* cells carrying pKM101 (21, 22).

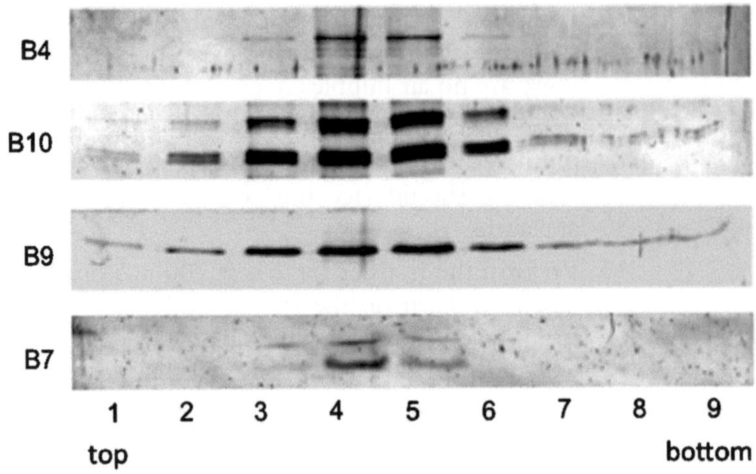

Fig. 1. Detection of enriched VirB complexes. Western blot of VirB/VirD4 T4SS machine subassemblies from *vir*-induced *A. tumefaciens* cells enriched by successive Strep-tag II affinity pull-down and CsCl density gradient centrifugation. Complexes were isolated from *A. tumefaciens* strain PC1010, a nonpolar *virB10* deletion mutant (31), engineered to produce N-terminally Strep-tagged VirB10. The western blot was developed successively with antisera specific for the VirB4, VirB7, VirB9, and VirB10 subunits. Lanes correspond to fractions recovered from the CsCl gradient. VirB subunits detected in the gradient fractions are listed at the left.

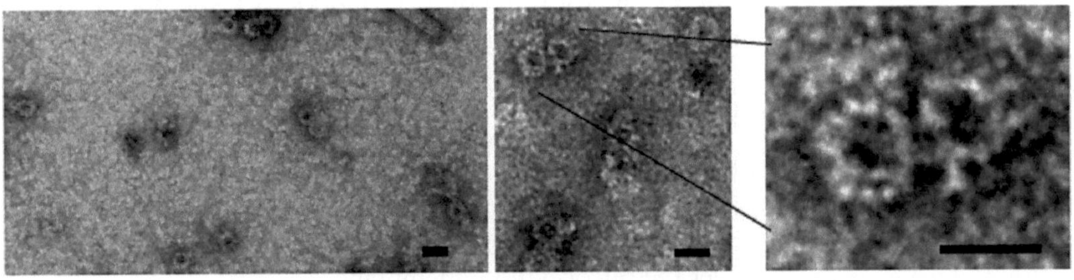

Fig. 2. Visualization of ring-shaped VirB complexes. Complexes in peak CsCl fractions 4 and 5 from Fig. 1 were pelleted by ultracentrifugation and resuspended in Lysate buffer (*see step 3* in Subheading 3.4). These samples were applied to a carbon-coated grid, stained with 1% uranyl acetate, and analyzed by transmission electron microscopy using a JEOL 1400 microscope. The ring-shaped complexes are similar in size and general architecture to the "core" complexes isolated from the *E. coli* pKM101 conjugation system by the Waksman laboratory (21, 22). Scale bar: 20 nm.

3.6. Transfer Immunoprecipitation (TrIP) Assay

This TrIP protocol results in recovery of DNA substrate—channel complexes by immunoprecipitation, as determined by PCR amplification of the immunoprecipitated material (see Fig. 3). In to our studies of the *A. tumefaciens* VirB/VirD4 T4SS, we have applied this protocol to characterize DNA substrate interactions with components of the *Enterococcus faecalis* pCF10 transfer system (32), and others have detected DNA substrate contacts with components of a conjugation machine in *Bacteroides fragilis* (33).

The Transfer ImmunoPrecipitation (TrIP) assay was adapted from the Chromatin ImmunoPrecipitation (ChIP) assay. For more information, on ChIP, see refs. 34–36.

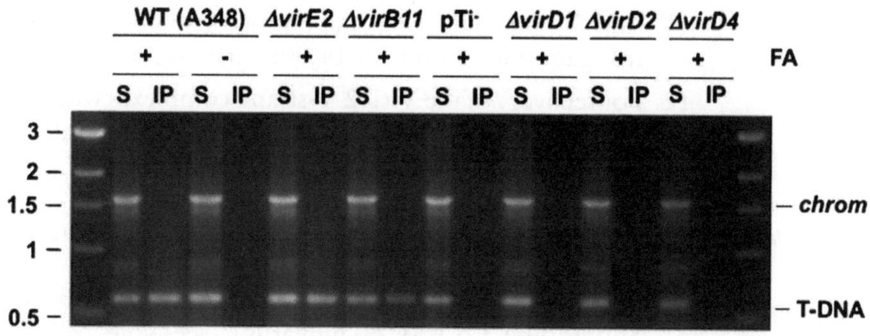

Fig. 3. Isolation of substrate DNA–channel subunit complexes with the TrIP assay. The translocating T-DNA substrate is formaldehyde cross-linked to channel subunits as described in Subheading 3.6.1. The resulting complexes are recovered by immunoprecipitation with antisera against the channel subunits as described in Subheading 3.6.2. The precipitated DNA substrate is detected by PCR amplification as described in Subheading 3.6.3. Here, anti-VirD4 antisera was used to precipitate the VirD4 substrate receptor following FA-cross-linking and disruption of the strains shown. Total soluble (S) and immunoprecipitated (IP) material were assayed for the presence of substrate T-DNA (T-DNA) or the *chvE* gene (chrom) carried on the chromosome. Strains: WT strain A348; pTi-, A348 lacking the pTi plasmid; nonpolar deletion mutants - Δ*virE2* (*virE2* codes for the exported effector VirE2; Δ*virD1*and Δ*virD2* (*virD1* and *virD2* code for relaxosomal subunits which are required for processing of the T-DNA substrate); Δ*virD4* (*virD4* codes for the substrate receptor). A precipitable VirD4-T-DNA complex is recovered from FA-cross-linked WT cells and mutants lacking the VirE2 effector and VirB11 ATPase, but not from WT cells without prior FA cross-linking or mutants lacking the T-DNA processing factors VirD1 and VirD2 or the substrate receptor VirD4 (23).

3.6.1. In Vivo Formaldehyde Cross-linking

1. Grow 1.5 ml of *A. tumefaciens* cells in MG/L medium to an OD_{600} = 0.5, pellet the cells by centrifugation, and resuspend in 6 ml of ABIM. Incubate for 14–16 h at 20°C with shaking.
2. Harvest the cells, wash with 6 ml of 20 mM sodium phosphate buffer pH 6.8, and suspend in 1 ml of cross-linking buffer.
3. Incubate the cells for 20 min at 18°C with shaking.
4. Add formaldehyde in 0.2% increments to reach a final concentration of 1% over a 15 min period, and then continue incubation for 40 min at room temperature without shaking.
5. Quench the formaldehyde by adding 141 μl of 1 M glycine per 1 ml of reaction mix (to reach a final concentration of 125 mM glycine) for 5–10 min at room temperature.
6. Pellet the cells and resuspend in 200 μl of Buffer A.
7. Incubate for 30 min at 37°C with shaking.
8. Add 900 μl of lysozyme solution, and continue incubating the mixture for 1–2 h on ice and then 30 min at 37°C with shaking.
9. Add Triton X-100 to a 4% final concentration and incubate the mixture for 15 min at room temperature rotating on a wheel.
10. Add the protease inhibitors cocktail to achieve a 5× dilution, and incubate the mixture with rocking for 15 min at 37°C and then for 2–3 h at 4°C on a wheel.

11. Add 3.2 ml of Buffer B and remove insoluble material by centrifugation for 15 min at 14,000 × *g*. Collect the supernatant.
12. For recovery of the VirD2-T-strand complexes with anti-VirD2 antibodies, treat the whole cells as described above in the absence of in vivo formaldehyde cross-linking.

3.6.2. Immunoprecipitation

1. For immunoprecipitation, first pre-incubate protein A-Sepharose CL4B (30 μl bed volume) with 1.1 ml of the detergent-solubilized, FA-cross-linked material from step 11 Subheading 3.6.1 for 60 min at RT (see Note 13).
2. Centrifuge at 5,000 × *g* and withdraw supernatant to another eppendorf tube. Discard pelleted Protein A-Sepharose and nonspecifically bound proteins.
3. Incubate the supernatant overnight at 4°C with 5 μl of antibody against a protein of interest and 30 μl of Protein A-Sepharose CL4B. Pellet the beads by centrifugation at 5,000 × *g* and carefully remove the supernatant ("S" fraction) to another eppendorf tube (see Note 14).
4. Wash the beads twice with buffer B supplemented with 1% Triton X-100 and once with buffer B supplemented with 0.1% Triton X-100.
5. Elute the material from the Sepharose beads by incubation for 20 min at 96°C in 20 μl of 10 mM Tris–Cl pH 6.8. This material is the "IP" fraction.
6. To resolve the proteins by SDS-PAGE, precipitate the material in the supernatant (S) fraction with 15% trichloroacetic acid. This is done by adding an equal volume of 30% TCA solution to the superanatant. Keep the sample on ice for at least 1 h. Then, centrifuge the samples for 15 min at 15,000 × *g* and 4°C. Remove the supernatant and wash the pellet twice with acetone. After each washing step, centrifuge the samples for 15 min at 15,000 × *g* and 4°C and remove the supernatant. Finally, dry the samples in the Speed Vac for 5 min, and then resuspend in sample loading buffer.
7. The immunoprecipitated (IP) material is resuspended directly in sample loading buffer.
8. Protein samples are boiled for 5 min prior to loading the gel.

3.6.3. PCR Amplification

1. Perform PCR in a 25 μl reaction volume containing 1/20 volume of the immunoprecipitates (IP fraction) or 1/100,000 vol. of the soluble (S) fraction.
2. Select the primers to amplify a ~300–500 bp fragment of the transferred T-DNA.
3. As a control, carry out PCR with a second set of primers designed to amplify a region of the chromosome.

4. Perform PCR amplification. An example of a typical reaction is as follows:

Immunoprecipitate ("IP" fraction)	1.25 μl
Taq PCR buffer	5 μl
dNTP's (2.5 mM stock)	1.6 μl
Forward primer (10 mM)	1 μl
Reverse primer (10 mM)	1 μl
Taq polymerase	0.2 μl
Water	15 μl

Reaction cycling conditions:

98°C for 45 s (Initial denaturation)

30 cycles of the following:

98°C for 10 s (denaturation)

45°C for 10 s (annealing)

72°C for 30 s (extension)

72°C for 10 min (final extension)

4°C or ice until the sample is visualized (see next step).

5. Separate the PCR products by electrophoresis through a 1.2% agarose gel and visualize by ethidium bromide staining (see Note 15).

4. Notes

1. Caution: Acrylamide is a neurotoxin and must be handled with gloves. Do not dispose of liquid acrylamide down the sink.
2. Caution: Formaldehyde is very toxic if inhaled, ingested, or absorbed through the skin.
3. Induction at 20°C is not required; however, efficient induction of the *vir* genes is best achieved at temperatures lower than ~23–25°C (37).
4. Induction of the *vir* genes in ABIM requires a minimum of 10 h incubation with vigorous aeration. Large scale-up of the cell culture without appropriate aeration, e.g., 1 l culture in a 2 l flask, is not advised as this will limit cell growth and diminish the yield of isolated machine subassemblies.
5. Cell pellets can be stored at −80°C for a few days prior to fractionation.

6. Cell lysis should be complete at this stage, and the cell culture should have a translucent appearance.
7. After addition of NaCl, the viscosity of the solution should considerably decrease.
8. We use a Beckman TLA100.3 centrifuge, with each centrifuge tube filled with ~3.5 ml of the cell lysate.
9. The suspension will become cloudy, but all flakes will not resuspend.
10. The use of sucrose gradients can be problematic for subsequent analysis by electron microscopy due to the presence of sucrose in the sample. The use of gel filtration chromatography offers promise for further purification, but in our experience requires scale-up of the starting culture to >10 l of induced cells for recovery of detectable levels of VirB subcomplexes upon fractionation.
11. The overlay prevents contact with atmospheric oxygen which inhibits acrylamide polymerization and also helps to level the resolving gel solution. The gel should polymerize within 15–30 min. If it has not polymerized within 1 h, prepare fresh AP solution and repeat the gel casting steps.
12. You need to be careful and efficient in quickly adding the stacking gel into the gel cassette. The gel combs need to be inserted immediately, as polymerization almost starts within a few minutes. Polymerization of stacking gel is complete within 15 min. After that, the combs can be taken out gently, and the wells washed gently and thoroughly for removing in air bubbles.
13. This step removes proteins that bind nonspecifically to the protein A-Sepharose beads.
14. The amount of antibody needed for immunoprecipitation of the T4SS subunits should be adjusted for each different batch of antibodies to achieve the best results. Too much antibody can cause background signal to occur.
15. This TrIP assay can generate either (1) no specific signal or (2) a high background level of nonspecific DNA, e.g., chromosomal DNA. When encountering either situation, we first verify that the antibodies successfully precipitate the target protein of interest. Although we typically employ polyclonal antibodies raised against the T4SS subunits, low signal or high background could be circumvented by use of monoclonal antibodies specific for precipitating epitope-tagged proteins of interest. If we encounter problems with the TrIP protocol, we first prepare or purchase new stocks of lysis buffer, formaldehyde, and reagents for PCR amplification. Further trouble shooting relies on strategies developed for enhancing specificity of DNA–protein cross-links recovered with the ChIP assay. These steps,

described in detail elsewhere (*see* refs. 34–36), include (1) adjustment of formaldehyde concentration and cross-linking time, (2) more extensive washing of the immunoprecipitates to remove nonspecifically bound DNA, (3) fragmentation of DNA by sonication prior to immunoprecipitation, (4) removal of antibodies and DNA-associated proteins by proteinase K digestion, phenol–chloroform extraction, and ethanol precipitation prior to PCR amplification, and (5) altering PCR amplification parameters.

Acknowledgments

We thank members of the Waksman, Galan, and Baron laboratories for sharing ideas regarding biochemical enrichments of T4SS subassemblies. We thank Drs. Sayyed Shah and Eric Cascales for initiating studies in this laboratory aimed at isolation of T4SS subassemblies and cross-linked DNA substrate–channel subunit complexes. We thank other members of the Christie laboratory for valuable discussions and technical expertise. This work was supported by NIH grant GM48746 to PJC.

References

1. Alvarez-Martinez CE, Christie PJ (2009) Biological diversity of prokaryotic type IV secretion systems. Microbiol Mol Biol Rev 73:775–808
2. Fernandez D, Spudich GM, Zhou XR, Christie PJ (1996) The *Agrobacterium tumefaciens* VirB7 lipoprotein is required for stabilization of VirB proteins during assembly of the T-complex transport apparatus. J Bacteriol 178:3168–3176
3. Hapfelmeier S, Domke N, Zambryski PC, Baron C (2000) VirB6 is required for stabilization of VirB5 and VirB3 and formation of VirB7 homodimers in *Agrobacterium tumefaciens*. J Bacteriol 182:4505–4511
4. Krall L et al (2002) Detergent extraction identifies different VirB protein subassemblies of the type IV secretion machinery in the membranes of *Agrobacterium tumefaciens*. Proc Natl Acad Sci U S A 99:11405–11410
5. Jakubowski SJ, Krishnamoorthy V, Christie PJ (2003) *Agrobacterium tumefaciens* VirB6 protein participates in formation of VirB7 and VirB9 complexes required for type IV secretion. J Bacteriol 185:2867–2878
6. Cascales E, Christie PJ (2004) *Agrobacterium* VirB10, an ATP energy sensor required for type IV secretion. Proc Natl Acad Sci U S A 101:17228–17233
7. Yuan Q et al (2005) Identification of the VirB4-VirB8-VirB5-VirB2 pilus assembly sequence of type IV secretion systems. J Biol Chem 280:26349–26359
8. Paschos A et al (2006) Dimerization and interactions of *Brucella suis* VirB8 with VirB4 and VirB10 are required for its biological activity. Proc Natl Acad Sci U S A 103:7252–7257
9. Jakubowski SJ et al (2009) *Agrobacterium* VirB10 domain requirements for type IV secretion and T pilus biogenesis. Mol Microbiol 71:779–794
10. Kerr JE, Christie PJ (2010) Evidence for VirB4-mediated dislocation of membrane-integrated VirB2 pilin during biogenesis of the *Agrobacterium* VirB/VirD4 type IV secretion system. J Bacteriol 192:4923–4934
11. Mossey P, Hudacek A, Das A (2010) *Agrobacterium tumefaciens* type IV secretion protein VirB3 is an inner membrane protein and requires VirB4, VirB7, and VirB8 for stabilization. J Bacteriol 192:2830–2838
12. Banta LM et al (2011) An *Agrobacterium* VirB10 mutation conferring a type IV secretion system gating defect. J Bacteriol 193:2566–2574

13. Sivanesan D, Baron C (2011) The dimer interface of *Agrobacterium tumefaciens* VirB8 is important for type IV secretion system function, stability, and association of VirB2 with the core complex. J Bacteriol 193:2097–2106
14. Yeo HJ, Savvides SN, Herr AB, Lanka E, Waksman G (2000) Crystal structure of the hexameric traffic ATPase of the *Helicobacter pylori* type IV secretion system. Mol Cell 6:1461–1472
15. Gomis-Ruth FX et al (2001) The bacterial conjugation protein TrwB resembles ring helicases and F1- ATPase. Nature 409:637–641
16. Hare S, Bayliss R, Baron C, Waksman G (2006) A large domain swap in the VirB11 ATPase of *Brucella suis* leaves the hexameric assembly intact. J Mol Biol 360:56–66
17. Yeo H-J, Yuan Q, Beck MR, Baron C, Waksman G (2003) Structural and functional characterization of the VirB5 protein from the type IV secretion system encoded by the conjugative plasmid pKM101. Proc Natl Acad Sci U S A 100:15947–15952
18. Bayliss R et al (2007) NMR structure of a complex between the VirB9/VirB7 interaction domains of the pKM101 type IV secretion system. Proc Natl Acad Sci U S A 104:1673–1678
19. Terradot L et al (2005) Structures of two core subunits of the bacterial type IV secretion system, VirB8 from *Brucella suis* and ComB10 from *Helicobacter pylori*. Proc Natl Acad Sci U S A 102:4956–4961
20. Bailey S, Ward D, Middleton R, Grossmann JG, Zambryski PC (2006) *Agrobacterium tumefaciens* VirB8 structure reveals potential protein-protein interaction sites. Proc Natl Acad Sci U S A 103:2582–2587
21. Fronzes R et al (2009) Structure of a type IV secretion system core complex. Science 323:266–268
22. Chandran V et al (2009) Structure of the outer membrane complex of a type IV secretion system. Nature 462:1011–1015
23. Cascales E, Christie PJ (2004) Definition of a bacterial type IV secretion pathway for a DNA substrate. Science 304:1170–1173
24. Jakubowski SJ, Krishnamoorthy V, Cascales E, Christie PJ (2004) *Agrobacterium tumefaciens* VirB6 domains direct the ordered export of a DNA substrate through a type IV secretion system. J Mol Biol 341:961–977
25. Jakubowski SJ, Cascales E, Krishnamoorthy V, Christie PJ (2005) *Agrobacterium tumefaciens* VirB9, an outer-membrane-associated component of a type IV secretion system, regulates substrate selection and T-pilus biogenesis. J Bacteriol 187:3486–3495
26. Christie PJ (2009) Structural biology: translocation chamber's secrets. Nature 462:992–994
27. Kubori T et al (1998) Supramolecular structure of the *Salmonella typhimurium* type III protein secretion system. Science 280:602–605
28. Sukhan A, Kubori T, Wilson J, Galan JE (2001) Genetic analysis of assembly of the *Salmonella enterica* serovar Typhimurium type III secretion-associated needle complex. J Bacteriol 183:1159–1167
29. Einhauer A, Jungbauer A (2001) The FLAG peptide, a versatile fusion tag for the purification of recombinant proteins. J Biochem Biophys Methods 49:455–465
30. Schmidt TG, Skerra A (2007) The Strep-tag system for one-step purification and high-affinity detection or capturing of proteins. Nat Protoc 2:1528–1535
31. Berger BR, Christie PJ (1994) Genetic complementation analysis of the *Agrobacterium tumefaciens virB* operon: *virB2* through *virB11* are essential virulence genes. J Bacteriol 176:3646–3660
32. Chen Y et al (2008) *Enterococcus faecalis* PcfC, a spatially localized substrate receptor for type IV secretion of the pCF10 transfer intermediate. J Bacteriol 190:3632–3645
33. Thomas J, Hecht DW (2007) Interaction of *Bacteroides fragilis* pLV22a relaxase and transfer DNA with *Escherichia coli* RP4-TraG coupling protein. Mol Microbiol 66:948–960
34. Strahl-Bolsinger S, Hecht A, Luo K, Grunstein M (1997) SIR2 and SIR4 interactions differ in core and extended telomeric heterochromatin in yeast. Genes Dev 11:83–93
35. Nelson JD, Denisenko O, Bomsztyk K (2006) Protocol for the fast chromatin immunoprecipitation (ChIP) method. Nat Protoc 1:179–185
36. Grably M, Engelberg D (2010) A detailed protocol for chromatin immunoprecipitation in the yeast *Saccharomyces cerevisiae*. Methods Mol Biol 638:211–224
37. Baron C, Domke N, Beinhofer M, Hapfelmeier S (2001) Elevated temperature differentially affects virulence, VirB protein accumulation, and T-Pilus formation in different *Agrobacterium tumefaciens* and *Agrobacterium vitis* strains. J Bacteriol 183: 6852–6861

Chapter 13

Production and Crystallization of Bacterial Type V Secretion Proteins

Hye-Jeong Yeo

Abstract

X-ray crystallography has become the most powerful approach to determine the three dimensional structures of proteins. The major bottleneck issues in protein crystallography are the availability of high-quality protein samples and the production of diffracting crystals. Since the type V secretion pathway involves unusually large substrate proteins (passenger domains or TpsA) and membrane proteins (β-barrel domains or TpsB), crystallography of type V secretion proteins deals with additional challenges in protein production and crystallization efforts. This chapter presents essential procedures used to generate successful crystals of type V secretion proteins beginning with different options for protein production. Following a description of the preparation and evaluation of crystallization experiments, optimization procedures of initial crystallization conditions are provided. A seeding protocol, employed to grow and obtain larger protein crystals, is also described.

Key words: Type V secretion, Autotransporters, Two-partner secretion, Crystallography

1. Introduction

The type V secretion pathway represents one of the largest groups of protein secretion systems in Gram-negative bacteria and includes autotransporters (ATs) and the functionally related two-partner secretion (TPS) systems (1, 2). The ATs are two domain proteins containing an N-terminal passenger domain that defines the specific function of the protein and a C-terminal β-domain that serves as a β-barrel translocator in the outer membrane. In TPS systems, the two domains are expressed as two separate proteins named TpsA (functional exoprotein) and TpsB (outer membrane translocator). In ATs, the inherent covalent link between the passenger and the translocator facilitates the translocation of the former without the need for a specific molecular recognition between the

Anne H. Delcour (ed.), *Bacterial Cell Surfaces: Methods and Protocols*, Methods in Molecular Biology, vol. 966,
DOI 10.1007/978-1-62703-245-2_13, © Springer Science+Business Media New York 2013

two modules, although AT secretion appears to be assisted by the outer membrane protein complex Omp85 (3). Remarkably, the TPS pathway appears to have solved the problem of "being unlinked" by specific recognition between TpsA (via N-terminal TPS domain) and TpsB (via periplasmic POTRA domains) (1). Most of known AT and TPS proteins are virulence factors associated with various diseases caused by pathogenic Gram-negative bacteria. Indeed, the AT passenger domains and the TpsA proteins are widely divergent both in sequence and in function. AT proteins include virulence factors of important human pathogens, such as the IgA1 proteases from *Neisseria gonorrhoeae*, *N. meningitidis*, and *Haemophilus influenzae*, the actin polymerization factor IcsA of *Shigella flexneri*, the AIDA-I adhesin of pathogenic *Escherichia coli*, the serum-resistant factor BrkA of *Bordetella pertussis*, and the cytotoxin VacA of *Helicobacter pylori* (4). TpsA proteins secreted by the TPS pathway are found in a wide variety of bacterial species and the best-studied TpsA/TpsB pairs include HMW1 and HMW1B of *H. influenzae*, FHA and FhaC of *B. pertussis*, HpmA and HpmB of *Proteus mirabilis*, and ShlA and ShlB of *Serratia marcescens* among others (1, 2).

X-ray crystallography is an essential technique in biochemistry to provide a framework for the structure–function relationship studies by revealing three-dimensional protein structures. Inevitably, crystallography of type V secretion proteins deals with unusually large substrate proteins (passenger domain or TpsA) and membrane proteins, often hampering protein production and crystallization efforts. Nonetheless, recent successes in several crystal structures of a few model systems represent significant progress in the field (5–14). However, many proteins with mostly putative functions continue to emerge as new members of type V secretion proteins, and they remain poorly understood. Continued advances in recombinant DNA technology, computational methodologies for structure determination including crystal data collection (not to mention the synchrotron beam sources) have made crystallization for challenging proteins attainable in almost any biochemistry laboratory setting. This chapter presents fundamental procedures used to generate successful crystals of type V secretion proteins beginning with different options for protein production. Following a description of the preparation and evaluation of crystallization experiments, optimization procedures of initial crystallization conditions are provided. A seeding protocol, employed to grow and obtain larger protein crystals, is also described. The protocols described here could be useful for a wide range of newly identified type V secretion proteins and provide starting strategies towards their structure determination.

2. Materials

2.1. Protein Production

1. Erlenmeyer flasks: 120 mL (for 50 mL culture) and 2 L (for 1 L culture).
2. Luria-Bertani (LB) broth: Dissolve 25 g of premix powder per 1 L water and autoclave.
3. *E. coli* strains containing appropriate expression vectors.
4. Culture incubator shakers.
5. Ampicillin stock solution (100 mg/mL): dissolve 2 g of ampicillin in 20 mL milli-Q water, filter through a 0.2 μm membrane filter, dispense in 4 and 1 mL aliquots, and store at −20°C.
6. Chloramphenicol stock solution (34 mg/mL): dissolve 0.68 g of chloramphenicol in 20 mL ethanol, filter through a 0.2 μm membrane filter, dispense in 5 mL aliquots, and store at −20°C.
7. Isopropyl β-D-1-thiogalactopyranoside (IPTG) stock solution (1 M): dissolve 2.38 g of IPTG in 10 mL milli-Q water, filter through a 0.2 μm membrane filter, dispense in 1 mL aliquots, and store at −20°C.
8. Ammonium sulfate.
9. Nylon membrane filters (0.2 μm) for chemical solution filtering.
10. Spectrophotometer.
11. Centrifuge and associated rotor.
12. Buffer S: 20 mM Tris-HCl pH 7.4 and 0.5 M NaCl.
13. Buffer A: 20 mM Tris–HCl pH 8.0 and 250 mM NaCl.
14. Buffer L: 20 mM Tris–HCl pH 8.0, 250 mM NaCl, and protease inhibitor.
15. EDTA-free Protease Inhibitor Cocktail Tablets: dissolve one tablet in 40 mL lysis buffer.
16. Sonicator with standard 3/8 in. disruptor horn.

2.2. Protein Purification and Sample Preparation for Crystallization

1. Ni-NTA agarose.
2. Ni-NTA superflow column.
3. Glutathione Sepharose 4 Fast Flow (GE Healthcare).
4. FPLC system.
5. SDS-PAGE system.
6. Vivaspin concentrators (15 and 4 mL).
7. HiTrapQ (5 mL) column.

8. 1 M imidazole pH 8.0: dissolve 34.04 g of imidazole in about 400 mL of water. Adjust to pH 8.0 by adding concentrated HCl, and then make up to 500 mL with water (see Note 1).
9. Buffer A: see item 13 Subheading 2.1.
10. Buffer W: 20 mM Tris–HCl pH 8.0, 250 mM NaCl, and 20 mM imidazole.
11. Buffer E: 20 mM Tris–HCl pH 8.0, 250 mM NaCl, and 250 mM imidazole.
12. Buffer I: 20 mM Tris–HCl pH 8.0, 250 mM NaCl, and 0.5 M imidazole.
13. GST cleavage buffer: 50 mM Tris–HCl pH 7.0, 150 mM NaCl, 1 mM EDTA, 0.2% Triton X-100, and 1 mM DTT.
14. PreScission protease (GE Healthcare).
15. Buffer Qa: 20 mM Bis-Tris pH 7.0.
16. Buffer Qb: 20 mM Bis-Tris pH 7.0 and 1 M NaCl.
17. Buffer G: 50 mM HEPES pH 7.0, 200 mM NaCl, 0.5 mM EDTA, and 5% Glycerol.
18. Glass columns (Kontes flex-column 1.5 × 20 cm).
19. Centrifuges.
20. Low protein binding filters (0.2 μm) for protein sample filtering.
21. Rotary shaker.
22. Phosphate Buffer Saline (PBS): dissolve one premix pouch (Sigma) per 1 L water and autoclave.
23. Gel filtration columns: HiPrep™ 16/60 Sephacryl™ S-100 HR for a large sample volume (up to 3 mL) or Superdex 75 10/300 GL for a small sample volume (0.5 mL).

2.3. Crystallization

1. 24-well plates (VDX plates, Hampton Research).
2. Sealant (Petroleum Jelly).
3. Siliconized cover slides.
4. Commercial crystallization screen solutions (Table 1).
5. Crystal plate incubator.
6. Filters (0.2 μm) and filtration system for stock solutions.
7. 20- or 30-mL syringe.
8. Light microscope.

2.3.1. Buffer Solutions at Different pHs (see Note 2)

Prepare 100 mL of 1 M buffer solutions pH 4.0–9.0. Filter each solution through a 0.2 μm membrane filter and keep at room temperature.

1. 1 M citric acid: dissolve 19.2 g in 80 mL water, adjust to pH 4.0, and make up to 100 mL.

Table 1
Representative screens for initial crystallization investigations

Screens (provider)	Descriptions
Crystal Screen (Hampton Research)	The original sparse matrix screen. 50 unique reagents
Index (Hampton Research)	A data-driven biased sparse matrix and grid screen. A 96 reagent crystallization screen.
pHClear Suite (Qiagen)	For systematic analysis of precipitant, concentration and pH A 96 reagents screen based on the eight most popular precipitants at various concentrations
JCGS suite (Qiagen)	An optimized sparse matrix screen based on the results of the Joint Center for Structural Genomics

2. 1 M sodium acetate: dissolve 8.2 g in 80 mL water, adjust to pH 5.0, and make up to 100 mL.
3. 1 M MES: dissolve 19.53 g in 80 mL water, adjust to pH 6.0, and make up to 100 mL (see Note 3).
4. 1 M HEPES: dissolve 23.83 g in 80 mL water, adjust to pH 7.0, and make up to 100 mL.
5. 1 M Tris–HCl: dissolve 12.11 g in 80 mL water, adjust to pH 8.0, and make up to 100 mL.
6. 1 M bicine: dissolve 16.32 g in 80 mL water, adjust to pH 9.0, and make up to 100 mL.

2.3.2. Salt Stock Solutions (see Note 4)

1. 2 M ammonium acetate: dissolve 6.16 g in water to a final volume of 40 mL.
2. 3.5 M ammonium sulfate: dissolve 115.6 g in water to a final volume of 250 mL.
3. 1 M calcium acetate: dissolve 6.32 g in water to a final volume of 40 mL.
4. 2 M lithium sulfate: dissolve 8.7 g in water to a final volume of 40 mL.
5. 4 M magnesium chloride: dissolve 15.2 g in water to a final volume of 40 mL.
6. 3 M sodium acetate: dissolve 9.84 g in water to a final volume of 40 mL.
7. 5 M sodium chloride: dissolve 29.2 g in water to a final volume of 100 mL.

2.3.3. Precipitant Stock Solutions (see Note 5)

1. 100% polyethylene glycol (PEG) 400.
2. 50% (w/v) PEG 4000.

3. 50% (w/v) PEG 6000.
4. 50% (w/v) PEG 8000.
5. 50% (w/v) PEG monomethyl ether (PEG-MME) 2000.
6. 50% (w/v) PEG-MME 5000.

3. Methods

The protocols described here are essential steps of the overall scheme required for the structure determination of proteins, as outlined in Fig. 1. Once a putative type V secretion protein is identified and cloned, production of the purified protein is necessary for its functional characterization and crystallization. While full-length passenger domains and full-length TpsA proteins are known to be extremely difficult to purify in sufficient amounts, one can define and purify (smaller) functional domains of the full-length proteins. Different expression systems (e.g., His-tagged proteins versus GST-fusion proteins among many other possibilities) often result in surprisingly positive outcomes.

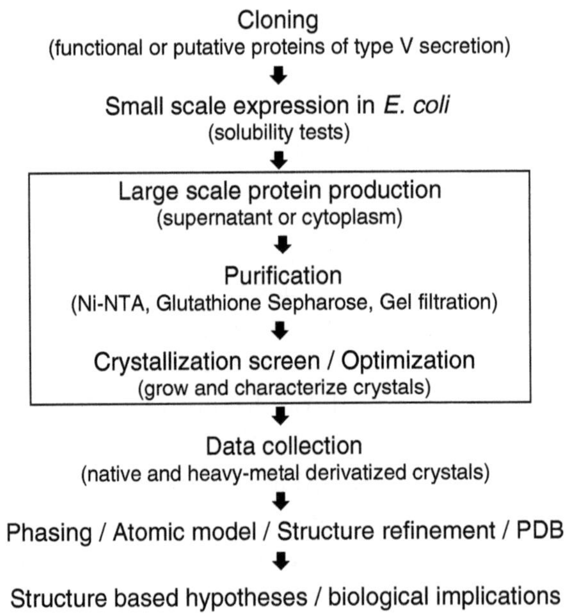

Fig. 1. Outline of key experimental steps required for obtaining the structure of proteins by X-ray crystallography. Following cloning of the genes encoding the passenger domains or TpsA proteins, mini-expression (5 mL) trials can be performed to optimize expression and test protein solubility. Subsequently, these conditions are applied on a large-scale expression (>500 mL for cytoplasmic overexpression or >2 L for cell culture supernatant expression). This chapter focuses on the bottleneck steps of the flowchart, indicated in the *box*: obtaining high quality protein sample and diffracting crystals.

3.1. Protein Production

For the large passenger domains or TpsA proteins, if they are inherently secreted to the extracellular medium, the most successful strategy to obtain protein solutions with high quality is to produce proteins in the cell culture supernatant (9, 11, 15, 16). With this observation in mind, the following protocols include different protein production strategies successfully employed for the type V secretion component proteins, and thus are easily applicable to other type V secretion proteins. Each production option requires a proper cloning vector system.

3.1.1. Protein Production in the Culture Supernatant

As an example, the HMW1-PP protein was produced using a DH5α strain containing pHMWB::HMWC (Cam^r) and pHMW$_{1\text{-}441}$::HAT (His-Affinity-Tag, Clontech) (Amp^r) (11). In this expression system, HMW1-PP is directed to the inner membrane through its signal sequence and to the outer membrane through its specific translocator HMW1B, and secreted in the cell culture medium.

1. In the evening of day 1, start a small-scale culture by inoculating a fresh colony (or frozen glycerol stock) of the expression strain into 50 mL of LB broth.
2. In the morning of day 2, add 10 mL (per liter) of the overnight culture to two 2 L flasks each containing 1 L of LB medium supplemented with proper antibiotics (e.g., for the HMW1-PP expression system, add 1 mL of chloramphenicol stock solution and 1 mL of ampicillin stock solution) and place them in a shaker (220 rpm) at 37°C.
3. When the culture reached an absorbance at 600 nm of ~0.8, add 200 μL of IPTG stock solution to each 1 L culture, and switch to a lower temperature (e.g., 23°C) and continue to incubate with shaking (180 rpm) for 4 more hours.
4. Spin down bacterial cells by centrifugation at 6,000×*g* for 20 min, and collect the supernatant.
5. Concentrate the supernatant by ammonium sulfate precipitation (e.g., 65% saturation) for over an hour. Subsequently, centrifuge at 15,000×*g* for 30 min, resuspend the protein pellets in 25 mL of buffer S and dialyze against buffer A. Now the protein sample is ready for the chromatography steps. Alternatively, concentrate the supernatant (2 L) to a small volume (less than 50 mL) using a concentrator filter (e.g., Vivaspin) and dialyze against buffer A.

3.1.2. Protein Production in the Cytoplasm

This protocol represents a standard overexpression system suitable for producing His-tagged proteins or GST-fusion proteins. This option has been useful for functional domains of type V secretion proteins (6, 11).

1. In the evening of day 1, start a small-scale culture by inoculating a fresh colony of the expression strain into 50 mL of LB supplemented with 50 μL ampicillin stock solution.
2. In the morning of day 2, add 10 mL (per liter) of the overnight culture to 1 L of LB medium supplemented with 1 mL ampicillin stock solution at 37°C.
3. When the culture reached an absorbance at 600 nm of ~0.8, add 400 μL of IPTG stock solution to the culture, and continue to incubate the culture with shaking at a lower temperature (e.g., 23°C) for 3 h (see Note 6).
4. Collect the bacterial cells by centrifugation at 5,000 × *g* for 20 min, discard the supernatant, resuspend the cell pellets in 80 mL of buffer L (see Notes 7 and 8), divide into 40 mL aliquots in 50 mL falcon tubes, and freeze in −20°C. The freezing step will help the subsequent cell lysis step.
5. On the day of purification, thaw the frozen cell stock (typically, an aliquot of 40 mL produces sufficient amount protein) at room temperature and sonicate the solution on ice at 40% cycle for 2 min and chill on ice for 3–5 min (repeat three times) (see Note 9).
6. Clarify the lysates by centrifugation at 40,000 × *g* for 20 min and collect the supernatant. Now the sample is ready for the chromatography step.

3.2. Protein Purification and Sample Preparation for Crystallization

3.2.1. Ni^{2+} Affinity Chromatography (Batch Purification)

1. Add 40 mL of the cleared lysate sample (from Subheading 3.1) to 5 mL of the 50% Ni-NTA agarose slurry and allow the proteins to bind for 40 min at 4°C by mixing gently using a rotary shaker.
2. Transfer the cell lysate and Ni-NTA agarose mixture into a column with the bottom cap closed.
3. Open the bottom cap and collect the flow-through. Now the column volume is about 2.5 mL.
4. Wash the column with 15 mL buffer W.
5. Elute the bound proteins four times with 2.5 mL buffer E and collect the eluate in four tubes.
6. Analyze the elution fractions by SDS-PAGE and assess the protein quality and quantity from the gel.
7. Carry out as soon as possible further purification steps, ion exchange column and/or gel filtration (see Notes 10 and 11).

3.2.2. Ni^{2+} Affinity Chromatography (FPLC Purification)

The most common practice of Ni^{2+} affinity chromatography is batch purification. In some cases, however, the purification works only if the Ni^{2+} affinity column step is performed using linear gradient elution via a FPLC system.

1. Apply the concentrated sample (less than 50 mL) to a Ni-NTA superflow column (5 mL) equilibrated with buffer A.

2. Following a washing step with 20 mL of buffer A mixed with 4% buffer I (i.e., 20 mM imidazole), elute the bound sample with a linear gradient of 20 mM to 0.5 M imidazole by combining buffer A and buffer I (total elution volume: 50 mL).
3. Analyze the fractions around the chromatogram peak by SDS-PAGE, and combine only desired fractions.
4. Proceed to further purification steps, ion exchange column and/or gel filtration, as soon as possible (see Notes 10 and 11).

3.2.3. GST Affinity Chromatography (Batch Purification)

1. Add 40 mL of the cleared protein sample (from Subheading 3.1.2) to 6 mL of the 50% Glutathione Sepharose 4 Fast Flow slurry and allow the proteins to bind for 2 h at 4°C by gentle stirring using a rotary shaker.
2. Sediment the lysate and Glutathione Sepharose mixture by centrifugation at 500×*g* for 5 min. Decant the supernatant (correspond to flow-through) to a falcon tube.
3. Wash the Glutathione Sepharose by adding 10 mL ice-cold PBS and mix gently by inverting the tube three or four times. Following centrifugation at 500×g for 5 min, decant the supernatant (corresponding to the wash). Repeat these steps to perform a total of three washes.
4. Proceed to cleave the GST moiety on the column by using 80 units PreScission protease in 3 mL of GST cleavage buffer overnight at 4°C.
5. Centrifuge the on-column reaction at 1,000×*g* for 5 min and then collect the supernatant (S1). Add 3 mL of GST cleavage buffer, mix gently, centrifuge at 1,000×*g* for 5 min, and collect the supernatant (S2). To recover the proteins (without tag) completely, repeat these steps two more times (S3 and S4).
6. After checking with SDS-PAGE, combine the highly pure fractions. Proceed with gel filtration, Subheading 3.2.5 (see Note 10).

3.2.4. Ion Exchange Chromatography (FPLC Purification)

Although ion exchange chromatography can be used as the first step of purification, where the sample is a crude cell lysate, this section describes a protocol used for the second step. This means the sample is already quite clean and usually contains a high concentration of salt.

1. Prepare the sample suitable for anion exchanger binding, by adjusting the salt concentration and buffer pH. This can be done by a simple dilution (two to fourfold) with buffer Qa (no salt).
2. After correctly connecting a 5 mL Hitrap Q column (for anion exchanger) to a FPLC system, inject the sample to the column equilibrated with buffer Qa mixed with 5% buffer Qb (i.e., 50 mM NaCl).

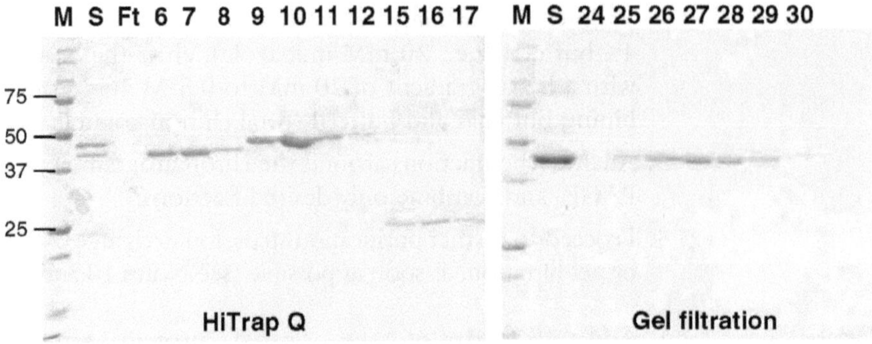

Fig. 2. SDS-PAGE analysis of HMW1-PP purification steps. The culture supernatant containing HMW1-PP was enriched by $(NH_4)_2SO_4$ precipitation and a Ni-NTA column step. Subsequently, an anion exchange column HiTrap Q was necessary to separate two protein fragments. The last purification step included gel filtration. *S* sample, *Ft* flow through.

3. Following a washing step with 10 mL of buffer Qa mixed with 5% buffer Qb, elute the proteins with a linear gradient of 50 mM –1 M NaCl by combining buffer Qa and buffer Qb (total elution volume: 50 mL).
4. Take small aliquots (20 μL for mini gel) from the peak fractions, and analyze them on SDS-PAGE (Fig. 2), and combine the desired fractions (now ready for gel filtration, Subheading 3.2.5).

3.2.5. Size Exclusion Chromatography

1. Equilibrate a gel filtration column with buffer G. Use a HiPrep™ 16/60 Sephacryl™ S-100 HR column for a large sample volume (up to 3 mL) or a high-resolution gel filtration column Superdex 75 10/300 GL for a small sample volume (0.5 mL).
2. Before injection into the gel filtration column, concentrate the sample using a Vivaspin concentrator filter to achieve an appropriate sample volume, as mentioned above.
3. After gel filtration, check the fractions for the presence and purity of the desired protein using SDS-PAGE (Fig. 2). Combine highly pure fractions only.

3.2.6. Protein Sample Preparation for Crystallization

The protein concentration is a very important parameter and protein crystallization requires a high protein concentration in general. Depending on the size and solubility of the protein, one can concentrate the sample to 5–20 mg/mL (large protein–small protein). For large type V secretion proteins, one may try crystallization at a lower concentration.

1. Place the protein sample from the final purification step in a Vivaspin concentrator (depending on the volume, use 15 mL or 4 mL), and centrifuge carefully according to the manufacturer's instructions by respecting the centrifugal force and time.
2. As the sample gets concentrated, carefully observe whether or not the sample starts to precipitate. If the sample is clear,

monitor the concentration using a UV spectrometer or colorimetric method (e.g., Bradford method).

3. Continue to concentrate the sample. If a large volume of concentrated protein sample is obtained, divided into small aliquots (100–200 μL) in eppendorf tubes, and proceed to crystallization with fresh sample (see Note 12). The remaining sample tubes can be stored frozen (−20°C) until use for crystallization.

3.3. Crystallization

3.3.1. Screening of Crystallization Conditions (see Note 13)

Crystallization of a protein is a controlled aggregation process. Among a few different crystallization techniques, the hanging drop vapor-diffusion method is the most popular approach, in which a droplet of a concentrated protein sample is mixed with an equal volume of crystallization solution and sealed above a reservoir of crystallization solution (Fig. 3a). By vapor diffusion over time, the concentration of components (protein and precipitant) in the

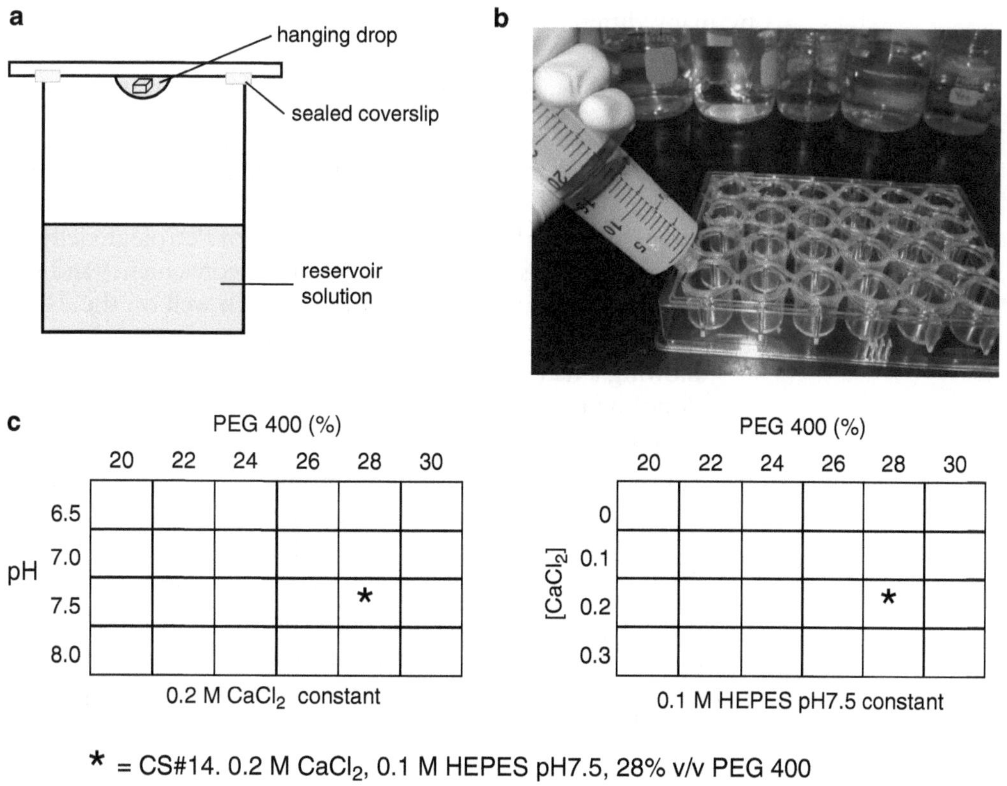

Fig. 3. Crystallization setup. (**a**) Schematics of the hanging drop vapor diffusion method. (**b**) A 4 × 6 well plate and crystallization plate preparation. Shown is an example of greasing the crystallization plate with Petroleum jelly. (**c**) Schematics of the optimization strategy. One of the crystallization conditions for HMW1-PP was found from Crystal Screen #14 (0.2 M $CaCl_2$, 0.1 M HEPES pH 7.5, 28% v/v PEG 400). To optimize and produce larger crystals, crystallization parameters were varied. Along a given axis, the concentration of one component of the well solution was altered while the others were kept constant. Shown are two plates that could be set up in the first round of optimization. The initial condition #14 is also incorporated in the plate (*star*).

droplet increases, in some cases causing the protein to precipitate. Precipitation in an orderly way can produce a protein crystal. Since it is impossible to predict in which condition a protein will crystallize, many different conditions have to be searched by trial-and-error, varying protein concentration, pH, temperature, and crystallization solution (three main components are precipitating agent or crystallant, salt, and buffer). The following protocols describe crystallization procedures using the hanging drop method that can be performed in almost any molecular biology laboratory settings, when a high-throughput robot system is not available (see Note 14). The most common approach for pursuing crystallization conditions is a screening matrix that is utilized to determine quickly which variables influence the protein solution behavior. Subsequently, one would refine the components of the hit (or promising) crystallization conditions by varying the concentrations of the components. Since the first commercial screen available as Crystal Screen (Hampton research Inc.), there has been a proliferation in the number of different crystallization screens provided by many different companies (Table 1). Although crystallization solutions can be prepared in the laboratory, it is most practical and efficient to utilize commercially available solutions to perform initial screening.

1. To be able to seal the cover slides on the plate, prepare crystallization plates by placing sealant (Petroleum jelly is cost effective and easy to handle). Put a good amount of Petroleum jelly into a medium size syringe (20 or 30 mL is convenient) and push down enough amounts to the rim of each well on the 24-well crystallization plate (Fig. 3b). This can be easily achieved by allowing a half circle of sealant and quickly moving the syringe along the row of the plate (you need to go back and forth for each row so that a full circle is greased).
2. Transfer 1 mL of each solution from the commercial crystallization kit to the 24-well plate. Practically, one can complete two to four plates (or 48–96 conditions) at a time. Without delay, proceed to set up hanging drops (see Note 15).
3. Place 1 μL protein sample onto the middle of a siliconized glass cover slide. It is convenient to make three to six cover slides in a row. Try to avoid air bubbles by carefully pipetting in and out.
4. Add 1 μL of each reservoir solution to each protein drop (use a new pipette tip for each reservoir transfer). Usually, two solutions (protein and reservoir) are mixed by diffusion, or you can pipette up and down once to mix. To obtain reproducible results, it is essential to keep the same technique. Take care not to introduce air bubbles into the droplet.

5. Flip the cover slide and place it on the corresponding well so that the droplet is hanging down (Fig. 3a). Seal the cover slide by gently pressing its edge over the sealant using a thin pipette tip. Adjust the coverslip so that the reservoir wells are completely sealed. Otherwise, droplets will dry out prematurely.
6. Once the plate is completed, minimize disturbance of the plate by lightly immobilizing the lid of the plate. This can be achieved by placing a small amount of clay underneath of the lid or using tape. Place the plates in a location that is ideally vibration free and at constant temperature (e.g., incubator 15°C) (see Note 16). Keep the plates undisturbed for a day.

3.3.2. Observation and Scoring

1. Using a light microscope (with >40× magnification), observe the crystallization drops starting the day following the setup of the crystallization screen. Make notes and score one of the outcomes: clear drops, precipitates, microcrystal-like, or crystals.
2. Make notes (e.g., size and shape) on any particular drops containing unusual crystal-like material. Observe if this materials changes with time (if crystals are present, they will grow).
3. Continue to observe the plates for any changes every day or every 2 days for a week. Subsequently, observe the plates twice a week for two additional weeks and afterwards once a week for a month.

3.3.3. Optimization

It is rather rare to obtain crystals suitable for data collection from the initial screens. When those crystals, if any, are too small, too thin (non-3D crystal), or clustered, optimization procedures can greatly improve the quality of crystals. Crystallization solutions are typically composed of a buffer, a salt, and a precipitant. Optimization of initial crystallization conditions is performed by varying one parameter along an axis of a 4×6 well plate while holding the others constant. On the other 4×6 well plate axis, a second parameter is varied resulting in a 24-well plate, each well containing a unique mixture of components (Fig. 3c). Each of the 24 wells contains a solution slightly different than the original condition used as a starting point for the optimization, and therefore usually drops will yield improved crystals (see Note 17). Optimization is commonly done manually using the hanging drop method as described above, even when the initial screening steps are carried out by high-throughput robotic systems.

1. For any crystallization drop showing crystals or signs of crystals, prepare a crystal plate that includes an initial one-dimensional grid to refine the concentration of precipitant, pH, or salt. For example, one crystallization condition for HMW1-PP was found from Crystal Screen #14 (0.2 M $CaCl_2$, 0.1 M

HEPES pH 7.5, 28% v/v PEG 400). The first optimization process would be varying two component parameters per plate, as shown in Fig. 3c.

2. Prepare reservoir solutions (1 mL each) by combining the correct amounts of precipitant, buffer, and salt from the concentrated stock solutions, and water. This can be done by adding the components directly to the plate wells followed by mixing on a shaker, or by adding them in an eppendorf tube first and then transferring the solution to the reservoir wells using a pipette (no need for additional mixing).
3. Proceed to make hanging drops by placing 1 μL of protein solution onto the coverslip, and adding 1 μL of reservoir solution. After completing the plate, place it in an incubator.
4. Monitor the crystal growth or the drop behavior under a microscope every other day for a week.
5. Based on the refined conditions, continue further rounds of optimization by repeating steps 1–4.

3.3.4. Additive Screening

Many chemicals including salts, small organic compounds, and other biologically active compounds are used as additives to help in protein crystal formation. The role of these compounds in the crystallization process varies from protein to protein. In some cases the additive is a ligand for the protein whose binding results in a change in protein conformation that helps the protein to crystallize. Additives can also serve to stabilize electrostatic interactions between the protein crystal subunits and consequently help in crystal formation. For instance, a kit composed of 96 different additives (Hampton Research) has been used to screen promising conditions from the optimization step.

1. Prepare four plates (for 96 additive screens) with each well containing the same reservoir crystallization solution. Usually, select a condition where one already observed crystals.
2. Proceed to make hanging drops. After forming the droplet of the protein sample and the reservoir solution as described above, add a small volume of additive to each droplet. For example, add 0.2 μL of additive agent to a 2 μL droplet.
3. Observe and analyze the drops.

3.3.5. Optimization by Crystal Seeding: Microseeding

For many challenging proteins, it is not uncommon that the crystals obtained from initial crystallization trials are inapt for data collection. Seeding technique can result in high quality crystals. When crystallization of a protein is very hard to reproduce, this technique is powerful as long as at least one crystal is available. There are several techniques including macroseeding and microseeding. It is important to assess a critical time for introducing the crystal seed. Generally, this is not known, and thus crystal seeds can be introduced in the crystallization drops at various time points. One of the most popular seeding techniques, microseeding, is described here. Microseeding involves controlling nucleation for crystals to grow. The idea is to provide only a few micro crystals in the pre-equilibrated drop such that good crystals form and grow around seed crystals.

1. Prepare pre-equilibrated drops 6–24 h before introducing seeds, as described above. When making reservoir solutions, use lower concentrations of precipitant than the concentration at which initial crystals formed in order to prevent self-nucleation.
2. Prepare seed stock. Under the microscope, carefully open and flip the coverslip containing crystals. Pick up one or few crystals from the drop and place in a small glass plate containing reservoir solution.
3. Transfer the crystals to a fresh stabilization solution (i.e., reservoir solution with slightly higher precipitant concentration), and crush the crystals using a mini pestle (or pipette tips). Subsequently, transfer 20 μL the crushed crystal solution into a new eppendorf tube containing 80 μL of the stabilization solution. Make tenfold and 100-fold serial dilutions with the stabilization solution.
4. Carefully open the cover slides from the pre-equilibrated plates and add 0.2 μL of seed solution by submerging the pipette tips into the drops.
5. Reseal the coverslip over the reservoir solution, and place in the same incubator.
6. Monitor every 2 days for the first week. Once crystals appear, do not touch the plate at least for a week.

4. Notes

1. It is convenient to prepare buffer solutions containing imidazole (various concentrations) from this 1 M stock solution. The 1 M imidazole solution is stable over a year at room temperature (do not expose the solution to the light).
2. The listed solutions are the most common buffers used for protein crystallization. Buffers should be used in the following ranges: citric acid (pH 2.5–6.5), sodium acetate (pH 3.6–5.6), MES (pH 5.5–6.7), HEPES (pH 3.0–6.5), Tris–HCl (pH 6.8–8.8), and bicine (pH 7.6–9.0). To optimize crystals, vary and refine the pH using the same buffer (i.e., if a protein crystallizes with reservoir solution containing MES pH 6.0, try out optimization using MES pH 5.6, 5.8, 6.2, 6.4, etc.) or one may

use different buffer for the same pH (i.e., sodium cacodylate instead of MES).

3. Due to the limited solubility of MES in water, making 1 M MES requires continual adjustments of pH with concentrated NaOH solution.
4. Concentrated salt solutions also serve as precipitants. In general, prepare highly concentrated solutions.
5. All solutions should be made with the high quality chemicals and the highest quality water available and cleaned by passing through a 0.2 μm filter system. In addition, many of common solutions (e.g., PEG 4000 and higher) utilized for crystallization reservoir solutions are highly viscous; one should be careful for proper preparations and pipetting in and out for reproducible results.
6. The most important factor for obtaining protein crystals suitable for crystallography is the protein sample itself. Many type V secretion proteins reveal a mediocre solubility in an *E. coli* expression system. One simple trial to improve solubility of recombinant proteins would be to induce protein expression at a lower temperature (room temperature or 18°C) and/or for a reduced time (e.g., 90 min instead of 3 h). As described, different expression and production systems can yield positive results.
7. Although our standard protocol uses 40 mL of lysis buffer for 500 mL culture pellets, one can adjust the volume of lysis buffer to improve solubility of proteins and the purification procedures.
8. When purifying soluble proteins (excluding membrane proteins) for the crystallization purpose, avoid including detergents such as TritonX-100 in buffer solutions, as exposing the protein samples to detergents at any step of purification appears to affect crystallization. However, in some cases, the presence of a tiny amount of TritonX-100 in lysis buffer may help solubility of the protein. Thus, if detergent free protocols fail to yield sufficient quantities of purified protein, one can turn to adding detergents in buffer solutions.
9. Sonication heats the lysate sample. By controlling the power, cycle and time, keep the lysate solution cold.
10. The entire process of purification, from the cell lysis step to the final concentration step, should be carried out promptly, which will result in a better yield and quality of protein. One can easily manage to complete two to three chromatography steps in a day.
11. Unless proteins are obtained from native sources, almost all recombinant proteins are designed to contain an affinity tag (e.g., His-tagged proteins and GST-fusion proteins), facilitating

purification. We observe that the His-tag of small proteins (<20 kDa) is often removed in order to generate high quality crystals. However, the short His-tag appears to have little influence for crystallization of the majority of medium to large proteins (>40 kDa) (17). Therefore, it is not crucial to remove the His-tag in the initial trials.

12. For manual screening (requiring 1 μl of protein per hanging drop), at least 100 μL of concentrated protein (5–10 mg/mL) would be reasonable to start crystallization trials, yielding about 100 conditions. Obviously, larger sets of screen will increase crystallization of a given protein.
13. Once a high quality concentrated protein samples are available, one can set up crystallization plates within a few days. Overall time required for observation, evaluation, and optimization of crystallization parameters vary from a few weeks to several months.
14. In recent years, high-throughput crystallization robotic systems have become very popular (almost every crystallography laboratory has a kind of robots), allowing for a very large-scale screen with much reduced time and reduced amounts of protein samples. However, for most of crystals, one needs to carry out a certain degree of optimization process manually before data collection. The protocols in this chapter serve as the basic methodologies for any molecular biologist to initiate and obtain crystals, promoting the bottleneck steps in protein crystallography. While our standard protocol uses 24-well plates for manual screening, recently there are many alternative plates provided by different companies, such as 48-well plate with sealant for hanging drop crystallization and (Hampton Research) and 96-well sitting drop plate (Corning/Hampton Research).
15. Any crystallization plates containing reservoir solutions (screen of optimization) should be used on the day they are prepared.
16. The temperature is an important parameter for protein crystallization and one may need to try out at different temperatures (4, 10, 23°C, etc.).
17. Crystallization experiments will show one of the following results: clear drops, precipitates, microscopic crystals, or crystals. If crystals are observed at least one condition, it is quite straightforward to produce sizable crystals suitable for X-ray diffraction through optimization of initial crystallization conditions. In addition to varying parameters of crystallization solutions, setting up larger drop volumes (e.g., 1.5 + 1.5, 2 + 2, etc.) is sometimes very effective in growing high quality crystals.

Acknowledgments

This work was supported in part by NIH grant AI068943 and Grant E-1616 from the Welch Foundation to H-JY

References

1. Jacob-Dubuisson F, Fernandez R, Coutte L (2004) Protein secretion through autotransporter and two-partner pathways. Biochim Biophys Acta 1694:235–257
2. Geme JW III, Yeo HJ (2009) A prototype two-partner secretion pathway: the *Haemophilus influenzae* HMW1 and HMW2 adhesin systems. Trends Microbiol 17:355–360
3. Ieva R, Bernstein HD (2009) Interaction of an autotransporter passenger domain with BamA during its translocation across the bacterial outer membrane. Proc Natl Acad Sci 106:19120–19125
4. Henderson IR, Nataro JP (2001) Virulence functions of autotransporter proteins. Infect Immun 69:1231–1243
5. Emsley P, Charles IG, Fairweather NF, Isaacs NW (1996) Structure of *Bordetella pertussis* virulence factor P.69 pertactin. Nature 381:90–92
6. Yeo HJ, Cotter SE, Laarmann S, Juehne T, St. Geme JW III, Waksman G (2004) Structural basis for host recognition by the *Haemophilus influenzae* Hia autotransporter. EMBO J 23:1245–1256
7. Oomen CJ, Van Ulsen P, Van Gelder P, Feijen M, Tommassen J, Gros P (2004) Structure of the translocator domain of a bacterial autotransporter. EMBO J 23:1257–1266
8. Clantin B, Hodak H, Willery E, Locht C, Jacob-Dubuisson F, Villeret V (2004) The crystal structure of filamentous hemagglutinin secretion domain and its implications for the two-partner secretion pathway. Proc Natl Acad Sci 101:6194–6199
9. Otto BR, Sijbrandi R, Luirink J, Oudega B, Heddle JG, Mizutani K, Park SY, Tame JR (2005) Crystal structure of hemoglobin protease, a heme binding autotransporter protein from pathogenic *Escherichia coli*. J Biol Chem 280:17339–17345
10. Meng G, Surana NK, St. Geme JW III, Waksman G (2006) Structure of the outer membrane translocator domain of the Haemophilus influenzae Hia trimeric autotransporter. EMBO J 25:2297–2304
11. Yeo HJ, Yokoyama T, Walkiewicz K, Kim Y, Grass S, St. Geme JW III (2007) The structure of the *Haemophilus influenzae* HMW1 pro-piece reveals a structural domain essential for bacterial two-partner secretion. J Biol Chem 282:31076–31084
12. Clantin B, Delattre AS, Rucktooa P, Saint N, Méli AC, Locht C, Jacob-Dubuisson F, Villeret V (2007) Structure of the membrane protein FhaC: a member of the Omp85-TpsB transporter superfamily. Science 317:957–961
13. Barnard TJ, Dautin N, Lukacik P, Bernstein HD, Buchanan SK (2007) Autotransporter structure reveals intra-barrel cleavage followed by conformational changes. Nat Struct Mol Biol 14:1214–1220
14. Weaver TM, Hocking JM, Bailey LJ, Wawrzyn GT, Howard DR, Sikkink LA, Ramirez-Alvarado M, Thompson JR (2009) Structural and functional studies of truncated hemolysin A from *Proteus mirabilis*. J Biol Chem 284:22297–22309
15. Meng G, Spahich N, Kenjale R, Waksman G, St. Geme JW III (2011) Crystal structure of the *Haemophilus influenzae* Hap adhesin reveals an intercellular oligomerization mechanism for bacterial aggregation. EMBO J 30:3864–3874
16. Khan S, Mian HS, Sandercock LE, Chirgadze NY, Pai EF (2011) Crystal structure of the passenger domain of the *Escherichia coli* autotransporter EspP. J Mol Biol 413:985–1000
17. Carson M, Johnson DH, McDonald H, Brouillette C, Delucas LJ (2007) His-tag impact on structure. Acta Crystallogr D Biol Crystallogr 63:295–301

Chapter 14

Assembly of Bacterial Outer Membrane Proteins

Jan Grijpstra, Martine P. Bos, and Jan Tommassen

Abstract

Various methods that are routinely used to study the subcellular localization of membrane proteins in wild-type Gram-negative bacteria fall short in genetic studies addressing the biogenesis of outer membrane proteins (OMPs). Here, we describe three biochemical methods that can be used in such studies to evaluate the proper assembly of OMPs into the outer membrane. The methods are based on (1) the differential electrophoretic mobility of folded and nonnative OMPs, (2) the intrinsically high protease resistance of folded OMPs, and (3) the observation that integral membrane proteins are not extracted from the membrane in solutions containing high concentrations of urea.

Key words: Outer membrane protein, β-barrel, Biogenesis, Bam complex, SDS-PAGE

1. Introduction

Whereas most integral proteins found in biological membranes span the membrane in the form of hydrophobic α-helices, most bacterial outer membrane proteins (OMPs) present an entirely different structure: they are β-barrels consisting of an even number of all antiparallel β-strands (1). OMPs are synthesized in the cytoplasm and have to be transported across the inner membrane and the aqueous periplasm to reach their destination. Research in the past decades has provided considerable insight in the biogenesis of these proteins (2, 3).

OMPs are synthesized as precursors with an N-terminal signal sequence, which marks them for transport across the inner membrane via the Sec system (4). During or shortly after transport, the signal sequence is cleaved off, and the nascent OMPs are bound by the chaperone Skp, which prevents their aggregation in the periplasm (2, 3). The OMPs are then targeted to the Bam complex in the outer membrane by virtue of a C-terminal signature sequence (5). There, folding takes place, probably assisted by periplasmic

Anne H. Delcour (ed.), *Bacterial Cell Surfaces: Methods and Protocols*, Methods in Molecular Biology, vol. 966,
DOI 10.1007/978-1-62703-245-2_14, © Springer Science+Business Media New York 2013

chaperones such as SurA, and the Bam complex mediates their insertion into the outer membrane (2, 3, 6). In their correctly assembled native state, OMPs are often extremely stable due to the existence of an extensive regular hydrogen-bond network between the transmembrane β-strands.

Studying the biogenesis of OMPs requires adequate methods to determine whether or not they are properly assembled into the outer membrane. Routine methods used to determine the localization of membrane proteins in Gram-negative bacteria include the separation of inner and outer membranes by isopycnic sucrose density-gradient centrifugation (7) or the selective solubilization of inner membrane proteins in detergents, such as Triton X-100 (8). However, such methods may fall short when the fate of recombinant OMPs is studied or when a protein that is required for their biogenesis is mutationally disrupted (9, 10). If the assembly of OMPs fails, the proteins may be degraded or they may form aggregates (inclusion bodies) in the periplasm. Such inclusion bodies co-sediment with the membranes during centrifugation, and, like correctly assembled OMPs, they are generally not solubilized in Triton X-100 and have a density very similar to that of outer membranes. Hence, such fractionation methods have frequently led to erroneous conclusions in the literature (for discussions, see refs. 9, 10).

Immunogold electron microscopy offers a reliable alternative to the cell fractionation methods mentioned above (11, 12). However, this method is tedious and requires highly specific antibodies and specialized equipment that is not available in every laboratory. Here, we describe three simple biochemical assays to evaluate the correct assembly of OMPs into the outer membrane, which have proven their suitability (13). Two methods are based on the extreme stability of correctly assembled OMPs. Usually these proteins do not denature in sample buffer if not heated prior to sodium dodecyl sulfate-polyacrylamide gel electrophoresis (SDS-PAGE). Independent of their oligomeric state, the native β-barrel structures of OMPs have a different electrophoretic mobility during SDS-PAGE than the denatured proteins, a phenomenon called heat-modifiability (14). Thus, by performing SDS-PAGE under non-denaturing condition (semi-native SDS-PAGE), it is fairly easy to determine whether an OMP has attained its native conformation, for which insertion into the outer membrane is a prerequisite. In addition, correctly assembled OMPs are usually highly resistant to proteases. The other method is based on the observation that inclusion bodies are readily solubilized in urea-containing solutions, while integral membrane proteins are not extracted under such conditions. We will describe the methods as we successfully used them to assess OMP assembly in *Neisseria meningitidis*, but we have also applied them in previous studies in *Escherichia coli*. Hence, with perhaps only minor modifications, they are likely to be applicable for OMP analyses also in other bacterial species.

2. Materials

2.1. Bacterial Culture

1. Spectrophotometer.
2. Humidified CO_2 incubator or candle jar (see Note 1).
3. GC-agar plates: Stir 18 g of GC-agar base in 500 mL of demineralized H_2O (dH_2O) (it will not dissolve). Autoclave for 15 min at 120°C and 2.7 kg/cm^2. After cooling to 60°C, add one bottle of Vitox supplement SR0090A and antibiotics as required for plasmid maintenance (see Note 2).
4. Tryptic Soy Broth (TSB): Dissolve 15 g of TSB in 500 mL of dH_2O (this will become a clear solution) and autoclave for 15 min at 120°C and 2.7 kg/cm^2.

2.2. Cell Envelope Preparation

1. Refrigerated benchtop centrifuge.
2. –80°C freezer.
3. Ultrasonic disintegrator Branson sonifier 450 (Branson Ultrasonics Corporation).
4. Ultracentrifuge.
5. 1 M Tris-base: Dissolve 121.14 g Tris-base in 1 L of dH_2O.
6. HCl for pH adjustments: Add 50 mL of 37% HCl to 50 mL of dH_2O in a fume hood.
7. TE buffer: Add 50 mL of 1 M Tris-base (final concentration: 50 mM) and 10 mL of a solution of 0.5 M ethylenediaminetetraacetic acid (EDTA) (final concentration: 5 mM), adjust to pH 8 with HCl and add dH_2O to a volume of 1 L.
8. 2 mM Tris–HCl (pH 7.6): Add 2 mL of 1 M Tris-base to 800 mL of dH_2O, adjust to pH 7.6 with HCl, and add dH_2O to a volume of 1 L.

2.3. Protease Digestion

1. 1 mg/mL trypsin in dH_2O (see Note 3).
2. 100 mM phenylmethanesulfonyl fluoride (PMSF): Dissolve 17.4 mg PMSF in 1 mL of ethanol. Caution: Wear gloves and protective clothing as PMSF is highly toxic and carcinogenic.

2.4. Urea Extraction

1. Rotary wheel.
2. Ultracentrifuge.
3. Microfuge.
4. 1 M Tris-base: See Subheading 2.2.
5. HCl for pH adjustments: See Subheading 2.2.
6. 5× stock solution Tris glycine buffer (100 mM Tris–HCl, 500 mM glycine): Dissolve 12.1 g Tris-base and 37.7 g of glycine in 1 L dH_2O. Adjust to pH 7.6 with HCl before adding all the dH_2O.

7. Urea extraction buffer (20 mM Tris–HCl, 100 mM glycine, 6 M urea, pH 7.6): Add 36 g of urea to 20 mL of 5× Tris glycine buffer and 33 mL of dH_2O. Let the urea dissolve by mixing for instance on a rotary wheel. Due to the hygroscopic properties of urea, the volume will increase. Adjust with dH_2O to 100 mL when the urea has completely dissolved.
8. 2 mM Tris–HCl (pH 7.6): See Subheading 2.2.
9. 100% (w/v) trichloroacetic acid (TCA): Dissolve 100 g of TCA in 50 mL of dH_2O, adjust to 100 mL with dH_2O.
10. Acetone ≥99% pure, cooled to 4°C.

2.5. Denaturing SDS-PAGE

1. SDS-PAGE gel system with glass plates and combs.
2. Power supply.
3. HCl for pH adjustments: *See* Subheading 2.2.
4. Running gel (RG) buffer: Dissolve 90.9 g Tris-base in 1 L of dH_2O, adjust to pH 8.8 with HCl.
5. Stacking gel (SG) buffer: Dissolve 30.3 g Tris-base in 1 L of dH_2O, adjust to pH 6.8 with HCl.
6. 40% Acrylamide/Bis solution 37.5:1.
7. 30% Acrylamide/Bis solution 29:1.
8. 1% (w/v) ammonium persulfate (APS): Dissolve 1 g APS in 100 mL of dH_2O.
9. 10% (w/v) SDS: Dissolve 10 g SDS in 100 mL of dH_2O. Caution: Wear a dust mask for protection against breathing SDS powder.
10. Tetramethylethylenediamine (TEMED).
11. Isopropanol ≥99.5% pure.
12. 1 M Tris-base: *See* Subheading 2.2.
13. 1 M glycine: Dissolve 75.07 g glycine in 1 L of dH_2O.
14. Running buffer: Mix 25 mL of 1 M Tris-base, 192 mL of 1 M glycine, and 10 mL of 10% SDS. Add dH_2O to a volume of 1 L; this should yield a pH of 8.3.
15. 1 M Tris–HCl (pH 6.8): Dissolve 12.114 g of Tris-base in 80 mL of dH_2O, adjust to pH 6.8 with HCl and the volume to 100 mL with dH_2O.
16. 2× sample buffer: Dissolve 2 mg bromophenol blue in 1.25 mL of 1 M Tris–HCl (pH 6.8), 2 mL of glycerol, 4 mL of 10% SDS, 0.5 mL of β-mercaptoethanol, and 2.25 mL of dH_2O. Store in aliquots at −20°C.
17. Molecular weight marker, e.g., a prestained marker.

2.6. Semi-native SDS-PAGE

Prepare materials as in Subheading 2.5 with the following modifications:

1. Step 14: The SDS concentration in the running buffer should be adjusted as required (see Note 4).
2. Step 16: Prepare 2× sample buffer without β-mercaptoethanol (substitute by dH_2O) and the SDS concentration should be adjusted as required (see Note 4).

2.7. Coomassie Brilliant Blue Staining

1. Scanner.
2. Fresh cling film.
3. Frame and clamps.
4. Staining solution: Dissolve 10 g of Coomassie Brilliant Blue (CBB) G250 in 400 mL of methanol and 100 mL of acetic acid and adjust to 1 L with dH_2O.
5. Destaining solution (10% acetic acid): Add 100 mL acetic acid to 900 mL of dH_2O. Destaining solution can be reused after removal of the CBB by filtration with activated charcoal.

2.8. Western Blotting

1. Blotting system.
2. Power supply.
3. Super RX Films.
4. Medical film processor.
5. 1 M Tris-base: See Subheading 2.2.
6. 1 M glycine: See Subheading 2.5.
7. 10% (w/v) SDS: See Subheading 2.5.
8. Transfer buffer: Mix 25 mL of 1 M Tris-base, 192 mL of 1 M glycine, 2 mL of 10% (w/v) SDS, and 200 mL of methanol. Add dH_2O to a volume of 1 L (see Note 5).
9. Protran nitrocellulose (NC) transfer membrane, Ø 0.45 μm.
10. Gel blotting paper.
11. Blot staining solution: Dissolve 100 mg of Ponceau S in 99 mL of dH_2O plus 1 mL acetic acid. This solution is reusable.
12. 1 M NaCl: Dissolve 58.44 g NaCl in 1 L of dH_2O.
13. HCl for pH adjustments: See Subheading 2.2.
14. Tris-buffered saline (TBS): Mix 20 mL of 1 M Tris-base and 150 mL of 1 M NaCl, adjust pH to 7.6 with HCl, and add dH_2O to a volume of 1 L.
15. Blocking buffer: Add 0.1 mL Tween-20 and 1 g nonfat milk powder, e.g., Protifar, to TBS in a final volume of 100 mL. Store the buffer at 4°C (see Note 6).
16. Antibody buffer: Add 0.1 mL Tween-20 and 0.1 g Protifar to TBS in a final volume of 100 mL. Store the buffer at 4°C.

17. Specific antisera against target OMPs appropriately diluted in antibody buffer (see Note 7).
18. Wash buffer: Add 0.1 mL Tween-20 to 100 mL of TBS.
19. Secondary antibodies, e.g., horseradish peroxidase-conjugated goat anti-rabbit IgG, appropriately diluted in antibody buffer (see Note 8).
20. Substrate for chemiluminescence reaction.

3. Methods

3.1. Bacterial Culture

1. Grow *N. meningitidis* overnight on GC-agar plates at 37°C in a humidified incubator with 5% (v/v) CO_2 or in a candle jar.
2. Resuspend bacteria from a GC plate in 3 mL of TSB (starting culture).
3. Measure the optical density of the starting culture at $\lambda = 550$ nm (OD_{550}).
4. Dilute the starting culture in 25 mL of TSB medium to an OD_{550} of 0.1, and grow for 5 h at 37°C with shaking.

3.2. Cell Envelope Isolation

1. After growing the bacteria in 25 mL of TSB for 5 h, measure the OD_{550}.
2. Collect the bacteria from the culture by centrifugation in a benchtop centrifuge (5,000 × *g*, 10 min, 4°C).
3. Resuspend the cell pellet in 15 mL of ice-cold TE buffer.
4. Store this bacterial suspension for at least 30 min at −80°C (see Note 9).
5. After thawing the bacterial suspension, perform ultrasonic disintegration in a Branson sonifier 450 for 5 min (duty cycle: 40%, output control: 7) using a macrotip. During sonication, leave the tube containing the bacterial suspension in a holder containing melting ice to maintain optimal cooling.
6. Remove unbroken cells by centrifugation in a benchtop centrifuge (12,000 × *g*, 15 min, 4°C).
7. Carefully transfer the supernatant to tubes suitable for ultracentrifugation.
8. Spin the supernatant in an ultracentrifuge (100,000 × *g*, 8 min, 4°C). Dissolve the pellet containing the cell envelopes in 2 mM Tris–HCl (pH 7.6). Store the cell envelopes until further use at −20°C (see Note 10).

3.3. Trypsin Digestion

1. Add 2.5 μL of 1 mg/mL trypsin to 97.5 μL of cell envelope suspension (final concentration of trypsin in the cell envelope suspension: 25 μg/mL).

2. Incubate overnight at room temperature (RT) (see Note 11).
3. Stop protease activity by adding 1 μL of 100 mM PMSF.
4. Analyze the samples by denaturing SDS-PAGE (see Subheading 3.5).

3.4. Urea Extraction

1. Add 30 μL of cell envelope suspension to 1 mL of urea extraction buffer in an Eppendorf tube.
2. Incubate for 1 h at RT on a rotary wheel.
3. Separate the insoluble material from the soluble fraction by ultracentrifugation (200,000 × *g*, 1 h, 4°C).
4. Dissolve the pellet in 30 μL of 2 mM Tris–HCl (pH 7.6) (see Note 10).
5. Add 100 μL of 100% (w/v) TCA to the supernatant (the solubilized proteins).
6. Mix vigorously with a vortex and incubate for 20 min on ice.
7. Spin for 10 min at 13,000 × *g* in a microfuge.
8. Remove the supernatant and add 500 μL of acetone (do not disturb the pellet).
9. Spin for 5 min at 13,000 × *g* in a microfuge.
10. Repeat steps 8 and 9 (see Note 12).
11. Remove the supernatant and air-dry the pellet.
12. Dissolve the pellet in 30 μL of 2 mM Tris–HCl (pH 7.6) and store at −20°C.
13. Analyze the samples by denaturing SDS-PAGE (see Subheading 3.5).

3.5. Denaturing SDS-PAGE

1. Prepare running gel solution for two gels by mixing 5 mL of RG buffer with x mL of 40% acrylamide/bis solution, $(4.5 - x)$ mL of dH_2O, 0.25 mL of 1% APS, 0.2 mL of 10% SDS, and 0.02 mL of TEMED (for x, see Note 13 and Table 1). Pour the running gel solution immediately after the addition of TEMED in between the glass plates until approximately 3 cm from the top. Fill the remaining space with isopropanol to prevent evaporation and contact with oxygen, which would slow down polymerization. The top layer of isopropanol will also result in a straight gel border.
2. Let the gels polymerize for 10–20 min.
3. Prepare the stacking gel solution for two gels by mixing 2.5 mL of SG buffer with 0.5 mL of 30% acrylamide/bis solution, 1.83 mL of dH_2O, 0.2 mL of 1% APS, 0.05 mL of 10% SDS, and 0.02 mL of TEMED.
4. Discard the top layer of isopropanol and pour the stacking gel solution on top of the polymerized running gel.

Table 1
Concentrations of SDS and amount of 40% acrylamide/bis used in semi-native SDS-PAGE to detect folded forms of various Neisserial OMPs

OMP	Sample buffer (%)	Running buffer (%)	40% Acrylamide/Bis (*x* in mL)
Omp85/BamA	1	0.025	$x = 1.75$ (7% final concentration)
Porins	0.1	0.1	$x = 2.5$ (10% final concentration)
Opa proteins	0.1	0.1	$x = 2.5$ (10% final concentration)
NspA	0.1	0.1	$x = 3.0$ (12% final concentration)
LptD	1	0.025	$x = 1.75$ (7% final concentration)

5. Immediately add a comb into the stacking gel (in between the two glass plates).
6. Let the gels polymerize for 20–30 min.
7. Assemble the gels in a gel system and add the running buffer to the buffer chambers.
8. Prepare samples by mixing preparations with an equal volume of 2× sample buffer and boiling for 10 min.
9. Load the samples and, in one lane, the molecular weight marker, and run the gels at ≤200 V at RT until the blue dye front reaches the bottom of the running gel.

3.6. Semi-native SDS-PAGE

Follow the procedure outlined in Subheading 3.5 with the following adjustments:

1. Steps 1 and 3: Replace the SDS solution with dH_2O.
2. Step 7: Use running buffer with adjusted SDS concentration (see Note 4).
3. Step 8: Use 2× sample buffer without β-mercaptoethanol and with adjusted SDS concentration (see Note 4). Do not boil the samples, but incubate on ice for 5–10 min.
4. Step 9: Run the gels at 12 mA at 4°C in a cold room, or by putting the gel unit in a bucket filled with ice (see Note 14).

3.7. CBB Staining

Some OMPs, for example the porins of *N. meningitidis*, are present in such large amounts in the cell envelopes that they can be visualized by staining of cell envelope protein patterns on gels with CBB G250.

1. De-assemble the gel unit and transfer the gels containing the separated cell envelope proteins into staining solution and leave for at least 30 min under shaking conditions at RT.

2. Remove the staining solution (can be reused).
3. Add destaining solution.
4. Leave under shaking conditions for 15 min and remove destaining solution.
5. Repeat steps 3 and 4 until the protein profile is clearly visible.
6. Transfer the gel into H_2O and scan the gel for further analysis.
7. If desired, also the original stained gel can be stored in a dried form in your laboratory notebook. To this end, assemble the gel in between two sheets of pre-wetted cellulose film and clamp the assembled gel in between two frames. To prevent the gel from breaking during the drying process, make sure that there are no air bubbles in between the film and the gel.
8. Dry the gel/cellulose film assembly at RT (~24 h).

3.8. Western Blotting

Most OMPs are lowly expressed and cannot be seen in the cell envelope protein patterns on a gel stained with CBB G250. In those cases, Western blotting can be used with specific antibodies to detect the proteins.

1. For semi-native gels, wrap the gels still contained between glass plates in aluminum foil and expose them to steam for 20 min prior to blotting. This will denature the proteins which will likely facilitate their recognition by antibodies (see Note 15).
2. Wear gloves at all stages, as fingerprints can become visible as background.
3. For each gel, cut one NC membrane and four gel blotting papers in the same or slightly bigger shape and size as the gel.
4. Soak the sponges of the blotting system, the gel, the NC membrane, and the gel blotting papers shortly in transfer buffer.
5. Assemble the transfer cassette, starting from the negative pole, with sponge, two gel blotting papers, the gel, the NC membrane, and again two gel blotting papers and a sponge. Roll a Pasteur pipette over the gel-NC assembly to remove any air bubbles.
6. Move the assembled transfer cassette to the tank, add a cooling device (shape filled with ice) and fill to the top with transfer buffer, connect it to a power supply, and carry out the transfer at 100 V for 1 h.
7. After the transfer is complete, remove the NC membrane carefully using forceps and in case no prestained markers were used, mark the protein side of the blot with a soft pencil.
8. The gel can be stained with CBB as described in Subheading 3.7 to assess whether all protein has been transferred.

9. Stain the NC membrane reversibly with the Ponceau S solution for 5 min at RT in order to confirm the transfer of the proteins from the gel. Air bubbles that may have been present during blotting are visible as empty spots on the blot. When a non-prestained marker was used, the marker bands can now be marked with a soft pencil.
10. Incubate the membrane with blocking buffer for 1 h at RT under continuous shaking in order to prevent nonspecific binding of the antibodies to the membrane.
11. After blocking, incubate the membrane for 1 h at RT under continuous shaking with the primary antiserum diluted in antibody buffer.
12. Wash the membrane three times for 5 min each at RT with wash buffer.
13. Incubate the membrane for 1 h at RT under continuous shaking with secondary antibody diluted in antibody buffer.
14. Wash the membrane three times for 5 min each at RT with wash buffer.
15. Apply the substrate solution for the chemiluminescence reaction to the blot, incubate for 5 min at RT, and discard the solution before placing the wet membrane between two plastic sheets in a film cassette.
16. Expose an RX film to the blot and develop the film using a medical film processor.

3.9. Example of Results

Figure 1 illustrates an example of the application of the biochemical methods described above. In the experiments shown, the effect of depleting cells of Omp85 (BamA), the central component of the Bam machinery, on OMP assembly was assessed (13). Compared are cell envelopes of control cells expressing Omp85 (lanes marked C) with cell envelopes isolated from mutant cells depleted of Omp85 (lanes marked M). *N. meningitidis* produces two different porins, PorA and PorB. The Omp85-depleted cells (Fig. 1a, lane 2) produce similar amounts of porins as the control cells (lane 1), which fractionate with the cell envelopes. However, when the samples are analyzed under semi-native SDS-PAGE conditions, the porins in the control sample remain in their native trimeric state (Fig. 1a, lane 3), while they appear as denatured monomers in the sample of the Omp85-depleted cells (lane 4). This lack of heat-modifiability, which is confirmed in the Western blots shown in Fig. 1c, indicates that the porins fail to be assembled into their native trimeric conformation in the Omp85-depleted cells. Heat-modifiability is also often used as a criterion to assess the assembly of monomeric OMPs. Usually, the native β-barrels migrate faster under semi-native SDS-PAGE conditions than the denatured forms of such proteins (see Note 16). This is illustrated for outer membrane phospholipase A in Fig. 1d, which migrates under semi-native conditions as a folded protein in the sample of the control cells, but accumulates as an unfolded protein in the Omp85-depleted cells.

The assembly of the porins in the Omp85-depleted cells was also assessed in protease-accessibility experiments and by urea extraction of cell envelopes. Figure 1a demonstrates that the porins are hardly degraded when cell envelopes of the control cells are treated with trypsin before denaturing SDS-PAGE (lane 5), while the porins accumulated in the Omp85-depleted cells are degraded under the same conditions (lane 6) (see Note 17). Furthermore, incubation of the cell envelope preparations in buffer containing urea did solubilize the porins in the sample from the Omp85-depleted cells, but not in that from control cells (Fig. 1b). These results confirm the lack of proper assembly of the porins in the Omp85-depleted cells (13).

4. Notes

1. *N. meningitidis* grows better with elevated levels of CO_2. When standard CO_2 incubators are not available, candle jars can be used.
2. Antibiotics most commonly used for selection in *N. meningitidis* are (with the final concentration shown in parentheses) chloramphenicol (25 μg/mL), kanamycin (80 μg/mL), and erythromycin (7 μg/mL). The development of plasmid vectors with the commonly used ampicillin-resistance marker is strongly discouraged, since penicillin or ampicillin is the method of choice for treatment of meningococcal infections, and the generation of resistant strains should be avoided.
3. In this protocol we used trypsin as the protease of choice. Note that other proteases can be used (e.g., chymotrypsin or proteinase K).
4. One should empirically test the optimal concentration of SDS in the semi-native sample and running buffer, since the amount of SDS that a native OMP can withstand without being denatured varies and is unpredictable (15). *See* Table 1 for examples. Note that SDS cannot totally be omitted. Many Neisserial OMPs have a high pI and may therefore not be negatively charged at the pH of 8.3 employed during SDS-PAGE. Therefore, at least some SDS remains essential to give them sufficient negative charge to enter the gel.
5. We add SDS to the transfer buffer since some OMPs of *N. meningitidis* have a high pI and are positively charged even at the pH of 8.3 of the transfer buffer; hence, these proteins,

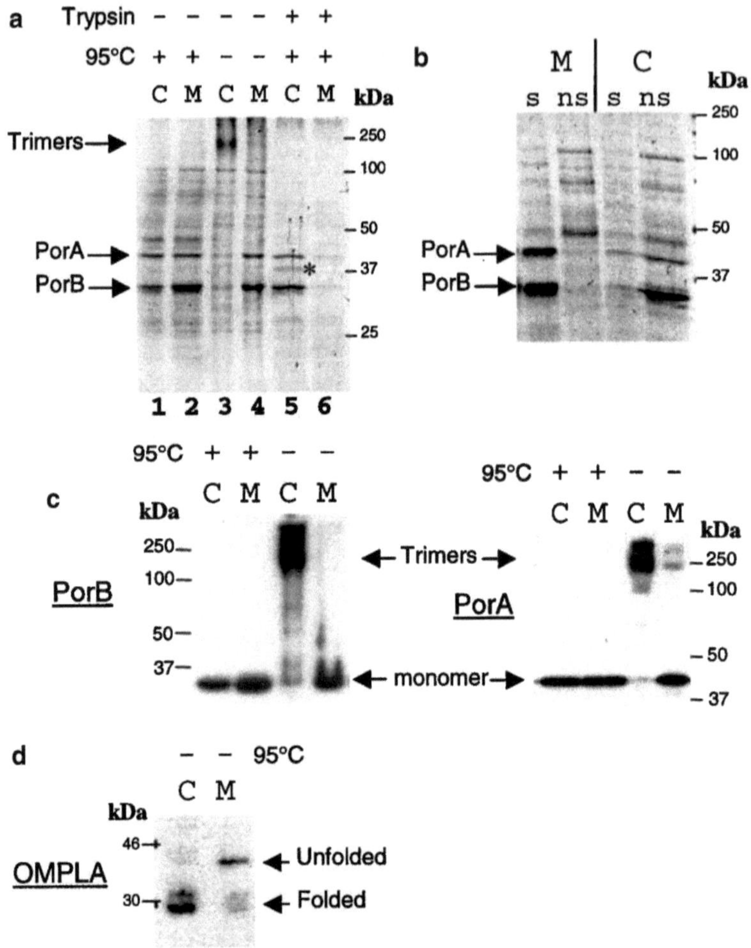

Fig. 1. Example of the implementation of the described methods in a study investigating the assembly of OMPs in a *N. meningitidis* strain depleted for Omp85 (BamA). Cell envelopes were isolated from control cells expressing Omp85 (C) and from cells depleted of Omp85 (M) and analyzed by (semi-native) SDS-PAGE with or without denaturation at 95°C as indicated. (**a**) CBB-stained semi-native SDS-PAGE gel. Where indicated, cell envelopes were treated with trypsin before denaturing SDS-PAGE. The *asterisk* in lane 5 marks a stable protease-digestion product of PorA. (**b**) CBB-stained SDS-PAGE gel of cell envelope samples extracted with urea; *s* fractions solubilized with urea, *ns* non-soluble fractions. (**c**, **d**) Immunoblot analyses after semi-native SDS-PAGE of cell envelopes with antibodies against PorB and PorA (panel **c**) or against outer membrane phospholipase A (OMPLA) (panel **d**). Molecular size markers (in kDa) are indicated. The figure is adapted from (13). Reprinted with permission from AAAS.

e.g., the Opa proteins, are not being transferred without SDS in the transfer buffer (M.P. Bos, unpublished results).

6. Depending on the antibodies used, other blocking agents might be needed to prevent nonspecific binding of the antibodies, e.g., 1% (w/v) bovine serum albumin (final concentration) instead of Protifar. In this case, also use 0.1% bovine

serum albumin instead of Protifar in the antibody buffer described in step 16.

7. Since antibody preparations vary in their levels of purity and specific binding properties, the working dilution has to be titrated for each antiserum.
8. Host species-specific secondary antibodies coupled to horseradish peroxidase are available from several manufacturers. Follow the manufacturer's instructions for the dilution to be used.
9. This step will both kill *N. meningitidis* and aid in cell fractionation. For nonpathogenic bacteria, this step may be omitted.
10. Resuspend the cell envelope pellet in 500 μL to 1 mL of Tris buffer depending on the amount of cells after growth (e.g., a wild-type strain grown to an OD_{550} of 3.5 can be dissolved in 1 mL of Tris buffer). Always dissolve the cell envelopes from different strains in the same ratio as the OD_{550} measured at step 1 of the procedure. To resuspend the cell envelopes, use a small magnetic stirrer at 4°C or leave the pellet in the buffer for several hours without stirring to dissolve. Pipetting up and down, to speed up solubilization, often results in foaming and poor solubilization.
11. Incubation time can be shorter (down to 15 min) depending on the target OMP; e.g., for Omp85 (BamA) incubation times of 1–4 h are sufficient to degrade nonnative complexes; the incubation time can be best tested empirically for each new OMP of interest.
12. This step serves to remove all traces of TCA. Residual TCA will prevent complete solubilization in sample buffer as it lowers the pH. Low pH is indicated by a change in color of bromophenol blue from blue to yellow. When samples appear yellow, some Tris buffer pH 8 may be added to restore optimal pH.
13. The concentration of acrylamide should be chosen such that the protein will migrate a fair distance into the gel. This depends on the size of the protein. It is advisable to use the lowest possible concentration as this was shown to increase the difference in migration between folded and non-folded OMPs (16). See Table 1 for examples.
14. During standard electrophoresis conditions, gels heat up. Therefore, running at low amperage and cooling are essential to prevent denaturation of the proteins during electrophoresis.
15. One has to consider the origin of the antibodies: Were they raised against denatured or native protein? In the former case, the antibody may not react with the protein in its native conformation and a heat-denaturing step may aid in antibody binding. For this reason, one cannot directly compare the amount of signal on blots from native and denatured OMPs.

16. In some exceptional cases of small monomeric β-barrel proteins, such as the meningococcal NspA protein (17), the correctly folded protein migrates more slowly under semi-native SDS-PAGE conditions than the heat-denatured form. For each OMP of interest, the electrophoretic mobility under semi-native SDS-PAGE conditions of the correctly folded protein relative to that of the heat-denatured protein should be determined before heat-modifiability can be used as a criterion to study the assembly of the protein in a mutant strain.
17. Due to their compact folding, the β-barrels of OMPs are highly resistant to proteases. Occasionally, a large cell-surface-exposed loop may be more flexible and therefore a target for proteases. In such a case, protease treatment of cell envelopes generates discrete protein fragments which can be detected on CBB-stained gels or Western blots (18). To some extent, this is also the case for PorA, which is cleaved in the long flexible loop L1 after prolonged incubation with trypsin, generating a large polypeptide that lacks a small N-terminal fragment (indicated with an asterisk in Fig. 1a, lane 5). In contrast, misfolded OMPs are totally degraded and do not yield fragments discernable on gels or blots (Fig. 1a, lane 6). Some OMPs, including Omp85, consist of two distinct domains, a membrane-embedded β-barrel and a second domain completely exposed to either side of the membrane. In such a case, the exposed domain can be completely degraded during protease treatment, but the β-barrel domain remains intact and can be detected on Western blots provided that the antiserum used recognizes this domain of the protein (19).

References

1. Koebnik R, Locher KP, Van Gelder P (2000) Structure and function of bacterial outer membrane proteins: barrels in a nutshell. Mol Microbiol 37:239–253
2. Bos MP, Robert V, Tommassen J (2007) Biogenesis of the gram-negative bacterial outer membrane. Annu Rev Microbiol 61:191–214
3. Tommassen J (2010) Assembly of outer membrane proteins in bacteria and mitochondria. Microbiology 156:2587–2596
4. du Plessis DJF, Nouwen N, Driessen AJM (2011) The Sec translocase. Biochim Biophys Acta 1808:851–865
5. Robert V, Volokhina EB, Senf F et al (2006) Assembly factor Omp85 recognizes its outer membrane protein substrates by a species-specific C-terminal motif. PLoS Biol 4:e377
6. Hagan CL, Silhavy TJ, Kahne D (2011) β-Barrel membrane protein assembly by the Bam complex. Annu Rev Biochem 80:189–210
7. Osborn MJ, Gander JE, Parisi E et al (1972) Mechanism of assembly of the outer membrane of *Salmonella typhimurium*. Isolation and characterization of cytoplasmic and outer membrane. J Biol Chem 247:3962–3972
8. Schnaitman CA (1974) Outer membrane proteins of *Escherichia coli*. III. Evidence that the major protein of *Escherichia coli* O111 outer membrane consists of four distinct polypeptide species. J Bacteriol 118:448–453
9. Tommassen J (1986) Fallacies of *E. coli* cell fractionations and consequences thereof for protein export models. Microb Pathog 1:225–228

10. Voulhoux R, Tommassen J (2004) Omp85, an evolutionary conserved bacterial protein involved in outer-membrane-protein assembly. Res Microbiol 155:129–135
11. Tommassen J, Leunissen J, van Damme-Jongsten M et al (1985) Failure of *E. coli* K-12 to transport PhoE-LacZ hybrid proteins out of the cytoplasm. EMBO J 4: 1041–1047
12. Bosch D, Leunissen J, Verbakel J et al (1986) Periplasmic accumulation of truncated forms of outer-membrane PhoE protein of *Escherichia coli* K-12. J Mol Biol 189: 449–455
13. Voulhoux R, Bos MP, Geurtsen J et al (2003) Role of a highly conserved bacterial protein in outer membrane protein assembly. Science 299:262–265
14. Nakamura K, Mizushima S (1976) Effects of heating in dodecyl sulfate solution on the conformation and electrophoretic mobility of isolated major outer membrane proteins from *Escherichia coli* K-12. J Biochem 80: 1411–1422
15. Kortekaas J, Müller SA, Ringler P et al (2006) Immunogenicity and structural characterisation of an in vitro folded meningococcal siderophore receptor (FrpB, FetA). Microbes Infect 8:2145–2153
16. Heller KB (1978) Apparent molecular weights of a heat-modifiable protein from the outer membrane of *Escherichia coli* in gels with different acrylamide concentrations. J Bacteriol 134:1181–1183
17. Vandeputte-Rutten L, Bos MP, Tommassen J et al (2003) Crystal structure of neisserial surface protein A (NspA), a conserved outer membrane protein with vaccine potential. J Biol Chem 278:24825–24830
18. Morona R, Tommassen J, Henning U (1985) Demonstration of a bacteriophage receptor site on the *Escherichia coli* K12 outer-membrane protein OmpC by the use of a protease. Eur J Biochem 150:161–169
19. Volokhina EB, Beckers F, Tommassen J et al (2009) The β-barrel outer membrane protein assembly complex of *Neisseria meningitidis*. J Bacteriol 191:7074–7085

Chapter 15

The Outer Membrane of Gram-Negative Bacteria: Lipid A Isolation and Characterization

Jessica V. Hankins, James A. Madsen, Brittany D. Needham, Jennifer S. Brodbelt, and M. Stephen Trent

Abstract

The isolation and characterization of the lipid A domain of lipopolysaccharide (LPS) are important methodologies utilized to gain understanding of the Gram-negative cell envelope. Here, we describe protocols often employed by our laboratory for small- and large-scale isolation of lipid A from bacterial cells. Additionally, we describe various methodologies including isolation of radiolabeled lipid A, thin layer chromatography, and various mass spectrometry methods. Tandem mass spectrometry is an integral tool for the structural characterization of lipid A molecules, and both coventional collision induced dissociation (CID) and new ultraviolet photodissociation (UVPD) methods are described.

Key words: Lipid A isolation, Bligh–Dyer, Thin layer chromatography, Lipopolysaccharide, Mass spectrometry, Collision induced dissociation, Photodissociation (PD)

1. Introduction

Lipopolysaccharide (LPS) covers the surface of Gram-negative bacteria and is anchored within the outer membrane by the lipid A moiety (Fig. 1) (1). The lipid A domain is the bioactive portion of LPS, activating the innate immune system via Toll-like receptor 4/ myeloid differentiation factor 2 (TLR4/MD2) (2). Substantial structural heterogeneity has evolved among Gram-negative lipid A species (Fig. 1), which is largely due to enzymes that modify lipid A following its well-conserved synthetic pathway (1, 3). Given this diversity and the fact that the biological activity of lipid A largely arises from its substituent pattern, the structural characterization of lipid A species is both an important and challenging task.

Jessica V. Hankins and James A. Madsen contributed equally to this work.

Anne H. Delcour (ed.), *Bacterial Cell Surfaces: Methods and Protocols*, Methods in Molecular Biology, vol. 966, DOI 10.1007/978-1-62703-245-2_15, © Springer Science+Business Media New York 2013

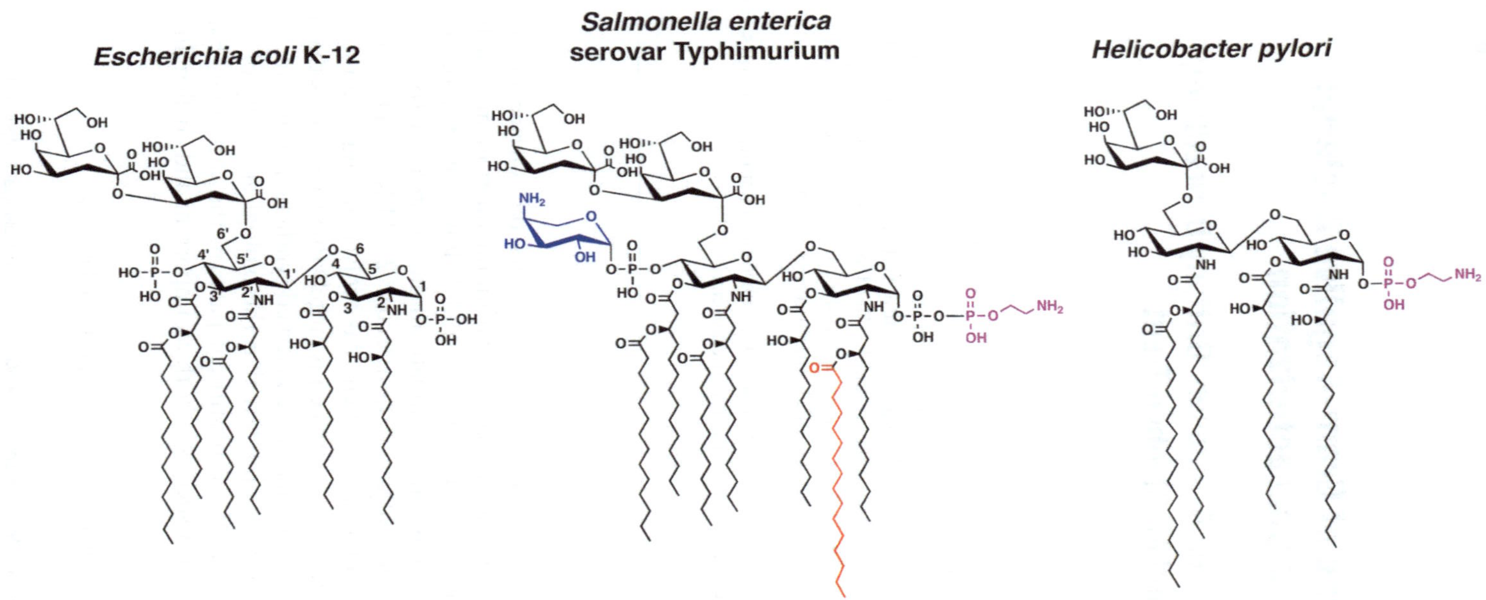

Fig. 1. Examples of the Kdo-lipid A domain of LPS from Gram-negative bacteria. The chemical structures of *Escherichia* coli K-12, *Salmonella enterica* serovar Typhimurium, and *Helicobacter pylori* are shown. The major lipid A species of *E. coli* is a hexa-acylated disaccharide of glucosamine with phosphate groups at the 1- and 4′-positions. The first sugar of the core oligosaccharide, Kdo (3-deoxy-d-*manno*-octulosonic acid), is attached at the 6′-position and serves as a bridge to link lipid A to the remaining carbohydrate domains of LPS. Although Gram-negative bacteria share a conserved pathway for lipid A biosynthesis similar to that of *E. coli* K12, there is a large amount of diversity in lipid A structures. For example, the phosphate groups of lipid A can be modified in some organisms (e.g., *S. enterica*) with the cationic sugar l-4-aminoarabinose or with a phosphoethanolamine residue. Additionally, an acyl chain can be added to *Salmonella* lipid A or in *H. pylori* acyl chains are removed.

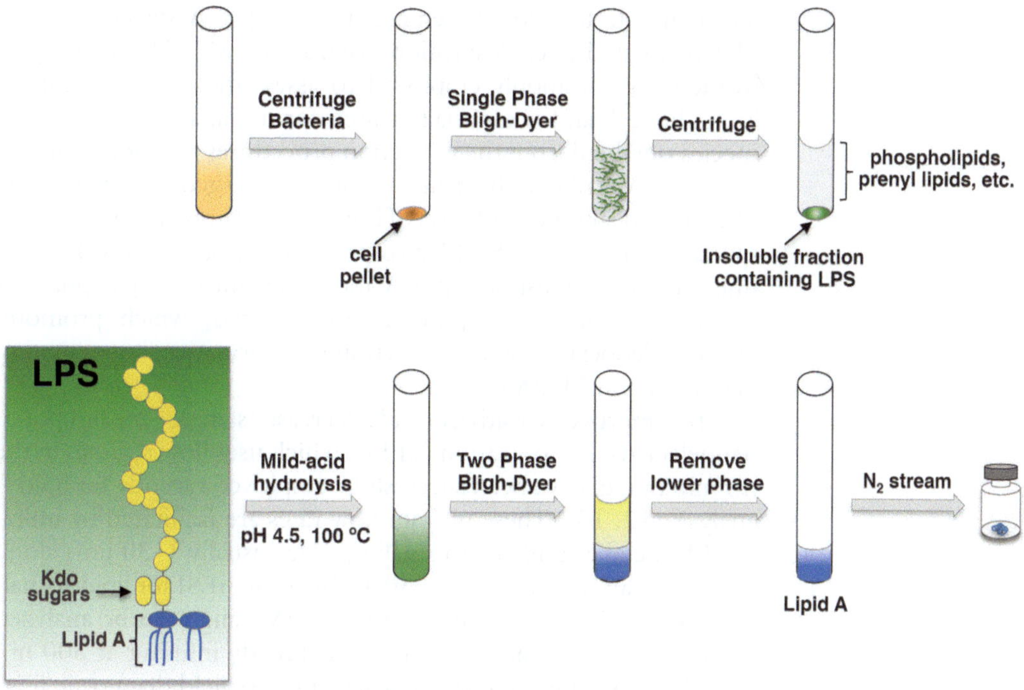

Fig. 2. Isolation of intact lipid A from whole cells. Schematic showing chemical hydrolysis of the lipid A moiety from whole LPS and Bligh–Dyer extractions of lipid A.

Various methodologies have been utilized for the extraction of LPS from the bacterial surface (4–7), the most common one being the hot phenol–water extraction procedure introduced by Westphal and Jann (8). From whole LPS, the lipid A domain can be released by mild-acid hydrolysis that selectively cleaves the linkage between the lipid region and the Kdo sugar (Fig. 1) (9). These LPS extraction methods take several days to complete, require degradation of protein and nucleic acids, and usually require the use of phenol (8). Also, LPS is naturally heterogeneous and some methods fail to extract specific types of LPS and the corresponding lipid A moiety.

Our laboratory has expanded the method proposed by Caroff (9) and Odegaard (10), which utilizes solvent extractions (Bligh–Dyer extractions) and mild-acid hydrolysis (Fig. 2). Bligh–Dyer extractions have been historically utilized for the isolation of whole lipid species from various types of preparations (e.g., animal tissue, plant tissue, etc.). The Bligh–Dyer method consists of multiple solvent extractions using chloroform, methanol, and water to extract the lipid species in the organic phase. Unlike other methods, this does not select for rough or smooth types of LPS, providing optimal recovery of lipid A species.

Several methods are employed to analyze isolated lipid A species. These include thin layer chromatography (TLC) and mass spectrometry techniques. Mass spectrometry (MS) has become an indispensable tool for the elucidation of lipid A structures (11). Matrix-assisted laser desorption/ionization (MALDI) mass spectrometry is commonly employed to assess molecular weights of intact lipid A and is especially suited for initial surveys of lipid A species obtained from the extraction procedures mentioned above; however, MALDI only generates singly charged ions which are often not ideal for elucidation of lipid A structures via tandem mass spectrometry (MS/MS). Electrospray ionization (ESI) is also routinely used to transport lipid A molecules into the gas phase as negatively charged ions, often multicharged, which promotes efficient dissociation and production of informative fragmentation patterns upon MS/MS.

To improve sensitivity and decrease sample consumption, nanoelectrospray ionization (nESI), which uses flow regimes in the nl/min range, have been successfully employed recently for lipid A analysis (12, 13). These nl/min flow rates are facilitated by utilizing ESI needles pulled to a small tip size (usually 1–30 μm) along with application of lower potentials. By using nESI for lipid A analysis, 20 μl of sample solution in the 1 μM range can be analyzed over an approximately 1 h period by directly infusing at 300 nl/min. The resulting ions are then activated to yield diagnostic fragmentation patterns (a process termed MS/MS). The default MS/MS method available on most commercial mass spectrometers is low-energy collision induced dissociation (CID). This method causes fragmentation upon collision of selected ions of interest with neutral bath gas molecules by application of a suitable potential to accelerate the ions. Simply stated, in this method, the internal energy imparted from the collisions causes fragmentation of the ions. CID has been vital in the characterization of lipid A and has been used to elucidate the lipid A structures of various bacterial species (14–25).

Although traditionally the most popular MS/MS method, CID generates only a limited series of product ions for lipid A molecules, mostly from C–O cleavages that result in neutral losses of phosphate groups and fatty acid chains. Ultraviolet photodissociation (UVPD) at 193 nm, an alternative MS/MS method, imparts energy by irradiating ions with high-energy photons via an excimer laser. UVPD has been shown recently to increase the depth of structural information for lipid A species by generating abundant product ions arising from cross-ring and inter-ring glucosamine cleavages, as well as cleavages between the amine and carbonyl groups on the 2′- and 2-linked primary acyl chains (12, 13). Furthermore, preferential cleavages of C–C bonds adjacent to carbonyl and hydroxyl groups have also been observed (12, 13). The combination of these diagnostic fragmentation pathways renders UVPD a valuable method that can be used to accurately identify subtle changes and modifications to lipid A structures.

2. Materials

Prepare all solutions using ultrapure water and store all reagents at room temperature. Measure all solvents in a glass graduated cylinder and store in glass solvent bottles with Teflon-lined caps. Teflon centrifuge tubes and rotary evaporator flasks should be rinsed with methanol and chloroform before use. Follow all waste disposal regulations when disposing of solvents and/or radioactive waste.

2.1. Lipid A Isolation

1. Growth medium (Luria broth or other media).
2. Orbital shaker.
3. Spectrophotometer.
4. Ultracentrifuge and associated rotors.
5. Phosphate buffered saline (PBS).
6. 250-ml Teflon centrifuge tubes.
7. Chloroform.
8. Methanol.
9. Single-phase Bligh–Dyer mixture: chloroform, methanol, water (1:2:0.8 v/v). Combine 200 ml of chloroform, 400 ml of methanol, and 160 ml of water in 1 l solvent bottle. Cap bottle and shake to mix. After shaking, loosen the cap to vent.
10. Hydrolysis buffer: 50 mM sodium acetate pH 4.5, 1% sodium dodecyl sulfate (SDS). First, prepare a 10% stock solution of SDS by weighing 10 g of SDS. Transfer the SDS (see Note 1) to a 100 ml graduated cylinder, add ~80 ml of water and stir. Once the detergent is in solution continue to add water to 100 ml. For 0.5 l of buffer, weigh 2.05 g of sodium acetate and transfer to a 500 ml beaker. Add water to a volume of ~350 ml and stir. Add 50 ml of 10% SDS. Mix and adjust pH to 4.5. Make the solution up to 0.5 l with water using a graduated cylinder.
11. Probe tip sonicator.
12. Water bath.
13. Rotary evaporator and flask.

2.2. Thin Layer Chromatography

1. TLC tank.
2. 20 × 20 cm Silica gel 60 TLC plates.
3. 3MM Chr chromatography paper (23 cm × 100 m).
4. TLC Solvent System: chloroform, pyridine, 88% formic acid, water (50:50:16:5 v/v). Combine 200 ml of chloroform, 200 ml of pyridine, 64 ml of 88% formic acid, and 20 ml of water in a 1 l solvent bottle. Cap bottle, shake to mix, and vent.

5. Chloroform–methanol (4:1, v/v): measure 100 ml of chloroform and transfer to a solvent bottle. Measure 25 ml of methanol and mix with chloroform.
6. Bath sonicator.
7. Nitrogen tank.
8. Nitrogen dryer.
9. Hot plate.
10. Glass chromatographic reagent atomizer.
11. 10% Sulfuric Acid–ethanol Mixture: Measure 10 ml of sulfuric acid and dilute to 100 ml with 100% ethanol. Carefully mix the solution and transfer to a glass chromatographic reagent atomizer.

2.3. Matrix-Assisted Laser Desorption/Ionization (MALDI-TOF) Mass Spectrometry

1. Chloroform–methanol (4:1, v/v): measure 4 ml of HPLC grade chloroform and transfer to a glass solvent bottle. Measure 1 ml of HPLC grade methanol and combine with chloroform. Mix.
2. Nitrogen tank.
3. MALDI plate.
4. Calibrant mixture: Sequazyme peptide mass standards (Applied Biosciences).
5. ATT Matrix: 6-aza-2-thiothymine in 50% acetonitrile. Add 50 μl of water and 50 μl of acetonitrile to a microcentrifuge tube. Add 6-aza-2-thiothymine so that the 50% acetonitrile is super-saturated. Vortex and centrifuge before use.
6. Mass spectrometer: MALDI-TOF/TOF ABI 4700 Proteomics Analyzer equipped with a Nd:YAG laser (355 nm) using a 200-Hz firing rate.

2.4. Nanoelectrospray Ionization and CID

1. Chloroform:methanol (1:1, v/v): Combine 2 ml of HPLC grade chloroform and 2 ml of HPLC methanol in a glass solvent bottle. Mix (see Note 2).
2. Bath sonicator.
3. Mass Spectrometer: Thermo Fisher Scientific LTQ XL (San Jose, CA) (see Note 3).
4. New Objectives PicoTip® online nanoESI emitter (30 μm tip size) (see Note 4).
5. Nanoelectrospray source: can be fabricated using any available power supply capable of delivering 1,000–2,500 V to a metal-coated New Objectives PicoTip® online nanoESI emitter.
6. Syringe pump: any pump capable of delivering flow rates down to 300 nl/min.

2.5. Ultraviolet Photodissociation

1. Chloroform:methanol (1:1, v/v): Combine 2 ml of HPLC grade chloroform and 2 ml of HPLC methanol in a glass solvent bottle. Mix (see Note 2).
2. 193 nm Coherent ExciStar XS excimer laser.
3. Modified vacuum manifold with a CaF_2 optical window to transmit photons into the ion trap.
4. 2-mm aperture: added prior to the ion trap after the optical window to prevent photons from irradiating the ion trap.
5. Laser trigger: TTL signal from mass spectrometer to a pulse/delay generator (Berkely Nucleonics Corporation), which triggers laser during MS/MS.
6. Thermo Fisher Scientific LTQ XL.

2.6. ^{32}P-Labeled Lipid A Isolation

1. Growth medium (Luria broth or other medium).
2. Orbital shaker.
3. Spectrophotometer.
4. Inorganic ^{32}P.
5. Glass centrifuge tubes with Teflon-lined caps.
6. Fixed angle clinical centrifuge.
7. PBS: see item 5, Subheading 2.1.
8. Chloroform.
9. Methanol.
10. Single-phase Bligh–Dyer mixture: see item 9, Subheading 2.1.
11. Hydrolysis buffer: see item 10, Subheading 2.1.
12. Bath sonicator.
13. Water bath.
14. Nitrogen gas.

2.7. Visualization of ^{32}P-labeled Lipid A Species via Thin Layer Chromatography

1. TLC tank.
2. 20 × 20 cm Silica gel 60 TLC plates.
3. 3MM Chr chromatography paper (23 cm × 100 m).
4. TLC Solvent System: see item 4, Subheading 2.2.
5. Chloroform–methanol (4:1, v/v): see item 5, Subheading 2.2.
6. Scintillation counter and vials.
7. Scintillation cocktail.
8. Plastic wrap.
9. PhosphorImager Screen.
10. Imaging analysis software.

3. Methods

3.1. Lipid A Isolation

An overall schematic of the Bligh–Dyer extractions is shown in Fig. 2.

1. Inoculate 5 ml of medium using a single colony. Grow overnight with shaking (220 rpm) at 37°C.
2. The next day, measure the OD_{600} and use the overnight culture to inoculate 200 ml of culture at a starting OD_{600} of 0.05 (see Note 5). Grow the cells with shaking (220 rpm) at 37°C until an OD_{600} of 0.8–1.0 is reached.
3. Harvest the cells via centrifugation at 10,000×*g* for 10 min (see Note 6). Wash the cell pellet with 100 ml of 1× PBS (see Note 7).
4. Resuspend the cells in 40 ml of 1× PBS and divide between two 250 ml Teflon centrifuge tubes (yielding 20 ml of suspension per tube) (see Note 8).
5. Add 25 ml of chloroform and 50 ml of methanol to each tube, making a single-phase Bligh–Dyer (chloroform, methanol, water; 1:2:0.8 v/v) (Fig. 2). Shake the bottles vigorously to mix.
6. Incubate at room temperature for 20 min to ensure cell lysis.
7. Centrifuge the mixture at 2,000×*g* for 20 min. The LPS will pellet along with proteins (Fig. 2); however, phospholipids and isoprenyl lipids will remain in the supernatant. Discard the supernatant.
8. Wash the LPS pellet with the single-phase Bligh–Dyer mixture (1:2:0.8 v/v).
9. Centrifuge at 2,000×*g* for 20 min. Discard the supernatant (see Note 9).
10. Suspend the LPS pellet in 27 ml of hydrolysis buffer (see Note 10).
11. Sonicate the sample using a probe tip sonicator at a constant duty cycle for 20 s at 50% output. Repeat sonication of the sample two times (20 s per burst).
12. Incubate the sample for 30 min in a boiling water bath (see Note 10). Remove from the water bath and allow the sample to cool to room temperature before proceeding.
13. To extract the lipids from the SDS solution, convert the solution into a two-phase Bligh–Dyer (Fig. 2) mixture by adding 30 ml of chloroform and 30 ml of methanol, yielding a chloroform, methanol, water (2:2:1.8, v/v) mixture. Mix by vigorously shaking the tube.

14. Centrifuge the sample for 10 min at 2,000×*g*. Extract the lower phase (chloroform portion) into a clean Teflon centrifuge tube using a glass pipet (see Note 11).
15. Perform a second extraction on the sample by adding 30 ml of chloroform to the upper phase from step 14. Vigorously mix the sample.
16. Centrifuge at 2,000×*g* for 10 min. Extract the lower phase, pooling it with the lower phase extracted in step 14.
17. Wash the pooled lower phases (60 ml total) by adding 60 ml of methanol and 54 ml of 1× PBS (see Note 12). This yields a two-phase Bligh–Dyer (chloroform, methanol, water; 2:2:1.8, v/v). Vigorously mix.
18. Centrifuge at 2,000×*g* for 10 min.
19. Remove the lower phase to a clean glass rotary evaporator flask and dry the sample using rotary evaporation (see Note 13).
20. Store the dried sample at -20°C for TLC analysis or MS analysis.

3.2. Visualization of the Lipid A Species via Thin Layer Chromatography

1. Use a TLC tank that will accommodate 20×20 cm plates for TLC analysis (see Note 14). Line the TLC tank with ~39 cm chromatography paper.
2. Prepare the TLC system by adding 200 ml of the TLC Solvent System to the tank. Allow the tank to pre-equilibrate for ~3 h.
3. Remove the rotary evaporator flask from the freezer (Subheading 3.1, step 20) and allow the flask to come to room temperature.
4. Add 5 ml of chloroform–methanol (4:1, v/v) to the sample, and bath-sonicate (30–60 s) to aid in removal of lipid from the sides of flask.
5. Using a glass transfer pipet, transfer the lipid sample to a clean glass tube.
6. Dry the sample under a stream of nitrogen using a nitrogen dryer.
7. Remove the silica from the top edge of the Silica gel 60 TLC plate. Using a dull pencil, draw a line 2 cm from the bottom of the plate, which goes across the entire TLC plate. This will be your reference line for spotting the samples. Additionally, mark 1 cm increments along the reference line so that samples will be spotted 1 cm apart (see Note 15).
8. Dissolve lipid (Subheading 3.2, step 6) in 250 μl of chloroform–methanol (4:1, v/v) and vortex.
9. Using a calibrated glass pipet, spot one-tenth of the volume (25 μl) onto the TLC plate. Allow samples to air-dry for 15 min.

10. Place the TLC plate containing the samples into the pre-equilibrated tank. Once the solvent front reaches the top of the TLC plate (see Note 16), remove the plate and air-dry (~30 min) (see Note 17).
11. While the plate is drying, turn on the hot plate at 250°C for charring (see Note 18).
12. In the fume hood, use a glass chromatographic reagent atomizer to spray the dried TLC plate with the 10% sulfuric acid–ethanol mixture. Make sure the spray evenly covers the TLC plate.
13. Place the TLC plate on the 250°C hot plate. Within a few minutes, you will be able to visualize the charred lipid samples as black/brown spots on the plate. Do not overexpose the plate, as this will cause the entire plate to turn brown in color and make the lipid A species difficult to visualize (see Note 19).

3.3. Determination of Lipid A Species via MALDI-TOF Mass Spectrometry

1. Resuspend the dried lipid A sample (Subheading 3.1, step 20) in ~200 μl of chloroform–methanol (4:1, v/v) and vortex.
2. Using a glass transfer pipet, transfer the lipid A sample to a glass vial (see Note 20) and dry under a stream of nitrogen.
3. Prepare the MALDI plate by adding 0.5 μl calibrant mixture to the MALDI plate on a spot near where the samples will be deposited (see Note 21).
4. Deposit 0.5 μl of ATT matrix onto the MALDI plate on each spot that a lipid A sample will be deposited.
5. To concentrate the sample for MS analysis, resuspend the dried lipid A sample in 20 μl of chloroform–methanol (4:1, v/v) to obtain a ~1–5 μg/μl solution, and vortex.
6. Once the matrix is dried, deposit 0.5 μl of sample onto the spot of ATT matrix and acquire spectra by scanning the sample for optimal ion signals (Fig. 3).

3.4. Nanoelectrospray Ionization and CID

1. Resuspend the dried lipid A (Subheading 3.1, step 20) in chloroform:methanol (1:1, v/v). Both methanol and chloroform should be HPLC grade to reduce clogging of nanoelectrospray tips.
2. Sonicate the diluted lipid A in a bath sonicator for 10 min to ensure that all materials are dissolved.
3. Set up the mass spectrometer for negative mode.
4. Directly infuse the diluted lipid A sample (see beginning of Subheading 2; Subheading 2.4) at 300 nl/min (see Note 22) using a syringe pump.
5. Slowly raise the nESI voltage from 1,000 to approximately 2,000 V until spray is stable and yields high ion signals.

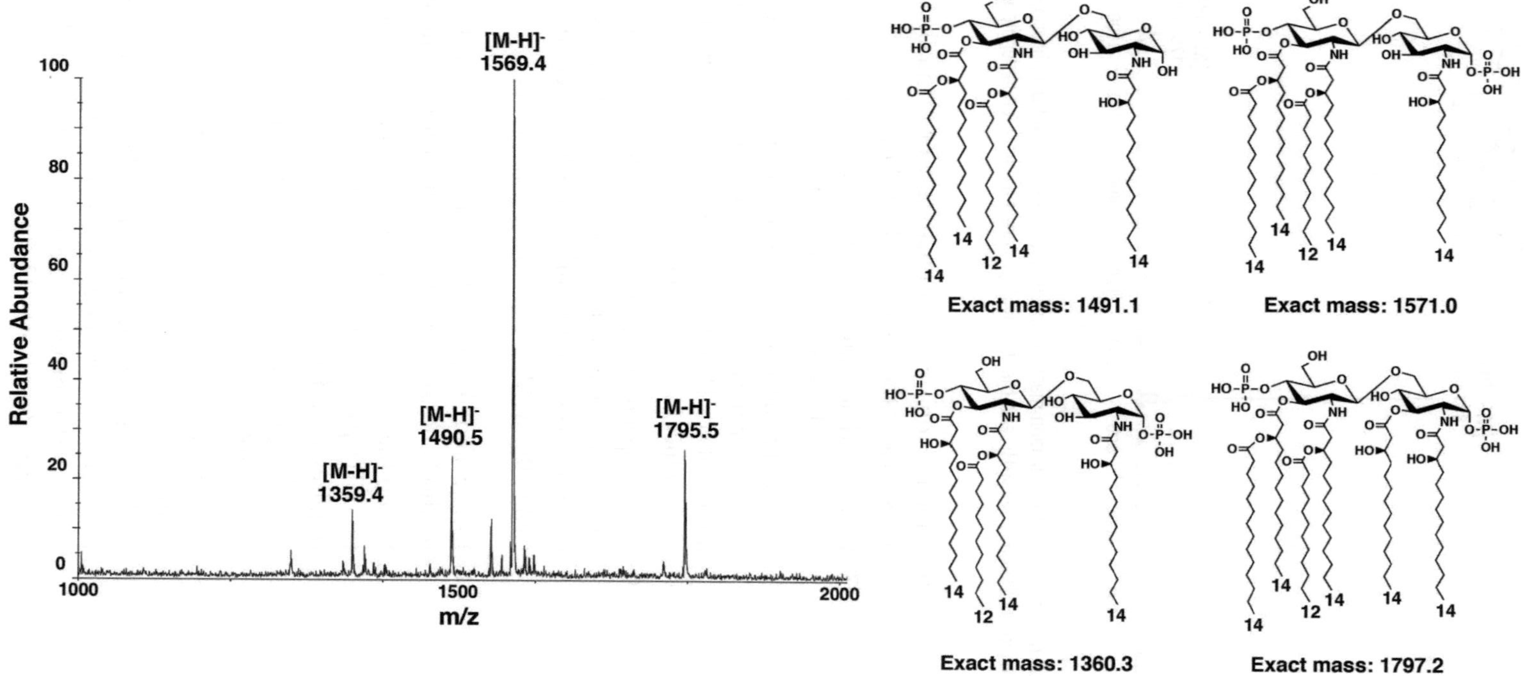

Fig. 3. Negative ion MALDI-TOF mass spectrum of a lipid A mixture. The singly deprotonated ions are denoted by $(M-H)^-$ and the predicted structure for each lipid is shown.

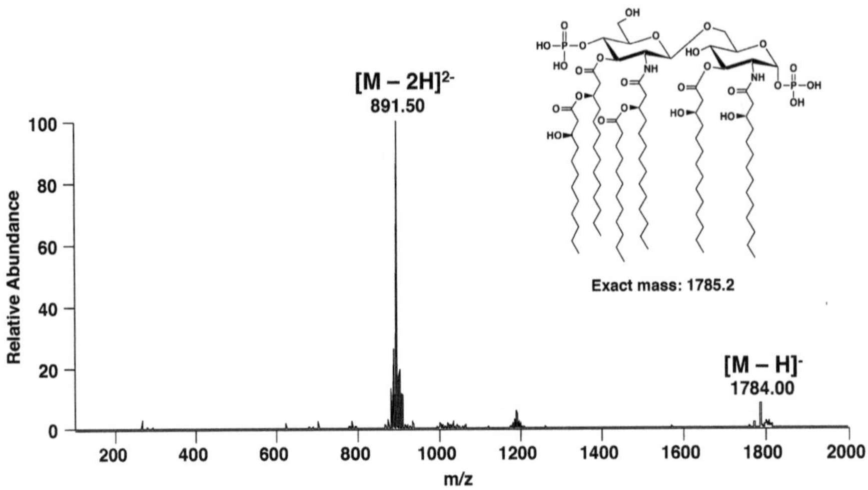

Fig. 4. Full negative mode nanoelectrospray ionization mass spectrum of a hexa-acylated *bis*-phosphorylated lipid A. The singly deprotonated lipid A ion is denoted by $(M\text{-}H)^-$ and the doubly deprotonated lipid A ion is denoted by $(M\text{-}2H)^{2-}$.

6. Tune the ion optics automatically or manually to enhance the ion signal of the lipid A analyte (Fig. 4).
7. Isolate and activate lipid A by selecting CID as the MS/MS method and isolating the precursor of interest.
8. Dissociate the selected precursor ion by increasing the CID voltage (or normalized collision energy) so that the precursor ion is only of 10% relative abundance as compared to the highest product ion (see Notes 23 and 24).
9. Collect spectra by averaging 3–300 scans or until high signal-to-noise is achieved for product ions (depends on the ion signal of the original precursor and how well the specific lipid A fragments) (Fig. 5a).

3.5. Ultraviolet Photodissociation

1. Isolate and fragment lipid A by selecting CID as MS/MS method and isolating the precursor ion of interest.
2. Set up the trigger so that the excimer laser only triggers when the CID parameter has no voltage applied (0% NCE). This requires a modest modification of the software.
3. Adjust the laser energy to between 1 and 8 mJ, and a repetition rate of 500 Hz.
4. Set the pulse generator to generate a laser pulse every 2 ms (500 Hz).
5. Activate the ion by increasing the laser energy and/or the number of laser pulses (thus increasing the total activation time) so that product ions are generated with high signal-to-noise ratio (see Notes 23–25).

6. Collect spectra by averaging 3–300 scans or until high signal-to-noise ratio is achieved for product ions (depends on the ion signal of the original precursor and how well the specific lipid A fragments) (Fig. 5b) (see Note 26).

3.6. ^{32}P-Labeled Lipid A Isolation

1. Inoculate 5 ml of medium using a single colony. Grow overnight at 37°C.
2. The next day, measure the OD_{600} and use the overnight culture to inoculate 7 ml of culture at a starting OD_{600} of 0.05. Add 2.5 μCi/ml of inorganic ^{32}P. Grow cells until an OD_{600} of 0.8–1.0 is reached.
3. Harvest the cells in glass centrifuge tubes with Teflon-lined cap using a fixed angle clinical centrifuge at 1,500 × *g* for 10 min (see Note 27). Discard the radioactive waste in an appropriate radioactive waste container.
4. Wash the cell pellet with 5 ml of 1× PBS. Centrifuge for 10 min at 1,500 × *g*. Discard the supernatant.
5. Resuspend the cells in 5 ml of single-phase Bligh–Dyer mixture (Fig. 2).
6. Vortex and incubate at room temperature for 20 min to ensure cell lysis.
7. Centrifuge in a clinical centrifuge at 1,500 × *g* for 20 min. Gently pour off the supernatant, which contains phospholipids and isoprenyl lipids.
8. Resuspend the LPS pellet in 1.8 ml of hydrolysis buffer via vortexing (see Note 10).
9. Sonicate the sample in a bath sonicator until the pellet is equally dispersed (~30 s).
10. Incubate the sample for 30 min in a boiling water bath (see Note 10).
11. Remove from the water bath and allow the sample to cool at room temperature for 5–10 min.
12. Convert the solution into a two-phase Bligh–Dyer mixture by adding 2 ml of chloroform and 2 ml of methanol, yielding a chloroform, methanol, aqueous (2:2:1.8 v/v) mixture (see Note 28).
13. Vortex and separate phases using a fixed angle clinical at 1,500 × *g* for 10 min. Extract the lower phase into a clean glass centrifuge tube using a glass pipet (first extraction).
14. Perform a second extraction on the sample by adding 2 ml of chloroform to the remaining upper phase from step 13.
15. Vortex and centrifuge at 1,500 × *g* for 10 min. Remove the lower phase and combine with the lower phase from the first extraction.

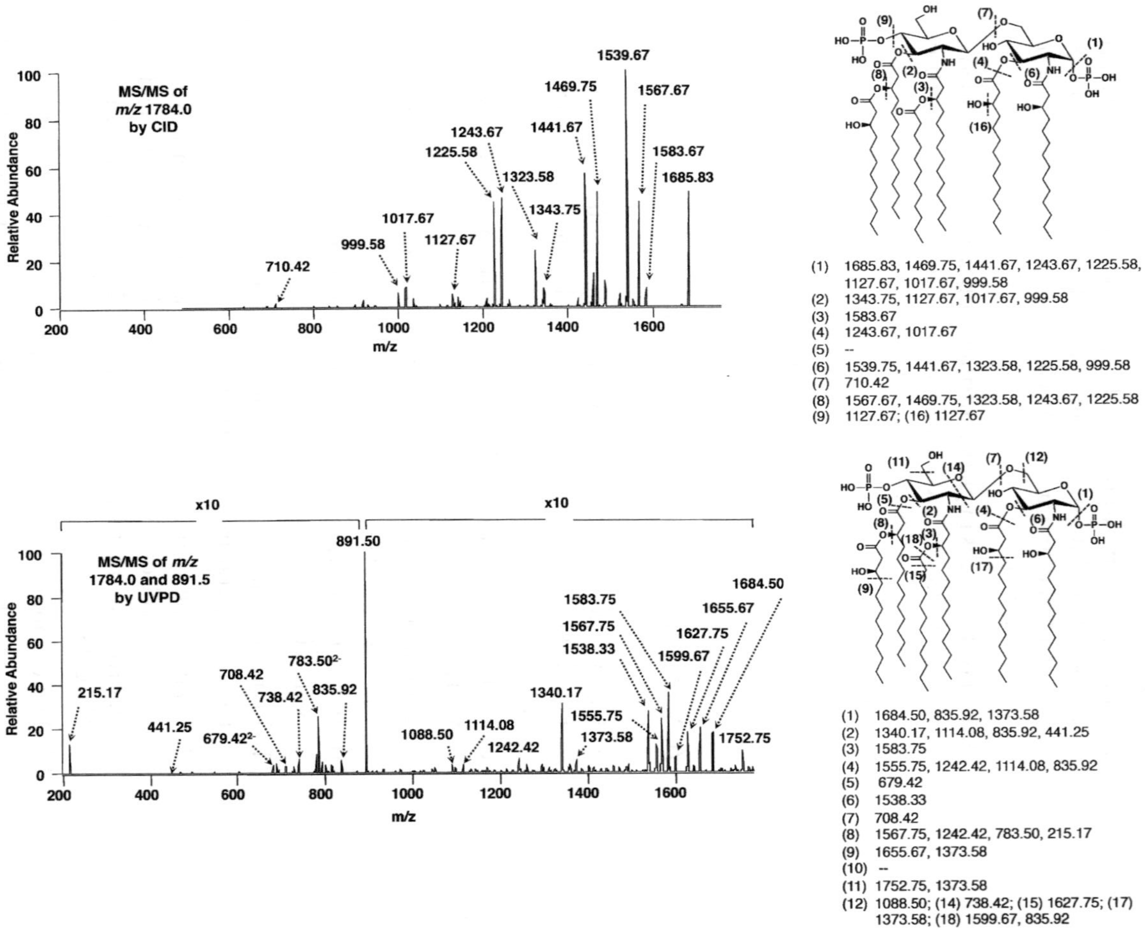

Fig. 5. (**a**) CID, (M-H)$^-$, and (**b**) 193 nm UVPD, (M-2H)$^{2-}$ mass spectra, and fragmentation maps for the hexa-acylated *bis*-phosphorylated lipid A shown in Fig. 4. (**a**) CID was performed on the deprotonated lipid, (M-H)$^-$, *m/z* 1,784, and (**b**) UVPD was undertaken on the doubly deprotonated lipid (M-2H)$^{2-}$, *m/z* 891.5. Fragmentation profiles are shown with *dashed lines* representing cleavage sites and are matched with the *m/z* values below the structure. The "×10" denotes a section of the spectrum that has been magnified ten times in order to more easily visualize product ions (Reproduced with permission from (3). Copyright (2011) American Chemical Society).

16. To the pooled lower phase (4 ml total volume), add 4 ml of methanol and 3.6 ml of water, yielding a two-phase Bligh–Dyer (chloroform, methanol, water; 2:2:1.8, v/v) (see Note 29).
17. Vortex and centrifuge at 1,500 × *g* for 10 min. Remove the lower phase to a clean glass tube and dry under a stream of nitrogen. Dried samples can be stored at −20°C until further use.

3.7. Visualization of ^{32}P-Labeled Lipid A Species via Thin Layer Chromatography

1. Prepare a TLC tank and TLC plates as described under Subheading 3.2.
2. Dissolve the ^{32}P-labeled sample in 500 μl of 4:1 chloroform:methanol (v/v). Vortex.
3. Add 5 μl of sample to a scintillation vial containing 5 ml of scintillation cocktail.
4. Count in scintillation counter and calculate the total CPMs of the sample.
5. Spot 10,000–20,000 cpm per sample on the TLC plate and allow the silica to dry (see Note 30).
6. Place the TLC plate containing the radiolabeled samples into the pre-equilibrated tank containing the TLC solvent system. Once the solvent front reaches the top of the TLC plate (~3 h), remove the plate and air-dry.
7. Wrap the plate in plastic wrap and expose to PhosphorImager Screen overnight.
8. The next morning, scan the screen to obtain the image (Fig. 6). Using imaging analysis software, this methodology also allows for quantitation of each lipid A species.

4. Notes

1. Move a balance to the fume hood and carefully weigh the SDS. SDS is harmful if inhaled.
2. Both methanol and chloroform should be HPLC grade to reduce clogging of nanoelectrospray tips.
3. Essentially, any mass spectrometer can be used as long as it has CID capabilities and can be outfitted with a nanoelectrospray ionization source.
4. Appropriate nanotips may vary between nanosources. Nanotips with smaller tip sizes (≤30 μm) can be used, but a higher rate of tip clogging may occur.
5. Culture volumes can be adjusted depending on the strain you are working with. For instance, *Vibrio cholerae* requires at least

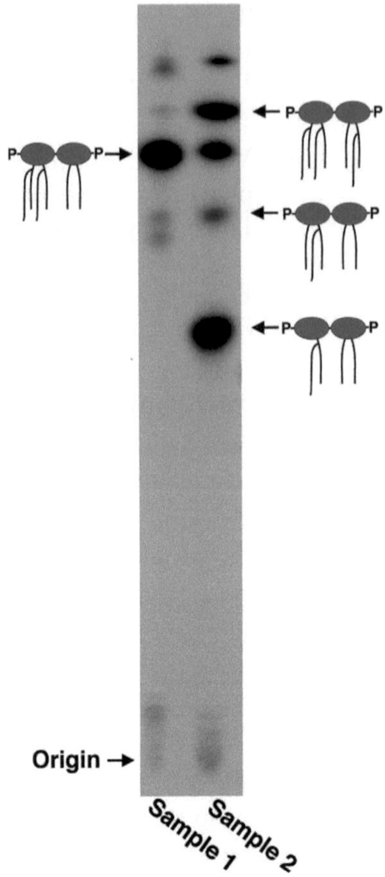

Fig. 6. TLC analysis of ^{32}P-labeled lipid A. Two different lipid A samples are shown. In Sample 1 a single lipid A species is present, whereas Sample 2 is heterogeneous. Cartoons are shown to indicate the lipid A structures.

200 ml of culture in order to obtain high quality mass spectra; however, the culture volume for *Escherichia coli* can be scaled down to 5 ml. More starting material is required for some bacterial species because the hydrolysis step that releases lipid A from the whole LPS is less efficient for some organisms. When scaling down the culture volume to 5 ml, split the sample into two glass centrifuge tubes. Cap the tubes with Teflon-lined caps and harvest cells using a fixed angle clinical centrifuge at 1,500×*g* for 10 min (see Subheading 3.6 for detailed instructions).

6. The length of centrifugation may need to be extended for bacterial cultures that are difficult to pellet.
7. If culture volume has been scaled down, wash the pellet in 10 ml of 1× PBS. Additionally, cell pellets can be stored at −20°C until further use. Alternatively, resuspend the pellet in 40 ml of 1× PBS and split into two 50 ml Falcon tubes (20 ml per tube). Store samples at −20°C.

8. If the culture volume has been scaled down, glass centrifuge tubes and a clinical centrifuge can be used for the lipid A isolation. Teflon-lined caps should be used with glass tubes. See Subheading 3.6 for detailed protocol using glass centrifuge tubes and the clinical centrifuge.
9. Additional wash steps may need to be included to reduce phospholipid contamination. When isolating lipid A from *E. coli* or *Salmonella*, only one wash is required; however, if isolating lipid A from other organisms (e.g., *V. cholerae* or *Helicobacter pylori*), additional wash steps are required.
10. By boiling the sample in 50 mM sodium acetate pH 4.5, 1% SDS, the lipid A domain is cleaved at the Kdo sugars from intact LPS (see Fig. 2). The boiling time should be increased to 1 h when isolating lipid A species from organisms (e.g., *V. cholerae*) which synthesize a phosphorylated-Kdo (26) domain to increase yield.
11. If isolating lipid A species of reduced hydrophobicity having fewer acyl chains (<5), an acidic Bligh–Dyer mixture can be used to increase yield. Add 225 μl of concentrated HCl to the SDS solution containing hydrolyzed lipid A followed by 30 ml of chloroform and 30 ml of methanol yielding a chloroform, methanol, 0.1 M HCl (2:2:1.8, v/v) mixture.
12. If isolating lipid A species of reduced hydrophobicity, use 60 ml of methanol and 54 ml of 0.1 M HCl to improve yield.
13. If using acidic Bligh–Dyers, add pyridine (1 drop of pyridine per 2 ml of final sample volume) to the sample before drying to neutralize the acid. Excess pyridine can lead to the removal of ester-linked fatty acids. Additionally, if lipid sample has difficulty drying, add a few milliliters of chloroform–methanol (4:1, v/v) to the flask and continue drying the sample to completion.
14. Two TLC plates (20 × 20 cm) can run in one TLC tank. One plate can be placed on one side of the ridge at the bottom of the tank. If it is necessary to run a second plate, place it on the other side of the ridge.
15. 10 × 20 cm TLC plates can also be used for charring lipid samples. If desired, the samples can be spotted along the 20 cm edge of the plate and placed in the tank, which shortens the run time from 3 h to ~45 min.
16. If possible, avoid spotting samples along the outside edges of the TLC plate. This helps solvent front and samples to migrate straight across plate.
17. Air dryers that have a cold and warm setting can also be used to dry the plates. If using a dryer, make sure to use the cold air

setting as the sample may degrade when exposed to hot air for extended periods of time.

18. Make sure hot plate is in fume hood prior to charring plate. Fumes from sulfuric acid–ethanol mixture are harmful.
19. If lipid spots are not visualized initially, allow plate to cool and spray again with sulfuric acid–ethanol mixture. Expose TLC plate to the hot plate and char again.
20. Our laboratory prefers using conical, "V"-shaped glass vials. This allows the lipid A sample to be concentrated in one location (in the bottom of the "V") as opposed to be spread out over the bottom of the tube. This is especially useful when suspending the dried sample in 20 μl of the chloroform–methanol mixture (4:1, v/v) for MS analysis.
21. Our laboratory utilizes Calibration mixture 1 from Sequazyme.
22. 20 μl of sample can be analyzed for up to approximately 1 h by directly infusing at 300 nl/min.
23. The characterization of lipid A structures is not necessarily impeded by incomplete annihilation of the precursor ion (as long as product ions are generated at a high signal-to-noise ratio).
24. Multiply charged ions dissociate more easily than singly charged ions due to coulombic repulsion. Thus, lower collision energies are generally used for higher precursor charge states.
25. UVPD is generally less sensitive than CID due to the generation of more product ions from ongoing secondary dissociation of primary product ions, thus subdividing the product ion current into more channels.
26. There are currently no commercially available automated spectral interpretation or database search programs to assist with the assignment of lipid A structures.
27. Do not centrifuge glass tubes above 1,500 × *g*, as higher speeds will cause glass tubes to shatter.
28. If isolating lipid A species of reduced hydrophobicity, an acidic Bligh–Dyer mixture can be used to increase yield. Add 15 μl of concentrated HCl to the SDS solution containing hydrolyzed lipid A followed by 2 ml of chloroform and 2 ml of methanol yielding a chloroform, methanol, 0.1 M HCl (2:2:1.8, v/v) mixture.
29. If isolating lipid A species of reduced hydrophobicity, use 4 ml of methanol and 3.6 ml of 0.1 M HCl to improve yield.
30. When visualizing radiolabeled lipid species, 10 × 20 cm or 20 × 20 cm TLC plates can be used. For 10 × 20 cm plates, samples should be spotted along the 10 cm edge of the plate to improve separation of lipid A species.

Acknowledgments

This work was supported by National Institutes of Health (NIH) Grants AI064184 and AI76322 to M.S.T. and Welch Foundation Grant F1155 and NIH RO1 GM103655 to J.S.B.

References

1. Trent MS et al (2006) Diversity of endotoxin and its impact on pathogenesis. J Endotoxin Res 12:205–223
2. Kim HM et al (2007) Crystal structure of the TLR4-MD-2 complex with bound endotoxin antagonist Eritoran. Cell 130:906–917
3. Raetz CR et al (2007) Lipid A modification systems in Gram-negative bacteria. Annu Rev Biochem 76:295–329
4. Yi EC, Hackett M (2000) Rapid isolation method for lipopolysaccharide and lipid A from gram-negative bacteria. Analyst 125:651–656
5. Darveau RP, Hancock RE (1983) Procedure for isolation of bacterial lipopolysaccharides from both smooth and rough Pseudomonas aeruginosa and Salmonella typhimurium strains. J Bacteriol 155:831–838
6. Galanos C, Luderitz O, Westphal O (1969) A new method for the extraction of R lipopolysaccharides. Eur J Biochem 9:245–249
7. Li J et al (1998) Development of an on-line preconcentration method for the analysis of pathogenic lipopolysaccharides using capillary electrophoresis-electrospray mass spectrometry. Application to small colony isolates. J Chromatogr 817:325–336
8. Westphal O, Jann K (1965) Bacterial lipopolysaccharide: extraction with phenol-water and further application of the procedure. Meth Carbohydr Chem 5:83–91
9. Caroff M et al (1991) Analysis of unmodified endotoxin preparations by 252Cf plasma desorption mass spectrometry: determination of molecular masses of the constituent native lipopolysaccharides. J Biol Chem 266: 18543–18549
10. Odegaard TJ et al (1997) Shortened hydroxyacyl chains on lipid A of Escherichia coli cells expressing a foreign UDP-N-acetylglucosamine O-acyltransferase. J Biol Chem 272:19688–19696
11. Banoub JH et al (2010) Structural investigation of bacterial lipopolysaccharides by mass spectrometry and tandem mass spectrometry. Mass Spectrom Rev 29:606–650
12. Madsen JA et al (2011) IR and UV photodissociation as analytical tools for characterizing Lipid A structures. Anal Chem (Washington, DC, US) 83:5107–5113
13. Hankins JV et al (2011) Elucidation of a novel *Vibrio cholerae* lipid A secondary hydroxy-acyltransferase and its role in innate immune recognition. Mol Microbiol 81:1313–1329
14. Chan S, Reinhold VN (1994) Detailed structural characterization of lipid A: electrospray ionization coupled with tandem mass spectrometry. Anal Biochem 218:63–73
15. Boue SM, Cole RB (2000) Confirmation of the structure of lipid A from Enterobacter agglomerans by electrospray ionization tandem mass spectrometry. J Mass Spectrom 35:361–368
16. Kussak A, Weintraub A (2002) Quadrupole ion-trap mass spectrometry to locate fatty acids on lipid A from Gram-negative bacteria. Anal Biochem 307:131–137
17. El-Aneed A, Banoub J (2005) Elucidation of the molecular structure of lipid A isolated from both a rough mutant and a wild strain of Aeromonas salmonicida lipopolysaccharides using electrospray ionization quadrupole time-of-flight tandem mass spectrometry. Rapid Commun Mass Spectrom 19:1683–1695
18. Murphy RC et al (2005) Mass spectrometry advances in lipidomica: collision-induced decomposition of Kdo_2-lipid A. Prostaglandins Other Lipid Mediat 77:131–140
19. Lee C-S et al (2004) Structural analysis of lipid A from Escherichia coli O157:H7:K- using thin-layer chromatography and ion-trap mass spectrometry. J Mass Spectrom 39:514–525
20. Wang Z, Li J, Altman E (2006) Structural characterization of the lipid A region of Aeromonas salmonicida subsp. salmonicida lipopolysaccharide. Carbohydr Res 341:2816–2825
21. Mikhail I et al (2005) Structural characterization of lipid A from nontypeable and type f Haemophilus influenzae: variability of fatty acid substitution. Anal Biochem 340:303–316
22. Madalinski G et al (2006) Gram-negative bacterial lipid A analysis by negative electrospray ion

trap mass spectrometry: stepwise dissociations of deprotonated species under low energy CID conditions. Int J Mass Spectrom 249/250:77–92

23. Silipo A et al (2008) Structural characterizations of lipids A by MS/MS of doubly charged ions on a hybrid linear ion trap/ orbitrap mass spectrometer. J Mass Spectrom 43:478–484
24. Jones JW et al (2008) Determination of pyrophosphorylated forms of lipid A in gram-negative bacteria using a multivaried mass spectrometric approach. Proc Natl Acad Sci U S A 105:12742–12747
25. Jones JW et al (2010) Comprehensive structure characterization of lipid A extracted from Yersinia pestis for determination of its phosphorylation configuration. J Am Soc Mass Spectrom 21:785–799
26. Hankins JV, Trent MS (2009) Secondary acylation of Vibrio cholerae lipopolysaccharide requires phosphorylation of Kdo. J Biol Chem 284:25804–25812

Chapter 16

Quantitative and Qualitative Preparations of Bacterial Outer Membrane Vesicles

Halima Chutkan, Ian MacDonald, Andrew Manning, and Meta J. Kuehn

Abstract

Gram-negative bacterial outer membrane vesicle production and function have been studied using a variety of quantitative and qualitative methods. These types of analyses can be hampered by the use of impure vesicle preparations. Here we describe a set of techniques that are useful for the quantitative analysis of vesicle production and for preparative yields of highly purified vesicles for studies of vesicle function or composition. Procedures and advice are also included for the purification of vesicles from encapsulated and low-yield strains.

Key words: Gram-negative bacteria, Outer membrane, Vesicle, Bleb, *Escherichia coli*, *Pseudomonas aeruginosa*, *Klebsiella pneumoniae*

1. Introduction

Outer membrane vesicles are spherical structures that bud naturally from all Gram-negative bacteria studied to date (Fig. 1) (1–4). Vesicles were first identified in the 1960s in cultures of *Escherichia coli* grown under lysine-limiting conditions but were thereafter observed in cultures of normally growing bacteria (5). Vesicle production, or vesiculation, is a natural process and is not a result of membrane instability (6). Vesiculation has been shown to occur in planktonic cultures, biofilms, and *in vivo* (7–9). Although the exact mechanism of vesicle formation is unknown, vesicles from a variety of species have been shown to play roles in survival and host interactions.

Understanding the composition and function and determining the quantity of vesicles require the isolation of vesicles from bacterial culture supernatants. Proper functional characterization as well as quantitative assessments of vesicle content and biogenesis rely

Anne H. Delcour (ed.), *Bacterial Cell Surfaces: Methods and Protocols*, Methods in Molecular Biology, vol. 966,
DOI 10.1007/978-1-62703-245-2_16,

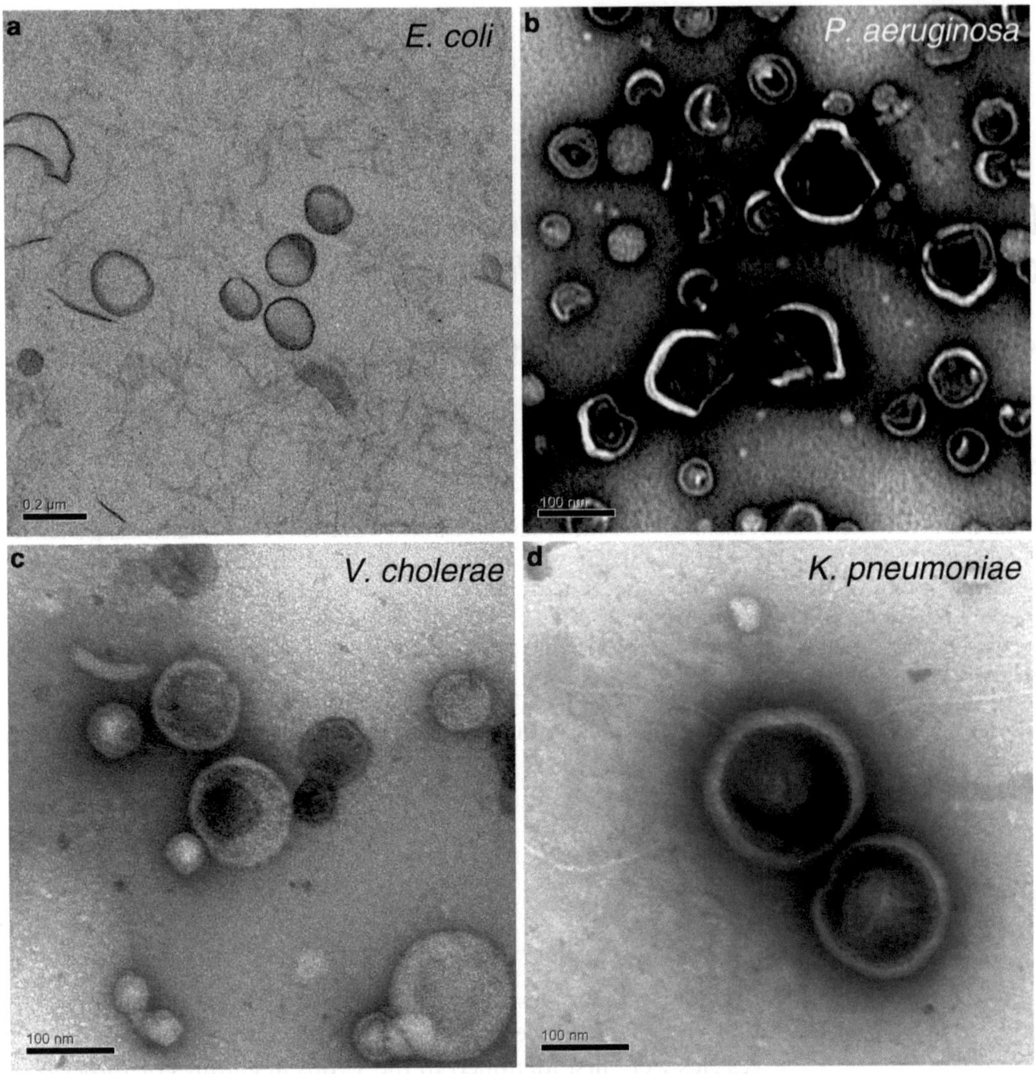

Fig. 1. Uranyl-acetate-stained electron micrographs of outer membrane vesicles purified from cultures of *E. coli*, *P. aeruginosa*, *Vibrio cholerae*, and *K. pneumoniae* (**a–d**, respectively). For electron microscopy procedure, *see* ref. 16. Size bar indicates 100 nm.

on the purity of the vesicles. If vesicles are simply sedimented by high-speed centrifugation from the culture supernatant they may be quite impure, depending on the propensity of other extracellular factors to co-sediment or co-precipitate with vesicles. These contaminants and other adherent extracellular material confuse studies of the vesicle proteome and host cell reactivity to vesicles. Differential density centrifugation of preparations generally allows the purification of vesicles away from such contaminants. However, as with any purification procedure, vesicle yield decreases as their purity increases. Therefore, other preparative and detection techniques must be used to acquire quantitative vesicle yield data.

Further complications in the preparation of vesicles arise from strains or species with low yields or which express capsule. In these cases, procedures can be modified to ensure sufficient vesicle yields.

For studies on the mechanism of vesicle formation or vesicle yield, a small-scale vesicle preparation is usually sufficient (6, 10). Vesicle quantity in these studies can be assessed by the specific analysis of outer membrane proteins and lipids in the pelleted supernatant and dividing by the amount of cells in the culture. These vesicle-specific probes would not detect contaminants, such as flagella, pili, and soluble factors, thus the vesicles produced from these small-scale preparations do not need to be further purified. Keeping the number of steps in quantitative vesicle analyses low reduces vesicle loss and error, increasing the accuracy and reproducibility of the data. By contrast, for studies on the interaction of vesicles with eukaryotic cells (e.g., toxicity or vesicle-mediated entry) or for vesicle content determination (e.g., proteomics or lipidomics), both a greater amount and a higher purity of vesicles are needed, and thus a large-scale vesicle preparation is required (11, 12). Density gradient centrifugation is commonly used to remove soluble and unassociated proteins from the vesicle fraction. The preparation of density-purified vesicles is particularly recommended for studies of vesicles from bacteria that secrete many other factors, such as *P. aeruginosa* (13). Although we generally recommend pooling the fractions which contain pure vesicles, it should also be considered that bacteria may produce vesicles of various densities with distinct functions. The nuances of the production and function of vesicles from the same organism, but with distinct densities, have yet to be explored.

Different protocols have also been used to isolate vesicles from different bacteria, including *Escherichia coli* (6), *Pseudomonas aeruginosa* (14), and *Neisseria meningitidis* (15). Here, we present protocols that have been used to efficiently isolate *E. coli* and *P. aeruginosa* vesicles for both quantitative (genetic) and qualitative (functional) studies (10, 11, 14). The *P. aeruginosa* protocol can be generally applied to bacteria that do not produce a lot of vesicles. Notes are included for capsulated strains (e.g., *Klebsiella pneumoniae*) since strongly associated cellular capsule can interfere with vesicle collection.

2. Materials

2.1. Small-Scale (Quantitative) Vesicle Preparation

2.1.1. General Preparation

1. 500 mL glass culture flask.
2. CFA (colonization factor antigen) broth: to approximately 3 L of water in a 4 L graduated cylinder or beaker, add 40 g casamino acids, 6 g yeast extract, 0.2 g $MgSO_4$, and 0.02 g $MnCl_2$. Adjust the pH to 7.4 using NaOH and then Q.S. (quantity

sufficient) to 4 L with water. Aliquot into culture flasks or bottles as necessary, autoclave, and store at room temperature.

3. LB broth: to 1 L of dH_2O in a 2 L culture flask, add 10 g of Bacto-tryptone, 5 g of yeast extract, and 10 g of NaCl. Autoclave and allow media to cool before inoculating.
4. DPBSS (Dulbecco's phosphate buffered saline supplemented with salt): to approximately 750 mL of water in a 2 L graduated cylinder or beaker, add 0.2 g KCl, 0.2 g KH_2PO_4, 11.70 g NaCl, 1.15 g Na_2HPO_4, 0.1 g $MgCl_2{\cdot}6H_2O$, and 0.1 g $CaCl_2$; Q.S. to 1 L with water.
5. Centrifuge, rotor, and 250 mL centrifuge bottles capable of spinning at 38,000 × *g*.
6. Vacuum filtration apparatus: glass filter holder and ground joint flask.
7. 0.45 μm PVDF Millipore filters, 90 mm.
8. 0.45 μm centrifugal filters.
9. LB agar plates: to approximately 750 mL of water in a 2 L beaker, add approximately 10 g of Bacto-tryptone, 5 g of yeast extract, and 10 g of NaCl and 15 g of agarose, and Q.S. to 1 L with water. Autoclave the solution and pour into plastic Petri plates. Allow the agar to set overnight. Store wrapped in plastic at 4°C.
10. Table top ultracentrifuge, rotor, and 1.7 mL tubes capable of spinning at 100,000 × *g*.
11. Protein concentration determination assay.
12. SDS-PAGE apparatus and gel, with Ruby or Coomassie stain.
13. FM4-64 fluorescent dye.
14. Spectrofluorometer.

2.1.2. Additional Material for Encapsulated Strains

1. Low-speed centrifuge with 250 and 500 mL polycarbonate centrifuge bottles.
2. 0.8 μm Cellulosic Acetate membrane filters, 90 mm.

2.2. Large-Scale Vesicle Preparation Using a Tangential Flow Concentrator

1. CFA or LB broth: see items 2 and 3, respectively, Subheading 2.1.1.
2. 3 L Fernbach culture flasks.
3. Low-speed centrifuge.
4. Tangential flow concentrator with a 100-kDa filter.
5. Concentrator wash buffer (1–2% Alconox): to approximately 2 L of water in a 4 L graduated cylinder or beaker, add approximately 45 g of Alconox, and Q.S. to 3 L with water.
6. Concentrator storage buffer (0.1 M NaOH): to approximately 2 L of water in a 4 L graduated cylinder or beaker, add 12 g of NaOH, mix, and Q.S. to 3 L with water.

7. 0.45-μm PVDF Millipore filter.
8. Glass vacuum filtration apparatus.
9. DPBSS: see item 4, Subheading 2.1.1.
10. LB agar: see item 9, Subheading 2.1.1.

2.3. Large-Scale Vesicle Preparation Without Concentrator

1. CFA or LB broth: see items 2 and 3, respectively, Subheading 2.1.1.
2. 3 L Fernbach culture flasks.
3. Low-speed centrifuge.
4. 0.45-μm PVDF Millipore filter.
5. Glass vacuum filtration apparatus.
6. DPBSS: see item 4, Subheading 2.1.1.

2.4. Small-Scale *P. aeruginosa* Vesicle Preparation

1. LB broth: see item 3, Subheading 2.1.1.
2. Amicon Ultra 10,000 MWCO centrifugation filters.
3. Slide-A-Lyzer® Dialysis Cassette 10,000 MWCO 3–12 m.
4. Low-speed centrifuge, rotor, 250 and 500 mL polycarbonate centrifuge bottles.
5. 50 mM Hepes, pH 7.5: to approximately 750 mL of water in a beaker add 11.9 g Hepes and adjust pH to 7.5 using NaOH. Q.S. to 1 L with water and filter-sterilize.
6. 0.45 μm PVDF Millipore filters.
7. Glass vacuum filtration apparatus.
8. 10 mL plastic syringe and needle.
9. Ammonium sulfate.

2.5. Purification of Vesicles by Density Gradient

1. OptiPrep diluent buffer: 0.85% NaCl in 10 mM Hepes, pH 7.4. Add 8.5 g NaCl and 2.38 g Hepes to approximately 750 mL of water in a graduated cylinder or beaker. Bring the pH to 7.4 with NaOH. Q.S. to 1 L with water and filter-sterilize.
2. OptiPrep solutions: dilute OptiPrep 60% stock solution in OptiPrep diluent buffer. The series of concentrations in the gradient steps depends on the strain of bacteria used, but a standard, recommended range of dilutions for the gradient is 40, 35, 30, 25, and 20% (final % Optiprep, v:v).
3. SDS-PAGE apparatus.
4. SDS-PAGE running buffer: to approximately 750 mL water in a graduated cylinder or beaker, add 14.4 g glycine, 3 g Tris base, and 1 g SDS. Q.S. to 1 L with water.
5. 15% SDS polyacrylamide gel: for the separating gel, mix 10.2 mL water with 7.5 mL 1.5 M Tris–HCl, pH 8.8, 0.15 mL 20% (w:v) SDS, and 12.0 mL Acrylamide/Bis-acrylamide

(30%/0.8% w:v). Add 0.15 mL 10% w:v ammonium persulfate (APS) and 0.02 mL tetramethylethylenediamine (TEMED).

6. Stacking gel (4% acrylamide): mix 3.075 mL water, 1.25 mL 0.5 M Tris–HCl, pH 6.8, 0.025 mL 20% w:v SDS, and 0.67 mL acrylamide/bis-acrylamide (30%:0.8%, w:v.) Add 0.025 mL 10% (w:v) APS and 0.005 mL TEMED.
7. 5× sample loading buffer: mix 0.1 M Tris–HCl, pH 6.8, 5 mL 100% glycerol, 0.5 g SDS, 1 mL β-mercaptoethanol, and 0.5 mL 2% bromophenol blue. Q.S. to 10 mL with dH_2O. Store aliquots at −20°C.
8. Ruby wash/fix solution: 7% acetic acid and 10% methanol in water. Add 70 mL acetic acid and 100 mL methanol to 830 mL dH_2O, and mix.
9. LB agar plates see item 9 Subheading 2.1.1.
10. 12.5-mL Ultraclear centrifuge tube.
11. Ultracentrifuge.
12. 0.45-μm PVDF Millipore centrifuge filters.
13. Tube rocker.

3. Methods

Perform all procedures at room temperature unless otherwise specified.

3.1. Small-Scale (Quantitative) Vesicle Preparation

1. Start an overnight culture in 5 mL of CFA or LB.
2. Inoculate 250 μL–2 mL of the overnight starter culture in each of up to six 500 mL-flasks containing 250 mL of CFA or LB (see Note 1).
3. Grow cultures at the desired temperature for the desired time and agitation (see Note 2).
4. Collect 1 mL, dilute 1,000- to 10,000-fold in sterile medium, and plate a tenfold dilution series to determine the amount of colony-forming units (CFU) in the culture (see Note 3).
5. To obtain vesicle-containing, cell-free supernatant for non-encapsulated strains, pour 240 mL of the growth culture into 250 mL bottles using a graduated cylinder and centrifuge at 10,000 × *g* for 10 min.
6. To obtain vesicle-containing, cell-free supernatant for encapsulated strains, follow these steps:
   (a) Centrifuge the cultures at 11,000 × *g* for 13 min in 500 mL polycarbonate bottles and collect the low-speed supernatant (see Note 4).

(b) Pour the low-speed supernatant into 250 mL polycarbonate bottles and centrifuge at 13,000 × *g* for 15 min and collect the mid-speed supernatant (see Note 5).

(c) Filter the mid-speed supernatant through a 0.8 μm filter, one sample at a time, and collect the filtrate that contains the vesicles. After each sample, remove the filter, wash the apparatus with 500 mL of dH_2O, replace with a new filter, and run the next sample through the apparatus.

7. Collect the supernatant in a sterile container, and filter it through a 0.45 μm Durapore PVDF Millipore filter using the glass vacuum filtration apparatus.
8. Centrifuge the filtered supernatant at 38,400 × *g* for 1–2 h (see Note 6).
9. Decant the supernatant (see Note 7).
10. Resuspend the vesicle-containing pellet in the residual supernatant remaining in the centrifuge bottle after decanting (approximately 1 mL) or resuspend in 1 mL fresh DPBSS. Transfer to 1.7 mL ultracentrifuge tube and centrifuge in tabletop ultracentrifuge at 100,000 × *g* for 1 h.
11. Decant the supernatant and resuspend the pellet in 100 μL DPBSS (see Note 8). The resuspended pellet is the vesicle preparation.
12. Filter-sterilize the vesicles using a 0.45 μm centrifugal filter. Check for sterility by plating on LB agar. Re-sterilize using 0.45 μm filters if needed.
13. To quantitate the vesicle yield in terms of mg vesicle protein/CFU, use a protein concentration determination assay to determine the total protein concentration in the vesicle preparation and divide the number by the CFU obtained from dilution plating of the culture at the time of harvest. In some cases, flagella and other non-vesicle proteins contaminate the preparation; however, further purification steps (e.g., using density gradient purification) prevent quantitative recovery. Therefore, vesicle yields often can be best compared using the quantity of vesicle-specific protein or lipid in the pelleted cell-free supernatant preparations. Here, an aliquot of the vesicle preparation is run on SDS-PAGE, stained for protein using Ruby or Coomassie, and either the total protein in each sample or the major outer membrane proteins (e.g., Omps F/C and A for *E. coli*) in each sample are determined by densitometry. Subsequently, the densitometry value is divided by the CFU and this vesicle yield compared between strains or treatments. Finally, vesicle yield can also be determined based on lipid content using FM4-64, a lipophilic fluorescent dye. For FM4-64-based measurements, 20 μL vesicle preparation is diluted in 560 μL of DPBSS, and 20 μL FM4-64 is added. Fluorescence

(RFU) is measured with an excitation of 506 nm and emission of 750 nm using a spectrofluorometer. Subsequently, the RFU value is divided by the CFU and this vesicle yield value compared between strains or treatments.

3.2. Large-Scale Vesicle Preparation Using a Tangential Flow Concentrator

1. Start an overnight culture in 10 mL of CFA or LB.
2. Autoclave 3 L culture flasks with 1.5 L of medium in each (12 L total).
3. Inoculate each flask with 750 μL of culture and grow at 37°C overnight.
4. Centrifuge the cultures at 15,000 × *g* for 15 min. Collect and pool the supernatants (see Note 9).
5. Drain the concentrator reservoir of concentrator storage solution and rinse with dH_2O.
6. Add the supernatants to the concentrator and turn on the flow through the filter. Take care to ensure that the volume in the reservoir does not go below 250 mL. Add the supernatant to the reservoir until all of the supernatant has been concentrated.
7. Once all of the supernatant has been concentrated down to 250 mL, backwash with 250 mL of medium through the filter for a total of 500 mL.
8. Drain the supernatant from the reservoir into centrifuge bottles, bypassing the filter, and centrifuge at 6,000 × *g* for 6 min.
9. Filter the supernatant through a 0.45-μm PVDF Millipore filter using the glass vacuum filtration apparatus.
10. Centrifuge the supernatant at 38,400 × *g* for 1–2 h and resuspend the pellet in a small volume of DPBSS (see Note 8).
11. Streak out a small amount of the resuspended vesicle-containing pellet on an LB plate to verify sample sterility. Re-filter if necessary.
12. Wash the concentrator filter with 2 L of concentrator wash buffer. Backwash approximately 500 mL of concentrator wash buffer through the filter into the reservoir. Bypassing the filter, drain the reservoir, and rinse with dH_2O until the reservoir is no longer sudsy. Drain 2.5 L dH_2O through the filter. Rinse the filter with 500 mL of concentrator storage buffer and store the reservoir with 2 L of concentrator storage buffer.

3.3. Large-Scale Vesicle Preparation Without Concentrator

1. Follow steps 1–4 of Subheading 3.2.
2. Filter each batch of supernatant through 0.45 μm PVDF Millipore filters using the glass vacuum filtration apparatus and store the cell-free supernatant at 4°C.
3. Once all of the cultures have been centrifuged and filtered, add the cell-free supernatant to six centrifuge bottles, and centrifuge at 38,400 × *g* for 1 h.

4. Decant the supernatant, add another aliquot of cell-free supernatant to the same bottles, and re-centrifuge until all of the cell-free supernatant has been pelleted.
5. Resuspend the vesicle pellets in five of the bottles and add the resuspended vesicles to the sixth bottle to pool all vesicle pellets.
6. Centrifuge the pooled vesicles at 38,000 × *g* for 1–2 h, decant the supernatant, and resuspend the vesicle pellet in the smallest possible volume of DPBSS.

3.4. Small-Scale *P. aeruginosa* Vesicle Preparation

1. Start an overnight culture in 5 mL of LB broth.
2. Inoculate a 1 L flask containing 250 mL of LB broth with 5 mL of culture and incubate at 37°C until an OD_{600} ~0.9–1.0 (see Note 10).
3. Pour the culture into a centrifuge bottle and centrifuge at 10,000 × *g* for 10 min. Collect the supernatant.
4. Filter the supernatant through a 0.45 μm PVDF Millipore filter using the glass vacuum filtration apparatus.
5. Pour the filtered supernatant into a 500 mL centrifuge bottle.
6. Add 129 g of ammonium sulfate to the filtered supernatant, add a stir bar, and dissolve at room temperature at moderate speed on a stir-plate.
7. Once dissolved, move to 4°C and continue stirring for at least 3 h to precipitate the vesicle-containing fraction from the supernatant.
8. Remove the stir bar and centrifuge the sample for 20 min at 10,000 × *g* (see Note 11).
9. Decant the supernatant carefully, without disrupting the pellet or the precipitate on the side of the bottle.
10. Resuspend the precipitate by gently rocking the bottle with 10 mL of 50 mM Hepes, pH 7.5 buffer.
11. Using a syringe and needle, inject the resuspended sample into 3–12 mL 10,000 molecular weight cut-off slide-a-lyzer dialysis cassettes. Remove the remaining air bubbles from the cassettes.
12. Dialyze the vesicle preparations in 50 mM Hepes, pH 7.5 buffer for at least 12 h at 4°C, with stirring. Use at least 100-fold the sample volume of buffer.
13. Collect, pool, and concentrate the samples to a final volume of 500 μL using 10,000 molecular weight cut-off Amicon Ultra centrifugation filters (see Note 12).

3.5. Purification of Vesicles by Density Gradient

1. Add OptiPrep (60% stock) to the resuspended vesicles at a ratio of 1:3 (by volume) to adjust the vesicle preparation to 45% OptiPrep (v:v) (see Note 13).

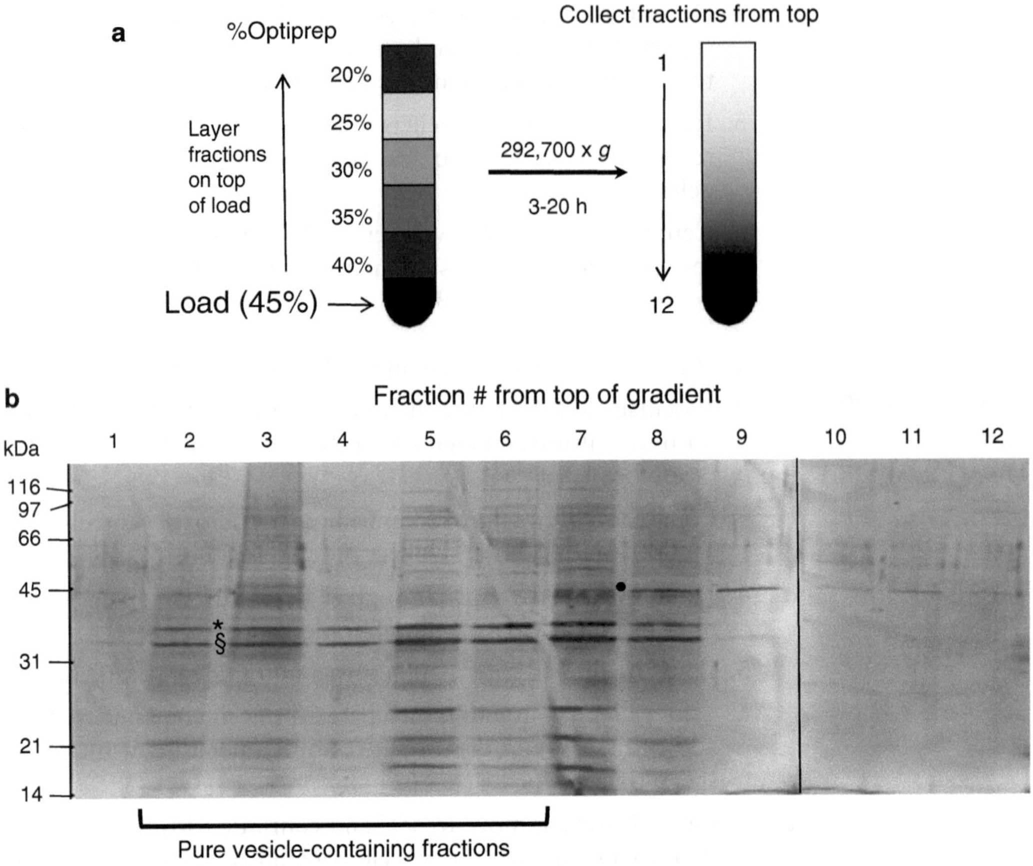

Fig. 2. (**a**) Schematic of ultracentrifuge tubes loaded with the vesicle preparation in the 45% Optiprep at the bottom of the tube, layered with decreasing density of Optiprep prior to and after centrifugation. After the gradient has reached equilibrium (3 h to overnight), fractions (typically 12 equal fractions of approximately 1 mL each) are removed sequentially from the top of the tube. (**b**) Optiprep gradient fractionation analysis by SDS-PAGE. Vesicles were collected from a culture of enterotoxigenic *E. coli* grown in LB, and the vesicle-containing pelleted supernatant was loaded on a 45–20% Optiprep gradient and ultracentrifuged. Twelve fractions (1 mL each) were collected and analyzed by Coomassie-stained 15% SDS-PAGE. The migration of molecular weight standards is indicated (kDa). Symbols * and § indicate the migration of OmpsF/C and OmpA, respectively, which are vesicle outer membrane proteins; • indicates the migration of a contaminating, non-vesicle protein that was present in the material originally pelleted at 38,400 × *g* from the culture supernatant. Pure vesicle-containing fractions are *bracketed*.

2. Pipette 2 mL of vesicles in 45% OptiPrep to the bottom of a 12.5-mL Ultraclear centrifuge tube.
3. Carefully layer 2 mL of each OptiPrep dilutions in descending order on the top of the preceding layer (see Note 14) (Fig. 2a).
4. Centrifuge the tube at 292,700 × *g* (120,000× *g* for *P. aeruginosa*) in an ultracentrifuge for 3 h to overnight.
5. Collect sequential fractions by placing the tip of a 1 mL pipette at the top of the gradient and carefully removing 1 mL (see Note 15).

6. Analyze a portion (12 μL) of each fraction using 15% SDS-PAGE and visualize the proteins in the gel using Coomassie or Ruby protein staining. Identify the vesicle fractions by the presence of outer membrane proteins (OmpF/C and OmpA for *E. coli* and OprF/D and OprH for *P. aeruginosa*) (Fig. 2b).
7. Pool the selected vesicle-containing fractions, and add them to a centrifuge bottle with at least tenfold the sample volume using DPBSS.
8. Centrifuge at 38,400 ×*g* for at least 3 h to remove the OptiPrep.
9. Resuspend the vesicle pellet in the smallest volume of DPBSS, and filter-sterilize the vesicles using 0.45-μm PVDF Millipore centrifuge filters (see Note 16).
10. Confirm sterility by plating an aliquot of vesicles on LB agar plates. Filter again through 0.45-μm PVDF Millipore centrifuge filters, if necessary.

4. Notes

1. If the preparations will be compared for quantitative yield (e.g., amount of vesicles/CFU), the preparation of duplicate samples is suggested to improve accuracy. Consequently, given the capacity limit of the rotor, a typical run can be used to compare three strains in duplicate at a time.
2. For encapsulated strains, stop growth incubation once the culture reaches an OD_{600} of 1.0–1.6 and place the culture on ice. Growth is restricted for capsulated strains to partially alleviate the problems of pouring off supernatants of heavily capsulated bacteria. The more you allow the strains to grow the harder it is to separate the capsule from the supernatant.
3. Plating of samples of the culture diluted 10^{-5}-, 10^{-6}-, and 10^{-7}-fold is recommended for determining an accurate bacterial count.
4. Encapsulated strains will have a sloppy pellet, meaning that as the supernatant is decanted, a substantial amount of capsule will mix with the supernatant after the first spin. This is why two pelleting steps are recommended. It is, however, important that a different centrifuge bottle is used for each spin (or at the least, that they are washed between spins) as the capsule will not stick to the bottle as well if there is already a pellet there.
5. This spin step is used to capture all of the capsule in the pellet of each bottle. Pour cautiously, and attempt to not disturb the pellet so as to mix the capsule back into the supernatant. If the

pellet is coming off the bottle wall, attempt to collect as much supernatant as possible (and avoid pouring off the pellet), but retain some of the supernatant, supplement with uncentrifuged cell-free supernatant to balance the bottles, and centrifuge again before filtration step.

6. Because the vesicle pellet may not be visible after this step, it is easiest to use a lab marker to designate the area on the outside of the centrifuge bottle where the pellet will form.
7. Vesicle pellets may not be visible, use caution when pouring off your supernatant and watch your pellet the entire time. If the sample appears to have disappeared, stop pouring and spin again. To ensure that no excess liquid leads to sample dilution, it may help to vacuum aspirate all liquid from the lip and wall of each centrifuge flask.
8. Generally 100 μL will resuspend the pellet without a problem. If the pellet is too large, resuspend in a more appropriate volume but do not forget to account for that change when comparing yields for samples. However, if the vesicle pellet is not visible after centrifugation, focus the pipetting of the 100 μL to the spot that was marked on the centrifuge bottle to identify the location of the vesicle pellet. Resuspending the pellet in DPBSS may result in a frothy solution. To ensure that the entire vesicle fraction is collected, take care to collect all of the bubbles.
9. The cell pellets may be very loose. Take care not to pour out the pellet with the supernatant. If the pellet is too loose, leave some supernatant in the bottom of the tube and re-centrifuge the bottle with fresh culture.
10. Cultures of *P. aeruginosa* are only grown to late log phase because degradative enzymes that could degrade outer membrane vesicles are produced during stationary phase.
11. If the precipitate remains suspended in the supernatant, re-centrifuge as necessary.
12. During concentration, the sample will become compacted at the bottom of the concentrator. Between spins, resuspend the concentrated sample by gentle pipetting.
13. OptiPrep is very viscous. We recommend using disposable plastic pipettes near a flame for all OptiPrep manipulations. When pipetting the OptiPrep solution, draw in approximately 1 mL more OptiPrep than necessary, and then during dispensing, ensure that the liquid drains down to the appropriate measurement.
14. Place the tip of a plastic pipette to the inside of the tube below the rim and slowly add each OptiPrep solution in a slow, steady stream. There should be a distinct boundary between each layer.

15. Because OptiPrep is so viscous, we recommend using a new tip for each tube, even for the same fraction number.
16. Sometimes, the vesicle suspension will be too concentrated to pass through the filter and may be retained on the filter. In these cases, a visual pellet will be visible on the filter. Resuspend the pellet in sterile DPBSS and re-sterilize the vesicles by filtering the suspension using more centrifugal filters.

Acknowledgements

This work was supported by NIH grants R01AI079068, R01AI064464, and R01GM099471-01. We thank Harold Meekel (University of North Carolina-Chapel Hill) for assistance with electron microscopy.

References

1. Kulp A, Kuehn MJ (2010) Biological functions and biogenesis of secreted bacterial outer membrane vesicles. Annu Rev Microbiol 64:163–184
2. Kuehn MJ, Kesty NC (2005) Bacterial outer membrane vesicles and the host-pathogen interaction. Genes Dev 19(22):2645–2655
3. Lee EY, Choi DS, Kim KP, Gho YS (2008) Proteomics in gram-negative bacterial outer membrane vesicles. Mass Spectrom Rev 27(6):535–555
4. Beveridge TJ (1999) Structures of gram-negative cell walls and their derived membrane vesicles. J Bacteriol 181(16):4725–4733
5. Bishop DG, Work E (1965) An extracellular glycolipid produced by Escherichia coli grown under lysine-limiting conditions. Biochem J 96(2):567–576
6. McBroom AJ, Johnson AP, Vemulapalli S, Kuehn MJ (2006) Outer membrane vesicle production by Escherichia coli is independent of membrane instability. J Bacteriol 188(15): 5385–5392
7. Ciofu O, Beveridge TJ, Kadurugamuwa J, Walther-Rasmussen J, Høiby N (2000) Chromosomal β-lactamase is packaged into membrane vesicles and secreted from Pseudomonas aeruginosa. J Antimicrob Chemother 45(1):9–13
8. Schooling SR, Beveridge TJ (2006) Membrane vesicles: an overlooked component of the matrices of biofilms. J Bacteriol 188(16): 5945–5957
9. Hellman J, Loiselle PM, Zanzot EM, Allaire JE, Tehan MM, Boyle LA, Kurnick JT, Warren HS (2000) Release of gram-negative outer-membrane proteins into human serum and septic rat blood and their interactions with immunoglobulin in antiserum to Escherichia coli J5. J Infect Dis 181(3):1034–1043
10. McBroom AJ, Kuehn MJ (2007) Release of outer membrane vesicles by Gram-negative bacteria is a novel envelope stress response. Mol Microbiol 63(2):545–558
11. Chutkan H, Kuehn MJ (2011) Context-dependent activation kinetics elicited by soluble versus outer membrane vesicle-associated heat-labile enterotoxin. Infect Immun 79(9): 3760–3769
12. Ellis TN, Leiman SA, Kuehn MJ (2010) Naturally produced outer membrane vesicles from Pseudomonas aeruginosa elicit a potent innate immune response via combined sensing of both lipopolysaccharide and protein components. Infect Immun 78(9):3822–3831
13. Bauman SJ, Kuehn MJ (2006) Purification of outer membrane vesicles from Pseudomonas aeruginosa and their activation of an IL-8 response. Microbes Infect 8(9–10):2400–2408
14. Bauman SJ, Kuehn MJ (2009) Pseudomonas aeruginosa vesicles associate with and are internalized by human lung epithelial cells. BMC Microbiol 9:26
15. Shoemaker DR, Saunders NB, Brandt BL, Moran EE, LaClair AD, Zollinger WD (2005) Intranasal delivery of group B meningococcal

native outer membrane vesicle vaccine induces local mucosal and serum bactericidal antibody responses in rabbits. Infect Immun 73(8): 5031–5038

16. Manning AJ, Kuehn MJ (2011) Contribution of bacterial outer membrane vesicles to innate bacterial defense. BMC Microbiol 11:258

Chapter 17

In Vitro Peptidoglycan Synthesis Assay with Lipid II Substrate

Jacob Biboy, Nhat Khai Bui, and Waldemar Vollmer

Abstract

Bacterial cell wall peptidoglycan is synthesized from lipid II precursor by two reactions. Glycosyltransferases polymerize the glycan chains and transpeptidases form the peptide cross-links. The bifunctional class A penicillin-binding proteins catalyze both of these reactions. Here, we describe an in vitro peptidoglycan synthesis assay utilizing radiolabeled lipid II substrate to monitor simultaneously peptidoglycan glycosyltransferase and transpeptidase activities. The products of the reaction are separated by high-pressure liquid chromatography and quantified by flow-through scintillation counting.

Key words: Peptidoglycan, Lipid II, Glycosyltransferase, Transpeptidase

1. Introduction

The essential peptidoglycan sacculus encases the cytoplasmic membrane in most bacteria to confer osmotic stability and to maintain cell shape (1). Peptidoglycan is composed of glycan chains of alternating *N*-acetylmuramic acid (Mur*N*Ac) and *N*-acetylglucosamine (Glc*N*Ac) residues which are cross-linked by short peptides. It forms a single, net-like layer in Gram-negative bacteria and a thick, multilayered structure with attached teichoic acid chains in Gram-positive bacteria. The pathway of peptidoglycan biosynthesis is targeted by important classes of antibiotics such as the β-lactams and glycopeptides.

Peptidoglycan is synthesized at the outer leaflet of the inner membrane from lipid II precursor by two enzymatic reactions. The glycan chains are polymerized by glycosyltransferases (2) and the peptide cross-links are formed by transpeptidases; the latter enzymes covalently bind β-lactams such as penicillin and are therefore called penicillin-binding proteins (PBPs) (3). The Gram-negative model bacterium *Escherichia coli* has six known

Anne H. Delcour (ed.), *Bacterial Cell Surfaces: Methods and Protocols*, Methods in Molecular Biology, vol. 966, DOI 10.1007/978-1-62703-245-2_17, © Springer Science+Business Media New York 2013

peptidoglycan synthases including the bifunctional glycosyltransferase-transpeptidases PBP1A, PBP1B, and PBP1C (also called class A PBPs), the transpeptidases PBP2 and PBP3 (class B PBPs), and the glycosyltransferase MtgA (4). PBP1A and PBP1B provide the main peptidoglycan synthesis activity in *E. coli* and have semi-redundant roles in the cell cycle. PBP2 is essential for cell elongation and PBP3 is essential for cell division, whereas the role of PBP1C and MtgA is not known (4).

During growth and cell division the sacculus is enlarged by the insertion of new peptidoglycan chains or patches into the existing layer presumably by the activities of membrane-attached, dynamic multienzyme complexes composed of peptidoglycan synthases and hydrolases (5). The precise composition of these complexes and the mechanisms by which they operate have remained unknown. Cytoskeletal proteins and associated cell morphogenesis proteins interact with peptidoglycan enzymes and position and/or control peptidoglycan synthesis complexes from inside the cell (6). The tubulin-like FtsZ forms a dynamic ring at mid-cell to which more than 12 essential cell division proteins including the peptidoglycan transpeptidase PBP3 localize. PBP3 interacts with the bifunctional peptidoglycan glycosyltransferase-transpeptidase PBP1B (7) and both synthases interact with the essential cell division protein FtsN (8). The actin-like MreB is required for peptidoglycan synthesis during cell elongation (9, 10). Filaments or patches of MreB move along the short axis of the cell depending on ongoing peptidoglycan synthesis (11). In *E. coli* and presumably other Gram-negative bacteria peptidoglycan synthases are regulated by outer-membrane anchored lipoproteins LpoA and LpoB (12, 13). LpoA specifically activates the transpeptidase activity of PBP1A, and LpoB activates PBP1B's transpeptidase and glycosyltransferase activities. Activation of peptidoglycan synthesis from the outer membrane could be a way to couple the rate of peptidoglycan growth with the cell's overall growth rate (14).

Deciphering the molecular mechanisms of peptidoglycan growth requires suitable in vitro assays for peptidoglycan synthases. There are several challenges to establish such assays. The PBPs are membrane proteins and thus require a detergent for solubilization. Many PBPs are notoriously difficult to purify and, once purified, are prone to denaturing. In addition, the lipid II substrate has a complicated structure and has not been commercially available until recently. In *E. coli* and other Gram-negative bacteria lipid II has the following structure: undecaprenylpyrophosphoryl-Mur*N*Ac(Glc*N*Ac)-L-Ala-γ-D-Glu-*meso*-Dap-D-Ala-D-Ala (*meso*-Dap, *meso*-diaminopimelic acid). A continuous assay with fluorescence labeled lipid II allows the characterization of glycosyltransferase activity in the absence of transpeptidation (15) and this assay has recently been adopted to microtiter format (16). Linear glycosyltransferase products, i.e., the glycan chain oligomers without peptide cross-links, have been separated by SDS-PAGE (17). Early studies used radioactive labeled lipid II to demonstrate peptidoglycan glycosyltransferase and/or transpeptidase activity of *E. coli* membrane preparations or with partially purified PBPs (18–20). Recently, the methods for the synthesis of lipid II have been improved, allowing preparation in larger amounts either semienzymatically (21) or by total chemical synthesis (22). Active PBPs can be purified in mg-amounts from overexpression strains (23, 24) and the reaction products can be separated by high-pressure liquid chromatography (HPLC) (25, 26).

Here, we describe the purification of *E. coli* PBP1B (Fig. 1) and the in vitro peptidoglycan synthesis assay performed in our laboratory (Figs. 2 and 3) (25). The protocol describes the purification of PBP1B from a 4 l culture of *E. coli* BL21 pDML924

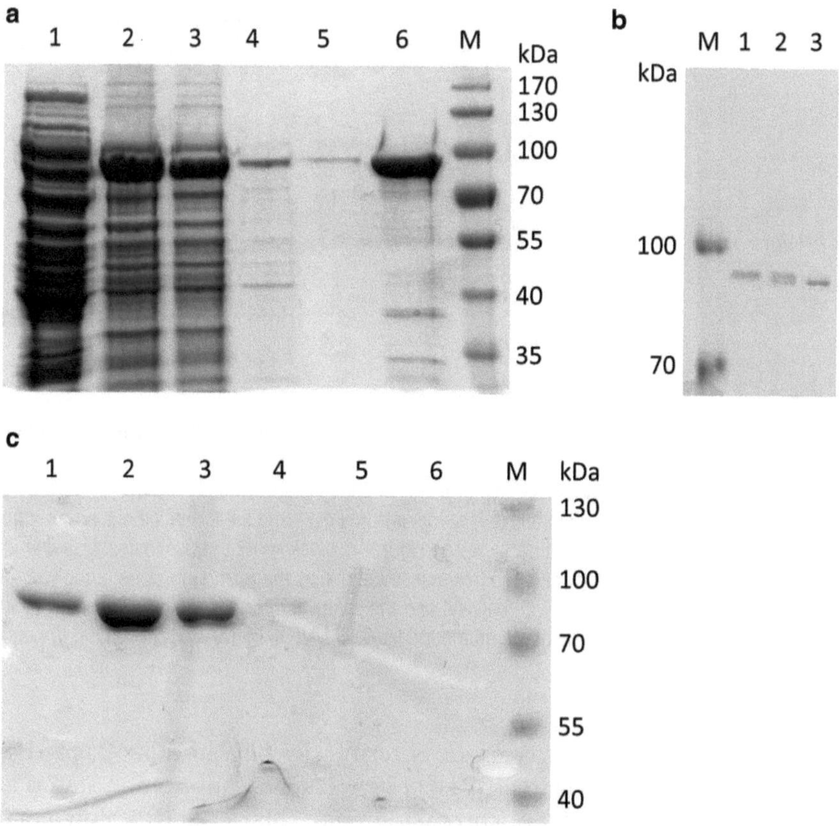

Fig. 1. Purification of PBP1B. (**a**) Coomassie Blue-stained 10% SDS-PAGE gel of 10 μl aliquots of the supernatant after centrifugation of the sonicated cells (**step 15** of Subheading 3.1.1; *lane 1*), the membrane extract (**step 19** of Subheading 3.1.1; *lane 2*), the flow-through after binding to Ni-NTA beads (**step 5** of Subheading 3.1.2; *lane 3*), the first wash fraction (**step 6** of Subheading 3.1.2; *lane 4*), the tenth wash fraction (**step 6** of Subheading 3.1.2; *lane 5*), the elution fraction (**step 7** of Subheading 3.1.2; *lane 6*), and the protein molecular weight marker (*lane 7*). (**b**) Coomassie Blue-stained 8% SDS-PAGE gel showing ~1 μg of His-PBP1B before thrombin cleavage (*lane 1*), thrombin-cleaved PBP1B (*lane 3*), and a mixture of both samples (*lane 2*). (**c**) Coomassie Blue-stained 10% SDS-PAGE gel of 10 μl aliquots of FPLC fractions 1–6 around the PBP1B elution peak (**step 7** of Subheading 3.1.3; *lanes 1–6*) and the protein molecular weight marker (*lane M*).

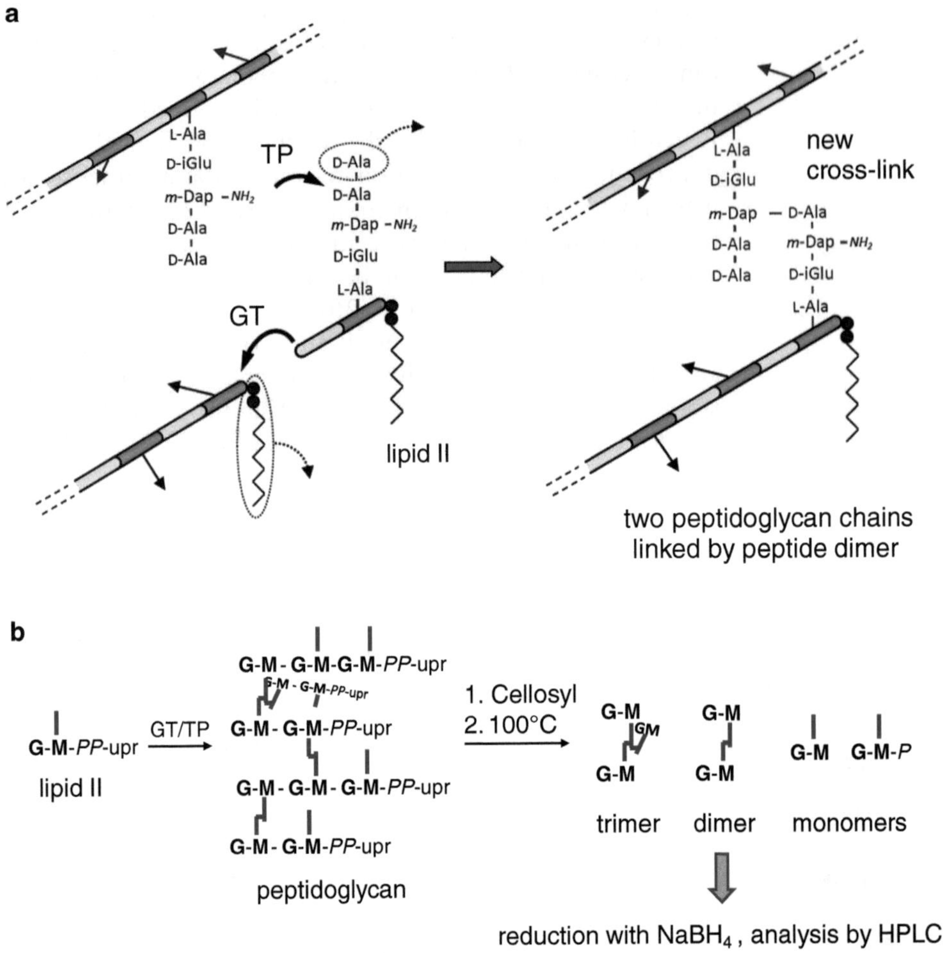

Fig. 2. Peptidoglycan synthesis reactions and principle of the peptidoglycan synthesis assay. (**a**) Cartoon showing the elongation of a glycan chain by glycosyltransferase (GT) reaction from lipid II and the formation of a new peptide cross-link by transpeptidation (TP). *light gray bar*, GlcNAc; *dark gray bar*, MurNAc, *black sphere*, phosphate; *zigzag line*, undecaprenyl lipid. (**b**) Principle of the in vitro peptidoglycan synthesis assay showing the formation of peptidoglycan from lipid II, followed by its degradation to muropeptides with cellosyl and the removal of undecaprenyl phosphate by incubation at 100°C. The latter leaves a phosphate residue at the muropeptide. *G* GlcNAc, *M* MurNAc, *P* phosphate, *upr* undecaprenyl.

which expresses an N-terminally oligohistidine-tagged version of PBP1B (His-PBP1B) (24). The first purification step is a Ni-NTA metal affinity chromatography using a gravity flow column. The oligohistidine tag is removed by the protease thrombin during the following dialysis step. The protein is then further purified by cation exchange chromatography by FPLC. Using this protocol we routinely obtain 12–20 mg of enzyme at a concentration of 0.8–1.5 mg/ml, which can be used for the peptidoglycan synthesis assay. In this assay we routinely use radioactively labeled lipid II, but the assay can be adopted for unlabeled lipid II which is now commercially available from the UK Bacterial Cell Wall Biosynthesis

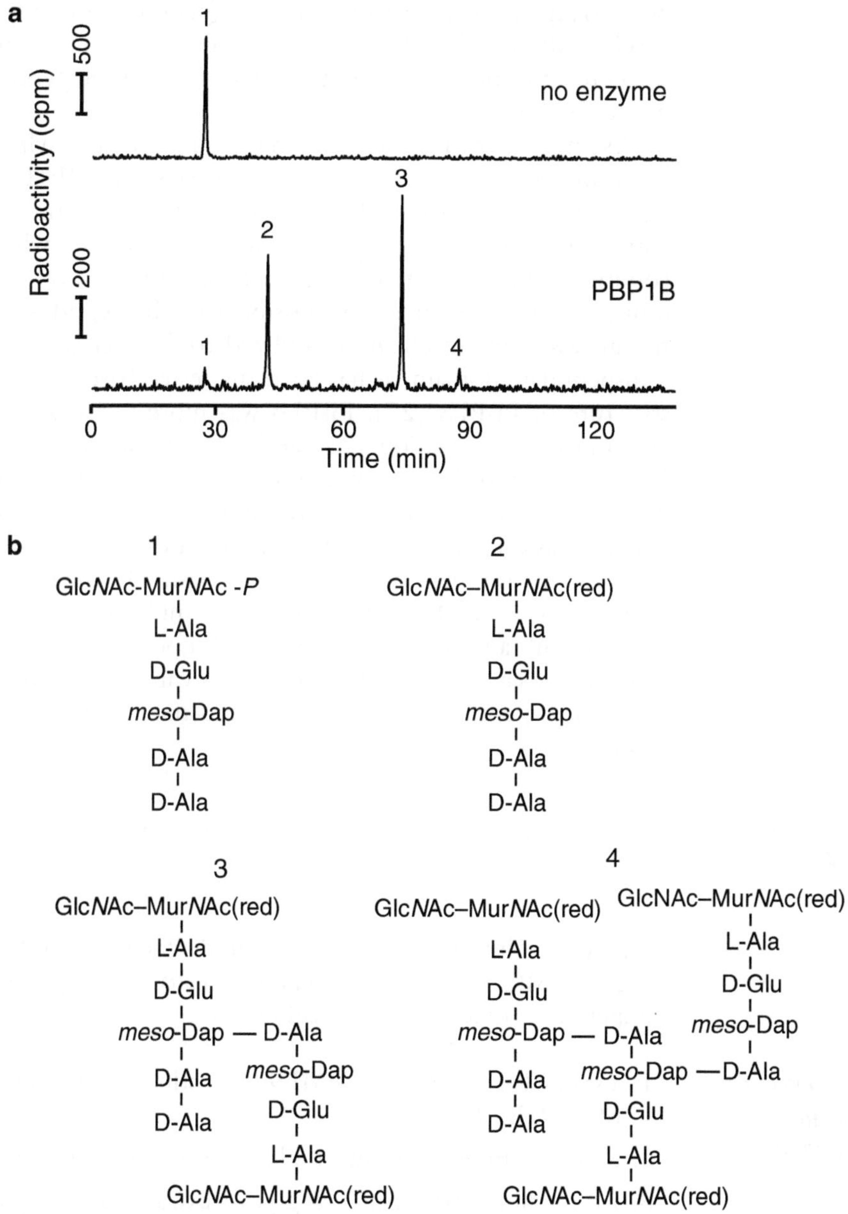

Fig. 3. Analysis of muropeptides by HPLC. (**a**) Lipid II was incubated with PBP1B followed by digestion with the muramidase cellosyl and reduction with sodium borohydride. The muropeptides were separated by reversed-phase HPLC and detected by flow-through scintillation counting. Fraction 1, monophosphorylated disaccharide pentapeptide; fraction 2, disaccharide pentapeptide (glycosyltransferase product); fraction 3, bis-disaccharide tetrapentapeptide (glycosyltransferase-transpeptidase product); fraction 4, tris-disacharide teratetrapentapeptide (glycosyltransferase-transpeptidase product). (**b**) Structures of the muropeptides separated in (**a**).

Network (UK-BaCWAN). Lipid II is utilized by PBP1B to form a cross-linked peptidoglycan product (Fig. 2). Digestion of this product with the muramidase cellosyl results in characteristic pattern of uncross-linked (monomeric) and cross-linked (dimeric and trimeric) muropeptides which we reduce with sodium borohydride prior to separation by HPLC (Fig. 3). Quantification of the muropeptides allows calculation of the average glycan chain length and the percent of peptides present in cross-links in the peptidoglycan synthesized in vitro.

We have used this assay to demonstrate the glycosyltransferase and transpeptidase activities of *E. coli* PBP1A and PBP1B (23, 25). For both bifunctional synthases we found that efficient transpeptidation requires ongoing glycosyltransferase reactions, whereas glycosyltransferase activity was independent of ongoing transpeptidation. Hence, it is possible that the peptides present in the growing glycan chain are utilized to form cross-links by the transpeptidase domain, as has been suggested based on the crystal structure of PBP1B (27). PBP1B was more active at conditions favoring its dimerization (25) and was stimulated by FtsN (8). PBP1A required the presence of oligomeric peptidoglycan chains or sacculi for transpeptidation reaction, and we have shown that PBP1A was capable of attaching newly synthesized peptidoglycan to sacculi by transpeptidation (23). Interestingly, the transpeptidation activity of PBP1A and PBP1B was stimulated by their cognate outer membrane activator, LpoA and LpoB, respectively (13). Thus, the assay described in this chapter can be used to study the activities of peptidoglycan synthases and the effects of interacting proteins.

2. Materials

Prepare all solutions with analytical grade chemicals and ultrapure water (18 MΩ cm at 25°C). Prepare and store all reagents at room temperature (unless indicated otherwise).

2.1. Purification of the Peptidoglycan Synthase PBP1B

2.1.1. Overproduction of PBP1B and Its Solubilization from Cell Membranes

1. *Escherichia coli* BL21 pDML924 producing His-tagged form of PBP1B (24).
2. Stock solution of 50 mg/ml kanamycin. Store at −20°C.
3. LB medium: 10 g NaCl, 5 g yeast extract, and 10 g tryptone per liter, pH 7.2.
4. Orbital shaker.
5. 2-l baffled flasks.
6. Spectrophotometer.
7. Centrifuge and associated rotor.
8. 0.5 M IPTG (isopropyl 1-thio-β-D-galactopyranoside) in water.
9. PBP1B basic buffer: 25 mM Tris–HCl pH 7.5, 10 mM $MgCl_2$, 1 M NaCl, 1 mM EGTA.

10. DNAse (freeze-dried).
11. 100 mM PMSF (phenylmethanesulfonylfluoride or phenylmethylsulfonyl fluoride) in ethanol.
12. 5 mg/ml Protease inhibitor cocktail in DMSO.
13. French press or probe sonicator.
14. Potter S Homogenizer 30 ml.
15. Light microscope.
16. Ultracentrifuge and associated rotor.
17. Triton X-100, purified for membrane research. Store at 4°C (see Note 1).
18. PPB1B extraction buffer: 25 mM Tris–HCl pH 7.5, 10 mM $MgCl_2$, 1 M NaCl, 20% glycerol, 2% Triton X-100.

2.1.2. Ni-NTA Affinity Chromatography

1. Ni-NTA-beads.
2. PPB1B extraction buffer: see item 18, Subheading 2.1.1.
3. Rotary wheel.
4. D-Tube Dialyzer Maxi; MWCO 6–8 kDa.
5. SDS-PAGE apparatus and buffers.
6. Coomassie Blue.
7. 1.8 U/μl Thrombin (restriction grade).
8. 5 M imidazole, pH 7.5.
9. Disposable columns (5 ml).
10. PPB1B dialysis buffer 1: 25 mM Tris–HCl pH 7.5, 10 mM $MgCl_2$, 1 M NaCl, 0.5 mM EGTA, 20% glycerol.

2.1.3. Ion Exchange Chromatography

1. PBP1B dialysis buffer 2: 10 mM sodium acetate pH 5.0, 10 mM $MgCl_2$, 0.02% NaN_3, 10% Glycerol, 1 M NaCl.
2. PBP1B dialysis buffer 3: 10 mM sodium acetate pH 5.0, 10 mM $MgCl_2$, 0.02% NaN_3, 10% Glycerol, 300 mM NaCl.
3. PBP1B dialysis buffer 4: 10 mM sodium acetate pH 5.0, 10 mM $MgCl_2$, 0.02% NaN_3, 10% Glycerol, 100 mM NaCl.
4. PBP1B dialysis buffer 5: 10 mM sodium acetate pH 5.0, 10 mM $MgCl_2$, 0.02% NaN_3, 10% Glycerol, 500 mM NaCl, 0.25% TX-100.
5. HiTrap SP HP 5 ml ion exchange column
6. PBP1B FPLC buffer A: 10 mM sodium acetate pH 5.0, 10 mM $MgCl_2$, 0.02% NaN_3, 10% Glycerol, 100 mM NaCl, 0.25% TX-100.
7. PBP1B FPLC buffer B: 10 mM sodium acetate pH 5.0, 10 mM $MgCl_2$, 0.02% NaN_3, 10% glycerol, 2 M NaCl, 0.25% TX-100.
8. Protein concentration determination assay.

2.2. Peptidoglycan Synthesis Reaction

1. The peptidoglycan precursor lipid II (undecaprenylpyrophosphoryl-*N*-acetylmuramic acid-(*N*-acetylglucosamine)-L-Ala-D-Glu-(γ)-*meso*-Dap-D-Ala-D-Ala) can be purchased from the synthetic facility of the UK Bacterial Cell Wall Biosynthesis Network (UK-BaCWAN), University of Warwick, UK (http://www2.warwick.ac.uk/fac/sci/lifesci/people/droper/bacwan/synthesis/). The facility also offers lipid II with L-Lys at position 3 of the peptide (instead of *meso*-Dap) and various versions of the UDP-linked precursors. In this work we use (^{14}C)GlcNAc-labeled lipid II (*meso*-Dap version, 0.15 mM, 10,000–12,000 dpm/ml) obtained from the laboratory of Dr. Eefjan Breukink (Department of Biochemistry of Membranes, University of Utrecht, The Netherlands). Lipid II is dissolved in chloroform/methanol 1:1 and stored at −80°C. Labeled and unlabeled versions of lipid II have been produced in the laboratories of Dr. Suzanne Walker (Harvard University, Massachusetts, MA, USA) and Dr. Shahriar Mobashery (University of Notre-Dame, Indiana, IN, USA).
2. Insert Micro Glass tubes (450 μl, 6 mm).
3. pH indicator paper: Acilit Strips, pH 0–6 and Neutralit Strips, pH 5–10.
4. PBP1B assay buffer (10×): 100 mM Hepes/NaOH, pH 7.5, 30 mM $MgCl_2$.
5. 1% Triton X-100 in 10 mM Hepes/NaOH, pH 7.5.
6. 1 M NaCl.
7. 20% Phosphoric acid.
8. Thermal shaker.
9. Dry bed heating block.

2.3. Muramidase Digestion and Reduction of the Muropeptides

1. The muramidase cellosyl is not commercially available. We use a cellosyl stock which has been provided by Höchst AG, Frankfurt, Germany. The commercially available muramidase mutanolysin (Sigma, Dorset, UK) has the same amino acid sequence as cellosyl and can be alternatively used at the same conditions. The enzyme is freeze-dried and stored long-term at 4°C. The muramidase is dissolved in 20 mM sodium phosphate, pH 4.8 and stored in 100 μl aliquots at −20°C. These stock solutions for daily use are active for several months.
2. Thermal shaker.
3. Dry bed heating block.
4. Microcentrifuge.
5. 0.5 M Sodium borate buffer pH 9.
6. Sodium borohydride (powder).

7. pH indicator paper: Acilit Strips, pH 0–6 and Neutralit Strips, pH 5–10.
8. 20% Phosphoric acid.

2.4. Separation of Muropeptides by HPLC

1. HPLC apparatus with binary pump system and radioactivity detector and/or UV detector.
2. Methanol.
3. HPLC buffer A: 50 mM sodium phosphate, pH 4.31 with 10 μl of 10% sodium azide per liter of buffer (see Note 2).
4. HPLC buffer B: 75 mM sodium phosphate, pH 4.75, 15% methanol (see Note 2).
5. HPLC column: Prontosil 120-3-C18 AQ reversed-phase column, 250×4.6 mm, 3 μm particle size.
6. Scintillation liquid Floscint III.

3. Methods

3.1. Purification of the Peptidoglycan Synthase PBP1B

3.1.1. Overproduction of PBP1B and Its Solubilization from Cell Membranes

1. Prepare 2× 50 ml of LB medium (with 50 μg/ml Kanamycin) in 250 ml flasks.
2. Inoculate each flask with a 10 μl loop from a glycerol stock of *E. coli* BL21 pDML924 cells. Incubate the culture on an orbital shaker at 37°C overnight for 14–16 h (see Note 3).
3. Prepare 10×400 ml of LB medium (with 50 μg/ml Kanamycin) in 2 l baffled flasks.
4. Inoculate each flask with 5 ml of overnight-grown culture from step 2.
5. Incubate the culture on an orbital shaker at 37°C to an OD_{578} of 0.7 (see Note 3).
6. Induce the overproduction of His-tagged PBP1B by adding 0.8 ml of a 0.5 M IPTG stock solution per flask and continue growing the cells for 5 h (see Note 4).
7. Place the flask on ice for 10 min.
8. Harvest the cells by centrifugation at 8,000×*g* for 20 min at 4°C (see Note 5).
9. Discard the supernatant and measure the wet weight of the cell pellet.
10. All further steps are either done on ice or in a cold room at 4°C.
11. Resuspend the cell pellet in PBP1B basic buffer (~5 ml for 1 g of wet cells weight).

12. Add a small amount of DNAse (~1–2 mg) to the resuspended cells.
13. Add the appropriate amounts of a 100 mM PMSF stock solution and of a 5 mg/ml Protease inhibitor cocktail to bring the final concentrations to 1 mM PMSF and 1 μg/ml of Protease inhibitor cocktail.
14. Disrupt the cells either by using a French press at 700 bar or by sonication with 6–8 pulses at 45 W (30 s "ON," 60 s "OFF"). Check for quantitative cell disruption by light microscopy.
15. Centrifuge the disrupted cell suspension at 130,000 ×*g* for 1 h at 4°C. Save 10 μl of the supernatant for SDS-PAGE analysis (step 8 in Subheading 3.1.2) and discard the rest.
16. Resuspend the pellet in 30 ml of PBP1B extraction buffer (see Note 6).
17. Stir the suspension at medium speed for at least 4 h at 4°C.
18. Centrifuge the suspension at 130,000 ×*g* for 1 h at 4°C. Reserve the supernatant and discard the pellet.
19. The His-tagged PBP1B and other membrane proteins are present in the supernatant. Save a 10 μl aliquot of the supernatant for SDS-PAGE analysis (**step 8** in Subheading 3.1.2).

3.1.2. Ni-NTA Affinity Chromatography

1. Take a 5 ml suspension of Ni-NTA beads (containing ~2.5 ml of beads) and wash them by centrifugation/suspension steps with 10 ml of water (three times) and 10 ml of PBP1B extraction buffer with 15 mM imidazole (three times).
2. Transfer the supernatant from step 18 in Subheading 3.1.1 to the pellet of Ni-NTA beads in a 50 ml tube.
3. Incubate the bead suspension overnight at 4°C on a rotary wheel.
4. Transfer the bead suspension into a disposable gravity column (see Note 7).
5. Let the beads settle on the bottom of the column and collect the flow-through. Take a 10 μl aliquot of each fraction for SDS-PAGE analysis (see Note 8).
6. Wash the beads with 5 ml of PBP1B extraction buffer with 15 mM imidazole. Repeat this wash step ten times. Collect the flow-through samples of each washing step and take a 10 μl aliquot of them for SDS-PAGE analysis (step 8 below).
7. Elute the His-tagged PBP1B by adding 5 ml of PBP1B extraction buffer with 400 mM imidazole. Repeat this step twice. Collect the eluates and take a 10 μl aliquot of each fraction for SDS-PAGE analysis (step 8 below).

8. Perform an 8 or 10% SDS-polyacrylamide gel electrophoresis to analyze samples from step 18 of Subheading 3.1.1 and steps 5–7 above. Stain the gel with Coomassie Blue and identify the PBP1B-containing elution fractions. Verify the absence of large amounts of PBP1B in the wash steps. Figure 1a shows an SDS-PAGE gel of samples from the Ni-NTA affinity chromatography.
9. Combine and transfer the PBP1B-containing elution fractions into a dialysis tube.
10. Dialyze the sample against 1 l of PBP1B dialysis buffer 1 for 1 h at 4°C. Repeat this step four times. Save an aliquot of the dialyzed sample for SDS-PAGE (step 14 below) (see Note 9).
11. Add 20 μl of 1.8 U/μl Thrombin into the dialysis tube and continue dialysing against 1 l of PBP1B dialysis buffer 1 for 1 h at 4°C.
12. Change the dialysis buffer and continue dialysing overnight.
13. Change the dialysis buffer and continue dialysing for 1 h.
14. Check for the completeness of the Thrombin cleavage by analyzing ~1 μg PBP1B containing aliquots of the pre-thrombin-cleaved sample (step 10 above) and of the post-thrombin-cleaved sample (step 13 above) by 8% SDS-PAGE using a long (20 cm) gel (*see* Fig. 1b for an example SDS-PAGE gel) (see Note 10).

3.1.3. Ion Exchange Chromatography

1. Dialyze the PBP1B solution for 45 min against 1 l of PBP1B dialysis buffer 2. Repeat this step two times.
2. Dialyze the protein solution overnight against 1 l of PBP1B dialysis buffer 3.
3. Dilute the protein solution with the same volume of PBP1B dialysis buffer 4.
4. Dialyze the protein solution for 1 h against 1 l of PBP1B dialysis buffer 4. Repeat this step three times every half an hour (see Note 11).
5. Prepare the HiTrap SP HP FPLC column by running at a flow rate of 2 ml/min ten column volumes of water, ten column volumes of PBP1B FPLC buffer B, and ten column volumes of PBP1B FPLC buffer A.
6. Reduce the flow rate to 1 ml/min and apply the protein solution from step 4 above to the column.
7. Apply a linear, 65-ml gradient from 100% PBP1B buffer A to 100% PBP1B buffer B at a flow rate of 1 ml/min. Detect proteins at 280 nm. Collect fractions of 5 ml.
8. Analyze the fractions by 8% SDS-polyacrylamide gel electrophoresis. Figure 1c shows an SDS-PAGE of PBP1B-containing fractions after ion exchange chromatography.

9. Combine fractions containing PBP1B (fractions 1–3, Fig. 1c) and transfer them into a dialysis tube.
10. Dialyze the PBP1B solution against 1 l of PBP1B dialysis buffer 5 for 1 h. Repeat this step three times.
11. Prepare 50–500 µl aliquot of the purified PBP1B solution and store them at −80°C.
12. Determine the concentration of the purified PBP1B using a method compatible with Triton X-100, $MgCl_2$, and glycerol.

3.2. Peptidoglycan Synthesis Reaction

The peptidoglycan synthesis reaction and sample preparation for HPLC are performed in an Insert Micro Glass vial placed into a 1.5-ml plastics reaction tube with lid. The total volume of the reaction is 50 µl. Carry out the following steps for each enzyme reaction.

1. Dry 1.2 nmol (11,000 dpm) of lipid II in vacuo and dissolve it in 3 µl of 1% Triton X-100 (see Note 12).
2. Prepare an assay buffer mixture calculated to obtain final concentrations of 10 mM Hepes/NaOH, pH 7.5, 3 mM $MgCl_2$, 150 mM NaCl, 0.18% Triton X-100 in a total assay volume of 50 µl. For the calculation take into account the volume of enzyme to be added in step 4.
3. Add the assay buffer mixture to the vial with the dried lipid II (step 1) and incubate the sample for 10 min on ice to dissolve the lipid II.
4. Add a sufficient volume of the purified PBP1B to obtain 0.94 µM (final concentration) to start the reaction.
5. Incubate the sample for 60 min at 37°C on a thermal shaker (see Note 13).
6. Stop the reaction by acidifying the sample to a pH value of 4.8–5.0 using 20% phosphoric acid (see Notes 14 and 15).
7. Boil the sample on a dry bed heating block at 100°C for 5–10 min and cool to ambient temperature.

3.3. Muramidase Digestion and Reduction of the Muropeptides

1. Add 10 µl of 0.5 mg/ml cellosyl to the sample and incubate for 60 min at 37°C on a thermal shaker (see Notes 13 and 16).
2. Stop the muramidase digestion by boiling for 10 min on a dry bed heating block at 100°C.
3. Centrifuge the sample for 15 min at maximum speed in a microcentrifuge and transfer the supernatant in a 1.5 ml microcentrifuge tube.
4. To reduce the muropeptides add an equal volume (~60 µl) of 0.5 M sodium borate, pH 9 and a small amount of solid sodium borohydride (tip of a spatula, approximately 0.5–1 mg) (see Notes 16–18).

5. Stop the reduction process after 30 min by adjusting the pH to a value of 4.0–4.5 with 20% phosphoric acid (see Notes 14, 15, and 17).
6. Centrifuge the sample for 15 min at maximum speed in a microcentrifuge and transfer the supernatant to an HPLC vial.

3.4. Separation of Muropeptides by HPLC

The HPLC systems for muropeptide analysis should have a binary pump system and should be connected to a flow-through radioactivity detector (see Note 19). Alternatively, or additionally, the reaction products can be visualized by a UV detector set at 205 nm (see Note 20). A C18 reversed-phase column is used to separate the reduced muropeptides according to Glauner (26). The flow rate of the HPLC system is set to 0.5 ml/min and the column temperature is at 55°C.

1. Prepare the HPLC column by passing 100% Methanol through the column for 15 min at ambient temperature.
2. Heat the column to 55°C and clean the system for another 30 min with 100% methanol.
3. Wash the column for 60 min with ultrapure water.
4. Wash the column with 15 ml of HPLC buffer B, followed by equilibration with 30 ml of HPLC buffer A.
5. Inject the sample and flush the injection loop for 2 min with HPLC buffer A.
6. Start the linear elution gradient for 135 min from 100% HPLC buffer A to 100% HPLC buffer B.
7. Re-equilibrate the column with 100% HPLC buffer A for 50 min for analysis of the next sample (see Note 21).

4. Data Analysis

Analyze the radioactivity profile with the software provided by the supplier of the HPLC system or the detector.

1. Identify the muropeptide fractions based on their retention time (Fig. 3a).
2. Determine the peak area of each muropeptide fraction and calculate the relative percentage of each muropeptide.
3. Calculate the percent peptides in cross-links and the average length of the glycan chains using the following formula:

$$\text{Peptides in cross links (\% of total)} = \%\text{peak } 3 + \%\text{peak } 4.$$

$$\text{Average glycan chain length (disaccharide units)} = 100\% / \%\text{peak } 1.$$

5. Notes

1. We recommend storing Triton X-100 at 4°C. At room temperature Triton X-100 accumulates peroxide radicals which can inactivate enzymes.
2. We prepare sodium phosphate HPLC running buffers according to Glauner (26) by dissolving the required amount of solid sodium hydroxide in water and adjusting the pH value with phosphoric acid.
3. Adjust the speed of the orbital shaker for good aeration of the liquid culture.
4. Cells should be harvested no longer than 5 h after the addition of IPTG because they may begin to lyse after this point.
5. Weigh the empty centrifuge tubes/buckets before centrifugation for determining the wet weight of the cell pellet.
6. Carefully homogenize the cell pellet with a Potter S Homogenizer.
7. Care should be taken to always cover the surface of the Ni-NTA beads with liquid when collecting them in the gravity column.
8. Avoid drying out the beads by fitting a small tubing with a clamp at the outlet of the column, which allows to stop the gravity flow.
9. This dialysis step is to remove the imidazole before Thrombin cleavage.
10. Use a long gel to separate the Thrombin-cleaved and uncleaved PBP1B versions.
11. The protein may precipitate at low NaCl concentration and therefore the dialysis steps are shorter than usual.
12. We recommend performing the peptidoglycan synthesis reaction in a glass vial because lipid II is prone to adhering to plastic tubes. Lipid II is stored in a chloroform/methanol (1:1) solvent. Prior to the enzyme assay the solvent is removed by vacuum drying lipid II in a SpeedVac apparatus for 5 min at room temperature.
13. The shaking speed should ensure good mixing of the reaction components during the incubation period.
14. The acid should not be added directly into the sample. Instead, small aliquots (0.5–1 μl) of the acid are added to the side wall of the tube which is then briefly centrifuged.
15. The pH is measured by taking a 0.5 μl aliquot of the sample onto the pH indicator paper.
16. The reaction product of lipid II is digested by muramidase resulting in muropeptides, the disaccharide peptide units of

peptidoglycan. Muropeptides are either without peptide cross-link (monomer) or contain one or more peptide cross-links (dimer, trimer, ...) (Fig. 3b).

17. Substantial amount of hydrogen (H_2) gas is formed and released during the reduction process and thereafter when the sample is acidified. We recommend punching a hole into the lid of the vial with a needle to prevent sudden opening of the lid and loss of sample.
18. The samples are reduced to convert the two anomeric forms of the Mur*N*Ac residues present in the muropeptides into muramitol, thereby improving muropeptide separation.
19. The flow cell volume of the scintillation detector may differ depending on the manufacturer. The scintillation liquid flow-rates should be set according to the manufacturer's instruction to obtain an optimal signal-to-noise ratio.
20. The sensitivity of UV detection of muropeptides is lower than for the radioactive lipid II used with high specific radioactivity.
21. The later period of the gradient (135–180 min) allows for equilibrating the column for the next run.

Acknowledgements

We thank Carolin Ewers for the SDS-PAGE analysis shown in Fig. 1b. This work was supported by the Biotechnology and Biological Sciences Research Council (BBSRC) and the EU Commission within the DIVINOCELL and EUR-INTAFAR projects.

References

1. Vollmer W, Blanot D, de Pedro MA (2008) Peptidoglycan structure and architecture. FEMS Microbiol Rev 32:149–167
2. Lovering AL, de Castro LH, Lim D, Strynadka NC (2007) Structural insight into the transglycosylation step of bacterial cell-wall biosynthesis. Science 315:1402–1405
3. Sauvage E, Kerff F, Terrak M, Ayala JA, Charlier P (2008) The penicillin-binding proteins: structure and role in peptidoglycan biosynthesis. FEMS Microbiol Rev 32:234–258
4. Vollmer W, Bertsche U (2008) Murein (peptidoglycan) structure, architecture and biosynthesis in *Escherichia coli*. Biochim Biophys Acta 1778:1714–1734
5. Höltje J-V (1998) Growth of the stress-bearing and shape-maintaining murein sacculus of *Escherichia coli*. Microbiol Mol Biol Rev 62:181–203
6. Cabeen MT, Jacobs-Wagner C (2010) The bacterial cytoskeleton. Annu Rev Genet 44:365–392
7. Bertsche U, Kast T, Wolf B, Fraipont C, Aarsman ME, Kannenberg K, von Rechenberg M, Nguyen-Disteche M, den Blaauwen T, Höltje J-V, Vollmer W (2006) Interaction between two murein (peptidoglycan) synthases, PBP3 and PBP1B, in *Escherichia coli*. Mol Microbiol 61:675–690
8. Müller P, Ewers C, Bertsche U, Anstett M, Kallis T, Breukink E, Fraipont C, Terrak M, Nguyen-Disteche M, Vollmer W (2007) The essential cell division protein FtsN interacts with the murein (peptidoglycan) synthase PBP1B in *Escherichia coli*. J Biol Chem 282:36394–36402
9. Daniel RA, Errington J (2003) Control of cell morphogenesis in bacteria: two distinct ways to make a rod-shaped cell. Cell 113:767–776

10. Uehara T, Park JT (2008) Growth of *Escherichia coli*: significance of peptidoglycan degradation during elongation and septation. J Bacteriol 190:3914–3922
11. van Teeffelen S, Wang S, Furchtgott L, Huang KC, Wingreen NS, Shaevitz JW, Gitai Z (2011) The bacterial actin MreB rotates, and rotation depends on cell-wall assembly. Proc Natl Acad Sci U S A 108:15822–15827
12. Paradis-Bleau C, Markovski M, Uehara T, Lupoli TJ, Walker S, Kahne DE, Bernhardt TG (2010) Lipoprotein cofactors located in the outer membrane activate bacterial cell wall polymerases. Cell 143:1110–1120
13. Typas A, Banzhaf M, van Saparoea VB, Verheul J, Biboy J, Nichols RJ, Zietek M, Beilharz K, Kannenberg K, von Rechenberg M, Breukink E, den Blaauwen T, Gross CA, Vollmer W (2010) Regulation of peptidoglycan synthesis by outer membrane proteins. Cell 143:1097–1109
14. Typas A, Banzhaf M, Gross CA, Vollmer W (2011) From the regulation of peptidoglycan synthesis to bacterial growth and morphology. Nat Rev Microbiol 10:123–136
15. Schwartz B, Markwalder JA, Seitz SP, Wang Y, Stein RL (2002) A kinetic characterization of the glycosyltransferase activity of *Escherichia coli* PBP1b and development of a continuous fluorescence assay. Biochemistry 41: 12552–12561
16. Offant J, Terrak M, Derouaux A, Breukink E, Nguyen-Disteche M, Zapun A, Vernet T (2010) Optimization of conditions for the glycosyltransferase activity of penicillin-binding protein 1a from *Thermotoga maritima*. FEBS J 277:4290–4298
17. Barrett D, Wang TS, Yuan Y, Zhang Y, Kahne D, Walker S (2007) Analysis of glycan polymers produced by peptidoglycan glycosyltransferases. J Biol Chem 282:31964–31971
18. Tamura T, Suzuki H, Nishimura Y, Mizoguchi J, Hirota Y (1980) On the process of cellular division in *Escherichia coli*: isolation and characterization of penicillin-binding proteins 1a, 1b, and 3. Proc Natl Acad Sci U S A 77: 4499–4503
19. Tomioka S, Ishino F, Tamaki S, Matsuhashi M (1982) Formation of hyper-crosslinked peptidoglycan with multiple crosslinkages by a penicillin-binding protein, 1A, of *Escherichia coli*. Biochem Biophys Res Commun 106: 1175–1182
20. van Heijenoort Y, Gomez M, Derrien M, Ayala J, van Heijenoort J (1992) Membrane intermediates in the peptidoglycan metabolism of *Escherichia coli*: possible roles of PBP 1b and PBP 3. J Bacteriol 174:3549–3557
21. Breukink E, van Heusden HE, Vollmerhaus PJ, Swiezewska E, Brunner L, Walker S, Heck AJ, de Kruijff B (2003) Lipid II is an intrinsic component of the pore induced by nisin in bacterial membranes. J Biol Chem 278: 19898–19903
22. Schwartz B, Markwalder JA, Wang Y (2001) Lipid II: total synthesis of the bacterial cell wall precursor and utilization as a substrate for glycosyltransfer and transpeptidation by penicillin binding protein (PBP) 1b of *Escherichia coli*. J Am Chem Soc 123:11638–11643
23. Born P, Breukink E, Vollmer W (2006) *In vitro* synthesis of cross-linked murein and its attachment to sacculi by PBP1A from *Escherichia coli*. J Biol Chem 281: 26985–26993
24. Terrak M, Ghosh TK, van Heijenoort J, Van Beeumen J, Lampilas M, Aszodi J, Ayala JA, Ghuysen JM, Nguyen-Disteche M (1999) The catalytic, glycosyl transferase and acyl transferase modules of the cell wall peptidoglycan-polymerizing penicillin-binding protein 1b of *Escherichia coli*. Mol Microbiol 34:350–364
25. Bertsche U, Breukink E, Kast T, Vollmer W (2005) *In vitro* murein peptidoglycan synthesis by dimers of the bifunctional transglycosylase-transpeptidase PBP1B from *Escherichia coli*. J Biol Chem 280:38096–38101
26. Glauner B (1988) Separation and quantification of muropeptides with high-performance liquid chromatography. Anal Biochem 172:451–464
27. Sung MT, Lai YT, Huang CY, Chou LY, Shih HW, Cheng WC, Wong CH, Ma C (2009) Crystal structure of the membrane-bound bifunctional transglycosylase PBP1b from *Escherichia coli*. Proc Natl Acad Sci U S A 106:8824–8829

Chapter 18

Extraction of Cell Wall-Bound Teichoic Acids and Surface Proteins from *Listeria monocytogenes*

Filipe Carvalho, María Graciela Pucciarelli, Francisco García-del Portillo, Didier Cabanes, and Pascale Cossart

Abstract

Gram-positive bacteria contain a cell wall consisting of a thick peptidoglycan layer decorated with surface proteins and polysaccharide-based polymers. The latter include the wall teichoic acids (WTAs), which are anionic glycopolymers covalently linked to the peptidoglycan matrix. They are constituted by a long backbone containing sugars of various sizes (trioses to hexoses) which can be reduced (polyols as in *Listeria*) or oxidized (uronic acids) and can undergo a variety of species- or often strain-specific modifications and substitutions. These confer unique biochemical properties to WTAs and any defect in the modification or substitution process can potentially affect their biological role in the overall cell wall physiology. Surface proteins can be associated to the cell wall by covalent bonds that anchor the protein to the peptidoglycan lattice. Due to the chemical nature of this bond, covalently bound proteins "co-purify" with peptidoglycan sacculi and are intrinsically insoluble at high temperatures and/or in the presence of ionic detergents. Analysis of this type of proteins therefore requires enzymatic digestion of peptidoglycan for the subsequent release of associated proteins. In contrast, proteins associated to the cell wall by non-covalent interactions are easier to isolate using ionic detergents. In this chapter, we describe methods for the extraction and analysis of (i) WTAs, (ii) covalently, and (iii) non-covalently cell wall-bound surface proteins from the Gram-positive pathogen *Listeria monocytogenes*.

Key words: Peptidoglycan, Wall teichoic acids, Surface proteins, Covalent anchorage, Digestion, Alkaline hydrolysis, PAGE, Alcian blue, Silver staining

1. Introduction

The cell wall of Gram-positive bacteria is a complex and dynamic structure composed largely by peptidoglycan, a giant macromolecule consisting of chains of alternate carbohydrate sugars cross-linked by peptidic side chains (1). Bacteria fabricate this molecule to ensure cell integrity in response to the high internal turgor pressure and to provide the cell with a defined morphology.

Anne H. Delcour (ed.), *Bacterial Cell Surfaces: Methods and Protocols*, Methods in Molecular Biology, vol. 966,
DOI 10.1007/978-1-62703-245-2_18, © Springer Science+Business Media New York 2013

Moreover, it serves as a scaffold to anchor different molecules, such as polysaccharides and proteins whose function is necessary and, in many cases, essential for the proper execution of important biological processes (2). In Gram-positive bacteria, these peptidoglycan-attached molecules constitute a major area of research in studies addressing the mechanisms by which these bacteria interact with the environment. In particular, many of these surface proteins have been demonstrated as implicated in bacterial pathogenesis (3).

Among the major glycopolymers that constitute the Gram-positive cell wall, teichoic acids are a group of anionic molecules, constituted by a chain of phosphodiester-linked polyol-phosphate repeats, which can be covalently attached either to the head group of membrane lipids (lipoteichoic acids, LTAs) or to the peptidoglycan (wall teichoic acids, WTAs). In comparison with LTAs, WTAs display a more diverse chemical composition, featuring a backbone of glycerol-, ribitol-, or even other sugar-phosphate species that can be substituted by D-alanine and/or a variety of glycosyl groups (2). This diversity confers WTAs with unique structural, chemical, and electrostatic properties that dictate their role in the bacterial surface physiology. Mutations interfering with WTA substitution pathways are expected to affect the overall stability of the native polymer. Such molecular changes can be assessed in vitro by employing a methodology that involves extraction of the WTAs, through chemical disruption of their inter-repeat linkages, followed by polyacrylamide gel electrophoresis (PAGE) analysis of the broken-down polymers (4). Effective extraction is performed with chemical treatment of purified cell walls with acid or alkali, which breaks the phosphodiester bonds between WTA repeats (5). The resulting mixture of differentially hydrolyzed polymeric species is then resolved electrophoretically in a tight-mesh polyacrylamide gel, under native conditions (4, 6, 7). To detect WTAs, the cationic dye Alcian blue (Fig. 1) is firstly used to specifically stain the polymers (through binding to the numerous phosphate groups), and silver staining is performed over the primary staining for contrasting purposes (8, 9). In this chapter, we use this approach to compare WTAs isolated from different strains of the intracellular human pathogen *Listeria monocytogenes*: a wild-type strain, an isogenic mutant defective in WTA glycosylation, and a complemented mutant that regained the ability to glycosylate WTAs.

Surface proteins also present different types of association to the cell wall. They can be covalently bound to the peptidoglycan in a reaction catalyzed by specialized enzymes named sortases, which recognize and cleave a sorting motif near the C-terminus of the protein (e.g., LPXTG proteins) (10). The high stability of this bond and the unique chemical properties of the peptidoglycan macromolecule, which remains insoluble even under stringent extraction conditions such as high temperatures and ionic detergents, dictate specific methods to isolate and characterize this class of surface

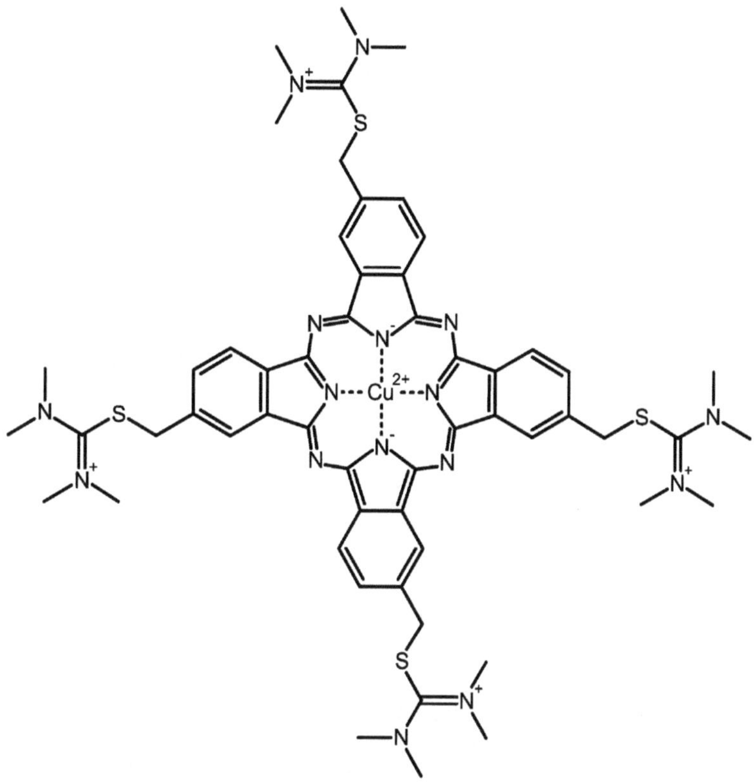

Fig. 1. The cationic dye Alcian blue, a copper phthalocyanine, binds WTAs through strong electrostatic interactions with the negatively charged phosphate groups present in each subunit of the polymer backbone.

proteins. In this chapter, we describe the purification and digestion of peptidoglycan material or associated proteins of *Listeria monocytogenes* for either individualized (biochemical) analyses on specific surface proteins or massive identification procedures (non-gel proteomics). These analyses were of interest given that *Listeria* spp. contain the largest number of genes encoding predicted surface proteins anchored covalently to the peptidoglycan. These methods have allowed the characterization of the profile of this group of proteins in different growing conditions, including the intracellular milieu of eukaryotic cells (11–13). In addition, methods involving purified peptidoglycan sacculi revealed that some of these proteins are detected differently when the peptidoglycan is digested from the "outside" (in intact bacteria) than when it is digested in its purified sacculi form (Fig. 2). Such evidence denotes probable topological differences within the group of *Listeria* surface proteins that are covalently bound to the peptidoglycan. Non-covalently bound surface proteins, on the other hand, are associated with cell wall components, such as LTAs, via repeated domains typically located in the C-terminus (14–16). The comparatively lower stability of such non-covalent interactions results in a labile cell wall anchorage of

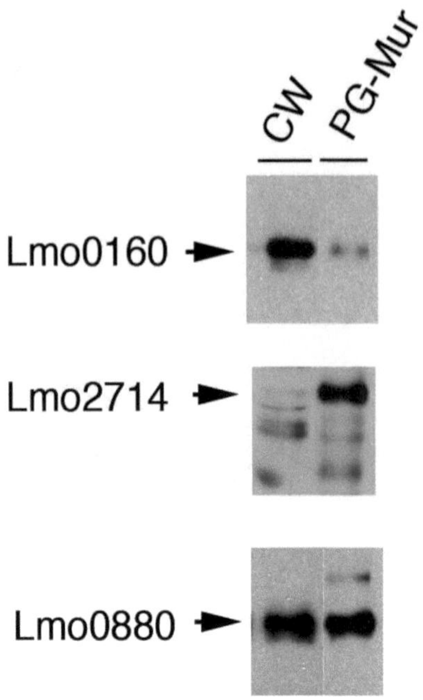

Fig. 2. Comparison of the amounts of three *Listeria monocytogenes* LPXTG surface proteins, covalently bound to peptidoglycan (Lmo0160, Lmo2714, and Lmo0880), by immunodetection after purification via two alternative methods involving the usage of muramidases: mutanolysin-mediated digestion of peptidoglycan in intact bacteria (CW, see Subheading 3.3.3) or cellosyl-mediated digestion of purified peptidoglycan sacculi (PG-Mur, see Subheading 3.3.2).

these proteins. As we also describe in this chapter, this association can be rapidly disrupted in the presence of ionic detergents that wash the proteins out of the bacterial surface.

2. Materials

Prepare solutions using analytical grade reagents and double-distilled water, and store at room temperature, unless indicated otherwise. Handle flammable, volatile, and/or highly odoriferous reagents under a ventilated chemical hood. Follow the recommended safety guidelines for adequate disposal of chemical wastes.

2.1. Extraction and Polyacrylamide gel Electrophoresis Analysis of *Listeria monocytogenes* WTAs

2.1.1. Extraction of WTA Polymers by Alkaline Hydrolysis

1. Brain–heart infusion (BHI) broth: Dissolve 37 g of BHI (Difco, BD) in 1 l of water. Autoclave at 121°C for 15 min.
2. Proteinase K (stock): 20 mg/ml in water. Store at −20°C.
3. TS buffer: 20 mM Tris–HCl, pH 8.0, 0.5% (w/v) sodium dodecyl sulfate (SDS). Dissolve 2.42 g of Tris and 5 g of SDS in 900 ml of water. Adjust the pH with 1 M HCl (see step 9) and complete 1 l with water.

4. Buffer 1: 50 mM 2-(*N*-morpholino)ethanesulfonic acid (MES), pH 6.5. Dissolve 10.86 g of MES sodium salt in 900 ml of water. Adjust the pH with 1 M HCl and make up to 1 l with water.
5. Buffer 2: 50 mM MES, pH 6.5, 4% (w/v) SDS. Dissolve 40 g of SDS per liter of Buffer 1 solution.
6. Buffer 3: 50 mM MES, pH 6.5, 2% (w/v) NaCl. Dissolve 20 g of NaCl per liter of Buffer 1 solution.
7. Buffer 4: 20 mM Tris–HCl, pH 8.0, 0.5% (w/v) SDS, 20 μg/ml proteinase K. Add 1 μl of proteinase K stock (see step 2) per ml of TS buffer (see step 3).
8. Hydrolysis solution: 0.1 M NaOH. Dissolve 0.4 g of NaOH in 100 ml of water. Prepare fresh before use.
9. 1 M HCl: Prepare 120 ml of solution by dilution of 10 ml of 37% (w/w) HCl stock in water.
10. 250-ml conical flask.
11. 30-ml Oak Ridge teflon centrifuge tubes.
12. 2-ml microcentrifuge tubes.
13. 2-ml screw cap microcentrifuge tubes.
14. Acid-washed glass beads, ≤106 μm.
15. Dialysis membrane (1,000-Da molecular mass cutoff).
16. High-speed refrigerated centrifuge (max. 30,000 × *g*), with 6 × 30 ml angle rotor.
17. Heating plate.
18. Benchtop microcentrifuge.
19. High-speed sample homogenizer (FastPrep®).
20. Thermomixer (Eppendorf®).
21. Vacuum concentrator (SpeedVac®).

2.1.2. Native PAGE Resolution of WTA Extracts

1. Native gel buffer: 3.0 M Tris–HCl, pH 8.5. Dissolve 363.4 g of Tris in 800 ml. Adjust the pH with HCl and complete 1 l with water.
2. Acrylamide solution: 30% (w/v) acrylamide/bis-acrylamide (29:1). Store at 4°C.
3. Ammonium persulfate (APS): 10% (w/v) in water (see Note 1).
4. *N*,*N*,*N*′,*N*′-tetramethylethylenediamine (TEMED).
5. Running buffer: 0.1 M Tris, 0.1 M tricine, pH 8.2. To prepare 1 l of 5× running buffer dissolve 60.6 g of Tris and 89.6 g of tricine in water (see Note 2). Store at 4°C.
6. Loading buffer (3×): 0.3 M Tris, 0.3 M tricine, pH 8.2, 30% (v/v) glycerol, 0.15% (w/v) bromophenol blue (BPB).

Mix 6 ml of 5× running buffer, 3 ml of glycerol, and 1 ml of water. Dissolve 1.5 mg of BPB. Prepare 1-ml aliquots and store at −20°C.

7. Isopropanol.
8. Vertical electrophoresis system (16×20 cm glass plates).
9. 250-ml Büchner flask with rubber bung.

2.1.3. Alcian Blue-Silver Staining of Native PAGE Gels

1. Alcian blue staining solution: 5% (v/v) acetic acid, 40% (v/v) ethanol, 0.1% (w/v) Alcian blue. Dissolve 0.2 g of Alcian blue 8GX in a mixture of 110 ml of water, 10 ml of glacial acetic acid, and 80 ml of absolute ethanol.
2. Alcian blue destaining solution: 10% (v/v) acetic acid, 40% (v/v) ethanol. Add 100 ml of glacial acetic acid and 400 ml of absolute ethanol to 500 ml of water.
3. Silver stain solution A: 2% (w/v) silver nitrate ($AgNO_3$), 2% (w/v) ammonium nitrate (NH_4NO_3). Dissolve 10 g each of silver nitrate and ammonium nitrate in 500 ml of water. Store in the dark (see Note 3).
4. Silver stain solution B: 10% (w/v) tungstosilicic acid. Dissolve 50 g of tungstosilicic acid in 500 ml of water.
5. Silver stain solution C: 3% (v/v) formaldehyde. Prepare fresh from 37% (w/v) formaldehyde. Store at 4°C (see Note 4).
6. Silver stain solution D: 5% (w/v) sodium carbonate (Na_2CO_3). Dissolve 50 g of sodium carbonate in 1 l of 50°C-heated water. Cool to room temperature before use.
7. Silver stain solution E: 5% (v/v) acetic acid. Add 50 ml of glacial acetic acid to 950 ml of water.
8. Orbital shaker.

2.2. Purification of Peptidoglycan Sacculi from Extracellular and Intracellular Listeria

2.2.1. Bacterial Growth in BHI Broth (Exponential Phase)

1. BHI broth: see item 1 of Subheading 2.1.1.
2. Phosphate-buffered saline (PBS): 137 mM NaCl, 2.7 mM KCl, 10 mM Na_2HPO_4, 2 mM KH_2PO_4, pH 7.4. Dissolve 8 g of NaCl, 0.2 g of KCl, 1.44 g of Na_2HPO_4, and 0.24 g of KH_2PO_4 in 800 ml of water. Adjust pH with 1 M HCl (see item 9 of Subheading 2.1.1), and add water to 1 l. Autoclave at 121°C for 15 min.
3. 25-ml and 1-l conical flasks.
4. High-speed refrigerated centrifuge.

2.2.2. Bacterial Growth in Chemically Defined Medium (Exponential Phase)

1. BHI broth: see item 1, Subheading 2.1.1.
2. 25- and 500-ml conical flasks.
3. PBS: see item 2, Subheading 2.2.1.

4. "A" salts (10×): 1.15 M Na_2HPO_4, 482 mM KH_2PO_4. Dissolve 154.8 g of $Na_2HPO_4 \cdot 7H_2O$ and 32.8 g of KH_2PO_4 in 500 ml of water. Autoclave.
5. "B" salts (100×): 166 mM $MgSO_4$. Dissolve 4.09 g of $MgSO_4 \cdot 7H_2O$ in 100 ml of water. Autoclave.
6. Glucose solution (20×): 20% (w/v) in water. Dissolve 20 g of glucose in 100 ml of water. Autoclave at 115°C for 15 min.
7. Amino acids solution (100×): 10 mg/ml each of L-leucine, DL-isoleucine, DL-valine, L-arginine, L-histidine, DL-methionine, L-tryptophan, and L-phenylalanine. Dissolve 200 mg of each amino acid in 20 ml of water (see Note 5). Sterilize through a 0.22-μm pore size filter.
8. "Cys-Glu" solution (100×): 10 mg/ml L-cysteine, 60 mg/ml L-glutamine. Dissolve 100 mg of L-cysteine and 600 mg of L-glutamine in 10 ml of water (see Note 5). Prepare fresh before each use. Sterilize through a 0.22-μm pore size filter.
9. Adenine solution (500×): 1.25 mg/ml adenine. Dissolve 25 mg of adenine in 20 ml of water. Sterilize using a 0.22-μm pore size filter.
10. Vitamin solution A: 0.5 mg/ml biotin, 0.5 mg/ml riboflavin. Dissolve 5 mg of each vitamin in 10 ml of water (see Note 6). Sterilize using a 0.22-μm pore size filter.
11. Vitamin solution B: 1 mg/ml calcium pantothenate, 1 mg/ml nicotinamide, 1 mg/ml pyridoxal-HCl. Dissolve 10 mg of each vitamin compound in 10 ml of water. Sterilize using a 0.22-μm pore size filter.
12. Vitamin solution C: 10 mg/ml *para*-aminobenzoic acid, 0.5 mg/ml α-lipoic acid in 1 ml of 95% (v/v) ethanol (see Note 7).
13. Vitamin stock (100×): Mix vitamin solutions A, B, and C with water for a final solution volume of 100 ml.
14. Ferric citrate solution (50×): 18 mM ferric citrate ($C_6H_5O_7Fe$). Dissolve 43.9 mg of $C_6H_5O_7Fe$ in 10 ml of water. Sterilize using a 0.22-μm pore size filter.
15. Improved minimal medium (IMM): Prepare this chemically defined growth medium by diluting each solution described in steps 1–9, 13, and 14 to a final concentration of 1×.
16. High-speed refrigerated centrifuge.

2.2.3. Bacterial Growth Inside Eukaryotic Cells (Intracellular Bacteria)

1. JEG-3 epithelial cell line from a human placental choriocarcinoma (ATCC HTB-36).
2. Eagle's minimal essential medium (MEM), commercially available.
3. Fetal bovine serum (FBS), commercially available.

4. Nonessential amino acids (NEAA) (100×), without L-glutamine, commercially available.
5. Sodium pyruvate solution (100×): 100 mM in MEM, commercially available.
6. L-glutamine solution (100×): 200 mM, commercially available.
7. JEG-3 complete medium: MEM supplemented with 10% (v/v) FBS, 1% (v/v) NEAA, 1 mM sodium pyruvate, and 2 mM L-glutamine.
8. BHI broth: see item 1, Subheading 2.1.1.
9. PBS: see item 2, Subheading 2.2.1.
10. Complete PBS: PBS, pH 7.4, 0.9 mM $CaCl_2$, 0.5 mM $MgCl_2$. Dissolve 99.8 mg of $CaCl_2$ and 101.6 mg of $MgCl_2.6H_2O$ in 1 l of PBS. Filter-sterilize.
11. Gentamicin solution (stock): 10 mg/ml gentamicin sulfate, commercially available. Store at 4°C.
12. Tissue culture lysis solution: 1% (v/v) phenol, 19% (v/v) ethanol, 0.4% (w/v) SDS. Add 10 ml of phenol and 190 ml of absolute ethanol to 700 ml of water. Dissolve 4 g of SDS and complete 1 l of solution with water.
13. 25-ml conical flask.
14. BioDish-XL plates.
15. Cell culture scrapers.
16. 1.5 ml microcentrifuge tubes.
17. High-speed refrigerated centrifuge.

2.2.4. Purification of Peptidoglycan Sacculi

1. PBS: see item 2, Subheading 2.2.1.
2. Protease inhibitors (50×): Dissolve one tablet of EDTA-free protease inhibitor cocktail in 1 ml of PBS.
3. DNase I (100×): Dissolve 10 mg of bovine pancreatic DNase I in 1 ml of water. Store aliquots at −20°C.
4. Lysis buffer: 2% (v/v) protease inhibitors, 100 μg/ml DNase I. Dilute the protease inhibitor and DNase I stock solutions in PBS to a final concentration of 1×.
5. SDS solution: 8% (w/v) in water (see Note 8).
6. Glass tube with screw top.
7. Glass beaker.
8. Heating plate with magnetic stirring.
9. FRENCH® Press Cell Disrupter, composed of motor-driven laboratory press and standard FRENCH® Cell (40,000 psi, 35 ml capacity).
10. High-speed refrigerated centrifuge.
11. Refrigerated ultracentrifuge (up to 300,000 × *g*).

2.3. Preparation of Listeria Surface Proteins Covalently Bound to the Cell Wall

2.3.1. Trypsin Digestion of Surface Proteins Bound to Purified Peptidoglycan Sacculi

1. Modified trypsin (see Note 9): 0.5 mg/ml. Suspend 20 μg of sequencing grade modified trypsin in 40 μl of "trypsin resuspension buffer" supplied by the manufacturer. Store aliquots at −20°C for up to 2 months.
2. Bicarbonate buffer: 50 mM $NaHCO_3$. Dissolve 10 mg of $NaHCO_3$ in 2.5 ml of ultrapure water. The solution is stable at 4°C for up to 2 months.
3. Trypsin digestion buffer: 50 mM $NaHCO_3$, 0.01% SDS (see Note 10). Add 3 μl of 8% (w/v) SDS (see item 5, Subheading 2.2.4) to 2.5 ml of bicarbonate buffer.
4. Refrigerated ultracentrifuge (up to 300,000×*g*).
5. Lyophilizer.

2.3.2. Muramidase Digestion of Purified Peptidoglycan Sacculi for Release of Cell Wall-Associated Proteins

1. Cellosyl from *Streptomyces coelicolor* (see Notes 11 and 12): 0.5 mg/ml in water. Store aliquots at −20°C.
2. Phosphate buffer solution A: 0.2 M NaH_2PO_4. Dissolve 27.8 g of NaH_2PO_4 in 1 l of water.
3. Phosphate buffer solution B: 0.2 M Na_2HPO_4. Dissolve 53.65 g of $Na_2HPO_4{\cdot}7H_2O$ in 1 l of water.
4. Cellosyl digestion buffer: 50 mM phosphate, pH 5.0. Mix 49 ml of phosphate buffer solution A and 1 ml of phosphate buffer solution B with 150 ml of water, to yield a 200-ml solution with pH 5.0.
5. PBS: see item 2, Subheading 2.2.1.
6. Heating block.
7. Benchtop microcentrifuge.

2.3.3. Muramidase Digestion of Peptidoglycan on "Intact Bacteria" for Release of Cell Wall-Associated Proteins

1. Mutanolysin from *Streptomyces globisporus* ATCC 21553 (≥4,000 units/mg protein): 2 mg/ml. Dissolve 12.5 mg of purified enzyme in 6.25 ml of TM buffer (see step 2). Store aliquots at −20°C (see Note 13).
2. TM buffer: 10 mM Tris–HCl, pH 6.9, 10 mM $MgCl_2$. Dissolve 1.21 g of Tris and 2.03 g of $MgCl_2{\cdot}6H_2O$ in 800 ml of water. Adjust pH with 1 M HCl (see item 9, Subheading 2.1.1) and complete 1 l of solution with water.
3. Mutanolysin digestion buffer: 10 mM Tris–HCl, pH 6.9, 10 mM $MgCl_2$, 0.5 M sucrose. Dissolve 85.5 g of sucrose in 400 ml of TM buffer. Complete 500 ml of solution with the same buffer.
4. RNase A (10×): Dissolve 2 mg of bovine pancreatic RNase A in 1 ml of water. Store aliquots at −20°C.
5. PBS: see item 2, Subheading 2.2.1.
6. Protease inhibitors: see item 2, Subheading 2.2.4.

7. TCA solution: 50% (w/v) trichloroacetic acid. Dissolve 250 g of trichloroacetic acid in water to a final volume of 500 ml.
8. Acetone. Store at −20°C.
9. 0.22-μm spin-centrifuge filter tubes.
10. Refrigerated benchtop microcentrifuge.

2.4. Extraction of Non-covalently Surface-Associated Proteins with Sodium Dodecyl Sulfate

1. BHI broth: see item 1, Subheading 2.1.1.
2. PBS: see item 2, Subheading 2.2.1.
3. Extraction buffer: 2% (w/v) SDS. Dissolve 2 g of SDS in 100 ml of PBS.
4. 0.22-μm spin-centrifuge filter tubes.
5. 100- and 250-ml conical flasks.
6. 1.5-ml microcentrifuge tubes.
7. High-speed refrigerated centrifuge.
8. Benchtop microcentrifuge.

3. Methods

Perform all procedures at room temperature, unless mentioned otherwise.

3.1. Extraction and Polyacrylamide gel Electrophoresis Analysis of Listeria monocytogenes WTAs

3.1.1. Extraction of WTA Polymers by Alkaline Hydrolysis

1. Inoculate 50 ml of BHI broth in a 250-ml conical flask with a single colony of freshly plated *Listeria monocytogenes*. Incubate the culture at 37°C with aeration (200 rpm) until the late stationary growth phase is reached (≥16 h) (see Note 14).
2. Collect bacteria from 30 ml of the overnight culture into an Oak Ridge tube by centrifugation (4,000 × *g*, 10 min).
3. Wash the bacterial pellet in 30 ml of buffer 1 and centrifuge (4,000 × *g*, 10 min).
4. Resuspend the bacteria in 30 ml of buffer 2. Incubate the suspension in a boiling water bath for 1 h (see Note 15).
5. After centrifugation (14,000 × *g*, 15 min, 4°C), resuspend the pellet in 2 ml of buffer 2 and centrifuge (14,000 × *g*, 15 min).
6. Wash the pellet in 1 ml of buffer 3 and centrifuge (14,000 × *g*, 15 min). Repeat this washing step twice in buffer 3, and twice in buffer 1, before making a final resuspension of the pellet in 1 ml of buffer 1.
7. Transfer the suspension to a 2-ml screw cap microcentrifuge tube containing 0.4 g of acid-washed glass beads. Break the cells in a high-speed homogenizer following a program comprising four cycles of 60 s at a speed of 4.0 m/s, interspaced with a 3-min incubation on ice.

8. After centrifugation (1,000×*g*, 5 min), transfer the supernatant to another tube. Wash the glass beads with 1 ml of water, centrifuge in the same conditions, and pool both supernatant fractions.
9. Collect the broken cell walls by centrifugation (14,000×*g*, 15 min). Resuspend the pellet in 1 ml of buffer 4, and incubate the suspension at 50°C for 4 h, without shaking (see Note 16).
10. Centrifuge the suspension (14,000×*g*, 15 min) and wash the pellet in 1 ml of buffer 3. Centrifuge in the same conditions and wash in 1 ml of water. Repeat this wash two more times.
11. Resuspend the pellet in 1 ml of hydrolysis solution (see Note 17). Shake vigorously in a thermomixer at room temperature for ≥16 h.
12. After centrifugation (14,000×*g*, 15 min), collect the supernatant containing hydrolyzed WTAs, and neutralize the solution with 0.1 ml of 1 M HCl.
13. Perforate the cap of a 2-ml microcentrifuge tube to create a large round hole. Transfer the WTA extract into the tube. Cut a small square of dialysis membrane and lay on top of the tube opening (large enough to cover it) and lock the cap into place. Float the inverted tube (with cap completely submerged) in 500 ml of water and perform dialysis for 4 h, with gentle agitation.
14. Lyophilize the dialyzed extract in a vacuum concentrator. Resuspend in water to achieve the desired concentration (see Note 18).

3.1.2. Native PAGE Resolution of WTA Extracts

1. Prepare a 20% separating gel by mixing 20 ml of native gel buffer and 40 ml of acrylamide solution, in a 250-ml Büchner flask.
2. Connect the lateral tube of the flask to a vacuum source and seal the flask mouth with a rubber bung. Turn on the vacuum and degas the mixture for 15 min, shaking regularly the flask (see Note 19). Cool the flask on ice (see Note 20).
3. Add 600 μl of APS and 60 μl of TEMED to the gel mixture, and cast the separating gel into a previously assembled and leak-free gel cassette. Save some space from the top of the glass plate to cast the stacking gel.
4. Quickly overlay the unpolymerized separating gel with isopropanol (see Note 21). After polymerization, wash out the isopropanol with water.
5. To prepare a 4% stacking gel, mix 3 ml of native gel buffer, 1.3 ml of acrylamide solution, 5.7 ml of water, 100 μl of APS, and 10 μl of TEMED. Quickly cast the gel over the polymerized separating gel to the top of the glass plate.

6. With the gel still unpolymerized, insert a 10-well comb, taking care not to introduce air bubbles.
7. After polymerization, remove the comb and mount the gel cassette in the electrophoresis system. Fill both the cathode and anode chambers with 1× running buffer.
8. Mix WTA extract samples with loading buffer. Do not heat. Load samples into the wells and electrophorese at 4°C, under a constant current of 20 mA, until the dye front (from the loading buffer) is almost at the bottom of the separating gel.
9. Remove the gel cassette from the electrophoresis system. Carefully, separate one of the glass plates from the gel, and then detach the gel from the other plate by holding the top corners and slowly pulling it up and towards the bottom (see Note 22).
10. Immerse the gel in a large container filled with ultrapure water and wash it for 5 min, under gentle agitation (40 rpm).

3.1.3. Alcian Blue-Silver Staining of Native PAGE Gels

Perform all incubations of the gel under gentle agitation (40 rpm), on an orbital shaker.

1. To detect the gel-resolved WTAs, incubate the gel in 200 ml of Alcian blue staining solution for 30 min.
2. Recover the staining solution (see Note 23). Remove excess of dye from the gel by washing it in 200 ml of Alcian blue destaining solution until there is no visible background staining.
3. Wash the gel twice in water for 15 min.
4. To perform silver staining, mix 5 ml each of silver stain solutions A, B, and C with 35 ml of water. Just before using, add 50 ml of silver stain solution D and incubate the gel with this developing mixture until a desired level of staining and contrast is reached (Fig. 3).
5. To stop further staining of the gel, add 100 ml of silver stain solution E to the gel and incubate for 15 min (see Note 24).
6. Wash the gel twice in water for 15 min, and store at 4°C in water.

3.2. Purification of Peptidoglycan Sacculi from Extracellular and Intracellular Listeria

The three procedures described below account for different experimental conditions in which bacteria are grown either in laboratory media or inside eukaryotic cells. The procedure steps from the collection of the bacteria until the purification of the peptidoglycan sacculi are the same.

3.2.1. Bacterial Growth in BHI Broth (Exponential Phase)

1. Inoculate 5 ml of BHI broth in a 25-ml conical flask with a single colony of freshly plated *Listeria monocytogenes*. Incubate overnight at 37°C with aeration (200 rpm).
2. Dilute 2.5 ml of overnight culture (see Note 14) in 250 ml of BHI broth and grow the culture at 37°C with aeration

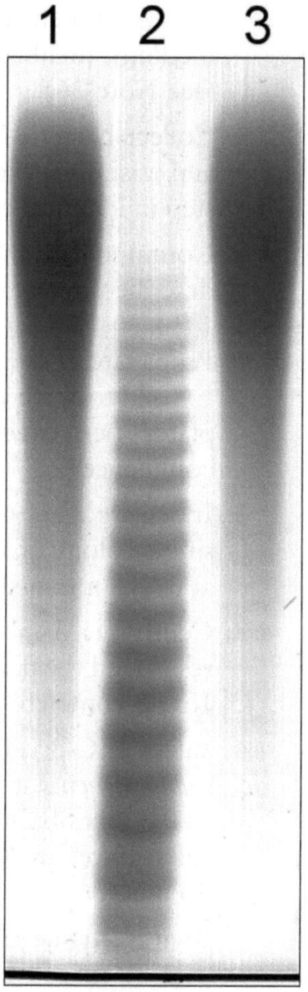

Fig. 3. *Listeria monocytogenes* WTA extracts analyzed by native PAGE and detected with a combination of Alcian blue and silver staining. WTAs isolated from a Δ*tam* strain, unable to perform WTA glycosylation (*lane 2*), show a gel migration profile distinct from that of WTAs extracted from the parental wild-type EGDe strain (*lane 1*), indicating a structural modification. A *trans*-complemented mutant strain that regained the ability to glycosylate WTAs (*lane 3*) yielded polymer extracts with a profile similar to the wild-type strain.

(200 rpm) until an optical density at 600 nm (OD_{600nm}) of about 0.25 (mid-exponential phase) (see Note 25).

3. Collect bacteria by centrifugation (10,000 × *g*, 10 min, 4°C) and wash the pellet in 20 ml of cold PBS. Repeat centrifugation step in identical conditions.
4. Continue with the final purification of peptidoglycan sacculi as described in Subheading 3.2.4.

3.2.2. Bacterial Growth in Chemically Defined Medium (Exponential Phase)

1. Inoculate 5 ml of BHI broth in a 25-ml conical flask with a single colony of freshly plated *Listeria monocytogenes* and incubate for 6 h at 37°C.
2. Centrifuge the bacteria (10,000 × *g*, 5 min) and resuspend the pellet in 5 ml of IMM medium (contains glucose as carbon

source). Inoculate a final culture volume of 100 ml of IMM medium. Incubate overnight at 37°C with aeration (200 rpm) (see Note 26).

3. Collect bacteria by centrifugation (10,000 × *g*, 10 min, 4°C) and wash the pellet in 20 ml of cold PBS. Repeat centrifugation step in identical conditions.
4. Continue with the final purification of peptidoglycan sacculi as described in Subheading 3.2.4.

3.2.3. Bacterial Growth Inside Eukaryotic Cells (Intracellular Bacteria)

1. Propagate JEG-3 epithelial cells in JEG-3 complete medium. Expand to reach surface area corresponding to five BioDish-XL plates per experiment (ca. 5.6×10^7 cells/plate at 80% confluence).
2. Inoculate 5 ml of BHI broth in a 25-ml conical flask with a single colony of *Listeria monocytogenes* and incubate overnight at 37°C in non-shaking conditions.
3. Centrifuge 5 ml of bacterial culture (10,000 × *g*, 5 min) and resuspend the pellet in the same volume of pre-warmed (37°C) JEG-3 complete medium.
4. Infect JEG-3 cells grown in BioDish-XL plates with *Listeria monocytogenes* at a multiplicity of infection of 10:1 (equivalent to 0.5 ml of bacterial suspension per tissue culture plate) when cell cultures are 80% confluent. Incubate at 37°C (5% CO_2) for 30 min.
5. Remove tissue culture medium and wash infected cells twice with 50 ml of complete PBS.
6. To each plate, add 50 ml of fresh JEG-3 complete medium containing 100 μg/ml of gentamicin to kill any remaining extracellular bacteria. Incubate for 90 min and then replace with fresh JEG-3 complete medium containing 10 μg/ml of gentamicin.
7. After 6 h of infection, wash the plates three times with 25 ml of complete PBS.
8. Lyse the infected cells in 20 ml of tissue culture lysis solution per plate. Detach the lysate with the help of a scraper and centrifuge (20,000 × *g*, 30 min, 4°C) to recover bacteria.
9. Wash each pellet of intracellular bacteria with 10 ml of cold PBS and centrifuge again (20,000 × *g*, 30 min, 4°C).
10. Collect the bacterial pellets in a small volume of PBS (100 μl per plate), transfer to 1.5-ml microfuge tubes, and centrifuge (10,000 × *g*, 10 min, 4°C).
11. Store bacterial pellets at −80°C. At this stage, it is necessary to pool a minimum of 20 pellets (coming from infection of 100 BioDish XL plates in 20 independent experiments) to obtain

the minimal amount of material required for final purification of peptidoglycan sacculi.

12. Continue with the final purification of peptidoglycan sacculi as described in Subheading 3.2.4.

3.2.4. Purification of Peptidoglycan Sacculi

1. Prepare bacterial pellets from growth in any of the three conditions described in Subheadings 3.2.1, 3.2.2, or 3.2.3.
2. Resuspend the bacterial pellets in 20 ml of cold lysis buffer. Maintain bacterial suspension on ice.
3. Break the bacteria in the FRENCH® press by three successive passages at 1,000 psi.
4. Remove unbroken cells by centrifugation (5,000 × *g*, 5 min, 4°C).
5. Recover the supernatant and ultracentrifuge at 140,000 × *g* (4°C) for 45 min.
6. Resuspend the pellet, containing membrane and peptidoglycan material, in 1.5 ml of cold PBS. Transfer the suspension to a glass tube with screw top. Insert the tube in a glass beaker and gently mix the suspension with 1.5 ml of boiling 8% (w/v) SDS (see Note 27). Maintain boiling conditions for 2 h and then overnight at 80°C.
7. Precipitate the peptidoglycan sacculi and associated proteins by ultracentrifugation (300,000 × *g*, 20 min, 25°C). Resuspend the pellet in 2.5 ml of pre-warmed (60°C) ultrapure water. Repeat the process three times and resuspend in either 300 μl of trypsin digestion buffer (see Subheading 2.3.1) or mutanolysin digestion buffer (see Subheading 2.3.3).

3.3. Preparation of Listeria Surface Proteins Covalently Bound to the Cell Wall

This section describes procedures for the preparation of proteins to be further analyzed either by proteomic techniques based on mass spectrometry (see refs. 3, 5) and thus cleaved into peptides by trypsin digestion (Subheading 3.3.1) or by biochemical characterization such as immunoblotting (see ref. 5) and thus released intact from the cell wall by muramidase digestion (Subheadings 3.3.2 and 3.3.3).

3.3.1. Trypsin Digestion of Surface Proteins Bound to Purified Peptidoglycan Sacculi

1. Resuspend peptidoglycan sacculi derived from 250 ml of an exponential bacterial culture ($OD_{600nm} \approx 0.25$) in 300 μl of trypsin digestion buffer.
2. Add 4 μl of modified trypsin and incubate overnight at 37°C with gentle agitation (50 rpm).
3. Centrifuge (300,000 × *g*, 20 min, 30°C) to release the peptides representative of surface proteins attached to the peptidoglycan.
4. Collect and lyophilize the supernatant fraction. Store at −20°C. This material is ready to be processed by proteomic techniques based on mass spectrometry.

3.3.2. Muramidase Digestion of Purified Peptidoglycan Sacculi for Release of Cell Wall-Associated Proteins

1. Resuspend peptidoglycan sacculi derived from 250 ml of an exponential bacterial culture ($OD_{600nm} \approx 0.25$) in 300 μl of cellosyl digestion buffer.
2. Add 30 μl of the cellosyl stock solution and incubate overnight at 37°C with gentle agitation (50 rpm).
3. Heat at 100°C for 10 min to inactivate the enzyme.
4. Centrifuge (15,000×*g*, 15 min, 4°C) to precipitate peptidoglycan-associated proteins (see Note 28).
5. Resuspend the pellet in 30 μl of PBS. The sample is ready to be further analyzed by immunoblotting techniques.

3.3.3. Muramidase Digestion of Peptidoglycan on "Intact Bacteria" for Release of Cell Wall-Associated Proteins

1. Collect bacteria grown in 250 ml of medium (BHI or IMM) to an $OD_{600nm} \approx 0.25$ by centrifugation (10,000×*g*, 10 min, 4°C).
2. Wash the bacterial pellet first in 20 ml of PBS and afterwards in 5 ml of mutanolysin digestion buffer. Finally, resuspend in 1.5 ml of mutanolysin digestion buffer.
3. Add 15 μl each of mutanolysin and RNase A stock solutions (final enzyme concentration: 20 μg/ml) and 30 μl of protease inhibitors. Incubate overnight at 37°C with gentle agitation (50 rpm).
4. Centrifuge (14,000×*g*, 15 min, 4°C) and filter the supernatant fraction in 0.22-μm spin-centrifuge tubes.
5. Add TCA solution to the filtrate to reach a final concentration of 15% (w/v) and incubate for 1 h at 4°C in chilled ice.
6. Centrifuge (20,000×*g*, 20 min, 4°C) and wash the pellet in cold acetone. Centrifuge again in same conditions. Discard the acetone and air-dry the pellet.
7. Resuspend in 60 μl of PBS. The sample is ready to be further analyzed by immunoblotting techniques.

3.4. Extraction of Non-covalently Surface-Associated Proteins with SDS

1. Inoculate 20 ml of BHI broth in a 100-ml conical flask with a single colony of *Listeria monocytogenes* and incubate overnight at 37°C with aeration (200 rpm).
2. Add 10 ml of the overnight culture to 90 ml of pre-warmed (37°C) BHI broth and incubate at 37°C (200 rpm) under aerobic conditions, until the culture reaches an $OD_{600nm} = 0.6–0.8$.
3. Harvest the bacteria by centrifugation (4,000×*g*, 10 min, 4°C).
4. Wash the bacterial pellet in 20 ml of cold PBS and centrifuge (4,000×*g*, 10 min, 4°C).
5. Resuspend in 1 ml of cold PBS, transfer to a 1.5-ml microfuge tube, and centrifuge (14,000×*g*, 5 min).
6. Resuspend the bacteria in 1 ml of extraction buffer and incubate under static conditions at 37°C for 30 min.

7. Spin down the suspension in a microcentrifuge (10,000 × *g*, 1 min) and collect the supernatant containing the extracted non-covalently bound surface proteins.
8. Filter the extract in a 0.22-μm spin-centrifuge tube. The filtrate is ready to be analyzed by SDS-PAGE or by immunoblotting techniques (see Note 29).

4. Notes

1. Ammonium persulfate is not stable in water and solutions withstand only a few hours at 4°C. It is advisable to prepare fresh solutions before each use.
2. An equimolar combination of both buffering agents is expected to yield a solution with a pH value very close (if not coincident) to 8.2, so there is usually no requirement for pH adjustment.
3. Silver ions are photosensitive and undergo reduction to solid silver particles with prolonged light exposure. It is recommended to store silver salt solutions in amber glass or foil-wrapped flasks.
4. Dilute formaldehyde solutions are unstable at room temperature, but can be stored at 4°C for a few days. We however recommend preparing fresh solutions from the stock.
5. Some of these amino acids are hydrophobic and the mixture should be prepared by heating the solution at 50°C to increase their solubility.
6. The vitamins biotin and riboflavin have low water solubility, so the solution must be prepared by heating at 50°C.
7. α-lipoic acid is also known as thioctic acid.
8. Concentrated SDS solutions should be maintained always at room temperature since solubility is extremely sensitive to temperature. An 8% (w/v) solution precipitates if stored at temperatures ≤10°C.
9. Trypsin hydrolyzes peptide bonds at lysine and arginine residues. Unmodified trypsin exhibits autocatalytic activity, generating peptides that can interfere in the HPLC or mass spectrometry analyses. Trypsin from Promega has been modified by reductive methylation, which makes it resistant to autolysis.
10. Addition of 0.01% (w/v) SDS is considered to favor trypsin digestion of proteins buried within the peptidoglycan lattice. Higher concentration can inactivate the enzyme.
11. Cellosyl is not commercially available. For our investigation, the enzyme was kindly gifted by Prof. Miguel A. de Pedro (Centro de Biología Molecular "Sevro Ochoa," Madrid, Spain).

12. Lysozyme hydrolyzes the β-1,4-glycosidic bond between *N*-acetylmuramic acid and *N*-acetylglucosamine residues in the peptidoglycan. However, in contrast to cellosyl, the activity of lysozyme drops drastically if the peptidoglycan presents natural modifications such as acetylation. This is the case of *Listeria* spp. Furthermore, lysozyme has been reported to exhibit unspecific "star" transpeptidase activity on purified peptidoglycan, generating artificial muropeptides.
13. Mutanolysin and cellosyl cleave the β-1,4-glycosidic bond between the *N*-acetylmuramic acid and *N*-acetylglucosamine residues in the peptidoglycan and both behave identically when used for in vitro digestion of purified peptidoglycan sacculi. Traditionally, mutanolysin has been more employed in digestion of the peptidoglycan exposed on intact bacteria.
14. Late stationary growth-phase cultures of *Listeria monocytogenes* in BHI broth typically display optical density values of 2.0–2.5 at 600 nm.
15. Incubation with SDS at high temperatures promotes a number of events: denaturation of surface autolytic proteins, which lose their cell wall-modifying activity; disruption and solubilization of the lipid-rich plasma membrane, facilitating cell death by loss of membrane integrity; and cell wall clearance of non-covalently linked proteins.
16. Incubation with proteinase K ensures that proteins or peptides still associated with the cell wall are fully degraded and removed.
17. It is crucial to use NaOH at the required concentration (to be sure, titrate or prepare fresh solution before each use) in order to ensure an efficient hydrolysis and high-yield extraction of the WTA polymers. In our hands, less concentrated NaOH solutions displayed a much lower extraction power.
18. To determine the amount of extracted WTAs, firstly weigh the empty tube in which the sample will be lyophilized; then, measure the weight of the same tube containing the sample after lyophilization, and finally subtract the weight of the tube alone.
19. Degassing the acrylamide mixture prevents the appearance of air bubbles within the gel matrix, and increases the speed of polymerization. It removes solubilized oxygen, which acts as an inhibitor of the formation of free radicals (from persulfate ions) that catalyze the acrylamide–bis-acrylamide co-polymerization reaction.
20. Because of the high acrylamide percentage in the resolving gel mixture and the degassing procedure, polymerization occurs quite fast once the initiators are added. Precooling the mixture gives enough time to properly cast the gel.

21. Isopropanol blocks contact between the resolving gel and atmospheric oxygen, and creates a hypoxic environment that promotes acrylamide polymerization. It also levels the surface of the resolving gel, creating an optimal interface with the stacking gel.
22. Handle high-percentage polyacrylamide gels with care (especially large and thin gels) since they tend to be more rigid and brittle due to their high monomer cross-linking degree.
23. This staining solution can be reused for an indefinite number of times.
24. The occurrence of effervescence is expected due to the release of carbon dioxide gas produced from the reaction between the acetic acid (in silver stain solution E) and carbonate ions (in the silver stain solution D). When the bubbling stops it is an indication that the staining reaction was fully halted.
25. It is optional to continue the cultivation of bacteria until early or late stationary phases. Further steps of bacteria processing are identical to those referred for bacteria collected in exponential phase.
26. Under these growth conditions, the overnight culture reaches a final optical density of ~3.0.
27. To ensure optimal mixing of the solution, it is recommended to use a magnetic stir inside the glass tube and continuously shake the solution.
28. The harsh treatment of isolating peptidoglycan sacculi by extensive boiling in SDS-containing solutions removes every protein not strongly associated to the peptidoglycan. Besides proteins covalently bound to the peptidoglycan, other surface proteins strongly bound to the peptidoglycan via diverse peptidoglycan-binding domains such as LysM or P60 (15) can also co-purify with peptidoglycan sacculi.
29. For protein quantification, do not use Coomassie dye-based methods (e.g., Bradford) because their reliability is limited by SDS concentrations ≥1% in the sample. As an alternative, employ SDS-insensitive quantification procedures, such as those based in the biuret test (e.g., bicinchoninic acid kits).

Acknowledgment

This work was supported by grants from the FCT (PTDC/SAU-MIC/111581/2009FCOMP-01-0124-FEDER-015844,PTDC/BIA-BCM/111215/2009FCOMP-01-0124-FEDER-014178, PTDC/BIA-BCM/100088/2008FCOMP-01-0124-FEDER-008860, and ERANet Pathogenomics Listress ERA-PTG/0003/2010 to

F.C. and D.C., and PhD fellowship SFRH/BD/61825/2009 to F.C.), from the Spanish Ministry of Economy and Competitiveness (BIO2010-18962 to M.G.P. and PIM2010EPA-00714 to F.G.P.), and from Institut Pasteur, Inserm, INRA ERC (Advanced Grant 233343) and Fondation les Mousquetaires to P.C.

References

1. Pucciarelli MG, Bierne H, García-del Portillo F (2007) The cell wall of *Listeria monocytogenes* and its role in pathogenicity. In: Goldfine H, Shen H (eds) *Listeria monocytogenes*: pathogenesis and host response. Springer, New York, pp 81–110. doi:DOI:10.1007/978-0-387-49376-3_5
2. Weidenmaier C, Peschel A (2008) Teichoic acids and related cell-wall glycopolymers in Gram-positive physiology and host interactions. Nat Rev Microbiol 6(4):276–287. doi:10.1038/nrmicro1861
3. Camejo A, Carvalho F, Reis O, Leitao E, Sousa S, Cabanes D (2011) The arsenal of virulence factors deployed by *Listeria monocytogenes* to promote its cell infection cycle. Virulence 2(5):379–394. doi:10.4161/viru.2.5.17703
4. Meredith TC, Swoboda JG, Walker S (2008) Late-stage polyribitol phosphate wall teichoic acid biosynthesis in *Staphylococcus aureus*. J Bacteriol 190(8):3046–3056. doi:10.1128/JB.01880-07
5. Endl J, Seidl HP, Fiedler F, Schleifer KH (1983) Chemical composition and structure of cell wall teichoic acids of staphylococci. Arch Microbiol 135(3):215–223
6. Pollack JH, Ntamere AS, Neuhaus FC (1992) D-alanyl-lipoteichoic acid in *Lactobacillus casei*: secretion of vesicles in response to benzylpenicillin. J Gen Microbiol 138(5):849–859
7. Pollack JH, Neuhaus FC (1994) Changes in wall teichoic acid during the rod-sphere transition of *Bacillus subtilis* 168. J Bacteriol 176(23):7252–7259
8. Min H, Cowman MK (1986) Combined alcian blue and silver staining of glycosaminoglycans in polyacrylamide gels: application to electrophoretic analysis of molecular weight distribution. Anal Biochem 155(2):275–285
9. Wolters PJ, Hildebrandt KM, Dickie JP, Anderson JS (1990) Polymer length of teichuronic acid released from cell walls of *Micrococcus luteus*. J Bacteriol 172(9):5154–5159
10. Marraffini LA, Dedent AC, Schneewind O (2006) Sortases and the art of anchoring proteins to the envelopes of gram-positive bacteria. Microbiol Mol Biol Rev 70(1):192–221. doi:10.1128/MMBR.70.1.192-221.2006
11. Calvo E, Pucciarelli MG, Bierne H, Cossart P, Albar JP, Garcia-Del Portillo F (2005) Analysis of the *Listeria* cell wall proteome by two-dimensional nanoliquid chromatography coupled to mass spectrometry. Proteomics 5(2):433–443. doi:10.1002/pmic.200400936
12. Pucciarelli MG, Calvo E, Sabet C, Bierne H, Cossart P, Garcia-del Portillo F (2005) Identification of substrates of the *Listeria monocytogenes* sortases A and B by a non-gel proteomic analysis. Proteomics 5(18):4808–4817. doi:10.1002/pmic.200402075
13. Garcia-del Portillo F, Calvo E, D'Orazio V, Pucciarelli MG (2011) Association of ActA to peptidoglycan revealed by cell wall proteomics of intracellular *Listeria monocytogenes*. J Biol Chem 286(40):34675–34689. doi:10.1074/jbc.M111.230441
14. Scott JR, Barnett TC (2006) Surface proteins of gram-positive bacteria and how they get there. Annu Rev Microbiol 60:397–423. doi:10.1146/annurev.micro.60.080805.142256
15. Bierne H, Cossart P (2007) *Listeria monocytogenes* surface proteins: from genome predictions to function. Microbiol Mol Biol Rev 71(2):377–397. doi:10.1128/MMBR.00039-06
16. Cabanes D, Dussurget O, Dehoux P, Cossart P (2004) Auto, a surface associated autolysin of *Listeria monocytogenes* required for entry into eukaryotic cells and virulence. Mol Microbiol 51(6):1601–1614

Chapter 19

Subfractionation and Analysis of the Cell Envelope (Lipo) polysaccharides of *Mycobacterium tuberculosis*

Anna E. Grzegorzewicz and Mary Jackson

Abstract

The cell envelope of *Mycobacterium tuberculosis*, the causative agent of tuberculosis in humans, is the source of carbohydrates of exceptional structure which play essential roles in the physiology of the bacterium and in its interactions with the host during infection. Much of what is known about their biosynthesis was derived from the phenotypic analysis of knockout or conditional knockout mutants of mycobacteria generated by random or specific insertional mutagenesis. Here, we describe the current techniques used to subfractionate *M. tuberculosis* cells and investigate major quantitative and qualitative changes in their cell envelope (lipo)polysaccharides.

Key words: *Mycobacterium tuberculosis*, Arabinogalactan, Lipoarabinomannan, Lipomannan, Glucan, Capsule

1. Introduction

The structure and composition of the cell envelopes of *Mycobacterium tuberculosis* and other mycobacteria differentiate these microorganisms from other prokaryotes (reviewed in ref. 1). In particular, it contains a number of unique (lipo)polysaccharides which have been the focus of much attention both from the perspective of drug development and understanding the host-pathogen interactions underlying the characteristic immunopathology of tuberculosis (1–5). In spite of being Gram-positive microorganisms, mycobacteria are endowed with an inner (IM) and an outer (asymmetrical) membrane (OM) separated by a periplasmic space (6, 7) (Fig. 1). Beyond the IM, a covalently linked complex consisting of mycolic acids, the heteropolysaccharide arabinogalactan (AG) and peptidoglycan (PG), constitutes the "core" of the cell wall. Interspersed somehow within this structure are the free (glyco) lipids of the OM with their fatty acyl chains intercalating into the

Anne H. Delcour (ed.), *Bacterial Cell Surfaces: Methods and Protocols*, Methods in Molecular Biology, vol. 966,
DOI 10.1007/978-1-62703-245-2_19, © Springer Science+Business Media New York 2013

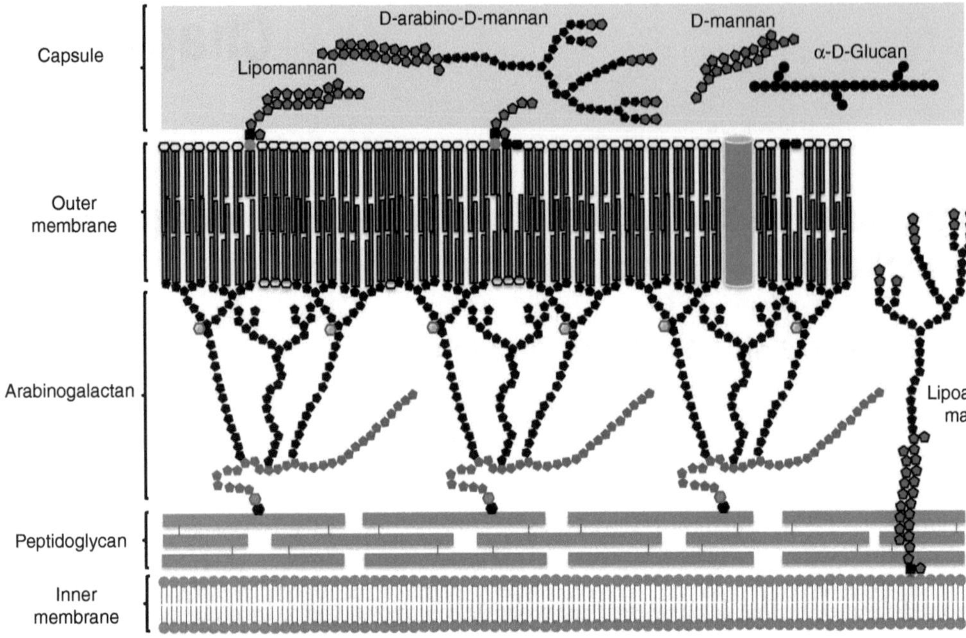

Fig. 1. Schematic representation of the *Mycobacterium tuberculosis* cell envelope (see text for details).

mycolic acid layer. Lipomannan (LM) and lipoarabinomannan (LAM), two lipopolysaccharides (also referred to as "lipoglycans") present in abundant quantities in the cell envelope of mycobacteria, possess a mannosylated phosphatidyl-*myo*-inositol moiety that mediates their non-covalent anchoring to the IM and OM (8) (Fig. 1). Finally, the outermost compartment of the cell envelope of *M. tuberculosis* consists of a loosely bound structure, referred to as "capsule," which is essentially composed of polysaccharides and proteins (9–11) (Fig. 1). The three major capsular polysaccharides are a high molecular weight α-D-glucan with a structure similar to that of glycogen (12) (see Note 1), a D-arabino-D-mannan (AM), and a D-mannan (1). The D-arabino-D-mannan and a D-mannan share structural resemblance with some segments of LM and LAM and are thought to be metabolically related (Fig. 2). Likewise, the arabinan domains of AG and that of LAM are structurally very similar and their biosynthesis involves the participation of common enzymes (Fig. 2).

The remarkable progress that was made over the last 15 years toward the elucidation of the biosynthetic pathways of *M. tuberculosis*' unique carbohydrates has heavily relied on genetic approaches to create mutants deficient in various aspects of their synthesis (reviewed in refs. 1, 5, 13). The phenotypic characterization of these mutants has allowed much of the glycosyltransferases involved in the biosynthesis of virulence-associated glycolipids, LM, LAM, and AG to be identified (1). How these macromolecules—which

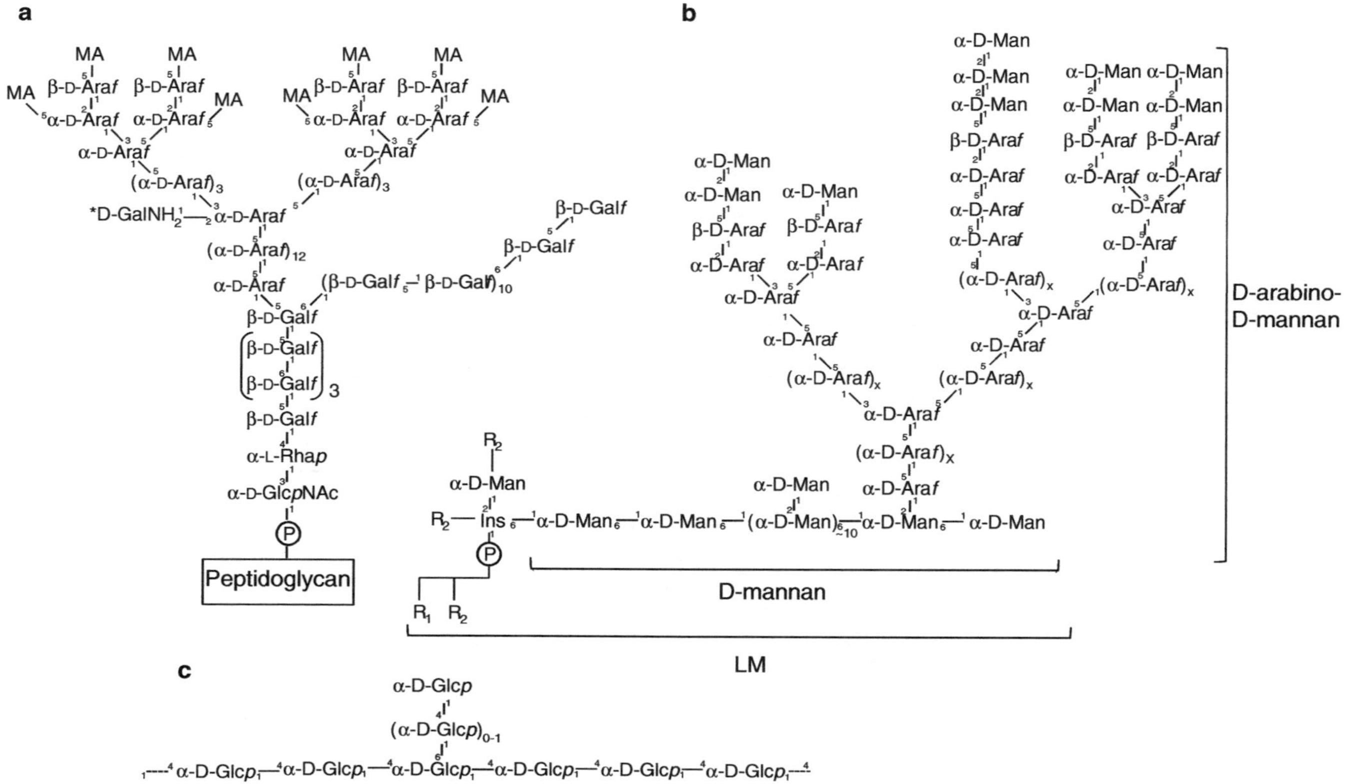

Fig. 2. Structures of the major cell envelope (lipo)polysaccharides from *Mycobacterium tuberculosis*. (**a**) Mycolyl arabinogalactan attached to peptidoglycan (mAGP); AG is made of a galactofuran of 30 Gal*f* residues and three arabinofurans (each made of 31 Ara*f* residues) of which only one is shown here. Some of the interior α-3,5,Ara*f* residues may be further substituted at position 2 with either galactosamine ($GalNH_2$) as shown herein or with a succinyl residue (15). AG is connected to PG via a linker unit (made of a rhamnosyl residue attached to a *N*-acetylglucosaminosyl-1-phosphate residue) at the reducing end and to mycolic acids (MA) at the nonreducing ends, all together forming the mycobacterial cell wall complex mAGP. (**b**) Lipoarabinomannan (LAM); the mannan moiety of LAM consists of approximately 20–30 Man*p* residues and the arabinan polymers of approximately 60 Ara*f* units; the precise number of arabinan chain(s) attached to mannan is still uncertain. Lipomannan (LM) is devoid of the arabinan chains of LAM. The capsular D-mannan and D-arabino-D-mannan are similar in structure to LAM from which they are thought to be derived. Note that both capsular polysaccharides are devoid of the phosphatidylinositol anchor of LAM. R_1-tuberculostearic acid, R_2-palmitic acid. (**c**) The repeating units of capsular α-D-glucan; the full-size molecule is >100,000 Da and represents about 80% of the capsular polysaccharides of *M. tuberculosis*.

are synthesized in part in the cytosol—reach their final periplasmic or OM location is less well documented and remains an area of intensive investigation. This chapter describes the current subfractionation and analytical methods that are used to investigate quantitative and qualitative changes in the capsular, cell wall- and IM/OM-associated (lipo)polysaccharides of *M. tuberculosis* (see Note 2). The analysis of *M. tuberculosis* (glyco)lipids and the detailed structural characterization of (lipo)polysaccharides are out of the scope of this review.

2. Materials

2.1. Isolation of the Capsular Material

1. *M. tuberculosis* H37Rv, frozen stock.
2. Gas medium: add 0.3 g of Bacto casitone, 0.05 g of Ferric ammonium citrate, 4.0 g of K_2HPO_4, 2.0 g of citric acid, anhydrous, 1.0 g of L-alanine, 1.2 g of $MgCl_2 \cdot 6\ H_2O$, 0.6 g of K_2SO_4, and 2.0 g of NH_4Cl to 800 ml of Milli-Q water. Adjust pH to 6.8 with 10 M NaOH. Add 10 ml of glycerol and bring volume to 1 l with Milli-Q water. Sterilize by autoclaving on liquid cycle at 121°C for 15 min.
3. Glass beads (4 mm-diameter).
4. Vacuum filtration system (0.22 μm).
5. 50-ml Falcon tube.

2.2. Preparation of Lipoglycans (LAM and LM)

1. *M. tuberculosis* H37Rv, frozen stock.
2. Gas medium (see Subheading 2.1, item 2).
3. Chloroform:methanol:water (10:10:3, v/v/v).
4. Endotoxin-free water (commercially available).
5. Endotoxin-free phosphate buffer saline (PBS) (commercially available).
6. 32% Triton X-114 solution in sterile endotoxin-free PBS.
7. Breaking buffer: add 70 μl of 1 mg/ml Pepstatin A, 50 μl of 1 mg/ml Leupeptin, 200 μl of 1 mg/ml PMSF, 150 μg of DNAse I, 165 μg of RNAse I to 100 ml of 8% Triton X-114.
8. Acid fast staining reagent.
9. Cold 95% ethanol.
10. Proteinase K solution: 2 mg/ml in 10 mM Tris–HCl (pH 7.5), 20 mM $CaCl_2$, 50% (v/v) glycerol.
11. Slide-A-Lyzer Casette 3.5 kDa MWCO (molecular weight cut off).
12. Vacuum concentrator, speed—vac.

13. Air bath.
14. Adjustable tilt rocking platform shaker.
15. French Press.
16. Oakridge Teflon FEP tubes
17. 7H10-OADC agar (commercially available).
18. 13 × 100 mm glass tubes.
19. 200-ml Falcon tubes.
20. 4-l conical flasks.
21. High- and low-speed centrifuges and rotors.

2.3. Separation of Lipoglycans (LAM and LM)

1. Lipoglycans extracted from *M. tuberculosis* H37Rv (~15 mg dry weight).
2. Column running buffer: 10 mM Tris–HCl, 0.2 M NaCl, 0.02% (w/v) sodium azide, 0.25% (w/v) deoxycholic salt, and 1 mM EDTA, pH 8.
3. Dialysis buffer: 10 mM Tris–HCl, 0.2 M NaCl, 0.02% (w/v) sodium azide, and 1 mM EDTA, pH 8.
4. 1 M NaCl.
5. Endotoxin-free water (commercially available).
6. 0.22 μm syringe filters.
7. 10 ml syringe.
8. High performance liquid chromatography system (HPLC).
9. Sephacryl S-100HiPrep 16/60 column.
10. Sephacryl S-200HiPrep 26/60 column.
11. Fraction collector.
12. Spectra/Por Dialysis Membranes (6–8 kDa MWCO and 12–14 kDa MWCO).

2.4. Tricine SDS/Polyacrylamide Gel Electrophoresis of Lipoglycans Followed by Periodic Acid Silver Staining

1. 5–10 μg of lipoglycans.
2. PageRuler Prestained Protein Ladder, 10–170 kDa.
3. 10–20% gradient Tricine SDS/polyacrylamide gel, commercially obtained.
4. 10× Tricine SDS running buffer, commercially obtained.
5. 2× Tricine SDS sample buffer and 10× Reducing agent, commercially obtained.
6. First fixative: 40% (v/v) methanol, 10% (v/v) acetic acid in distilled water.
7. Second fixative: 5% (v/v) methanol, 7% (v/v) acetic acid in distilled water.
8. 0.7% (w/v) periodic acid solution in the first fixative.

9. 2.5% (v/v) glutaraldehyde solution in distilled water.
10. 0.0025% (w/v) α-dithiothreitol solution (DTT) in distilled water.
11. 0.1% (w/v) silver nitrate solution in distilled water.
12. 3% (w/v) sodium carbonate solution in distilled water.
13. 37% (v/v) formaldehyde solution in distilled water
14. Developer solution: to 200 ml of 3% sodium carbonate add 100 μl of 37% formaldehyde.
15. 50% (w/v) citric acid solution in distilled water.
16. Adjustable tilt rocking platform shaker.

2.5. Preparation of the Mycolate-Arabinogalactan Peptidoglycan (mAGP) Complex

1. Cell walls from *M. tuberculosis* H37Rv (i.e., remaining pellet after lipoglycans extraction, see Subheading 3.2, step 18).
2. Oakridge Teflon FEP tube.
3. 2% SDS in PBS.
4. Proteinase K solution: add 10 mg in 1 ml of 10 mM Tris–HCl (pH 7.5), 20 mM $CaCl_2$, 50% (v/v) glycerol.
5. Acetone.
6. Magnetic stir/heat plate.
7. High-speed centrifuge and rotors.

2.6. Preparation of Arabinogalactan (AG)

1. Purified mAGP (50–150 mg dry weight).
2. Oakridge Teflon FEP tube.
3. 0.5% KOH in methanol.
4. Methanol.
5. Chloroform.
6. 0.05 M sulfuric acid.
7. Barium carbonate.
8. Vacuum concentrator, speed—vac.
9. Air bath.
10. Magnetic stir/heat plate.
11. 16 × 100 mm glass tubes.
12. High-speed centrifuge and rotors.

2.7. Analysis of Monosaccharide Composition by Gas Chromatography/Mass Spectrometry

1. 100 μg/ml of monosaccharide standards stock solutions (arabinose, fucose, galactose, glucose, mannose, rhamnose, ribose, xylose and *myo*—inositol) in Milli-Q water.
2. 100 μg/ml of 3-*O*-methyl-glucose (internal standard) in Milli-Q water.
3. 13 × 100 mm glass tubes with Teflon lined lids.

4. 2 M Trifluoroacetic acid (TFA) in Milli-Q water.
5. Reducing agent: 10 mg/ml sodium borodeuteride in 1 M aqueous ammonium hydroxide/95% ethanol (1:1, v/v).
6. Acetic acid, glacial.
7. Acetic anhydride.
8. Chloroform.
9. Methanol.
10. Vacuum concentrator, speed—vac.
11. Heat block.
12. Air bath.
13. CP 3800 gas chromatograph equipped with an MS320 mass spectrometer and fitted with a DB 5 column (30 m×0.20 mm i.d.).
14. Low speed centrifuge and rotor.

3. Methods

3.1. Isolation of the Capsular Material

To isolate the capsule, i.e., the outermost compartment of the mycobacterial cell envelope, a gentle extraction with glass beads is applied. The mechanical treatment of mycobacterial cells with glass beads disperses the cells by extracting the amorphous material without causing cell disruption (10). The monosaccharide composition of the capsule (9, 10) is determined by gas chromatography/mass spectrometry analysis of alditol acetate derivatives.

1. *M. tuberculosis* H37Rv is grown in 200 ml of Gas medium at 37°C with shaking to late-log phase (see Note 3).
2. Harvest the cells by centrifugation at 3,000×*g* for 15 min at 4°C and combine all pellets in a 50 ml Falcon tube.
3. To isolate the capsule, add 10 g of glass beads per 2 g of the wet bacterial pellet and gently shake for 2 min.
4. Suspend the bacterial pellet and beads in 50 ml of Milli-Q water and spin down at 3,000×*g* for 15 min at 4°C. Collect the supernatant and filter immediately through a 0.22 μM pore-size-filter.
5. Freeze-dry the filtered supernatant (capsule material) by lyophilization.
6. Weigh the lyophilized material and dissolve in Milli-Q water.
7. Use a 50 μg aliquot of this preparation for alditol acetate preparation for monosaccharide composition determination (see Subheading 3.7).

3.2. Preparation of Lipoglycans (LAM and LM)

To prepare mycobacterial lipoglycans suitable for structural analyses, the cells are first delipidated. The delipidation step removes all peripheral, noncovalently bound lipids. The delipidated cells are disrupted mechanically by French Press and lipoglycans are extracted with detergent and precipitated with ethanol. Proteinase K treatment is applied to digest and remove proteins. This procedure yields a mixture of LAM and LM (see Note 4).

1. *M. tuberculosis* is grown in 8 l (2×4-l conical flasks) of Gas medium to late-log phase (see Note 3).
2. Harvest the cells by centrifugation at 3,000×*g* for 15 min at 4°C. Wash the cell pellet three times with PBS and transfer to an Oakridge Teflon FEP tube.
3. Delipidate the cells with 10 ml of chloroform:methanol:water (10:10:3, v/v/v), for every 1 g of wet cells. After rocking the tubes for 2 h, centrifuge at 27,000×*g* for 20 min at 15°C.
4. The delipidation step is repeated two more times (see Note 5).
5. To remove residual organic solvents, the cells are dried on the air bath at room temperature overnight (see Note 6).
6. Grind the dried cell pellet to achieve a fine powder.
7. Delipidated cells are suspended in a minimal amount of breaking buffer and freeze-thawed three times to achieve complete suspension before passing over French Press eight times at 1,500 psi (see Note 7).
8. Transfer the suspension of broken cells to Oakridge Teflon FEP tubes and remove any remaining unbroken cells by centrifugation at 3,000×*g* for 5 min.
9. Add breaking buffer to the supernatant (broken cells) at a ratio of 1:1 (v/v) and rock overnight at 4°C.
10. Centrifuge the broken cells (from step 9) at 27,000×*g* for 1 h at 4°C (see Note 8).
11. Transfer the clear supernatant to Oakridge Teflon FEP tubes and keep the pellet at 4°C.
12. Incubate the supernatant at 37°C to separate the aqueous layer from the detergent layer (see Note 9).
13. After the supernatant is partitioned into two layers, centrifuge at 27,000×*g* for 15 min at 25°C.
14. Carefully collect the aqueous layer (upper layer) into the tubes containing the pellet from step 11. Pool the detergent layer and store at 4°C.
15. Add Triton X-114 to the aqueous layer containing the pellet from step 11 to achieve the final concentration of 8% (v/v).
16. Rock the tubes gently at 4°C for 2 h.

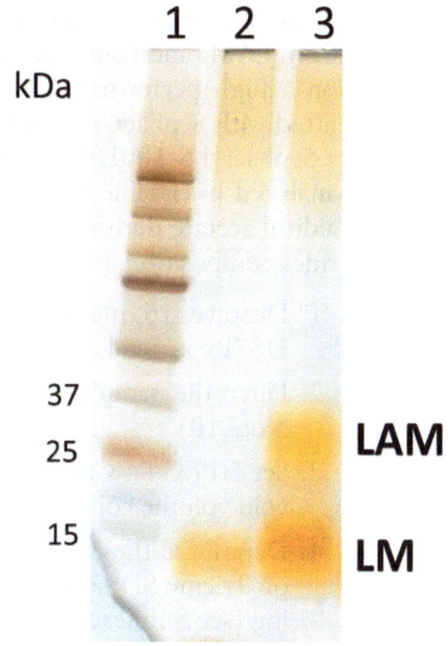

Fig. 3. Tricine SDS polyacrylamide gel analysis of lipoglycans. Purified lipoglycans were run on a Tricine SDS polyacrylamide gel and visualized by PAS staining. *Lane 1* molecular weight marker, *Lane 2* Purified LM, *Lane 3* LAM and LM.

17. Repeat detergent extraction two more times (steps 10–16).
18. Combine all detergent extracts and keep the pellet (later referred to as "cell wall pellet"; see Subheading 3.5) for the preparation of mAGP.
19. To precipitate the extracted lipoglycans, add cold 95% ethanol to the combined detergent layers at a ratio 10:1 (v/v) and leave at −20°C overnight.
20. Centrifuge at 27,000 × *g* for 20 min at 4°C.
21. Collect the precipitate in one tube and dry using a vacuum concentrator.
22. Add proteinase K at a concentration 2 mg/ml and digest at 37°C overnight.
23. Dialyze the digest for 24 h against the endotoxin-free water using a 3.5 kDa MWCO Slide-A-Lyzer Casette.
24. Transfer the sample to 13 × 100 mm glass tubes and dry using a vacuum concentrator.
25. To determine if the sample contains protein-free soluble lipoglycans, run a Tricine SDS/polyacrylamide gel followed by PAS (Periodic Acid Silver staining) (see Subheading 3.4 and Fig. 3). If proteins are still present in the sample, repeat the proteinase K digestion step.

3.3. Separation of Lipoglycans (LAM and LM)

A size-exclusion chromatography system is used to separate LM from LAM (and contaminating PIMs; see Note 4). It is performed on a high performance liquid chromatography (HPLC) system fitted with Sephacryl S-200HiPrep 26/60 column in tandem with a Sephacryl S-100HiPrep 16/60 column. Purified lipoglycans are analyzed by Tricine SDS/Polyacrylamide gel electrophoresis and alditol acetate derivatives are prepared to determine monosaccharides composition and ratios.

1. Dissolve 15 mg lipoglycans extracted from *M. tuberculosis* H37Rv in 5 ml of column running buffer.
2. Filter the sample through 0.22-μm syringe filter units (see Note 10).
3. Set HPLC flow rate at 1 ml/min. After collecting the column void volume collect about 100 fractions of 2 ml.
4. Determine the elution profile by running every other fraction on Tricine SDS/polyacrylamide gel and develop by PAS staining (see Subheading 3.4).
5. Based on the results of PAS staining make pools of pure LAM and LM.
6. Dialyze against the dialysis buffer at 37°C for 24 h. Use the following membranes for each pool: 12–14 kDa MWCO for LAM and 6–8 kDa MWCO for LM.
7. Change the dialysis buffer to 1 M NaCl and continue dialysis at room temperature for another 24 h.
8. Dialyze against endotoxin-free water at room temperature for 24 h.
9. Transfer the dialyzed fractions to 100 ml round-bottom flask and freeze-dry by lyophilization.

3.4. Tricine SDS/Polyacrylamide Gel Electrophoresis of Lipoglycans Followed by Periodic Acid Silver Staining

Lipoglycans are run on a Tricine SDS/polyacrylamide gel and visualized by PAS staining.

1. Dissolve 5 μg of purified lipoglycans in 8 μl of Milli-Q water.
2. Add 10 μl of 2× Tricine SDS Sample buffer, 2 μl of 10× reducing agent to the sample and incubate at 100°C for 5 min.
3. Load 10 μl of PageRuler Prestained Protein Ladder and the sample onto a gel.
4. Run the gel at constant voltage of 125 V. The expected current should be 80 mA/gel at the beginning and 40 mA/gel at the end of the run.
5. Transfer the gel to a dish containing 100 ml of the first fixative and rock for 45 min (see Note 11).
6. Discard the first fixative and add 100 ml of 0.7% periodic acid solution. Rock for 7 min (see Note 12).

7. Discard the periodic acid and add 100 ml of the second fixative and rock for 5 min.
8. Discard the second fixative, add enough 2.5% glutaraldehyde to cover the gel and rock for 5 min.
9. Remove the glutaraldehyde and wash the gel with 100 ml of distilled water for 10 min. Repeat washing step as needed (see Note 13).
10. Add the 0.0025% DTT solution for 6 min.
11. Discard the DTT solution and after adding 100 ml of 0.1% silver nitrate rock for 5 min.
12. Decant off the silver nitrate solution and rinse the gels briefly with Milli-Q water three times.
13. Develop the gel in 200 ml of developer solution. Start by incubating the gel in about 50 ml of the developer. After the solution becomes brown/yellow, discard the developer and replace with the fresh 50 ml of developer. Incubate for 10–20 s. Replace again with 100 ml developer and incubate until the bands are as dark as desired.
14. Stop development by adding 10 ml of 50% citric acid for 10 min and wash the stained gels with water.

3.5. Preparation of the Mycolate-Arabinogalactan Peptidoglycan (mAGP) Complex

To prepare the mAGP complex, the cell wall pellet is extracted with 2% SDS, which disrupts membranes and hydrophobic interactions. Proteinase K digestion is applied to digest and remove any protein contaminants. Finally, the mAGP pellet is washed with acetone, dried, and alditol acetate derivatives are prepared to determine the monosaccharide composition of mAGP.

1. Transfer the cell wall pellet (see Subheading 3.2, step 18) to an Oakridge Teflon FEP tube.
2. Resuspend the cell wall pellet in 30 ml of 2% SDS solution and stir with a magnetic stir bar at room temperature for 30 min (see Note 14).
3. Centrifuge at 27,000 × *g* for 10 min at 25°C. Decant the supernatant.
4. Repeat steps 2–3 two more times.
5. Add 26 ml of 2% SDS and 4 ml of Proteinase K solution to the pellet. Stir with a magnetic stir bar at room temperature for 30 min.
6. Centrifuge at 27,000 × *g* for 10 min at 25°C.
7. Discard the supernatant and add 30 ml of 2% SDS solution to the pellet.
8. After stirring with a magnetic stir bar at 90°C for 1 h, centrifuge at 27,000 × *g* for 10 min at 25°C.

9. Repeat steps 7–8 two more times.
10. Discard the supernatant, add 30 ml of water to the pellet and stir at room temperature for 30 min.
11. Centrifuge at 27,000 × *g* for 10 min at 25°C.
12. Discard the supernatant and after adding 5 ml of acetone to the pellet stir at room temperature for 10 min.
13. Centrifuge at 27,000 × *g* for 10 min at 25°C, decant acetone and dry the pellet in a chemical fume hood.
14. Weight 1 mg of purified mAGP and perform alditol acetate derivatization for monosaccharide composition determination (see Subheading 3.7).

3.6. Preparation of AG

To isolate AG, the mAGP complex is treated with 0.5% KOH. The basic hydrolysis removes the cell wall bound mycolic acids from the mAGP complex. The released mycolic acids are extracted with organic solvents. At this step, the pellet consists mainly of AGP. AG is released from AGP by mild acid hydrolysis. The supernatant, which contains the solubilized AG is treated with barium carbonate to neutralize the acid. Alditol acetates are prepared to analyze the monosaccharide composition and purity of AG.

1. Transfer mAGP to an Oakridge Teflon FEP tube and resuspend in 30 ml of 0.5% KOH in methanol and stir with a magnetic stir bar at 37°C for 4 days.
2. Centrifuge at 27,000 × *g* for 20 min.
3. Discard the supernatant and resuspend the pellet in 30 ml of methanol.
4. Centrifuge at 27,000 × *g* for 20 min.
5. Resuspend the pellet in 20 ml of chloroform and mix well using a stir bar for 5–10 min at room temperature. Then add 10 ml of methanol and centrifuge at 27,000 × *g* for 20 min (see Note 15).
6. Resuspend the pellet in 30 ml of methanol and centrifuge at 27,000 × *g* for 20 min.
7. Discard the supernatant and dry the pellet on the air bath. The pellet now consists of rather pure AGP.
8. Transfer the AGP pellet to a 16 × 100 mm tube and add 5 ml of 0.05 M sulfuric acid.
9. Stir with a magnetic stir bar at 37°C for 4 days.
10. Centrifuge at 3,000 × *g* for 15 min at 25°C.
11. Transfer the supernatant to a new 16 × 100 mm tube (see Note 16).
12. To neutralize the acid, add a few grains of barium carbonate and incubate overnight at room temperature.

13. Centrifuge at 3,000 × *g* for 15 min at 25°C.
14. Transfer the supernatant to a new 16 × 100 mm tube and dry using a vacuum concentrator.
15. Dissolve in 5 ml of Milli-Q water and use 5 μl for alditol acetate preparation for monosaccharide composition determination (see Subheading 3.7).

3.7. Analysis of Monosaccharide Composition by Gas Chromatography/Mass Spectrometry

To analyze the monosaccharide composition of the sample, it is necessary to derivatize the glycosyl residue to alditol acetates. This procedure involves three major steps: hydrolysis, reduction, and acetylation.

1. Combine 10 μl of monosaccharides stock solutions and 10 μl of the internal standard in 13 × 100 mm glass tube (see Note 17).
2. Add your sample and 10 μl of the internal standard to another 13 × 100 mm glass tube.
3. Completely dry the samples using a vacuum concentrator.
4. Add 250 μl of TFA and heat at 120°C for 2 h.
5. Completely dry the content of each tube on the air bath.
6. Add 100 μl of methanol to each tube and dry on air bath.
7. Repeat step 6 one more time.
8. Add 200 μl of the reducing agent and let it stand at room temperature for 2 h (see Notes 18 and 19).
9. Stop the reaction by adding two drops (~50 μl) of glacial acetic acid and dry the samples on air bath.
10. Add 100 μl of 10% glacial acetic acid in methanol and dry the samples on air bath.
11. Repeat step 10 three more times.
12. Add 100 μl of methanol and dry on air bath.
13. To acetylate, add 100 μl of acetic anhydride and heat at 120°C for 2 h in a heating block.
14. Cool the tubes at room temperature and dry the samples on the air bath under a gentle stream of air.
15. Add 1 ml of Milli-Q water and 2 ml of chloroform and mix thoroughly.
16. Centrifuge at 2,500 × *g* for 5 min at 4°C.
17. Discard the upper aqueous layer.
18. Wash the organic layer with 1 ml of Milli-Q water.
19. Centrifuge at 2,500 × *g* for 5 min at 4°C.
20. Transfer the lower organic layer to 13 × 100 mm glass tube and dry on air bath under a gentle stream of air (see Note 20).
21. Dissolve each sample in 50 μl of chloroform.

22. Inject 1 μl of the sample into the gas chromatograph-mass spectrometer. Use an initial temperature of 50°C for 1 min, increasing to 150°C at 30°C/min, followed by a 5°C/min increase to 275°C.

4. Notes

1. Although mycobacterial glycogen is very similar in structure to capsular α-D-glucan, glycogen is localized intracellularly and the capsular α-D-glucan will thus readily be separated from glycogen using the subfractionation procedure with glass beads described under Subheading 3.1. A description of how intracellular glycogen may be prepared can be found in ref. 12.
2. Although this chapter refers of *M. tuberculosis*, the same procedures can be used to analyze the cell envelope (lipo)polysaccharide content of other *Mycobacterium* species.
3. All steps involving live *M. tuberculosis* H37Rv must be performed under Biosafety Level III conditions.
4. Remaining phosphatidylinositol mannosides (PIMs), particularly the more polar (i.e., more mannosylated) forms of them, may contaminate this preparation.
5. Delipidated bacterial cell pellet may be brought outside the BSL-3 containment laboratory at this point since, if applied under the conditions described here, *M. tuberculosis* should be killed by the treatment with chloroform/methanol. To ensure that no viable bacteria remain, plate some of the non-diluted suspension on 7H10-OADC agar and incubate the plates for 3–4 weeks at 37°C.
6. Cover tubes with foil and dry the sample under a gentle stream of air.
7. You may perform an acid fast staining (14) on the suspension to ensure that at least 90% of the cells have been broken.
8. If the supernatant is not clear, repeat centrifugation until a clear supernatant is obtained. It is very important to obtain a clear supernatant devoid of any residual contamination with mAGP.
9. Triton X-114 is miscible with water at 4°C, however it separates from water at 37°C and a biphasic suspension forms. The detergent layer is the lower layer.
10. If the sample is too concentrated and doesn't easily go through the 0.2-μm filter, it can first be filtered through a 0.8 μm filter.
11. The gel can be kept in the first fixative and water indefinitely.

12. Periodic acid enhances the staining of carbohydrates and shouldn't be used when the goal is to verify the contamination of lipoglycans with proteins.
13. Background is reduced with increased washing with water. An overnight washing works well.
14. Before each centrifugation step, remove the magnetic stir bar from the tube.
15. The organic extract/supernatant contains mycolic acids.
16. The pellet consists of peptidoglycan.
17. 3-*O*-methyl-glucose is used as an internal standard to determine quantitatively the amount of monosaccharides in the sample.
18. Reducing agent must be freshly made prior to use.
19. The reduction reaction gives better results when performed overnight.
20. It is important not to contaminate the organic layer with the water layer.

Acknowledgments

Research on mycobacterial (lipo)polysaccharides in the authors' laboratory is supported by the National Institutes of Health/ National Institute of Allergy and Infectious Diseases grant AI064798. Dr. Shiva Kumar Angala is gratefully acknowledged for his assistance in the preparation of Fig. 3.

References

1. Kaur D, Guerin ME, Škovierová H et al (2009) Biogenesis of the cell wall and other glycoconjugates of *Mycobacterium tuberculosis*. Adv Appl Microbiol 69:23–78
2. Daffé M, Draper P (1998) The envelope layers of mycobacteria with reference to their pathogenicity. Adv Microb Physiol 39:131–203
3. Daffé M, Etienne G (1999) The capsule of *Mycobacterium tuberculosis* and its implications for pathogenicity. Tuber Lung Dis 79:153–169
4. Brennan PJ, Crick DC (2007) The cell-wall core of *Mycobacterium tuberculosis* in the context of drug discovery. Curr Top Med Chem 7:475–488
5. Gilleron M, Jackson M, Nigou J et al (2008) Structure, activities and biosynthesis of the Phosphatidyl-*myo*-Inositol-based lipoglycans. In: Daffé M, Reyrat JM (eds) The Mycobacterial Cell Envelope. ASM Press, Washington, DC, pp 75–105
6. Hoffmann C, Leis A, Niederweis M et al (2008) Disclosure of the mycobacterial outer membrane: cryo-electron tomography and vitreous sections reveal the lipid bilayer structure. Proc Natl Acad Sci U S A 105:3963–3967
7. Zuber B, Chami M, Houssin C et al (2008) Direct visualization of the outer membrane of mycobacteria and corynebacteria in their native state. J Bacteriol 190:5672–5680
8. Pitarque S, Larrouy-Maumus G, Payré B et al (2008) The immunomodulatory lipoglycans, lipoarabinomannan and lipomannan, are exposed at the mycobacterial cell surface. Tuberculosis 88:560–565
9. Lemassu A, Daffé M (1994) Structural features of the exocellular polysaccharides of

Mycobacterium tuberculosis. Biochem J 297: 351–357

10. Ortalo-Magné A, Dupont MA, Lemassu A et al (1995) Molecular composition of the outermost capsular material of the tubercle bacillus. Microbiology 141:1609–1620
11. Sani M, Houben ENG, Geurtsen J et al (2010) Direct visualization by cryo-EM of the mycobacterial capsular layer: a labile structure containing ESX-1-secreted proteins. PLoS Pathog 6:e1000794
12. Dinadayala P, Sambou T, Daffé M et al (2008) Comparative structural analyses of the alpha-glucan and glycogen from *Mycobacterium bovis*. Glycobiology 18: 502–508
13. Berg S, Kaur D, Jackson M et al (2007) The glycosyltransferases of *Mycobacterium tuberculosis*- roles in the synthesis of arabinogalactan, lipoarabinomannan, and other glycoconjugates. Glycobiology 17:35R–56R
14. Ebersole LL (1992) Acid-fast staining procedures. In: Isenberg HD (ed) Clinical microbiology procedures handbook, vol. 1. ASM, Washington DC., pp 3.5.1–3.5.11
15. Bhamidi S, Scherman MS, Rithner CD et al (2008) The identification and location of succinyl residues and the characterization of the interior arabinan region allows for a model of the complete primary structure of *Mycobacterium tuberculosis* mycolyl arabinogalactan. J Biol Chem 283:12992–13000

Chapter 20

Protein Disulfide Bond Formation in the Periplasm: Determination of the In Vivo Redox State of Cysteine Residues

Katleen Denoncin, Valérie Nicolaes, Seung-Hyun Cho, Pauline Leverrier, and Jean-François Collet

Abstract

Many proteins secreted to the bacterial cell envelope contain cysteine residues that are involved in disulfide bonds. These disulfides either play a structural role, increasing protein stability, or reversibly form in the catalytic site of periplasmic oxidoreductases. Monitoring the in vivo redox state of cysteine residues, i.e., determining whether those cysteines are oxidized to a disulfide bond or not, is therefore required to fully characterize the function and structural properties of numerous periplasmic proteins. Here, we describe a reliable and rapid method based on trapping reduced cysteine residues with 4′-acetamido-4′-maleimidylstilbene-2,2′-disulfonic acid (AMS), a maleimide compound. We use the *Escherichia coli* DsbA protein to illustrate the method, which can be applied to all envelope proteins.

Key words: Redox state, *Escherichia coli*, Periplasm, DsbA, Disulfide bond, Cysteine, Maleimide, AMS labeling

1. Introduction

Disulfide bonds are vital for the correct folding of many proteins that are secreted to the bacterial cell envelope. In *Escherichia coli*, about 40% of the periplasmic proteins possess at least two cysteine residues and can therefore form disulfide bonds (1, 2). Well-characterized proteins whose folding involves the formation of at least one disulfide bond include alkaline phosphatase (3), ribonuclease A (4), the outer membrane lipoprotein RcsF (5), and the essential β-barrel protein LptD (6–8).

Although disulfide bonds can form spontaneously in the presence of molecular oxygen, air oxidation is a rather slow process. In order to ensure high rate of protein folding and secretion, catalysts

Anne H. Delcour (ed.), *Bacterial Cell Surfaces: Methods and Protocols*, Methods in Molecular Biology, vol. 966,
DOI 10.1007/978-1-62703-245-2_20,

of disulfide bond formation are present in most living cells. The first bacterial catalyst of disulfide bond formation, DsbA, was identified in the periplasm of *E. coli* by the group of Jon Beckwith (9). DsbA, like many other oxidoreductases, belongs to the thioredoxin (Trx) superfamily (10). Proteins from that family have a canonical CXXC catalytic motif located in a highly conserved fold (11). The catalytic CXXC motif of DsbA is maintained oxidized in vivo, which enables DsbA to react with proteins entering the periplasm to oxidize them: upon interacting with a newly secreted protein, the disulfide bond is transferred from the catalytic site of DsbA to the substrate protein (9). Concomitantly, two electrons are transferred from the substrate to DsbA, releasing DsbA in a reduced state. In order to remain active in the periplasm, DsbA needs to be reoxidized. The protein that recycles DsbA is the inner membrane protein DsbB, which transfers the electrons received from DsbA to the electron transport chain (12, 13).

DsbA is a powerful oxidant (14) but lacks proofreading activity. Thus, proteins that require formation of disulfides between cysteine residues that are nonconsecutive in the sequence are often incorrectly oxidized by DsbA. Therefore, the *E. coli* periplasm contains an isomerization system to catalyze the rearrangement of nonnative disulfide bonds (15).

The central player of the *E. coli* isomerization pathway is the protein disulfide isomerase DsbC, a soluble homodimeric protein characterized by a V-shaped structure (16). Each monomer of DsbC presents a C-terminal Trx-like domain with a CXXC active site that is found predominantly in the reduced state in vivo (note that only reduced DsbC can rearrange nonnative disulfide bonds) (16, 17). The protein that maintains DsbC reduced in the periplasm is the inner membrane protein DsbD (18), which transfers electrons from the cytoplasmic Trx system to the periplasm (19), by a yet unknown mechanism. The Trx system, which involves Trx and Trx reductase, maintains cytoplasmic proteins reduced at the expense of NADPH (11).

The *E. coli* disulfide bond system has now become the paradigm of oxidative protein folding pathways in bacteria. However, bioinformatic analyses of whole-genome sequencing data have revealed that there is a significant diversity of bacterial disulfide bond-forming systems (20). The majority of these bacterial systems are still not characterized, and further investigation is required to understand in detail how the oxidative folding pathways catalyze disulfide bond formation in bacteria other than *E. coli*.

Here we describe a method that can be used to determine whether two cysteine residues present in a protein are oxidized to a disulfide bond. This is an important question to address when dealing with periplasmic proteins. For instance, determination of the in vivo redox state of the catalytic cysteine residues of a periplasmic oxidoreductase is often a key element indicating whether the protein functions in an oxidizing or in a reducing pathway. Proteins that catalyze disulfide formation, such as DsbA, are indeed maintained oxidized in vivo, whereas protein disulfide isomerases, such as DsbC, are maintained in the reduced state. If the protein of interest is not an oxidoreductase, determining the redox state of its cysteine residues will provide useful information regarding the folding pathway and structural stability of this protein (see Note 1).

In the method that we describe, proteins are first denatured under acidic conditions by the addition of trichloroacetic acid (TCA). This treatment prevents nonspecific thiol-disulfide exchange reactions that could influence the final result. Then, denatured proteins are solubilized in a SDS-containing buffer in the presence of a thiol-labeling reagent. Commonly used reagents are modified forms of *N*-ethylmaleimide (NEM), such as 4-acetamido-4′-maleimidylstilbene-2,2′-disulfonic acid (AMS). AMS is a 0.49 kDa labeling reagent that consists of a fusion between a maleimide moiety and another small size molecule. Maleimide-based molecules react with the free thiolate groups of reduced cysteine residues and covalently modify them (see Fig. 1). In contrast, cysteine residues that are involved in a disulfide bond are not modified. Thus, oxidized proteins migrate with the expected size on SDS-PAGE, whereas reduced proteins are shifted towards higher molecular weights. Here, we use the *E. coli* DsbA–DsbB system to illustrate the method and determine the in vivo redox state of DsbA both in a wild-type strain (where it is oxidized) and in a *dsbB* deletion mutant strain (where it is reduced).

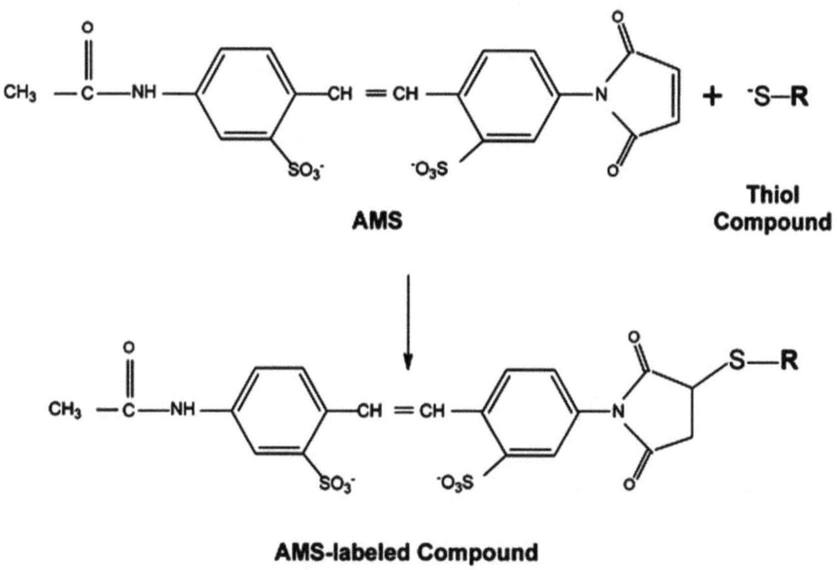

Fig. 1. Maleimide compounds such as AMS react with reduced cysteine residues and covalently modify them.

2. Materials

2.1. Equipment

1. Dry block heating shaker.
2. Microcentrifuge.
3. Gel-filtration (desalting) columns.
4. UV-Visible Spectrophotometer.
5. Gel electrophoresis system.
6. Semi-dry protein transfer system.
7. Western blot developer.

2.2. Bacterial Strains and Growth Media

1. Luria Bertani (LB) agar plates: dissolve 5 g of Bacto yeast extract, 10 g of Bacto tryptone, 15 g of Bacto agar, and 10 g of NaCl to a final volume of 1 L with water (see Note 2). Autoclave and store at room temperature.
2. LB broth medium: mix 5 g of Bacto yeast extract, 10 g of Bacto tryptone, 10 g of NaCl and dissolve to a final volume of 1 L with water (see Note 3). Autoclave and store at room temperature.
3. Three *E. coli* strains are used in our study: wild-type MC4100 (21) and two deletion strains, prepared in MC4100, lacking either *dsbA* (*dsbA* strain) or *dsbB* (*dsbB* strain). These genes were deleted by the standard P1 transduction procedure (22).

2.3. Solutions

1. DTT solution: to prepare a 1 M solution, dissolve 0.77 g of DTT in water to a final volume of 5 mL. Aliquot in 1 mL portions and store at −20°C.
2. GSSG solution: dissolve 613 mg of GSSG in 1 mL of buffer-containing 50 mM of potassium phosphate buffer (KPi) (see Note 4), pH 7 and 0.1 mM EDTA. Store the stock solution at −20°C.
3. TCA solution: to avoid weighing out this highly corrosive reagent, the whole content of the bottle (1 kg) is dissolved by adding 454 mL of water to obtain a 100% solution. Store at 4°C.
4. AMS solution: because powder of AMS (4-acetamido-4′-maleimidylstilbene-2,2′-disulfonic acid, disodium salt) sticks to the bottle, resuspend the whole content of the bottle (25 mg) in 2.33 mL of 50 mM Tris–HCl, pH 7.5, 0.1% SDS (v/v), 10 mM EDTA to obtain a final 20 mM AMS solution. Store at −20°C. Protect solution from light.
5. 5× Laemmli buffer: 10% SDS (v/v), 0.05% bromophenol blue (w/v), 61% glycerol (v/v), 300 mM Tris–HCl, pH 6.8. Store at −20°C.
6. 10× SDS-PAGE running buffer: mix 144 g of glycine, 30 g of Tris base, and 50 mL of 20% SDS. Bring to 1 L with water. Dilute ten times before use.

7. 1× Western blot transfer buffer: mix 5.8 g of Tris base, 1.875 mL of 20% SDS, 200 mL of methanol, and 2.93 g of glycine. Bring to 1 L with water.
8. 20× Tris-buffered saline (TBS): mix 160 g of NaCl, 4 g of KCl, and 60 g of Tris base. Adjust to pH 7.4 with HCl. Bring to 1 L with water.
9. 1× TBS-T: dilute the 20× TBS solution to 1× and mix with 0.05% (v/v) of Tween-20 detergent.
10. Blocking buffer: 5% (w/v) skim milk powder (see Note 5) in 1× TBS.
11. Primary antibody solution: anti-DsbA, raised in rabbit. Immunological serums are stored at −80°C. Alternatively, they can be stored at 4°C when sodium azide is added (0.02%). Working solutions for primary antibody incubations are prepared at 1:10,000 dilution in 1× TBS containing 1% Bovine Serum Albumin (BSA) (w/v). This solution should be prepared freshly.
12. Secondary antibody solution: goat anti-rabbit IgG (whole molecule)-peroxidase is stored at 4°C. Working solutions for secondary antibody incubations are prepared freshly at a 1:5,000 dilution in 1× TBS containing 1% BSA (w/v). This solution should be prepared freshly.
13. Chemiluminescence substrate: stored at 4°C. Mix equal volumes of solution A (Luminol/Enhancer) and solution B (Stable peroxide buffer). This solution should be prepared freshly.

3. Methods

3.1. Determination of the In Vivo Redox State of an Oxidoreductase

In the method described below, we determine the in vivo redox state of *E. coli* DsbA using AMS as a thiol-labeling reagent. AMS is a maleimide derivative compound of 490 Da that covalently modifies free thiols (see Note 6). The reaction proceeds via attack by the thiolate anion ($-S^-$) of a reduced cysteine residue on the electron poor C=C bond of the maleimide moiety (see Note 7) in a Michael-type reaction (see Fig. 1). This results in the formation of a stable thioether linkage between AMS and the cysteine residue.

Thus, in wild-type cells, where DsbA is oxidized, AMS will not modify the protein. In contrast, the catalytic cysteine residues of DsbA are modified by AMS when the protein is expressed in cells lacking DsbB (remember that DsbB maintains DsbA oxidized). This leads to the addition of ~1 kDa to the protein (2 × 0.49 kDa), causing a visible shift in SDS-PAGE (see Fig. 2). Here, we detect DsbA expressed from the chromosome using a specific antibody. Alternatively, the protein of interest can be expressed from a plasmid fused to a tag for which a specific antibody is commercially available.

We recommend using a *dsbA* mutant (or an empty plasmid) as a negative control to unambiguously identify the protein. Moreover, when purified preparations of the protein of interest are available, oxidized and reduced samples should be prepared by incubating the protein with GSSG or DTT, respectively. These samples will then be incubated with AMS and serve as controls (see below).

3.1.1. Sample Preparation

1. Wild-type, *dsbA*, and *dsbB* strains are streaked onto LB plates. Incubate the plates overnight at 37°C.
2. Pick up a single colony of the wild-type, *dsbA*, and *dsbB* strains and grow each of them in 5 mL of fresh LB broth medium at 37°C overnight with constant shaking at 200 rpm.
3. Take a 1 mL sample of each preculture. Measure OD_{600} in a spectrophotometer.
4. Inoculate 30 mL of fresh LB broth medium with 300 μL of the overnight wild-type preculture. Proceed similarly with the *dsbA* and *dsbB* mutants (see Note 8).
5. Grow the bacterial cells at 37°C with constant shaking at 200 rpm to an OD_{600} of 0.5 (see Note 9).
6. When OD_{600} reaches 0.5, take two 1.8 mL aliquots of the wild-type and *dsbB* cultures and one 1.8 mL aliquot of the *dsbA* culture (see Note 10). Precipitate the cellular material by mixing each aliquot with 200 μL of 100% ice-cold TCA (already prepared in a 2 mL microcentrifuge tube stored on ice). The final TCA concentration will be 10% v/v.
7. Also, when OD_{600} reaches 0.5, take 5 mL of the wild-type culture and add 50 μL of 1 M DTT (final DTT concentration will be 10 mM). Incubate 5 min at 37°C, with constant shaking (200 rpm) (see Note 11). After incubation with DTT, take a 1.8 mL aliquot and proceed like in step 6 (see Note 12).
8. Incubate on ice for 30 min (see Note 13).
9. Centrifuge at 16,000 × *g* at 4°C for 5 min.
10. Remove supernatant carefully, leaving protein pellet intact (see Note 14).
11. Wash the pellet with 200 μL of cold acetone (see Note 15). Do not vortex the pellet.
12. Centrifuge at 16,000 × *g* at 4°C for 5 min.
13. Remove supernatant carefully, leaving protein pellet intact.
14. Dry pellet by placing the tube at 37°C for 15 min to remove acetone.

3.1.2. Preparation of the Controls Using the Purified Protein

If a preparation of the purified protein of interest is available, we recommend applying the AMS-trapping technique both to a fully reduced and to a fully oxidized sample of the protein. These samples will be used as controls to facilitate the identification of the oxidized and reduced forms of the protein of interest on the SDS-PAGE gel.

1. To fully reduce the protein, incubate 500 μL of purified DsbA (up to 300 μM) (see Note 16) with 10 mM DTT for 1 h at room temperature.
2. To fully oxidize the protein, incubate another 500 μL of purified DsbA (up to 300 μM) with 100 mM GSSG (see Note 17) for 1 h at 37°C.
3. Remove the excess of DTT or GSSG using an NAP-5 gel-filtration column equilibrated with 50 mM Tris–HCl, pH 7.5. Elute with 700 μL of 50 mM Tris–HCl, pH 7.5 (see Note 18).
4. Determine the concentration of the oxidized and reduced proteins by measuring the UV absorbance at 280 nm using a spectrophotometer. Calibrate to zero absorbance with the buffer solution (here, 50 mM Tris–HCl, pH 7.5) (see Note 19).
5. Add 100 μL of 20% ice-cold TCA to 100 μL of a solution containing 50 mM Tris–HCl, pH 7.5 and 2 μM of either the oxidized or reduced form of the DsbA protein (see Note 20).
6. Follow steps 7–14 in Subheading 3.1.1.

3.1.3. Thiols Labeling with AMS

1. Resuspend one of the pellets prepared for each strain (wild-type, *dsbA,* and *dsbB* strains) in Subheading 3.1.1, step 6, in 30 μL of a solution containing 50 mM Tris–HCl, pH 7.5, 0.1% SDS (v/v) and 10 mM EDTA (AMS solution without AMS). These samples will serve as controls.
2. Resuspend the pellets prepared for the second 1.8 mL aliquot of the wild-type and *dsbB* strains (see Subheading 3.1.1, step 6), as well as the pellet prepared for the wild-type in Subheading 3.1.1, step 7, in 30 μL of the AMS solution.
3. Resuspend the pellet prepared in Subheading 3.1.2 from the reduced protein in 30 μL of the AMS solution.
4. Resuspend the pellet prepared in Subheading 3.1.2 from the oxidized protein in 30 μL of a solution containing 50 mM Tris–HCl, pH 7.5, 0.1% SDS (v/v) and 10 mM EDTA (AMS solution without AMS).
5. Use a vortex and mix vigorously for 1 min (see Note 21).
6. Incubate the samples in the dark at 37°C for at least 60 min (see Note 22) under constant shaking (1,400 rpm) by using a Thermomixer.
7. Add 10 μL of 5× Laemmli buffer to each sample. The buffer color should remain blue. If TCA was not properly removed, the buffer color turns to yellow.
8. Heat the samples at 95°C for 15 min to solubilize proteins (see Note 23).

3.1.4. Detection of the Labeled Proteins by Western Blot Analysis

1. Load 10 μL of the samples prepared above onto a 12% SDS-Tris-Glycine PAGE gel (see Note 24) following the loading order shown in Fig. 2.
2. Run SDS-PAGE gel at 200 V until the dye front has reached the bottom of the gel (see Note 25).
3. Transfer the proteins to a nitrocellulose membrane at 180 mA for 60 min in transfer buffer using a semi-dry protein transfer system.
4. Immerse the membrane in the blocking solution for 60 min on a rocking platform at room temperature (see Note 26).
5. Rinse the membrane three times in TBS-T (10 min each) and once in TBS at room temperature on a rocking platform.
6. Incubate the membrane in the primary antibody solution for 60 min on a rocking platform at room temperature (see Note 27).
7. Remove the solution containing the primary antibody and wash the membrane three times in TBS-T (10 min each) and once in TBS at room temperature on a rocking platform.
8. Incubate the membrane in the secondary antibody solution for 60 min on a rocking platform at room temperature.
9. Remove the solution containing the secondary antibody and wash the membrane three times in TBS-T (10 min each) and once in TBS at room temperature on a rocking platform.
10. Add the chemiluminescence substrate and incubate for 1 min without agitation.
11. Discard the chemiluminescence substrate, place the membrane in a transparent sheet protector, being careful not to trap air bubbles (see Note 28).
12. Expose the membrane to a medical X-Ray film for appropriate time interval to obtain an optimal detection of the protein of interest (see Note 29).
13. Figure 2 shows the in vivo redox state of *E. coli* DsbA in a wild-type (lanes 1, 2, and 3) and in a *dsbB* strain (lane 4 and 5). The samples were incubated (lane 1) or not (lanes 2, 3, and 4) with 10 mM DTT prior to the incubation with (lanes 1, 2, and 4) or without (lanes 3 and 5) AMS. The oxidized and reduced samples of the purified DsbA preparation were loaded as controls (lane 7 and 8). The identification of the band corresponding to DsbA was unambiguously established by loading a sample prepared from the *dsbA* strain (lane 6).

4. Notes

1. When the protein of interest is a putative DsbA substrate, we recommend using a bioinformatic approach prior to the experimental procedure in order to determine whether the cysteine

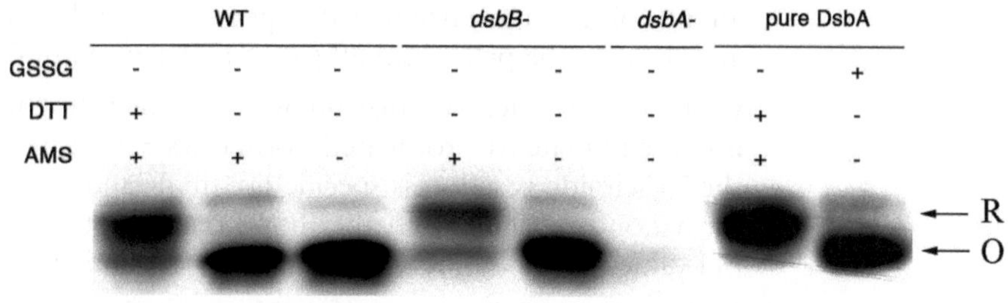

Fig. 2. Determination of the in vivo redox state of DsbA in wild-type and *dsbB* strains. *Lanes* 1, 2, and 3 correspond to the samples prepared from the wild-type strain. *Lanes* 4 and 5 correspond to the samples prepared from the *dsbB* mutant. These samples were incubated with DTT and/or AMS as indicated on the figure. *Lanes* 2 and 4 show the in vivo redox state of DsbA in the wild-type (oxidized) and in the *dsbB* mutant (reduced), respectively. The sample prepared from the *dsbA* knockout strain was loaded in *lane* 6. *Lanes* 7 and 8 show the fully reduced (+DTT/+AMS) and fully oxidized (+GSSG) purified DsbA, respectively.

residues present in the protein are conserved in homologous proteins. A BLAST search is commonly used to find sequences homologous to the protein of interest in sequence databases (http://blast.ncbi.nlm.nih.gov/Blast.cgi). Then, multiple alignments of the homologous sequences are performed using ClustalW2 to identify regions of conservation (http://www.ebi.ac.uk/Tools/msa/clustalw2/). The conservation of a pair of cysteine residues in a periplasmic protein strongly suggests that these cysteines are involved in the formation of a disulfide bond.

2. Alternatively, a commercially available preparation of LB agar medium can be used (MP Biomedicals). Solubilize 40 g in 1 L of water. Autoclave. Store powder at room temperature and plates at 4°C.

3. An alternative is to use a commercial preparation of LB broth medium (Sigma). Solubilize 20 g of powder in 1 L of water. Autoclave. Store at room temperature.

4. Potassium phosphate buffer (KPi) consists of a mixture of monobasic dihydrogen phosphate (KH_2PO_4) and dibasic monohydrogen phosphate (K_2HPO_4). To prepare around 300 mL of a 0.5 M stock solution of KPi buffer at pH 7, titrate 185 mL of a 0.5 M solution of K_2HPO_4 with a 0.5 M solution of KH_2PO_4 to reach pH 7. Store at room temperature or at 4°C.

5. Purchase from any supermarket.

6. Alkylating reagents with higher molecular weight, such as Mal-PEG (5 kDa) or PEG-(Mal)2 (2 kDa), can be used. The trapping protocol is similar to the one described here. However, PEG-containing molecules may interfere with the running of the samples in the gel and/or with the transfer of the proteins onto the nitrocellulose membrane. Therefore, we recommend removing the PEG molecules by one more TCA precipitation

after thiol labeling. Then repeat steps 6–14 in Subheading 3.1.1 and dissolve the pellet into 30 μL of 1× Laemmli buffer.

7. Cysteines can react with maleimides only when the thiol group is in the thiolate (deprotonated) form (Cys-S^-). At pH 6.5–7.5, the maleimide reaction is specific for thiols and results in the formation of a stable thioether bond that is not reversible. At pH values above 8.5, reactions with primary amines can occur.
8. If the OD_{600} of the overnight wild-type and mutant cultures are significantly different, standardize the inoculation volume for each strain, to reach an initial OD_{600} of about 0.03.
9. Stopping the bacterial growth at a higher OD_{600} (during the stationary phase instead of the exponential phase) is not recommended. During stationary phase, cysteine residues can undergo oxidation due to the accumulation of reactive oxygen species (ROS), which may influence the result of the analysis.
10. In this particular case, 1.8 mL of the bacterial culture is precipitated by the addition of TCA. This volume should be adapted depending on the protein abundance in vivo and on the antibody sensitivity.
11. Incubation with DTT, a reducing agent, leads to the reduction of the cysteine residues that are involved in a disulfide bond or are oxidized to a sulfenic acid. The DTT-reduced samples will serve as controls for the fully reduced form of the protein of interest.
12. For disulfides that are difficult to reduce, we recommend to dissolve the pellet (prepared in step 14—Subheading 3.1.1) into 30 μL of 1× Laemmli buffer-containing 10 mM DTT. After a 30 min incubation at 37°C with shaking, proceed to a second TCA precipitation by adding 3 μL of 100% TCA to the tube. Then, proceed from step 8.
13. The incubation on ice can be longer than 30 min. Overnight incubations are recommended to improve the solubilization of compact pellets.
14. TCA pellets are not always visible. Align the tubes in the rotor so that the cap hinges point outward and remove carefully the supernatant without touching the side of the tube under the hinge.
15. Washing with acetone is useful to remove traces of acid. Proceed with care (do not detach the pellet). We recommend washing twice with acetone.
16. Purified proteins are stored at −20°C in 50 mM Tris–HCl, pH 7.5, 100 mM NaCl.
17. Alternatively, oxidation of DsbA can be achieved by incubating the protein with 10 mM diamide for 30 min at 4°C.
18. We use an elution volume smaller than the volume suggested by the supplier to prevent any contamination of the protein fraction by either GSSG or DTT.

19. Following the Beer-Lambert law, protein concentration (M) = (Absorbance at 280 nm)/(absorbance coefficient) for a path length of 1 cm. The absorbance coefficient of the protein of interest can be calculated using ProtParam (http://web.expasy.org/protparam).
20. In this particular case, 2 μM of the reduced and oxidized forms of the DsbA protein were precipitated by the addition of TCA. This concentration should be adapted depending on the antibody sensitivity.
21. Make sure that the pH of each sample is above 7. Remaining traces of TCA can significantly decrease the pH (see Note 15). If the pH is too acidic, we recommend using an AMS solution containing 300 mM Tris–HCl, pH 7.5 instead of 50 mM.
22. Incubation in the dark under constant shaking can be longer than 60 min, and up to overnight.
23. After this step, the samples can be stored at –20°C.
24. The percentage of the separating gel depends on the molecular weight of the protein of interest: 8% or 10% SDS-PAGE gels are better for resolving high molecular weight proteins, whereas 12% or 16% SDS-PAGE gels are ideally suited for the separation of smaller proteins.
25. Running gels at voltage below 200 V increases resolution. 150 V is a longer but better alternative to observe clear shifts between the oxidized and reduced forms of the proteins.
26. Alternatively, membrane can be stored overnight or even up to 3 days in the blocking solution at 4°C without agitation.
27. Alternatively, the membrane can be incubated in the primary antibody solution overnight at 4°C without agitation.
28. Make sure that you discard all the remaining chemiluminescence substrate from the membrane to obtain nicer images.
29. Test various times of exposure, from very short (a few seconds) to long ones (more than 30 min). The optimal exposure time will depend both on the antibody sensitivity and on the protein concentration in the samples.

Acknowledgments

This work was supported by the European Research Council (FP7/2007–2013) ERC independent researcher starting grant 282335—Sulfenic to J.F.C. J.F.C. is a Chercheur Qualifié and PL a Chargé de Recherches of the Belgian FNRS. KD and VN are supported by a grant from the FRIA.

References

1. Dutton RJ, Boyd D, Berkmen M, Beckwith J (2008) Bacterial species exhibit diversity in their mechanisms and capacity for protein disulfide bond formation. Proc Natl Acad Sci U S A 105:11933–8
2. Vertommen D, Depuydt M, Pan J, Leverrier P, Knoops L, Szikora JP, Messens J, Bardwell JC, Collet JF (2008) The disulphide isomerase DsbC cooperates with the oxidase DsbA in a DsbD-independent manner. Mol Microbiol 67:336–49
3. Sone M, Akiyama Y, Ito K (1997) Differential *in vivo* roles played by DsbA and DsbC in the formation of protein disulfide bonds. J Biol Chem 272:10349–52
4. Messens J, Collet JF, Van Belle K, Brosens E, Loris R, Wyns L (2007) The oxidase DsbA folds a protein with a nonconsecutive disulfide. J Biol Chem 282:31302–7
5. Leverrier P, Declercq JP, Denoncin K, Vertommen D, Hiniker A, Cho SH, Collet JF (2011) Crystal structure of the outer membrane protein RcsF, a new substrate for the periplasmic protein-disulfide isomerase DsbC. J Biol Chem 286:16734–42
6. Denoncin K, Vertommen D, Paek E, Collet JF (2010) The protein-disulfide isomerase DsbC cooperates with SurA and DsbA in the assembly of the essential beta-barrel protein LptD. J Biol Chem 285:29425–33
7. Ruiz N, Chng SS, Hiniker A, Kahne D, Silhavy TJ (2010) Nonconsecutive disulfide bond formation in an essential integral outer membrane protein. Proc Natl Acad Sci U S A 107:12245–50
8. Braun M, Silhavy TJ (2002) Imp/OstA is required for cell envelope biogenesis in *Escherichia coli*. Mol Microbiol 45:1289–302
9. Bardwell JC, McGovern K, Beckwith J (1991) Identification of a protein required for disulfide bond formation in vivo. Cell 67:581–9
10. Martin JL, Bardwell JC, Kuriyan J (1993) Crystal structure of the DsbA protein required for disulphide bond formation *in vivo*. Nature 365:464–8
11. Collet JF, Messens J (2010) Structure, function, and mechanism of thioredoxin proteins. Antioxid Redox Signal 13:1205–16
12. Bardwell JC, Lee JO, Jander G, Martin N, Belin D, Beckwith J (1993) A pathway for disulfide bond formation *in vivo*. Proc Natl Acad Sci U S A 90:1038–42
13. Bader M, Muse W, Ballou DP, Gassner C, Bardwell JC (1999) Oxidative protein folding is driven by the electron transport system. Cell 98:217–27
14. Grauschopf U, Winther JR, Korber P, Zander T, Dallinger P, Bardwell JC (1995) Why is DsbA such an oxidizing disulfide catalyst? Cell 83:947–55
15. Rietsch A, Belin D, Martin N, Beckwith J (1996) An *in vivo* pathway for disulfide bond isomerization in *Escherichia coli*. Proc Natl Acad Sci U S A 93:13048–53
16. McCarthy AA, Haebel PW, Torronen A, Rybin V, Baker EN, Metcalf P (2000) Crystal structure of the protein disulfide bond isomerase, DsbC, from *Escherichia coli*. Nat Struct Biol 7:196–9
17. Joly JC, Swartz JR (1997) *In vitro* and *in vivo* redox states of the *Escherichia coli* periplasmic oxidoreductases DsbA and DsbC. Biochemistry 36:10067–72
18. Katzen F, Beckwith J (2000) Transmembrane electron transfer by the membrane protein DsbD occurs via a disulfide bond cascade. Cell 103:769–79
19. Rietsch A, Bessette P, Georgiou G, Beckwith J (1997) Reduction of the periplasmic disulfide bond isomerase, DsbC, occurs by passage of electrons from cytoplasmic thioredoxin. J Bacteriol 179:6602–8
20. Shouldice SR, Heras B, Walden PM, Totsika M, Schembri MA, Martin JL (2011) Structure and function of DsbA, a key bacterial oxidative folding catalyst. Antioxid Redox Signal 14: 1729–60
21. Casadaban MJ (1976) Transposition and fusion of the *lac* genes to selected promoters in *Escherichia coli* using bacteriophage lambda and Mu. J Mol Biol 104:541–55
22. Miller JH (1992) A short course in bacterial genetics: laboratory manual. Cold Spring Harbor, New York

Chapter 21

Using Reporter Genes and the *Escherichia coli* ASKA Overexpression Library in Screens for Regulators of the Gram Negative Envelope Stress Response

Julia L. Wong, Stefanie L. Vogt, and Tracy L. Raivio

Abstract

We describe methods for screening the *E. coli* ASKA overexpression library for clones that lead to altered expression of reporter genes. First, a promoter of interest is cloned upstream of either the *lacZ* or *luxCD-ABE* genes to yield reporter genes in which transcription is proportional to the levels of β-galactosidase or luminescence produced by strains carrying the reporter. The ASKA library is then condensed into two 96-well plates resulting in mixed preparations of 12 plasmids in each well. The plasmids in each well are transformed into the reporter strain and transformants are screened for either altered β-galactosidase or light production. The genes contained in ASKA clones that result in altered reporter gene expression are amplified and sequenced and the ASKA clone for the gene identified is retransformed into the parent reporter strain to confirm the effect. We have used screens like this one to look for new *E. coli* genes that, when over-expressed, result in the altered expression of promoters that are regulated by the envelope stress response. The identity of the clones can yield information about the nature of inducing cues and/or additional regulatory molecules. The techniques are broadly applicable to any microbial function of interest.

Key words: Reporter gene, *E. coli* ASKA overexpression library, β-Galactosidase, Luminescence, *lacZ*, *Lux*

1. Introduction

The creation of libraries of knockout mutants and inducible constructs representative of the entire *Escherichia coli* genome has been a huge boon to the *E. coli* community (1, 2). Both libraries have been used extensively in large-scale genetic screens and selections designed to investigate specific research questions (180 combined PubMed citations for both libraries). In these screens and selections, the alteration, elimination, or creation of a specific phenotype by a knockout allele or overexpression clone is used to identify genes and conditions involved in discrete cellular processes of

Anne H. Delcour (ed.), *Bacterial Cell Surfaces: Methods and Protocols*, Methods in Molecular Biology, vol. 966,
DOI 10.1007/978-1-62703-245-2_21, © Springer Science+Business Media New York 2013

interest. In addition, individual Keio library mutants and ASKA overexpression clones have accelerated the research of many. Others have exploited these libraries to search for synthetic phenotypes indicative of interactions that can serve as the basis for the construction of biological networks (e.g., (3)). In our lab, we have used the ASKA overexpression library in combination with transcriptional reporter genes to screen for new genes and cellular processes that impact the *E. coli* envelope stress response.

E. coli (and other Gram-negative organisms) devotes a large repertoire of regulators to monitoring and responding to changes in the envelope (4), the cellular compartment including the inner membrane, the periplasm, the cell wall, the outer membrane, and all of the resident macromolecules. It is logical that the envelope is so closely monitored since it is the first cellular compartment to experience environmental changes that may impact bacterial growth. Amongst the regulators devoted to measuring and reacting to changes in the envelope is a suite of overlapping envelope stress responses that have been demonstrated to respond to a variety of conditions that impact some or all of the separate components of the envelope (4). Genetic screens and large-scale transcriptomic studies indicate that the adaptations conferred by these envelope stress responses intersect and influence multiple aspects of the cell envelope (e.g., (5–12)). Our group has focused on the study of two envelope stress responses in *E. coli* and other Gram-negative pathogens—the Cpx and Bae envelope stress responses. Both are controlled by a two-component signal transduction pathway made up of a sensor kinase (CpxA or BaeS) and a response regulator (CpxR or BaeR). CpxA and BaeS utilize a periplasmic sensing domain to detect envelope stress signals, which are conveyed to a cytoplasmic kinase/phosphatase domain that controls phosphorylation of CpxR and BaeR, respectively. The phosphorylated response regulators bind DNA and alter gene expression to institute adaptation. Genetics, bioinformatics, and transcriptomics have been used successfully to define the genes regulated by CpxR and BaeR, and thus the nature of the adaptation conferred (see ref. 4 for a review). Analysis of upstream events, particularly the nature of inducing signals, has been more difficult, as with other two-component signaling pathways. Similarly, an understanding of the mechanisms that tie envelope stress responses to each other and other cellular regulatory pathways is just beginning to arise.

In this chapter, we describe a genetic approach that can facilitate the identification of cellular processes that may be detected by a given regulatory pathway as well as connector proteins that link regulatory pathways together to forge networks. We are using this approach to study the regulation of the CpxA and BaeS sensor kinases specifically, but the methods are broadly applicable to other regulators. Our methods initially involve the creation of a reporter gene that will inform on the activity of the regulator of interest. The transcriptional control region of a gene controlled by the regulator is cloned upstream of either the *lacZ* gene or the *luxCDABE* genes encoding bacterial luciferase. A strain bearing this reporter is then transformed with pools of plasmids representative of the entire ASKA overexpression library (2) and levels of β-galactosidase or luciferase are measured using indicator media or a luminometer. The plasmids from transformants with altered reporter gene expression are isolated and ultimately sequenced to identify the gene of interest. The identity of the genes uncovered in this type of screen can be used as a basis for hypotheses about the nature of inducing signal(s) and/or potential auxiliary regulators involved in network formation or signal sensing.

2. Materials

2.1. Construction of Luminescent Transcriptional Reporter Genes

2.1.1. Preparing Vector DNA for Cloning

1. Bacterial strains: glycerol stocks of *E. coli* TOP10 harboring luminescence reporter plasmids pNLP10 and pJW15 (13, 14) (Fig. 2) and stocks of bacteria containing a promoter of interest to act as a PCR template (in the example, *E. coli* strain MC4100). Store at –80°C.
2. LB agar: 1% (w/v) Bacto tryptone, 0.5% (w/v) yeast extract, 0.5% (w/v) sodium chloride, 1.5% (w/v) agar.
3. LB broth: 1% (w/v) Bacto tryptone, 0.5% (w/v) yeast extract, 0.5% (w/v) sodium chloride.
4. Kanamycin stock: 50 mg/mL in water. Stored at 4°C.
5. Plasmid miniprep kit.
6. Culture tubes.
7. Incubator shaker.
8. Autoclaved deionized water.
9. 10× Buffer K: 200 mM Tris–HCl, pH 8.5, 100 mM magnesium chloride, 10 mM dithiothreitol (DTT), 1,000 mM potassium chloride. Provided with *Bam*HI restriction enzyme.
10. 10× Buffer H: 500 mM Tris–HCl, pH 7.5, 100 mM magnesium chloride, 10 mM DTT, 1,000 mM sodium chloride. Provided with *Eco*RI restriction enzyme.
11. *Bam*HI restriction enzyme stock (see Note 1): 60 U/μL. The enzyme stock is in storage buffer: 10 mM Tris–HCl, pH 7.5, 400 mM potassium chloride, 0.1 mM EDTA, 1 mM dithiothreitol, 0.15% Triton X-100, 0.01% BSA, 50% (v/v) glycerol.
12. *Eco*RI restriction enzyme stock: 20 U/μL. The enzyme stock in storage buffer: 10 mM Tris–HCl, pH 7.5, 100 mM potassium chloride, 0.1 mM EDTA, 1 mM dithiothreitol, 0.15% Triton X-100, 0.01% BSA, 50% (v/v) glycerol.

13. 6× DNA loading dye: 0.25% (w/v) bromophenol blue, 0.25% (w/v) xylene cyanol FF, 30% (v/v) glycerol in water.
14. Agarose.
15. Ethidium bromide.
16. Agarose gel electrophoresis setup.
17. Gel extraction kit.

2.1.2. PCR Amplification of the Promoter Region

1. Thermocycler.
2. LB agar: see item 2 in Subheading 2.1.1
3. Platinum *Taq* DNA Polymerase High Fidelity (Invitrogen), 5 U/μL. The enzyme stock is supplied in storage buffer: 20 mM Tris–HCl, pH 8.0, 0.1 mM EDTA, 1 mM dithiothreitol, stabilizers (unspecified by manufacturer), and 50% (v/v) glycerol.
4. 50 mM magnesium sulfate: provided with the Platinum *Taq* DNA Polymerase High Fidelity.
5. 10× High Fidelity buffer: 600 mM Tris-sulfate, pH 8.9, 180 mM ammonium sulfate, provided with the Platinum *Taq* DNA Polymerase High Fidelity.
6. 5 mM dNTP solution (1.25 mM each dNTP). Store at −20°C.
7. Betaine solution: 5 M betaine.
8. Template DNA: 3–4 MC4100 colonies resuspended in 500 μL of deionized water and boiled for 5 min.
9. PCR primers to amplify the *micF* promoter, resuspended in deionized water to a concentration of 10 μM: PmicFForEcoRI, TTTTGAATTCACAGATCTAATTTGTTGCC; PmicFRevBamHI, TTTTGGATCCTAAAGTTAATGATGAT AGCG.
10. PCR purification kit.
11. Agarose.
12. Ethidium bromide.
13. Agarose gel electrophoresis setup.

2.1.3. Cloning of the Promoter Region

1. Agarose.
2. Ethidium bromide.
3. Agarose gel electrophoresis setup.
4. 6× DNA loading dye: 0.25% (w/v) bromophenol blue, 0.25% (w/v) xylene cyanol FF, 30% (v/v) glycerol in water. Store at room temperature.
5. T4 DNA ligase: 1 U/μL. T4 DNA ligase stock is supplied in storage buffer: 10 mM Tris–HCl, pH 7.5, 50 mM potassium chloride, 1 mM dithiothreitol, 50% (v/v) glycerol.

6. 5× DNA ligase reaction buffer: 250 mM Tris–HCl, pH 7.6, 50 mM magnesium chloride, 5 mM ATP, 5 mM DTT, 25% (w/v) polyethylene glycol-8000; provided with the T4 DNA ligase.
7. One Shot TOP10 chemically competent *E. coli* cells (Invitrogen).
8. SOC medium: 2% (w/v) tryptone, 0.5% (w/v) yeast extract, 10 mM sodium chloride, 2.5 mM potassium chloride, 10 mM magnesium chloride, 10 mM magnesium sulfate, 20 mM glucose. Provided with One Shot TOP10 chemically competent *E. coli* cells.
9. LB agar: see item 2 in Subheading 2.1.1.
10. Kanamycin.
11. LB broth: see item 3 in Subheading 2.1.1.
12. Incubator.
13. Plasmid miniprep kit.
14. Incubator shaker.

2.1.4. Confirming the Presence of an Insert by PCR

1. Recombinant *Taq* DNA polymerase, 5 U/μL. Recombinant *Taq* is provided in storage buffer: 20 mM Tris–HCl, pH 8.0, 0.1 mM EDTA, 1 mM dithiothreitol, 50% (v/v) glycerol, stabilizers (unspecified).
2. 10× PCR buffer: 200 mM Tris–HCl, pH 8.4, 500 mM potassium chloride. Provided with the Recombinant *Taq* DNA polymerase.
3. 50 mM magnesium chloride. Provided with the Recombinant *Taq* DNA polymerase.
4. 5 mM dNTP solution (1.25 mM each dNTP). Store at −20°C.
5. Betaine solution: 5 M betaine.
6. PCR primers to amplify inserts from pNLP10 or pJW15, resuspended in deionized water to a concentration of 10 μM; pNLP10F: GCTTCCCAACCTTACCAGAG; pNLP10R: CA CCAAAATTAATGGATTGCAC.
7. Agarose.
8. Ethidium bromide.
9. Agarose gel electrophoresis setup.
10. Luminometer.

2.2. Pooling the ASKA Overexpression Library Plasmids

1. Glycerol Stocks of *E. coli* AG1 containing ASKA plasmids in 96-well plates covered with foil sealing film, from the ASKA library (2). Store at −80°C.
2. Foil sealing film.

3. LB broth: 1% (w/v) Bacto tryptone, 0.5% (w/v) yeast extract, 0.5% (w/v) sodium chloride.
4. LB agar: 1% (w/v) Bacto tryptone, 0.5% (w/v) yeast extract, 0.5% (w/v) sodium chloride, 1.5% (w/v) agar.
5. 25 mg/mL chloramphenicol in 95% ethanol. Stored at 4°C.
6. 96-well microtiter plates.
7. Multi-channel micropipetor.
8. 96-pronged plastic, disposable replicators.
9. Autoclaved 50% glycerol in water.
10. Sterile wooden dowels.
11. Plasmid miniprep kit.
12. Autoclaved deionized water.

2.3. Genetic Screens of ASKA Overexpression Plasmid Pools

1. Glycerol stocks of *E. coli* MC4100 harboring a *lacZ* or *lux* reporter, i.e., *cpxP–lacZ* on λRS88 incorporated into genome (TR50) or *cpxP–lux* on pJW25 (NLP104), respectively. MC4100 λRS88 *cpxP–lacZ* (pCA24N) or MC4100(pJW25, pCA24N). Store at −80°C.
2. LB agar: 1% (w/v) Bacto tryptone, 0.5% (w/v) yeast extract, 0.5% (w/v) sodium chloride, 1.5% (w/v) agar.
3. LB broth: 1% (w/v) Bacto tryptone, 0.5% (w/v) yeast extract, 0.5% (w/v) sodium chloride.
4. MacConkey Agar: 50 g in 1 L water.
5. Kanamycin: 50 mg/mL in water. Stored at 4°C.
6. Chloramphenicol: 25 mg/mL in 95% ethanol. Stored at 4°C.
7. Isopropyl-β-1-D-thiogalactopyranoside (IPTG): 0.1 M in sterile deionized water. IPTG stored at −20°C.
8. Incubator shaker.
9. 250-mL culture flasks.
10. Spectrophotometer.
11. 50-mL conical tubes.
12. Centrifuge and associated rotor.
13. Pooled ASKA plasmids collected as per Subheading 3.2 below.
14. Autoclaved deionized water. Store at 4°C.
15. Sterile 10% glycerol in water. Store at 4°C.
16. 0.2 mL Eppendorf tubes.
17. Electrocuvettes.
18. Electroporation machine.
19. 1.5 mL microcentrifuge tubes.
20. Sterile glass Pasteur pipettes.

21. Sterile toothpicks.
22. Sterile wooden dowels.
23. Luminometer.
24. Excel software.

2.4. Identifying ASKA Clones of Interest

1. LB broth: 1% (w/v) Bacto tryptone, 0.5% (w/v) yeast extract, 0.5% (w/v) sodium chloride.
2. Chloramphenicol: 25 mg/mL in 95% ethanol. Stored at 4°C.
3. Plasmid miniprep kit.
4. *Taq* DNA Polymerase, 500 U/mL. The enzyme stock is supplied in storage buffer: 20 mM Tris–HCl, pH 8.0, 0.1 mM EDTA, 1 mM dithiothreitol, stabilizers (unspecified by manufacturer), and 50% (v/v) glycerol.
5. 50 mM magnesium chloride, provided with the *Taq* DNA Polymerase.
6. 10× PCR buffer: 200 mM Tris–HCl pH 8.4, 500 mM potassium chloride (without magnesium chloride), provided with the *Taq* DNA Polymerase. Store at −20°C.
7. PCR primers to amplify ASKA library clones from pCA24N, resuspended in deionized water to a concentration of 100 mM. pCA-F CGTCTTCAC CTC GAGAAATC; pCA-R, CTGAACA AATCCAGATGGAG. Store at −20°C.
8. Autoclaved deionized water.
9. 5 mM dNTP solution: 1.25 mM each dNTP in sterile, deionized water. Store at −20°C.
10. Betaine solution: 5 M betaine. Store at 4°C.
11. Thermocycler.
12. Agarose.
13. Agarose gel electrophoresis setup.
14. 6× DNA loading dye: see item 4 Subheading 2.1.3.
15. PCR purification kit.
16. BigDye ET Sequencing Kit (Applied Biosystems): BigDye Terminator sequencing buffer (5×) diluted to 1× in 200 mM Tris–HCl pH 9.0 and 5 mM magnesium chloride and stored at −20°C; BigDye Terminator premix containing enzyme and nucleotides; purification buffer.
17. BigDye Terminator sequencing buffer (5×): 400 mM Tris, 10 mM $MgCl_2$, pH 9.0, provided in the BigDye ET Sequencing Kit.
18. Purification buffer: 1.5 M sodium acetate, pH 8, 250 mM EDTA, provided in the BigDye ET Sequencing Kit.
19. 1.5 mL Eppendorf tubes.

20. Microfuge.
21. 95% Ethanol stored at –20°C.
22. 70% Ethanol stored at –20°C.
23. ABI 3730 Genetic Analyzer for sequencing (Applied Biosystems).

2.5. Confirming Phenotypes

1. Glycerol stocks of *E. coli* MC4100 harboring a *lacZ* reporter (*cpxP–lacZ* on λRS88 incorporated into genome) (TR50) or a *lux* reporter (*cpxP–lux* on pJW25) (NLP104). Store at –80°C.
2. Glycerol Stocks of *E. coli* AG1 containing ASKA plasmids in 96-well plates covered with foil sealing film from the ASKA library (5). Store at –80°C.
3. Foil sealing film.
4. LB agar: 1% (w/v) Bacto tryptone, 0.5% (w/v) yeast extract, 0.5% (w/v) sodium chloride, 1.5% (w/v) agar.
5. LB broth: 1% (w/v) Bacto tryptone, 0.5% (w/v) yeast extract, 0.5% (w/v) sodium chloride.
6. MacConkey Agar: 50 g in 1 L water.
7. Kanamycin: 50 mg/mL in water. Stored at 4°C.
8. Chloramphenicol: 25 mg/mL in 95% ethanol. Stored at 4°C.
9. Isopropyl-β-1-D-thiogalactopyranoside (IPTG): 0.1 M in water. Stored at 4°C.
10. Culture tubes.
11. Rotatory shaker.
12. Sterile 10% glycerol in water. Store at 4°C.

3. Methods

Luminescent or *lacZ* reporter constructs contain a promoter of interest (see Note 2) and are used to screen plasmid pools derived from the *E. coli* ASKA overexpression library (2) for genes that alter light production or lactose metabolism. Plasmids of interest are isolated and the cloned gene is identified by nucleotide sequencing. In our lab, we use this protocol to screen for genes that, upon overexpression, lead to altered expression of envelope stress responsive promoters. The identification of these genes can yield insights into what cellular processes and/or regulators are involved in the control of envelope stress adaptation (Fig. 1).

3.1. Construction of Luminescent Transcriptional Reporter Genes

To create a luminescent transcriptional reporter for expression of the gene of interest, the region encompassing nucleotides –500 to +100 with respect to the translational start site of the gene of interest is cloned into the plasmids pNLP10 and/or pJW15 (see Note 3) (Fig. 2).

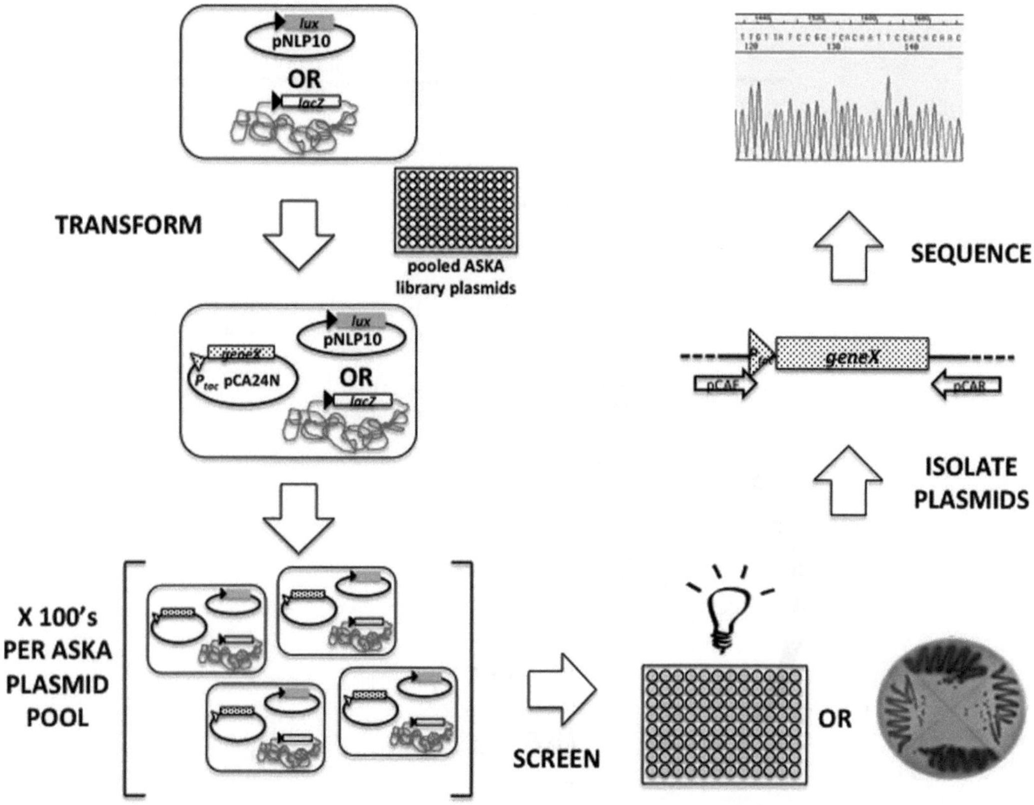

Fig. 1. Overview of method for using *E. coli* ASKA overexpression clone library to screen for genes affecting expression of promoter X. An *E. coli* strain carrying a pNLP10-based luminescent reporter or a β-galactosidase reporter gene incorporated into the chromosome is successively transformed with pools of plasmids representing the *E. coli* ASKA overexpression clone library. Transformants are plated either to lactose-containing MacConkey agar to detect β-galactosidase or to LB plates, each carrying the appropriate antibiotics. Approximately 100 colonies are examined per pool of plasmids. Transformants showing altered β-galactosidase activity are streaked for isolated colonies to a second lactose-containing MacConkey plate to confirm the altered phenotype. Transformants carrying luminescent reporter genes are picked into 96-well plates containing LB broth and antibiotics and luminescence is measured using a luminometer. Any transformants with altered luminescence relative to the parent strain are stored as frozen stocks. Genes of interest from ASKA clones are identified by sequencing using primers homologous to the ASKA vector pCA24N and either isolated colonies or plasmid preps from strains of interest as template. Stippled arrowhead/bars indicate promoter/gene of interest, *lux* refers to the *luxCDABE* genes responsible for luminescence, *lacZ* refers to the gene for β-galactosidase, *geneX* indicates a particular clone from the ASKA overexpression library, pCAF and pCAR indicate primers used to sequence inserts in the pCA24N plasmid.

Previous studies have demonstrated that, even if the transcriptional start site of a gene is unknown, all of the regulatory protein binding sites are typically located within 500 bp of the translation initiation codon (15). The promoter region is PCR-amplified from a bacterial colony, with primers containing restriction enzyme sites appropriate for directional cloning of the insert (see Note 1 for more information about choosing restriction enzymes). The purified PCR product and the target vector are then restriction digested and ligated together. After transformation into *E. coli* TOP10 competent cells, transformants are analyzed by PCR and sequencing to confirm that

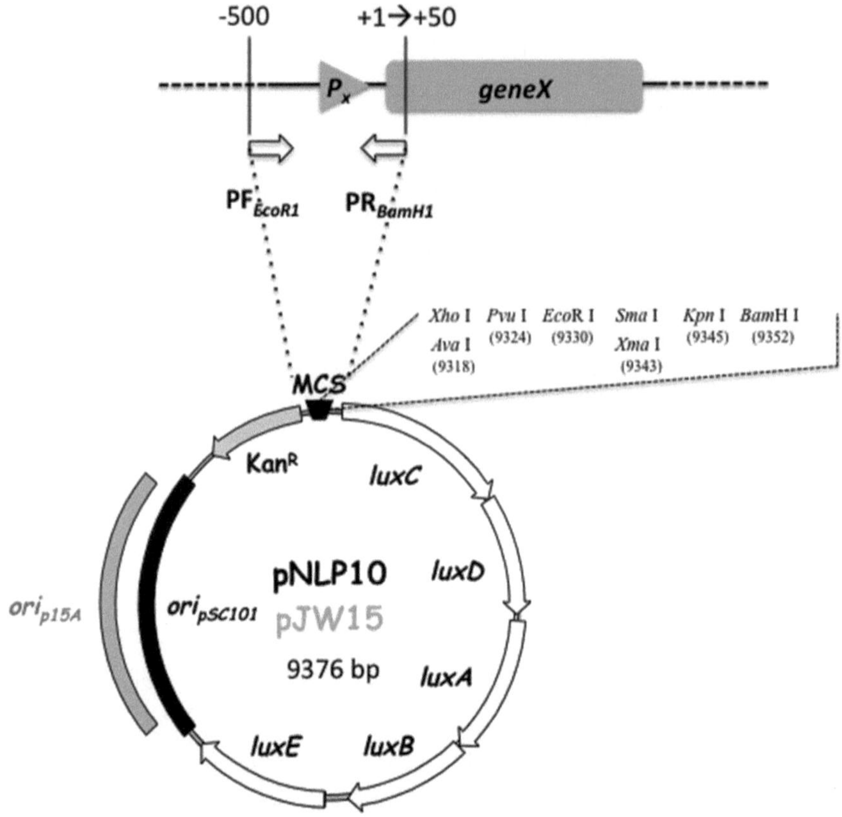

Fig. 2. Diagram of plasmids pNLP10 and pJW15 and the construction of luminescent reporter genes. PCR with restriction tagged primers is used to amplify a region corresponding to approximately 500 bp upstream of the translational start site. The PCR product is digested with *Eco*RI and *Bam*HI and cloned into the same sites on either the pNLP10 (ori_{pSC101}) or pJW15 (ori_{p15A}) luminescent reporter plasmids. Positive clones are confirmed by measuring their ability to confer luminescence and through DNA sequencing.

reporter plasmids contain the desired promoter region. When properly inserted into the reporter plasmid, the promoter of the gene of interest drives transcription of the *luxCDABE* operon, causing bacteria harboring the reporter to produce luminescence proportional to the level of expression from the promoter. As an example, we describe the construction of a reporter for expression of the gene *micF* from *E. coli* strain MC4100.

3.1.1. Preparing Vector DNA for Cloning

1. Streak *E. coli* TOP10 cells harboring pNLP10 or pJW15 from the frozen glycerol stock onto LB agar containing 30 μg/mL kanamycin and incubate at 37°C overnight.
2. Inoculate an isolated colony into each of two test tubes containing 5 mL LB broth with 30 μg/mL kanamycin. Incubate cultures at 37°C with aeration at 225 rpm overnight.
3. Purify plasmid DNA using a plasmid miniprep kit. Use an entire 5 mL culture per miniprep (make two minipreps in total).

Elute the DNA from each column with 30 μL sterile deionized water. Pool the two minipreps for a total of 60 μL purified plasmid.

4. Digest 40 μL of the plasmid in a 50-μL reaction containing 1.25 μL of each restriction enzyme (in this case *Eco*RI and *Bam*HI) and 5 μL of the appropriate universal buffer (buffer K). Incubate overnight at 37°C.
5. Add 10 μL of 6× DNA loading dye to the reaction, then load the mixture into two wells of a 0.9% agarose gel.
6. Run the gel at 80 V for approximately 1 h.
7. Stain the gel with ethidium bromide.
8. Excise the 9.3-kb bands from the gel and purify DNA using a gel extraction kit, purifying both gel slices using one column. Elute the DNA from the column in 30 μL of sterile deionized water.

3.1.2. PCR Amplification of the Promoter Region

1. Prepare the template for PCR of the promoter region by streaking bacteria harboring the gene of interest (in this case, *E. coli* strain MC4100) from the frozen stock onto LB agar and incubating overnight at 37°C.
2. To amplify the *micF* promoter from *E. coli* strain MC4100 by PCR, prepare a 50-μL PCR reaction containing 5 μL of 10× High Fidelity Buffer, 2 μL of 50 mM magnesium sulfate, 8 μL of 5 mM dNTPs, 5 μL of betaine solution, 1 μL of primer PmicFForEcoRI, 1 μL of primer PmicFRevBamHI, 26.5 μL of deionized water, 1 μL of template DNA, and 0.5 μL of Platinum *Taq* DNA Polymerase High Fidelity.
3. Program the thermocycler as follows: 94°C for 2 min; 30 cycles of 94°C for 30 s, 55°C for 30 s, 68°C for 30 s; 68°C for 10 min.
4. Load 10 μL of the reaction onto a 1.5% agarose gel to confirm the presence of a 456-bp band.
5. Purify the PCR product using a PCR purification kit. Elute DNA in 50 μL of deionized water.

3.1.3. Cloning the Promoter Region

1. Digest 20 μL of the purified PCR product in a 50-μL reaction as described in step 4 Subheading 3.1.1.
2. Add 10 μL of 6× DNA loading dye to the reaction then load the mixture into two wells of a 1.5% agarose gel. Excise the ~450-bp bands from the gel and purify as described in step 8 in Subheading 3.1.1.
3. Set up a ligation reaction containing 5 μL of purified promoter DNA, 3 μL of purified vector DNA, 2 μL of 5× DNA ligase reaction buffer, and 1 μL of T4 DNA ligase. Incubate at 16°C overnight.

4. Transform 2 μL of the ligation into TOP10 competent cells. Thaw one aliquot of TOP10 cells on ice for 10 min, then add 2 μL of the ligation reaction and incubate on ice for 30 min. Heat-shock the cells at 42°C for 30 s, then return to ice for 2 min. Add 250 μL of SOC broth, and incubate the cells in a 37°C shaker for 1 h.
5. Plate transformants on LB agar containing 30 μg/mL kanamycin and incubate at 37°C overnight. Expect 10–100 colonies, with the majority containing the desired insert.
6. To purify the plasmid DNA from transformants, restreak six transformants on LB agar with 30 μg/mL kanamycin for single colonies. Inoculate a single colony from each purified transformant into six tubes containing 5 mL LB with 30 μg/mL kanamycin and incubate at 37°C with aeration at 225 rpm overnight. The following day, isolate plasmid DNA from the six cultures using a plasmid miniprep kit.

3.1.4. Confirming the Presence of an Insert by PCR

1. Prepare a 50-μL PCR reaction containing 5 μL of 10× PCR Buffer, 2 μL of 50 mM magnesium chloride, 8 μL of 5 mM dNTPs, 5 μL of betaine solution, 1 μL of primer pNLP10F, 1 μL of primer pNLP10R, 26.5 μL of deionized water, 1 μL of template (a 1:30 dilution of plasmid DNA in deionized water), and 0.5 μL of recombinant *Taq* DNA polymerase. The *Taq* polymerase should be added to the reaction only after it has been heated to 96°C to prevent the formation of nonspecific PCR products.
2. Program the thermocycler as follows: 96°C for 2 min; 30 cycles of 96°C for 45 s, 50°C for 30 s, 72°C for 1 min; 72°C for 10 min.
3. Load 10 μL of the reaction onto a 1% agarose gel to confirm the presence of a ~750-bp band (~300 bp of flanking DNA plus size of the promoter insert). Further confirmation of the presence of an insert containing a promoter can be obtained by sequencing the PCR product using primers pNLP10F and pNLP10R, or by measuring luminescence of TOP10 cells harboring the construct in a luminometer.

3.2. Pooling the ASKA Overexpression Library Plasmids

The ASKA library is divided in half (plates 1–24, plates 25–47) and ASKA plasmids are extracted from every strain in the library. Plasmids are pooled by well number in either the first or second half of the library. For example, plasmid pool A1-1 represents plasmids pooled from every well labeled A1 in plates 1–24 of the ASKA library.

1. Aliquot 100 μL of LB broth containing 30 μg/mL of chloramphenicol into each well of two 96-well plates using a multichannel micropipetor.

2. Remove frozen plates 1–3 of the ASKA library from the freezer. Thaw at room temperature for 2 min (see Note 4).
3. For each plate, pull back the foil cover and scratch the surface of each well in a circular motion using a sterile 96-pronged replicator. Inoculate one of the 96-well microtiter plates by plunging the pronged replicator into the first of the prepared plates and swirling the media gently.
4. Repeat steps 2 and 3 for all plates 1–24 in the ASKA library until all of the wells have been inoculated 24 times each.
5. Repeat steps 2 and 3 for plates 25–47, but inoculate the second prepared 96-well microtiter plate.
6. Replace the lids on both 96-well microtiter plates. Grow statically at room temperature for 18–30 h.
7. Add 100 µL of sterile 50% glycerol into each well using a multichannel micropipetor. Cover the plate with a foil seal. Store at −80°C.
8. Using a sterile wooden dowel, scrape the surface of an individual glycerol stock and make many dense streaks across the surface of an agar plate containing 25 µg/mL chloramphenicol. Incubate at 37°C overnight.
9. Scrape the surface of the plate clean using sterile wooden dowels (approximately 3–4) without removing an excess of agar. Inoculate 5 mL of LB broth with bacteria collected from a single plasmid pool. Repeat for each plasmid pool.
10. Extract plasmids using a plasmid prep kit. Elute in 50 µL sterile deionized water. Store at −20°C in 96-well microtiter plates or microfuge tubes.

3.3. Genetic Screens of ASKA Overexpression Plasmid Pools

This section describes the screening of ASKA overexpression plasmid pools for clones that impact reporter gene expression. *E. coli* cells harboring an appropriate reporter (i.e., *lacZ* or *lux*) are made electrocompetent and transformed by electroporation to ensure efficient transformations of plasmid pools. When using a *lacZ* reporter, transformants are plated on indicator media containing appropriate antibiotics and colonies with the desired phenotype are isolated and confirmed. When using a *lux* reporter, transformants are plated to LB media containing appropriate antibiotics, colonies are picked and transferred to 96-well microtiter plates, grown to mid-exponential phase, and luminescence is measured in a luminometer. As an example, we give the procedure for screening overexpression of the ASKA library in *E. coli* MC4100 λRS88 *cpxP–lacZ* (TR50) (20) and *E. coli* MC4100(pJW25) (*cpxP–lux*) (NLP104) (13).

1. To prepare electro competent cells, streak *E. coli* MC4100 λRS88 *cpxP–lacZ* or MC4100(*cpxP–lux*) from frozen glycerol stocks onto LB agar or LB agar containing 30 µg/mL kanamycin, respectively, and incubate at 37°C overnight.

2. Select a single colony and inoculate into 5 mL LB broth (MC4100 λRS88 *cpxP–lacZ)* or LB broth containing 30 μg/mL kanamycin (MC4100(pJW25)) and incubate at 37°C with aeration at 225 rpm overnight.
3. Subculture 250 μL of the overnight culture into each of two 250 mL flasks, each containing 125 mL of LB broth (and 30 μg/mL kanamycin for MC4100(pJW25)). Grow the flask cultures at 37°C with aeration at 225 rpm for 1 h or until the OD measured at 600 nm reaches 0.5–0.6.
4. Aliquot the flask cultures into five 50 mL conical tubes, 50 mL of culture in each, and centrifuge for 10 min at 12, 200 × *g* at 4°C. Discard the supernatant.
5. Resuspend the cells in each tube from the previous step with 50 mL of sterile water at 4°C to wash. Centrifuge for 10 min at 12, 200 × *g* at 4°C. Discard supernatant.
6. Repeat step 5, but resuspend the bacterial pellet in 25 mL of sterile water at 4°C.
7. Repeat step 5 but resuspend the bacterial cell pellet in 500 mL of sterile 10% glycerol in water at 4°C.
8. Repeat step 5, but resuspend the cells with 100 μL of sterile 10% glycerol in water at 4°C. Aliquot 50 μL of cells into 0.2 mL Eppendorf tubes. To use immediately, incubate on ice and proceed to step 9. To store for future use (up to 10 days), store at −80°C.
9. To transform the competent cells with plasmid DNA from each plasmid pool (see Note 5), add 1 μL of pooled plasmid preparation to a 50 μL aliquot of cells. Tap gently to mix. Incubate 2 min on ice. All steps from this point forward are done for each plasmid pool.
10. Transfer the cells from the Eppendorf tubes to sterile electrocuvettes. Electroporate the plasmids into the cells using an electroporation machine.
11. Immediately add 1 mL of LB broth to the reaction.
12. Transfer the transformation mix from the cuvette to a 1.5 mL microcentrifuge tube using a sterile glass Pasteur pipette. Incubate for 2–3 h at 37°C with aeration at 225 rpm.
13. Serially dilute the transformation reactions into LB broth, using dilutions of 10^{-2}–10^{-6}. Add 100 μL of transformation reaction to 900 μL of fresh LB broth and vortex to mix. Repeat but add 100 μL of diluted culture to 900 μL of fresh LB broth and repeat until desired dilutions have been achieved (see Note 6).
14. Spread 100 μL of each dilution onto MacConkey agar plates containing 25 μg/mL chloramphenicol if using a *lacZ* reporter. Plate on LB agar containing 30 μg/mL kanamycin

and 25 μg/mL chloramphenicol if using a strain carrying a *lux* reporter. Incubate at 37°C overnight.

15. Streak out the parent reporter strain carrying the ASKA library empty vector pCA24N (MC4100 λRS88 *cpxP–lacZ* (pCA24N) or MC4100(pJW25, pCA24N)) for isolated colonies to a MacConkey agar plate containing 25 μg/mL chloramphenicol (for *lacZ* reporters) or to an LB agar plate containing 30 μg/mL kanamycin and 25 μg/mL chloramphenicol (for *lux* reporters).
16. If screening for altered levels of β-galactosidase production, proceed to step 17. If screening for altered levels of luminescence, proceed to step 21.
17. Examine a MacConkey agar plate from each plated dilution to identify those plates containing 100–200 well isolated colonies per plate. From these plates, use visual inspection and comparison to isolated colonies of the parent strain carrying the vector control pCA24N to identify colonies that exhibit a noticeably different color compared to the parent strain.
18. Select single colonies of interest using a sterile toothpick. Streak these for isolated colonies onto a sectored MacConkey agar plate containing 25 μg/mL chloramphenicol (see Note 7). We routinely streak out five colonies on each sectored plate together with the parent strain carrying the empty ASKA vector control pCA24N for comparison (see Note 8).
19. Incubate all plates at 37°C overnight.
20. Examine each MacConkey agar plate. If the isolated colonies of the streaked transformants retain the phenotype of interest, relative to the parental control strain (MC4100 λRS88 *cpxP–lacZ* (pCA24N)), then proceed to Subheading 3.4 (see Note 9).
21. Identify the dilution series plate on which 100–300 well-isolated transformants are growing. From that plate, use sterile toothpicks to pick individual colonies into separate wells of a 96-well microtiter plate containing 200 μL of LB plus 30 μg/mL of kanamycin and 25 μg/mL of chloramphenicol (see Note 7). In each plate, be sure to inoculate one well with the vector control strain MC4100(pJW25, pCA24N). Since there are 24 plasmids per ASKA clone pool, we routinely screen around 100 colonies per transformation to ensure good representation of different transformants (i.e., approximately 4× coverage).
22. Grow the 96-well plates for 3–4 h at 37°C, shaking at 225 rpm, or until the cultures have reached an optical density measured at 600 nm in the mid-log range (0.4–0.6) (see Note 10).
23. Measure and record luminescence in a luminometer as well as the optical density at 600 nm to confirm the growth stage.
24. Transfer the luminescence and OD data derived from the 96-well plates to an Excel (or similar) file.

25. Divide the level of luminescence of each transformant by the OD at 600 nm to standardize light levels to growth.
26. Graph the normalized level of luminescence on the *y*-axis and the well number of each transformant on the *x*-axis. The amount of luminescence should fluctuate closely around that of the control strain (MC4100(pJW25, pCA24N)).
27. Identify transformants that lie well above or below the "noise" of the majority of the transformants and the level of luminescence of the control strain.
28. Streak the cultures from the corresponding wells to sectored LB agar plates containing 25 μg/mL chloramphenicol and 30 μg/mL kanamycin for isolated colonies (see Note 7). Incubate overnight at 37°C. Freeze strains in glycerol (see Note 9).
29. To confirm the initially observed luminescence phenotype, inoculate three individual, isolated colonies from each plated transformant into 200 μL of LB containing 30 μg/mL of kanamycin and 25 μg/mL of chloramphenicol. Inoculate three isolated colonies of the parent vector control strain, MC4100 (pJW25, pCA24N) in the same fashion.
30. Grow these cultures for 4–6 h at 37°C shaking at 225 rpm and measure and graph luminescence as in steps 23–26.
31. If the cultures of the isolated colonies of the streaked transformants retain the phenotype of interest, and it is stable (i.e., the average luminescence of the three transformants of interest is significantly different than that of the control strain), then proceed to Subheading 3.4.

3.4. Identifying ASKA Clones of Interest

The genes contained on ASKA plasmids isolated from transformants of interest are identified by using colony PCR to amplify them and then subjecting them to sequence analysis.

1. Prepare a 100-μL PCR reaction containing 5 μL of 10× PCR Buffer, 2 μL of 50 mM magnesium chloride, 8 μL of 5 mM dNTPs, 5 μL of betaine solution, 1 μL of primer pCA-F, 1 μL of primer pCA-R, and 76.5 μL of deionized water.
2. Resuspend single colonies of transformants of interest (obtained in Subheading 3.3, steps 18 or 28) by picking the colony from the plate with the tip of a disposable 200 μL micropipette tip and then pipetting up and down in the PCR reaction mix several times to resuspend.
3. Boil the reactions for 5 min and then allow them to cool to room temperature for about 5 min.
4. Add 0.5 μL of recombinant *Taq* DNA polymerase.
5. Program the thermocycler as follows: 96°C for 2 min; 30 cycles of 96°C for 45 s, 50°C for 30 s, 72°C for 1 min; 72°C for 10 min.

6. Load 10 μL of each PCR reaction onto a 1.5% agarose gel to determine the size of the inserts. Any plasmids bearing the same size insert are likely to carry the same clone. One representative plasmid of each size insert should be chosen for sequencing.
7. Purify the PCR products using the PCR purification kit. Elute the DNA in 30 μL of deionized water.
8. Sequence the purified PCR product using BigDye ET by preparing the following reaction mixture: 6 μL BigDye Terminator sequencing, 2 μL BigDye enzyme premix, 1 μL of either pCA-F or pCA-R, 2 μL of purified PCR template DNA from step 7 above (50–400 ng DNA), and 9 μL of deoionized water in a 0.5 mL eppendorf tube.
9. The PCR cycling reaction is as follows: 96°C for 2 min followed by 30 cycles of 96°C for 10 s, 50°C for 5 s, 60°C for 4 min. Reactions are stored at −20°C until ready for use.
10. Transfer the sequencing reaction to a 1.5 mL eppendorf tube containing 2 μL of purification buffer. Pipet up and down slowly to mix.
11. Add 80 μL 95% ethanol stored at −20°C and vortex briefly to mix. Incubate on ice for 15 min.
12. Pellet the DNA by spinning the reaction mixture for 15 min at max speed in a microfuge. Discard the supernatant.
13. Add 500 μL 70% ethanol at −20°C. Vortex thoroughly to mix. Incubate on ice for 5 min.
14. Pellet the purified DNA by spinning for 5 min at max speed in a microfuge. Aspirate the supernatant carefully to avoid the pellet, which may not be visible.
15. Dry the pellet in the fumehood for 1–2 h or in a speed vac on high for 20 min. Store at −20°C until ready to proceed to step 16.
16. Sequencing is performed using an ABI 3730 Genetic Analyzer.
17. To confirm the identity of genes of interest from the DNA sequence, perform a nucleotide BLAST (Basic Local Alignment Search Tool) at the EcoCyc database (http://ecocyc.org/blast.html). Copy and paste your DNA sequence into the query box, select "nucleotide" for both "type of query sequence" and "type of database to search against." The search results should reveal the identity of the clone of interest.

3.5. Confirming Phenotypes

Phenotypes are confirmed by transforming known ASKA library clones corresponding to the genes identified in Subheading 3.4 into the reporter strain to verify the phenotype initially observed in the screen.

1. Streak out *E. coli* containing the ASKA clone of interest from the library onto LB agar containing 25 μg/mL chloramphenicol. Incubate at 37°C overnight.
2. To isolate the ASKA clones of interest, inoculate a single isolated colony into a tube containing 5 mL LB with 25 μg/mL chloramphenicol and incubate at 37°C with aeration at 225 rpm overnight. The following day, isolate plasmid DNA from the culture using a plasmid miniprep kit.
3. Transform the plasmid into *E. coli* MC4100 λRS88 *cpxP–lacZ* or MC4100(pJW25) as in steps 1–14 of Subheading 3.3.
4. Plate 100 μL of each dilution of the transformation mix onto MacConkey agar containing 25 μg/mL chloramphenicol (*lacZ* reporters) or 30 μg/mL kanamycin (if using a *lux* reporter). Incubate at 37°C overnight.
5. To confirm the phenotype of the transformants, streak individual transformant colonies for isolated colonies to sectored lactose-containing MacConkey plates containing 25 μg/mL chloramphenicol (for *lac* reporters) or sectored LB plates containing 25 μg/mL of chloramphenicol and 30 μg/mL kanamycin (for *lux* reporters). On each plate, also streak the vector control parent strain for comparison (either MC4100 λRS88 *cpxP–lacZ* (pCA24N) or MC4100(pJW25, pCA24N)). Incubate at 37°C overnight.
6. The following day, examine the MacConkey plates to ensure that the transformants yield the same phenotype as initially observed during the screen in Subheading 3.3. For *lux* reporters, follow steps 29 and 30 of Subheading 3.3 to confirm that the transformants exhibit the same phenotype as that originally identified in Subheading 3.3.

4. Notes

1. When choosing restriction enzymes for cloning promoters into pNLP10 or pJW15, there are two important considerations. First, you must choose two enzymes whose recognition sequences are not present in the promoter region to be cloned. Assuming you have access to the DNA sequence of the promoter region, the presence of restriction enzyme sites can be determined using one of numerous online tools (such as RestrictionMapper, http://www.restrictionmapper.org/). Second, in order to ensure proper orientation of your insert, the restriction site located further from the *luxCDABE* operon in pNLP10/pJW15 must be included in the forward primer (*Eco*RI in the example), while the restriction site located closer

to the *luxCDABE* operon is included in the reverse primer (*Bam*HI in the example). The plasmid map in Fig. 2 shows which enzymes have sites in the MCS of pNLP10 and pJW15.

2. We have used *lacZ* reporter genes extensively in screens like this. We employ *lacZ* reporter genes that are contained on a λ phage that is integrated into the bacterial chromosome. Their construction and use has been described extensively (16, 17), and so we will not repeat it here. These reporters are extremely useful in laboratory strains of *E. coli* K-12 that are susceptible to infections with commonly used bacteriophages such as λ and P1, but cannot be employed in the study of other strains resistant to such infections. Because of this, we have modified existing luminescent reporter plasmids (18) with origins of replication that are compatible in a broad host range (Fig. 2). The resulting plasmids, pNLP10 and pJW15 ((13, 14), Fig. 2), contain large multiple cloning sites that facilitate the cloning of transcriptional control regions of interest upstream of the *lux-CDABE* genes which encode products that result in light production.
3. pNLP10 and pJW15 are identical plasmids except for their origins of replication (Fig. 2). Both contain the same multiple cloning site and encode a promoterless *luxCDABE* operon as well as a kanamycin resistance cassette. However, pNLP10 contains the pSC101 origin (very low copy number, ~5 copies/cell), while pJW15 contains the p15A origin (low copy number, ~10–12 copies/cell). Use of the lower copy plasmid pNLP10 is generally desirable, with two exceptions: first, when cloning a very weak promoter, the higher copy pJW15 will give increased sensitivity; and second, pJW15 is often easier to transform into bacterial strains that contain restriction–modification systems, such as enteropathogenic *E. coli* (19).
4. It is important to work with small numbers of plates at this step and to minimize the time that the plates are at room temperature in order to avoid complete thawing of the library, as successive freeze-thaw cycles will result in significant loss of viability.
5. This method must be repeated from this point on for every pool of ASKA library plasmids. We have found it most practical to work with 2–4 plasmid pools at a time.
6. The optimal number of colonies for screening β-galactosidase activity on Maconkey agar or for picking transformants into 96-well plates is 100–200 colonies per plate. Once an optimal serial dilution for each batch of competent cells is determined, this dilution can be used for every transformation until a new batch of competent cells is required.
7. IPTG can be included at a concentration of 10–100 micromolar to induce the expression of the ASKA clones. We have found

that there is generally overexpression of the ASKA open reading frames in the absence of induction, and often induction can lead to a defect in growth which might impact expression of the reporter gene. For this reason, we often perform our genetic screens in the absence of IPTG.

8. Occasionally, "jackpot" transformations occur in which multiple plasmids in one plasmid pool influence the expression of the reporter gene. In this case, we routinely choose ten to further characterize, since we find that this number generally results in the identification of the same clone(s) multiple times, indicating that this level of sampling is sufficient.
9. To preserve colonies of interest, inoculate 5 mL of LB broth with an isolated colony that possesses the phenotype of interest. Grow culture at 37°C with aeration at 225 rpm overnight. Prepare a cryo-tube containing 750 μL of sterile 10% glycerol. Aliquot 750 μL of overnight culture into the tube and seal with lid. Store at −80°C.
10. The growth phase at which luminescence is measured is user-defined. If searching for genes that impact processes that occur at specific growth phases, it may be desirable to screen the transformants accordingly.

Acknowledgments

The authors acknowledge Shannon Leblanc and Anna Nevesinjac for technical notes on condensing the ASKA library into plasmid pools. Work in the Raivio lab is funded by operating grants from the Canadian Institutes of Health Research (MOP 199847) and the Natural Sciences and Engineering Research Council (NSERC). TLR is supported by a Senior Scholar Award from Alberta Innovates Health Solutions. SLV is supported by scholarships from NSERC, Alberta Ingenuity, and the Killam Trust.

References

1. Baba T, Ara T, Hasegawa M, Takai Y, Okumura Y, Baba M, Datsenko KA, Tomita M, Wanner BL, Mori H (2006) Construction of Escherichia coli K-12 in-frame, single-gene knockout mutants: the Keio collection. Mol Syst Biol 2:2006.0008
2. Kitagawa M, Ara T, Arifuzzaman M, Ioka-Nakamichi T, Inamoto E, Toyonaga H, Mori H (2005) Complete set of ORF clones of Escherichia coli ASKA library (a complete set of E. coli K-12 ORF archive): unique resources for biological research. DNA Res 12(5):291–299
3. Babu M, Diaz-Mejia JJ, Vlasblom J, Gagarinova A, Phanse S, Graham C, Yousif F, Ding H, Xiong X, Nazarians-Armavil A et al (2011) Genetic interaction maps in Escherichia coli reveal functional crosstalk among cell envelope biogenesis pathways. PLoS Genet 7(11):e1002377
4. MacRitchie DM, Buelow DR, Price NL, Raivio TL (2008) Two-component signaling and gram negative envelope stress response systems. Adv Exp Med Biol 631:80–110
5. Batchelor E, Walthers D, Kenney LJ, Goulian M (2005) The Escherichia coli CpxA-CpxR

envelope stress response system regulates expression of the porins ompF and ompC. J Bacteriol 187(16):5723–5731
6. Bury-Mone S, Nomane Y, Reymond N, Barbet R, Jacquet E, Imbeaud S, Jacq A, Bouloc P (2009) Global analysis of extracytoplasmic stress signaling in Escherichia coli. PLoS Genet 5(9):e1000651
7. Button JE, Silhavy TJ, Ruiz N (2007) A suppressor of cell death caused by the loss of sigmaE downregulates extracytoplasmic stress responses and outer membrane vesicle production in Escherichia coli. J Bacteriol 189(5): 1523–1530
8. Connolly L, De Las PA, Alba BM, Gross CA (1997) The response to extracytoplasmic stress in *Escherichia coli* is controlled by partially overlapping pathways. Genes Dev 11:2012–2021
9. Danese PN, Silhavy TJ (1997) The sigma(E) and the Cpx signal transduction systems control the synthesis of periplasmic protein-folding enzymes in Escherichia coli. Genes Dev 11(9):1183–1193
10. Gerken H, Charlson ES, Cicirelli EM, Kenney LJ, Misra R (2009) MzrA: a novel modulator of the EnvZ/OmpR two-component regulon. Mol Microbiol 72(6):1408–1422
11. Jubelin G, Vianney A, Beloin C, Ghigo JM, Lazzaroni JC, Lejeune P, Dorel C (2005) CpxR/OmpR interplay regulates curli gene expression in response to osmolarity in Escherichia coli. J Bacteriol 187(6): 2038–2049
12. Raffa RG, Raivio TL (2002) A third envelope stress signal transduction pathway in Escherichia coli. Mol Microbiol 45(6):1599–1611
13. MacRitchie DM, Ward JD, Nevesinjac AZ, Raivio TL (2008) Activation of the Cpx envelope stress response down-regulates expression of several locus of enterocyte effacement-encoded genes in enteropathogenic Escherichia coli. Infect Immun 76(4):1465–1475
14. Price NL, Raivio TL (2009) Characterization of the Cpx regulon in Escherichia coli strain MC4100. J Bacteriol 191(6):1798–1815
15. Shimada T, Makinoshima H, Ogawa Y, Miki T, Maeda M, Ishihama A (2004) Classification and Strength Measurement of Stationary-Phase Promoters by Use of a Newly Developed Promoter Cloning Vector. J Bacteriol 186(21): 7112–7122
16. Silhavy TJ, Berman ML, Enquist LW (1984) Experiments with gene fusions. Cold Spring Harbor Laboratory Press, Cold Spring Harbor, NY
17. Simons RW, Houman F, Kleckner N (1987) Improved single and multicopy *lac*-based cloning vectors for protein and operon fusion. Gene 53:85–96
18. Beeston AL, Surette MG (2002) pfs-dependent regulation of autoinducer 2 production in Salmonella enterica serovar Typhimurium. J Bacteriol 184(13):3450–3456
19. Hobson N, Price NL, Ward JD, Raivio TL (2008) Generation of a restriction minus enteropathogenic Escherichia coli E2348/69 strain that is efficiently transformed with large, low copy plasmids. BMC Microbiol 8:134
20. Raivio TL, Silhavy TJ (1997) Transduction of envelope stress in Escherichia coli by the Cpx two-component system. J Bacteriol 179: 7724–7733

Chapter 22

Isolation of Bacteria Envelope Proteins

Shu Quan, Annie Hiniker, Jean-François Collet, and James C.A. Bardwell

Abstract

Proteomic analysis on cell envelope proteins from Gram-negative bacteria requires specific isolation techniques. We found that conventional extraction methods such as osmotic shock cause extracts to be heavily contaminated with soluble cytoplasmic proteins. These cytoplasmic protein contaminants constitute the major signal in proteomic analysis and can overwhelm the signals coming from genuine envelope components. After extensive testing of various protocols for the preparation of envelope contents, we found that a modified version of the method of Oliver and Beckwith (1) consistently produces the cleanest extract of periplasmic and outer membrane proteins (2). We have designated this very simple method TSE extraction because it uses a Tris-sucrose solution supplemented with EDTA. Cytoplasmic and inner membrane protein contaminants are not evident on 1D SDS polyacrylamide gels and contribute to less than 6% of total signal in very sensitive mass spectrometry analysis. This straightforward method is therefore ideal for analyzing specific proteomic changes in the cell envelope.

Key words: Gram-negative bacteria, Periplasmic extraction, Outer membrane extraction, Envelope proteins

1. Introduction

Bacterial cells are protected by a complex, multilayered structure called the cell envelope. In Gram-negative bacteria, the cell envelope is composed of an inner cytoplasmic membrane, a thin peptidoglycan cell wall, and a lipopolysaccharide-containing outer membrane that surrounds the peptidoglycan layer; the inner and outer membranes delimit an aqueous space termed the "periplasm" (3). The periplasm has gained increasing attention as a compartment for recombinant protein expression, in part because it favors disulfide bond formation. Additionally, the periplasm only contains a small proportion of the total cellular protein; therefore, expression into this compartment results in a substantial purification upon periplasmic extraction. Many methods have been established to mechanically or chemically disrupt the outer membrane and cell wall in order

Anne H. Delcour (ed.), *Bacterial Cell Surfaces: Methods and Protocols*, Methods in Molecular Biology, vol. 966, DOI 10.1007/978-1-62703-245-2_22, © Springer Science+Business Media New York 2013

to release periplasmic contents, either for purification of recombinant proteins or for proteomic studies. Some of these methods, such as osmotic shock, lysozyme-EDTA treatment, polymyxin digestion, and chloroform extraction, showed high enrichment of recombinant proteins in the extracts and up to 90% recovery of the target proteins (4–6). However, we found that these periplasm extraction methods, presumably optimized for maximal recovery of recombinant proteins, also release a substantial amount of cytoplasmic protein. Periplasmic proteins are estimated to constitute only ~4–16% of the total cellular protein in *Escherichia coli* (7, 8). Thus, cytoplasmic proteins outnumber and out-mass periplasmic contents by approximately ten-fold. As a consequence, relatively minor release of cytosolic contents will result in substantial contamination of a periplasmic extract. This release can be due to cell lysis during the repetitive pelleting and resuspension steps of the osmotic shock procedure, or it can result from destabilization of the inner membrane by chemicals such as chloroform and polymyxin, which are utilized in other extraction procedures (9, 10).

During a study examining substrate specificity of the periplasmic oxidoreductases in *E. coli* (2), we found that the osmotic shock and polymyxin procedures failed to produce a clean periplasmic fraction suitable for proteomic analysis. Indeed, when the most abundant "periplasmic" spots on two-dimensional polyacrylamide gels were analyzed by mass spectrometry, we found that many actually corresponded to cytoplasmic proteins. In addition, we noticed that the amount of cytoplasmic proteins released by these procedures varied substantially from experiment to experiment, making the accurate quantification of periplasmic contents extremely difficult.

To circumvent these problems, we tested a large number of periplasmic extraction procedures. We found that modification of a periplasmic extraction method described by Oliver and Beckwith (1), which we renamed TSE extraction, gave the cleanest periplasmic preparation by far. We note that this methodology contains many outer membrane proteins in addition to periplasmic proteins but has very few contaminating cytosolic or inner membrane proteins (11). As outlined below, outer membrane proteins can be removed by an additional ultracentrifugation step. Mass spectrometry identification of a typical extract prepared from *E. coli* by this method (omitting the ultracentrifugation step) contains detectable quantities of ~130 periplasmic proteins (34% of total), 50 outer membrane proteins (13% of total), 2 extracellular proteins, ~180 cytoplasmic proteins, and ~20 inner membrane proteins (altogether 53% of total) (Table 1 and Fig. 1). Although the number of cytoplasmic and inner membrane proteins detected seems substantial, the average signal coming from cytoplasmic and inner membrane proteins is only two spectra counts, 13-fold lower than the average signal of outer membrane proteins and 23-fold lower than that of

Table 1
Classification of protein identified in *E. coli* MC4100 envelope extract prepared by the TSE method according to their subcellular localization and relative intensity determined by mass spectrometry

Subcellular localization	Relative intensity (spectral counts)			
	Total no. of proteins	High (>20)	Medium (3–20)	Low (1–2)
Periplasmic	129	36	63	30
Outer membrane	50	9	24	17
Extracellular	2	1	1	0
Inner membrane	21	0	2	19
Cytoplasmic	183	0	31	152
Total	385	46	121	218

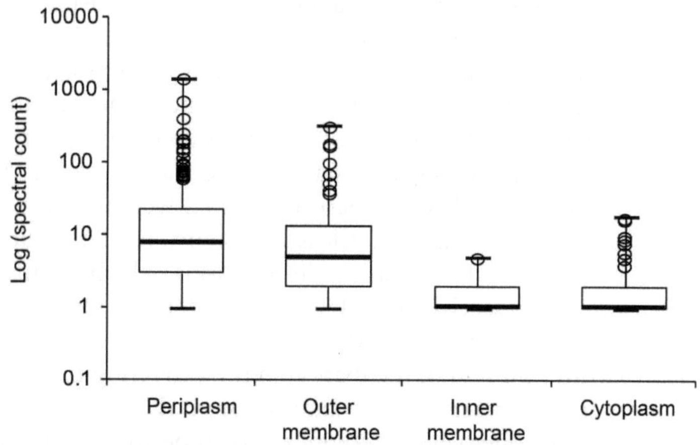

Fig. 1. Intensity distribution of proteins identified by mass spectrometry analysis on a typical extract prepared from *E. coli* using the TSE method as measured by spectral count (SC). The number of SC for a protein is the total number of MS/MS spectra taken on peptides from this protein in a given 2D-LC-MS/MS analysis. This value is linearly correlated with the protein abundance. Proteins are grouped according to their subcellular localization. The *lower boxes* represent the first quartile (25%) and the *upper boxes* represent the third quartile (75%). The *thick lines* across each of the boxes show the medians. Note that the first quartiles in inner membrane and cytoplasmic samples equal the medians in these samples. *Short horizontal lines* show the maximal and minimal values, and *circles* represent the outliers, which are three times larger than the likely range of variation.

periplasmic proteins. Among the 100 most abundant proteins, only eight are cytoplasmic; the rest are either outer membrane proteins (21%) or periplasmic proteins (71%). These eight cytoplasmic protein contaminants are among the most abundant *E. coli* proteins, with copy numbers higher than 1,000 per cell (12), and include thioredoxin and EF-Tu, which are known to be very easily extracted in standard periplasmic extraction procedures (13). Assuming that overall spectral signal corresponds to protein abundance, the TSE extract contains 78% periplasmic proteins, 16% outer membrane proteins, 0.4% inner membrane proteins, and 5% cytosolic proteins. We note that mass spectrometry analysis is extremely sensitive, and the majority of the cytoplasmic proteins are unlikely to be visible using techniques with lower sensitivity, such as SDS-PAGE or 2D gel electrophoresis. A recent periplasmic proteome study in *Pseudomonas aeruginosa* compared extractions prepared by three methods, one of which is similar to our method except for their use of lysozyme in the extraction buffer and a higher concentration of EDTA (14). This similar method resulted in the highest number of unique spots on 2D gels corresponding to periplasmic proteins and the lowest level of cytoplasmic contaminations.

In this simple TSE extraction procedure, cells are first pelleted to remove the media, resuspended in a concentrated solution of sucrose in Tris buffer supplemented with EDTA, incubated for 30 min on ice, and then recentrifuged. The supernatant is the TSE periplasm/outer membrane extract. Sucrose in the extraction buffer increases the extracellular osmolality, causing the cells to shrink and release water and periplasmic contents into the surrounding medium. Sucrose can go through the outer membrane to enter the periplasmic space but is unable to penetrate the inner membrane. The presence of sucrose at the water–inner membrane interface presumably helps to stabilize the membrane (15) and therefore prevents cell lysis. Sucrose is also a well-known protein stabilizer (16), protecting the released proteins. EDTA facilitates periplasmic extraction by chelating divalent ions, which normally stabilize the lipopolysaccharide (LPS) in the outer membrane, resulting in LPS release and increased permeability of the outer membrane (17). After 30 min of incubation in the extracting buffer, the centrifugation step acts to efficiently separate the soluble envelope proteins from other cellular components. If desired, an additional ultracentrifugation step at 100,000 × *g* for 1 h can be performed to separate outer membrane proteins (which pellet at 100,000 × *g*) from periplasmic proteins, which remain soluble (11). This method is an effective way to extract soluble envelope proteins and additionally is less labor intensive than many other methods. We named it TSE, standing for Tris, Sucrose, and EDTA, the components of the extraction buffer.

We compared extracts purified by TSE extraction with those prepared by either polymyxin extraction or a modified osmotic shock procedure (Fig. 2). Polymyxin and modified osmotic shock were chosen as the gold standards of periplasmic purification because we found that they produce much purer periplasmic extracts than other methods (lysozyme-EDTA, traditional osmotic shock, and chloroform extractions). The purity of subcellular extracts in *E. coli* has traditionally been estimated by analyzing for protein markers of known subcellular localization.

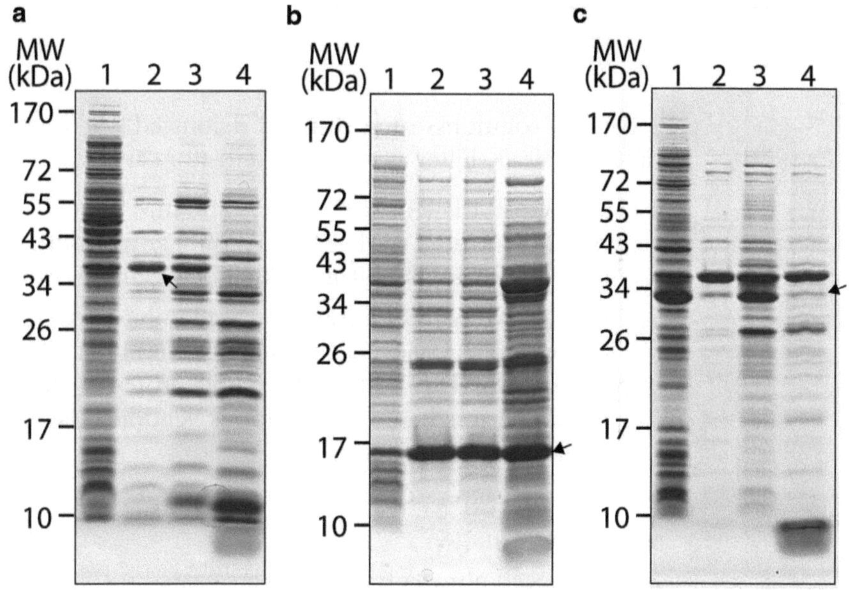

Fig. 2. Comparison of the envelope extracts prepared using the TSE procedure and two other methods. Periplasmic proteins were separated via SDS-PAGE using 14% Tris-glycine gels. For all gels, *lane 1* whole cell lysate, *lane 2* TSE extract, *lane 3* modified osmotic shock extract, *lane 4* polymyxin extract. *Panel* (**a**) shows the extraction of the *E. coli* K12 strain MG1655. The *arrow* indicates enrichment of the outer membrane proteins OmpC and OmpF. *Panel* (**b**) shows the extraction of the *E. coli* K12 strain MG1655 over-expressing a 16 kDa periplasmic protein, Spy (marked by an *arrow*). Over-expression of Spy makes the cells vulnerable to lysis during the periplasmic extraction procedures. However, the TSE method still produces a cleaner periplasm than the polymyxin method (compare *lanes* 2 and 4). *Panel* (**c**) shows the extraction of the *E. coli* BL21 strain over-expressing a 33 kDa cytoplasmic protein, Hsp33 (marked by an *arrow*). Although the TSE (*lane* 2) and polymyxin methods (*lane* 4) resulted in a slight release of Hsp33 into the periplasm, the modified osmotic shock method (*lane* 3) released a much larger fraction of Hsp33 into the periplasm. In conclusion, the TSE method produces the cleanest periplasmic extract under all the conditions we have tested.

Often, β-galactosidase serves as a cytoplasmic marker and plasmid-encoded β-lactamase serves as a periplasmic marker (18). However, we speculate that this method may overestimate the purity of the preparations since β-galactosidase, which is over 100 kDa, is likely to be much harder to "shock out" than smaller cytosolic proteins like thioredoxin. A simpler and more realistic way to estimate the periplasmic purity is to compare the band pattern of periplasmic extracts on 1D SDS-PAGE gels with the band pattern of whole cell extracts (see Fig. 2). Since envelope proteins constitute only about 4–16% of the proteome, a clean extract is expected to produce a distinctive band pattern with very little overlap with that generated by whole cell extraction. Substantial contamination with cytosolic proteins can be demonstrated by the appearance of dominant bands from the total cell extract appearing as major components of the so-called periplasm. We found that for various strain backgrounds (*E. coli* K12 and B strains) including those that over-express various proteins, the TSE method generates the cleanest envelope pattern of all methods tested (Fig. 2).

2. Materials

Prepare all solutions using distilled deionized water and reagent grade chemicals. Filter all solutions and store at 4°C.

1. Tris-sucrose-EDTA (TSE) buffer: 200 mM Tris–HCl, pH 8.0, 500 mM sucrose, 1 mM EDTA. Aliquot an appropriate amount and add protease inhibitor cocktail before use.
2. Complete Mini, EDTA-free Protease Inhibitor Cocktail Tablets (Roche Applied Science, Indianapolis, IN, USA). Use 1 tablet per 10 mL of TSE buffer. Alternative protease inhibitor cocktails can be purchased from various vendors. Use the dosage specified by the manufacturer.
3. Centrifuge and associated rotor.
4. Microcentrifuge.
5. High speed ultracentrifuge and associated rotor.
6. Polymyxin buffer: 50 mM Tris–HCl, pH 7.5, 5 mM EDTA, 150 mM NaCl, 1 mg/mL polymyxin B sulfate.
7. Ice-cold water.
8. Reducing SDS sample buffer: 60 mM Tris–HCl, pH 7.0, 2% SDS, 1.32 M glycerol, 0.01% bromophenol blue, 1% 2-mercaptoethanol.

3. Methods

The protocols described below are to be performed on a 100 mL of a fresh bacteria culture at an optical density of 1.0 at 600 nm. For cell cultures of different volumes or different optical densities, adjust the volume of reagents proportionately (see Note 1). Carry out all procedures on ice unless otherwise specified.

3.1. TSE Periplasmic Extraction

1. Harvest bacterial cells by centrifugation at 3,000 × *g* for 20 min at 4°C (see Note 2).
2. Discard the supernatant and carefully remove the last few drops of liquid with a pipette (see Note 3).
3. Gently resuspend the pellet in 1 mL of TSE buffer using a wire loop (see Notes 4–7).
4. Incubate the cells in TSE buffer on ice for 30 min (see Note 8).
5. Transfer the cell suspension to a microcentrifuge tube and centrifuge at 16,000 × *g* (maximal speed) for 30 min at 4°C.
6. Transfer the supernatant to a new microcentrifuge tube; this supernatant constitutes the envelope extract.

7. (Optional) If separation of outer membrane proteins from periplasmic proteins is desired, centrifuge the supernatant from step 6 at 100,000 × *g* for 1 h at 4°C. The pellet from this step contains the outer membrane proteins; the supernatant contains the periplasmic proteins.

3.2. Polymyxin Treatment Control (see Fig. 2)

This procedure is adapted from Jonda et al. (see ref. 19).

1. Follow steps 1 and 2 in Subheading 3.1 to pellet cells.
2. Gently resuspend the pellet in 1 mL of 1 mg/mL polymyxin buffer and incubate cells on ice for 1 h.
3. Continue with steps 5 and 6 in Subheading 3.1.

3.3. Modified Osmotic Shock Control (see Fig. 2)

This procedure is adapted from the protocol for PeriPreps™ Periplasting Kit (Epicenter Biotechnologies, Madison, WI, USA).

1. Follow steps 1 and 2 in Subheading 3.1 to pellet cells.
2. Gently resuspend the pellet in 0.5 mL TSE buffer and incubate on ice for 15 min.
3. Add 0.5 mL of ice-cold water, mix, and incubate on ice for an additional 15 min.
4. Continue with steps 5 and 6 in Subheading 3.1.

3.4. Whole Cell Lysate Control (see Fig. 2)

1. Follow steps 1 and 2 in Subheading 3.1 to pellet 1 mL of cells.
2. Resuspend the pellet in 0.1 mL reducing SDS sample buffer and heat at 99°C for 10 min.

4. Notes

1. We have routinely used cell amounts as low as 6×10^8–2×10^9 cells (the equivalent of 2 mL at an optical density of 1.0 at 600 nm). We do not recommend using fewer cells because resuspension of cells in less than 20 μL of TSE buffer is difficult.
2. Centrifugation at higher *g* forces may produce tight pellets, which are more difficult to resuspend, thus increasing the risk of cell lysis.
3. Any remaining liquid will dilute the TSE buffer and therefore decrease the efficiency of extraction.
4. Cell pellets must be freshly prepared. Old or frozen pellets should not be used as they are prone to cell lysis, particularly during resuspension.
5. Avoid vigorous pipetting because this may cause cell lysis resulting in contamination of the extract with cytoplasmic proteins.

6. If cells are difficult to resuspend, increase the amount of TSE buffer to 2.5 mL.
7. If the concentration of target protein is low, decrease the amount of TSE to 0.5 mL to produce a more concentrated periplasmic extract. Extracts can also be concentrated with centrifugal filter devices or precipitated by 10% TCA. TSE buffer shows no interference with TCA precipitation.
8. The release of envelope proteins in the presence of sucrose and EDTA is solely dependent on diffusion. Therefore, we use prolonged incubation and centrifugation steps to ensure substantial release of envelope proteins.

References

1. Oliver DB, Beckwith J (1982) Regulation of a membrane component required for protein secretion in *Escherichia coli*. Cell 30:311–319
2. Hiniker A, Bardwell JC (2004) *In vivo* substrate specificity of periplasmic disulfide oxidoreductases. J Biol Chem 279:12967–12973
3. Silhavy TJ, Kahne D, Walker S (2010) The bacterial cell envelope. Cold Spring Harb Perspect Biol 2:a000414
4. Chen YC et al (2005) Comparison of various methods for periplasmic release of recombinant creatinase from Escherichia coli. J Chin Inst Chem Eng 36:527–532
5. Chen YC et al (2004) Modified osmotic shock for periplasmic release of a recombinant creatinase from *Escherichia coli*. Biochem Eng J 19: 211–215
6. Donohuerolfe A, Keusch GT (1983) *Shigella dysenteriae*-1 cytotoxin: periplasmic protein releasable by polymyxin B and osmotic shock. Infect Immun 39:270–274
7. Ames GF, Prody C, Kustu S (1984) Simple, rapid, and quantitative release of periplasmic proteins by chloroform. J Bacteriol 160: 1181–1183
8. Nossal NG, Heppel LA (1966) The release of enzymes by osmotic shock from *Escherichia coli* in exponential phase. J Biol Chem 241:3055–3062
9. Folch J, Lees M, Sloane Stanley GH (1957) A simple method for the isolation and purification of total lipides from animal tissues. J Biol Chem 226:497–509
10. Teuber M, Bader J (1976) Action of polymyxin B on bacterial nembranes. Binding capacities for polymyxin B of inner and outer membranes isolated from *Salmonella typhimurium* G30. Arch Microbiol 109:51–58
11. Vertommen D et al (2009) Characterization of the role of the *Escherichia coli* periplasmic chaperone SurA using differential proteomics. Proteomics 9:2432–2443
12. Ishihama Y et al (2008) Protein abundance profiling of the *Escherichia coli* cytosol. BMC Genomics 9:102
13. Berrier C et al (2000) Elongation factor Tu and DnaK are transferred from the cytoplasm to the periplasm of *Escherichia coli* during osmotic downshock presumably via the mechanosensitive channel MscL. J Bacteriol 182:248–251
14. Imperi F et al (2009) Analysis of the periplasmic proteome of Pseudomonas aeruginosa, a metabolically versatile opportunistic pathogen. Proteomics 9:1901–1915
15. Strauss G, Hauser H (1986) Stabilization of lipid bilayer vesicles by sucrose during freezing. Proc Natl Acad Sci U S A 83:2422–2426
16. Lee JC, Timasheff SN (1981) The stabilization of proteins by sucrose. J Biol Chem 256: 7193–7201
17. Leive L (1965) Release of lipopolysaccharide by EDTA treatment of *E. coli*. Biochem Biophys Res Commun 21:290–296
18. Dixon RA, Chopra I (1986) Leakage of periplasmic proteins from *Escherichia coli* mediated by polymyxin B nonapeptide. Antimicrob Agents Chemother 29:781–788
19. Jonda S et al (1999) Complementation of DsbA deficiency with secreted thioredoxin variants reveals the crucial role of an efficient dithiol oxidant for catalyzed protein folding in the bacterial periplasm. EMBO J 18:3271–3281

Chapter 23

Patch Clamp Electrophysiology for the Study of Bacterial Ion Channels in Giant Spheroplasts of *E. coli*

Boris Martinac, Paul R. Rohde, Charles G. Cranfield, and Takeshi Nomura

Abstract

Ion channel studies have been focused on ion channels from animal and human cells over many years. Based on the knowledge acquired, predominantly over the last 20 years, a large diversity of ion channels exists in cellular membranes of prokaryotes as well. Paradoxically, most of what is known about the structure of eukaryotic ion channels is based on the structure of bacterial channels. This is largely due to the suitability of bacterial cells for functional and structural studies of biological macromolecules in a laboratory environment (1). Development of the "giant spheroplast" preparation from *E. coli* cells was instrumental for functional studies of ion channels in the bacterial cell membrane. Here we describe detailed protocols used for the preparation of giant spheroplasts as well as protocols used for the patch-clamp recording of native or heterologously expressed ion channels in *E. coli* spheroplast membrane.

Key words: Bacteria, Archaea, Cephalexin, Lysozyme, Mechanosensitive channels, K^+ channels, Patch clamp, Laplace's law

1. Introduction

In the past, the classical electrophysiological techniques of the current clamp and voltage-clamp based on impaling cells with two glass microelectrodes were not applicable to bacteria because of their very small size. The patch-clamp recording technique established in 1981 by Erwin Neher, Bert Sakmann, and their coworkers (2) provided a means to overcome this shortcoming. For the first time, this technique allowed electrophysiologists to study the electrical properties of very small cells, including bacteria, by examining ionic currents flowing through individual ion channels in their cellular membranes in situ. Gaining access to ion channels embedded in the bacterial cell membrane, however, poses a few challenges. Firstly there is the bacterial cell wall which creates a partition that is not readily surmountable. Another problem, particularly for researchers

Anne H. Delcour (ed.), *Bacterial Cell Surfaces: Methods and Protocols*, Methods in Molecular Biology, vol. 966,
DOI 10.1007/978-1-62703-245-2_23, © Springer Science+Business Media New York 2013

desiring to undertake patch clamp electrophysiological experiments of ion channels, is the relative dimensions of the bacteria themselves. The dimensions of an *E. coli* cell, a Gram-negative rod-shaped bacterium, is about 0.8 μm wide and 2 μm long, which is almost the same as a diameter of a typical patch pipette. Therefore, a larger bacterial object (with cell wall removed) is necessary for patching the inner membrane where the ion channels are found. The preparation of "giant spheroplasts" (see Note 1) presented a major technical advancement in this direction by allowing, for the first time, the electrophysiological investigation of *E. coli* cell membranes (3, 4), and created the opportunity for structure and function studies of ion channel proteins in prokaryotes (5).

Here we describe the established procedure(s) in which giant spheroplasts can be formed from bacterial cells where the bacterial cell wall has been removed (4, 6). We follow this procedure with a protocol for recording ion channel activities in bacterial spheroplasts, with a special interest in mechanosensitive (MS) ion channels (7, 8).

2. Materials

All solutions should be prepared using ddH_2O and analytical grade reagents. The solutions are stored in the fridge at 4°C. Some of the reagents (EDTA, HEPES, sucrose, Tris-HCl, $MgCl_2$) are stored at room temperature, whereas antibiotics (cephalexin) and enzymes (DNase, lysozyme) are stored in the fridge (4°C) or freezer (−20 to −30°C), respectively.

2.1. Giant Spheroplast Preparation

1. Autoclaved Erlenmayer flasks for culture: two 100-ml, one 200-ml.
2. Lysogeny broth (LB): 10 g/l tryptone, 5 g/l yeast extract, 10 g/l NaCl, pH 7.4 (adjusted with NaOH), autoclaved before use.
3. Selection antibiotics, if required (see Note 2).
4. Cephalexin solution: 10 mg/ml in water. Filter-sterilize with a 0.22 μm syringe filter. Make at least 2 ml to allow for filter sterilization. Freshly made.
5. DNase (from bovine pancreas) solution: 5 mg/ml in water, freshly made.
6. Lysozyme (from chicken egg white) solution: 5 mg/ml in water freshly made.
7. 1 M $MgCl_2$: 20.33 g $MgCl_2 \cdot 6H_2O$ in 100 ml water.

8. 1 M Tris–HCl (Trizma® base, minimum), pH 7.2: 12.11 g/100 ml water, adjust pH with HCl (see Note 3).
9. 125 mM EDTA: 4.65 g/100 ml, pH 7.8, adjusted pH with NaOH (see Note 4).
10. Fresh 0.8 M sucrose solution (see Note 5). It is convenient to measure 27.4 g sucrose into a 100 ml Schott™ bottle or similar, and fill with water to around three quarters full. Fasten lid and shake to dissolve the sucrose. Top up with water to the 100 ml capacity groove. Refrigerate if the procedure continues onto the next day.
11. Stop Solution: 2,100 μl of 0.8 M sucrose, 228 μl of water, 48 μl of 1 M $MgCl_2$, 24 μl of 1 M Tris–HCl pH 7.2. Make fresh.
12. Dilution Solution: 20 ml of 0.8 M sucrose, 200 μl of 1 M $MgCl_2$ (10 mM final), 200 μl of 1 M Tris–HCl pH 7.2 (10 mM final). Make fresh and chill on ice.
13. Shaker incubator for 37°C, and another for 42°C. If a second incubator is not available, one will readily suffice.
14. Bench top centrifuge suitable for 50 ml Falcon™ type tubes.

2.2. Patch-Clamp Recording from Giant Spheroplasts

2.2.1. Solutions

1. Spheroplast preparation: Ion channel recording from giant spheroplasts are carried out using either freshly prepared giant spheroplasts or spheroplasts that have been stored up to several months in a −20 or −30°C freezer.
2. Pipette solution: 200 mM KCl, 40 mM $MgCl_2$, 5 mM HEPES, pH 7.2 (adjusted with KOH).
3. Bath solution: 250 mM KCl, 90 mM $MgCl_2$, 5 mM HEPES, pH 7.2 (adjusted with KOH) (see Note 6).

2.2.2. Patch-Clamp Setup (Basic Components)

1. Faraday cage (homemade or commercial).
2. Phase contrast inverted microscope, with common phase ring 10× to 40×, and PHP objectives (see Note 7).
3. Experimental chamber (homemade or commercial).
4. Micromanipulator.
5. Patch-clamp amplifier (AxoPatch 200B amplifier, Molecular Devices).
6. Digitizer.
7. Computer plus data acquisition and analysis software.
8. Piezoelectric pressure transducer, with a pressure range of ±15 psi (±775 mmHg).
9. High-Speed Pressure Clamp-1 apparatus (optional).
10. Borosilicate glass pipettes.

3. Methods

3.1. Preparation of Giant Spheroplasts

All strains used for the preparation of giant spheroplasts are derivatives of *E. coli* K12. The spheroplast preparations are made from *wild-type* (WT) strains (AW405 (4), AW737 (9), Fragl (10)) or mutant strains having deletions either in major porins OmpC and OmpF or MS channel genes (one to four genes deleted), which can also be used for plasmid expression of MS or K^+ channels from various bacterial and archaeal species. The mutant strains include AW740 (*ompC, ompF*) (11, 12), AW737-KO (*mscL::Cm*) (13), MJF367 (Δ*mscL*), MJF451 (Δ*yggB*), MJF379 (Δ*kefA*), MJF453 (Δ*kefA*, Δ*mscL*), MJF429 (Δ*kefA*, Δ*yggB*), MJF455 (Δ*mscL*, Δ*yggB*), MJF455 ((Δ*mscL*, Δ*yggB*)/*pyggB2*), MJF465 (Δ*kefA* Δ*yggB* Δ*mscL*) (10, 14–16) and PB114 (*recA::Tn10, mscS, mscK::Kan, yjeP*) (17), MJF611 (*mscK, mscS, ybdG*) and MJF612 (*yggB, mscL, mscK, ybdG*) (16). Deletion mutant strains are utilized for patch clamping so that a specific channel may be studied without interference from other channels that may be activated under similar conditions. Knockout strains are also utilized when studying an altered (mutated) channel expressed on a plasmid within a strain with the wild-type channel deleted. Following the original giant spheroplast preparation (3, 4) several variations of this method are also widely used (18–23) (see Note 8).

The procedure for *E. coli* spheroplast production will differ slightly depending on whether the ion channel to be studied is constitutionally expressed (from natural genome expression) or whether it is induced, typically from an expression plasmid. Both situations are described together below, with additional notes where required when an expression plasmid is used. Plasmid expression protocols may need to be individually optimized. This may be dependent on the *E. coli* expression plasmid used, the background strain used, the induction level, the nature of the expressed ion channel protein, the original species of the protein, as well as the desired expression level. The basic procedure given below may be adequate, but may be modified depending on the success of the culture growth, the quality of the spheroplasts, and ultimately the "patchability" of the system.

Up to 3 days may be required to make a preparation of giant spheroplasts.

3.1.1. Preculture

1. Add 5 μl or an aliquot of a glycerol stock of the desired strain (or colony pick from an agar plate) to 10 ml LB media in a 50 or 100 ml flask. If expressing a potentially toxic gene or a gene prone to recombination, use of fresh transformants from a colony is favored over use of glycerol stocks.
2. Add the appropriate antibiotics, if any (see Notes 2 and 9).

3. Grow overnight at 37°C with 240 rpm shaking for ¾ in. stroke shaker incubator or 180 rpm for a 2 in. stroke shaker incubator.

3.1.2. Main Culture

1. The main culture starts early in the morning. Take 200 μl of the above overnight culture and add it to 20 ml LB in a 100 ml flask. Add any appropriate selection antibiotics (see Note 9).
2. Set up 54 ml LB culture in a ≥225 ml flask, and place it in a shaker incubator at 42°C (if available, if not use 37°C), to be pre-warmed and ready for use when the OD of the main culture has reached the required value.
3. Grow the main culture at 37°C with shaking (same rpm as the overnight culture) until the (optical density) OD_{600} is approx. 0.4 (better) to 0.5.
4. Make fresh 0.22-μm filtered 10 mg/ml cephalexin solution while the main culture grows.

3.1.3. Elongation Culture

1. Set incubator to 42°C if not already done (see Note 10).
2. Add 6 ml of the above exponential growth culture to the pre-warmed 54 ml LB.
3. Opt out of using selective antibiotics, keep solutions sterile.
4. Initiate the elongation growth of "snakes" (Fig. 1a) by adding 360 μl of the cephalexin solution (60 μg/ml final). Reduce the rpm of incubator shaker to 180 rpm for a ¾ in. stroke incubator or 120 rpm for a 2 in. stroke incubator at 42°C.
5. During this time make fresh 5 mg/ml lysozyme (at least 120 μl) and 5 mg/ml DNase (at least 100 μl) solutions.
6. At intervals, take 20 μl or so aliquots to view under a light microscope. Continue the growth of snakes until they reach around 100–150 μm (see Note 11); the length can be ~25 % less if the next step below is required.

3.1.4. Induced Expression Culture

The expression steps addressed here are only required if an induced expression system is being used, typically when a plasmid expression system is being utilized. If channels native to a knockout or wild-type strain are of interest only (and thus no expression plasmid is needed), this section will be skipped. Actual induction criteria depend on the expression of interest (see Note 12).

1. Induce the expression system. Typically this is done chemically; the most widespread systems utilize isopropyl β-D-1-thiogalactopyranoside (IPTG), though this may differ depending on the plasmid used (see Note 12).
2. Continuing shaking, allow the expression of the channel for half an hour or to 1 h (see Note 13).

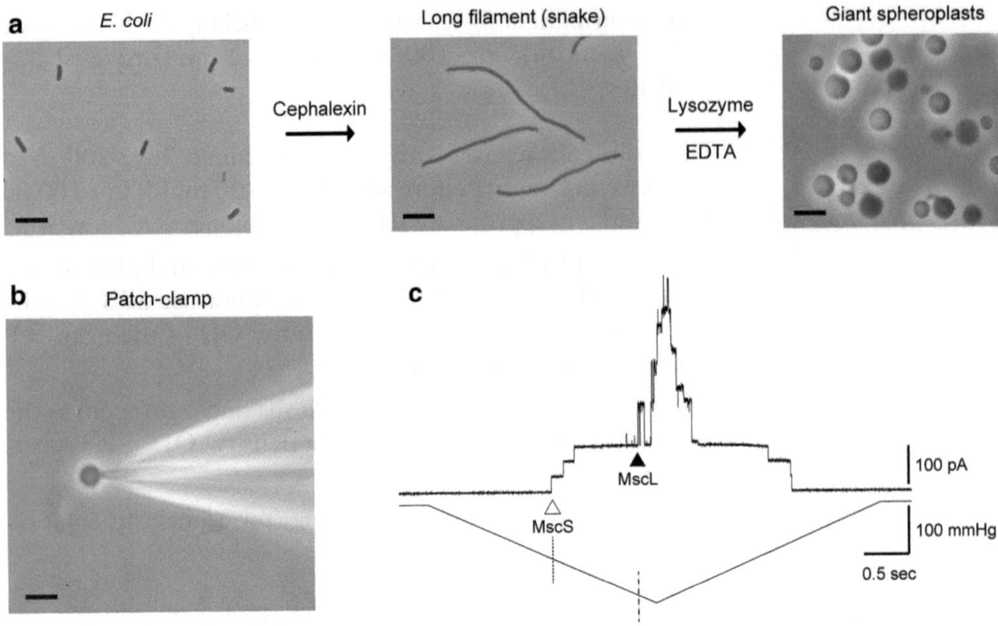

Fig. 1. Generation of and patch-clamp recording from giant spheroplasts. (**a**) A method for generating giant spheroplasts from *E. coli*. The scale bar in all images represents 5 μm. (**b**) The patch pipette is shown with a giant spheroplast attached at the tip. (**c**) Mechanosensitive channel activity recorded in an inside-out membrane patch from a giant spheroplast of the AW737 wild-type *E. coli* strain. Single MscS (*open triangle*) and MscL (*filled triangle*) channels are shown (*upper trace*). The corresponding negative pressure applied to a patch pipette is shown in the trace below the channel current trace. The activation threshold of MscS in this particular patch was −114.5 mmHg (*dotted line*), whereas the activation threshold of MscL was −195.2 mmHg (*dashed line*) giving the activation threshold ratio of 1.7, in good agreement with the reported results (33). Pipette potential was held at +30 mV.

3.1.5. Spheroplasting

The usual practice is to continue with the giant spheroplasts preparation, below. However, if desired, the preparation may be continued the next day. In this case the culture is stored at 4°C for next day's use. Experience has shown that in some instances, quality ("patchability"), of giant spheroplasts may be increased if the culture is stored overnight, but this may not be universal (see Note 14).

1. Split the "snake" culture evenly into two 50 ml Falcons™ tubes (see Note 15).
2. Centrifuge each tube 5 min at 3,000 × *g*.
3. Discard each supernatant completely.
4. Resuspend each pellet with 2.5 ml of 0.8 M sucrose by swirling only, not pipetting. If a pellet is troublesome to resuspend, vigorous/fast swirling is allowable, as is a few quick wrist flicks if necessary.
5. Repeating the centrifugation and resuspension in 2.5 ml of 0.8 M sucrose can improve spheroplast quality.
6. Add 150 μl of 1 M Tris–HCl pH 7.2 to each resuspended pellet solution. Swirl.

7. Then add 120 μl of the lysozyme solution to each resuspended pellet solution. Swirl.
8. Then add 50 μl of the DNase I solution to each resuspended pellet solution. Swirl.
9. Then add 150 μl of 0.125 M EDTA to each resuspended pellet solution. Swirl.
10. The time after the addition of EDTA is critical. Take a 15 μl sample to observe spheroplast formation under the phase-contrast microscope every minute. This helps to determine if spheroplast formation is occurring faster (or slower) than expected and thus the Stop Solution may be added at different times than suggested in the next two steps (see Notes 7 and 15).
11. 5 min after EDTA addition, add 1 ml Stop Solution to one tube only and swirl immediately.
12. Seven and a half minutes after EDTA addition add 1 ml Stop Solution to the remaining tube and swirl immediately.
13. Add 7 ml cold Dilution Solution to each sample, and swirl (see Note 16).
14. Choose the best sample (the one at 5 min or at 7.5 min, or both) and aliquot 50 μl into 200 μl PCR tubes (in a rack on ice), to a desired number of aliquots. The 200 μl tubes can be contained within, for example, a 30 ml collection vial and stored at −20 or −30°C. Discard excess.
15. Spheroplasts are ready for use. Keep the tube of spheroplasts in an ice bucket during the experiment. Once an aliquot is thawed, it will need to be discarded after use.

3.2. Patch-Clamp Recording from Giant Spheroplasts

Giant spheroplast preparations have predominantly been used to record activities of MS channels from bacteria and archaea. The channels include the MscS and MscL channels of *E. coli* (4, 21, 24), *Bacillus subtilis, Erwinia carotovora, Pseudomonas fluorescens, Hemophilus influenzae, Clostridium perfrigens, Staphylococcus aureus, Synechocystis* (25), and *Corinebacterium glutamicum* (26), the MscMJ and MscMJLR channels of the archaeon *Methanococcus jannashii* (15, 27), MscSP of the marine bacterium *Silicibacter pomeroyi* (28) as well as the MscK (29) and MscM (16) channels of *E. coli*. Giant spheroplast preparations have also been used to characterize functionally K^+ channels of the bacterium *Listeria monocytogenes* (22) and the archaeon *Methanobacterium thermoautotrophicum* (30).

3.2.1. Patch-Clamp Pipettes

1. Patch-clamp micropipettes are formed from borosilicate glass microcapillaries by using a pipette puller. Recording pipettes should be pulled to a ~1 μm in diameter, corresponding to a pipette resistance in the range of 2.5–4.9 MΩ in Bath solution (bubble number = 4.0–5.0; see Note 17).

2. To reduce electrical noise, pipette tips can be coated using Sylgard 184 (31) or transparent nail enamel (4), but is usually not required when recording from bacterial MS channels whose conductance is large and single channel currents are ≥10 pA at voltages usually applied in patch clamp experiments (i.e., ≥±10 mV).

3.2.2. Recording

1. A 2–5 μl aliquot of giant spheroplasts is introduced into the recording chamber filled with the bath solution at 22°C. Recording pipettes are backfilled with the pipette solution filtered through a 0.20 μm filter.
2. Lower the patch pipette in the recording chamber and try to catch a floating spheroplast at the tip of the pipette by applying slight suction (see Notes 18 and 19). Form a giga-ohm seal (>1 GΩ) by applying further suction to the patch pipette by mouth or a syringe. Suction is halted when a sudden decrease in pipette current occurs. MscL and MscS activities are usually recorded in inside-out excised patches, which are obtained by briefly exposing the pipette tip to air.
3. Negative pressure (suction) recorded in mmHg is applied to patch pipettes using a syringe or High-Speed Pressure Clamp apparatus and is monitored using a piezoelectric pressure transducer.
4. Ion currents arising from activation of MS channels using negative pipette pressure are recorded with a patch-clamp amplifier. Currents are usually filtered at 2 kHz and digitized at 5 kHz for offline analysis.
5. Single channel recordings can be analyzed using software such as pCLAMP (Axon Instruments) or in-house applications.
6. For MS ion channels it is preferable to measure membrane tension thresholds for channel opening rather than simply measuring applied pressure thresholds because the resulting tension in the membrane patch depends on its geometry, i.e., the same pressure could produce larger tensions in membrane patches of larger diameter and vice versa. Membrane tension measurements can be obtained when the patch is actually visible in the pipette by the use of fluorescence confocal microscopy or differential interference contrast (DIC) microscopy (32) (see Note 20, Fig. 2c, d).

4. Notes

1. Spheroplasts of *E. coli* have a double membrane, as opposed to protoplasts from Gram-positive bacteria and eukaryotic microbes (e.g., yeast).
2. Expression plasmids, or regulatory plasmids, if required, should be maintained with the appropriate antibiotic(s). Many of the

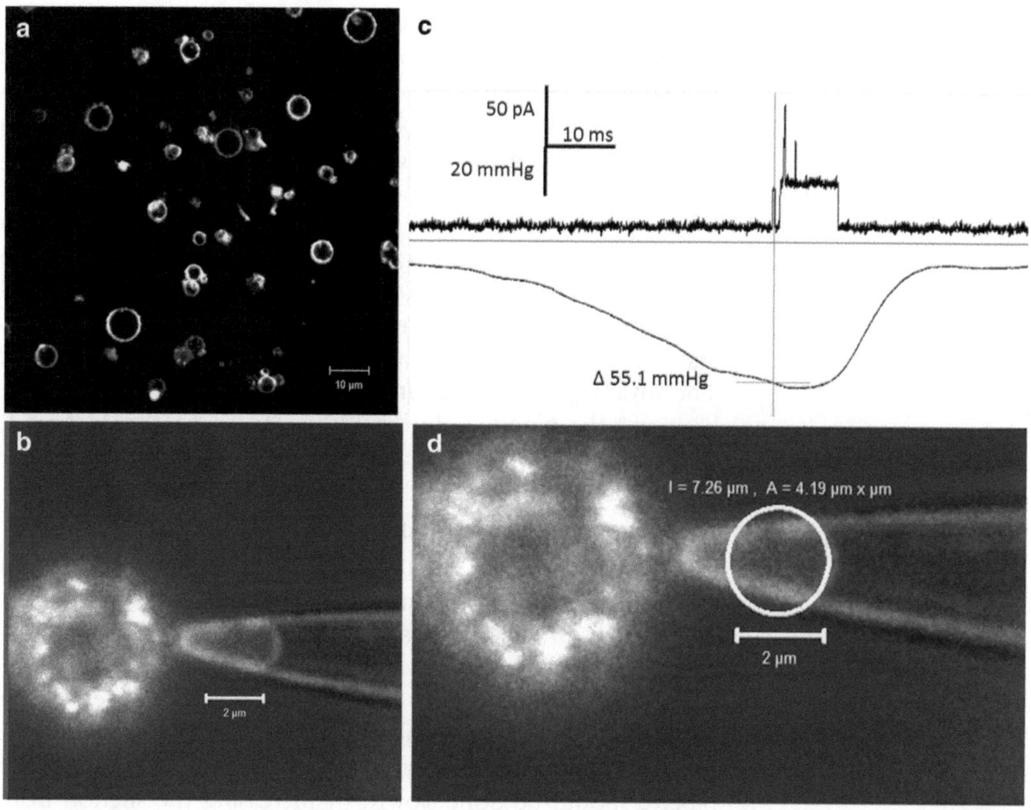

Fig. 2. Patch-clamp recording from fluorescently labeled giant spheroplasts. (**a**) Giant spheroplasts labeled with DI-8-ANEPPSDHQ fluorescent dye and viewed by a confocal microscope. Scale bar corresponds to 10 μm. (**b**) A giant spheroplast shown at the tip of the patch pipette in a cell-attached recording mode. A membrane patch can be seen inside the patch pipette. Scale bar corresponds to 2 μm. (**c**) A 2 s segment of a recording showing activity of two MscS channels (*upper trace*) and a pressure recording (*lower trace*). The channels were activated at a negative pressure applied to the patch pipette of −55.1 mmHg. (**d**) Same figure as in (**b**) shown on an enlarged scale. Measurement of the membrane patch diameter allows calculation of membrane tension corresponding to the negative pressure applied to the patch pipette using Laplace's law (see Note 20). In this particular case, the tension required for activation of MscS channels would be ~4.2 mN/m in agreement with published results (34).

knockout strains have inherent antibiotic resistance common to certain expression or regulatory plasmids. If an intended plasmid gives the same antibiotic resistance as the inherent resistance of a strain, then the strain and plasmid are not compatible to each other, as the plasmid cannot be selected after transformation, nor would it be maintained.

3. For certain "gain of function" mutants, disruption to spheroplast formation may be reduced if Tris–HCl is replaced with PIPES.
4. EDTA chelates calcium ions thus rendering the outer membrane more fragile.
5. 0.8 M sucrose counteracts the high internal turgor pressure of bacteria.

6. Using 250 mM KCl/90 mM $MgCl_2$ in the recording bath solution is important because of the osmolarity of the spheroplast suspension (~800 mOsm, which is approximately the same as the osmolarity inside giant spheroplasts). However, if it is important to keep the ion concentration the same in the bath and pipette solutions, and thus use 200 mM KCl/40 mM $MgCl_2$ in the bath solution, the osmolarity of the bath solution can be adjusted by adding sucrose, sorbitol, or mannose to achieve the osmolarity of 800 mOsm. It is better to use sorbitol or mannose rather than sucrose because sucrose is more viscous and it may cause problems with formation of the giga-ohm seal. The advantage of maintaining an osmotic gradient between the bath and the pipette solution (200 mM KCl/40 mM $MgCl_2$ ~520 mOsm) is that it helps drawing a spheroplast into the pipette and thus facilitates formation of the giga-ohm seal.
7. It is advantageous to use a phase-contrast microscope for recording channel activities from giant *E. coli* spheroplasts. Using a phase-contrast microscope allows one to contrast three different populations of spheroplasts according to their optical appearance, i.e., "shiny," "black," and "transparent." The best spheroplasts for obtaining a giga-ohm seal are the shiny ones. These spheroplasts have two membranes (outer and inner one, since *E coli* is a Gram-negative bacterium). Applying suction to the pipette should be done slowly, by increasing the suction from time to time and waiting until one observes a "jump" in the current trace on the computer or oscilloscope screen. This "jump" indicates a breakage of the outer membrane and the beginning of seal formation between the patch pipette and the inner cytoplasmic membrane where the ion channels are located.
8. In recent literature on bacterial MS channels the *yggb* and *kefA* genes have been renamed to *mscS* and *mscK* since they were shown to encode the MscS and MscK channels.
9. One may consider using appropriate antibiotics inherent to any strain. This is not required if the sterile technique used is considered satisfactory, especially so if another antibiotic is used for any plasmid retention.
10. 42°C incubation is used to suppress flagella formation. Avoidance of flagella formation is to aid ease of a giga-ohm seal formation.
11. As a guide for distance, the diameter of the microscope field of view can be calculated as: 1,000 μm × (eyepiece field number/the objective magnification number). The eyepiece field number (FN) should be listed after the eyepiece magnification (×) number and should not be confused with it.
12. If an expressed ion channel has not been successfully patched from giant spheroplasts, a useful starting guide would be to

emulate conditions for protein production of the channel, if they exist, or of a similar channel. Conditions to consider are expression temperature, amount of chemical inducer, any additives to culture media such as glycerol, or static pause steps in culture. Note, for patching, expression time or amount of inducer can be much less than if a protein was to be purified. Generally, induction of the protein does not need to occur for more than half an hour to 1 h. Note that excessive shaking should be avoided. Excessive shaking may break the elongated "snakes" of *E. coli*. With respect to the incubation temperature for flagella inhibition, 42°C may not be the optimal or suitable temperature for expressing a particular channel. In these instances a lower suitable expression temperature would be used, though forming a seal when patching may become more difficult.

13. Certain toxic proteins (e.g., Gain of function (GOF) mutants of MscL) may only allow expression for 15 min without causing cellular damage. With only 15 min of expression, most patches will unfortunately be devoid of the desired channel. If single channel recordings are required, less expression rather than a greater expression time will be utilized. For pQE plasmids (Qiagen), only very low levels of IPTG roughly correlate with expression level, so it is best to use medium (500 μM) to high amounts of IPTG while varying the expression time. Other systems, such as pBAD plasmids (Invitrogen) have a tight correlation between expression levels and inducer (L-arabinose) concentration, so varying the inducer concentration rather than time, or both, may be best employed.

14. A helpful "trick" for generating giant spheroplasts amenable to patch-clamp recording is to leave "snakes" overnight at 4°C (cold room or lab fridge) and make the spheroplasts the next day. This trick is useful to obtain soft spheroplasts after 4–5 min upon addition of EDTA, which are amenable to formation of a giga-ohm seal and single-channel recording.

15. The sample is split (at the "snake" pelleting stage) so that two different times of EDTA treatment can be used (5 min vs. 7.5 min). The 5 and 7.5 min are two suggested time points for those inexperienced with giant spheroplasts production. Conditions or strains may dictate lesser times, or perhaps greater times. For complete novices, note also that the procedure does not convert every *E. coli* cell (or string of non-septated cells, "snakes"), to a spheroplast. Patchable spheroplasts are minor elements formed against a background of non-converted material and other cell debris. Spheroplasts will form by ballooning out and from certain snakes.

16. The spheroplasts are diluted so that debris do not contaminate and stick to patch pipettes when a spheroplast is targeted for patching (contaminants stuck to the pipette will block seal

formation). Methods exist where spheroplasts are isolated via a sucrose gradient. However, the purification via a sucrose gradient is not great and much material is lost. A simpler approach is to merely dilute the sample. A sucrose gradient can be tried on the material before the Dilution Solution is utilized/added.

17. A "bubble number" is defined as a reading on a 10 cm^3 syringe connected via a piece of silicon tubing to a glass micropipette emerged in 100 % EtOH to which positive pressure is applied by the syringe plunger starting from the 10 cm^3 setting. The cc number (1–10 cm^3) reached by the plunger rubber head at which the first air bubbles are observed emerging from the tip of the pipette in EtOH is read as the "bubble number" for the particular micropipette and is characteristic of the pipette tip opening size. The smaller the bubble number, the smaller the opening of the pipette tip is. A small pipette tip diameter corresponds to a high resistance of the patch pipette (measured in MΩ) when the pipette is filled with the pipette solution and is immersed in the bath solution, while connected to the electrical circuit of the patch-clamp amplifier.
18. Higher spheroplast activation pressures when patch-clamping might imply that the EDTA treatment was not long enough. However, longer snakes require less EDTA treatment, as their cell wall is less substantial, resulting in more fragile spheroplasts, which may not be suitable for single channel recording because their membranes tend to be leaky.
19. After being introduced in the recording chamber (<1 ml volume), the spheroplasts settle on the bottom of the chamber. Usually the bottom of the chamber is made of glass (a thin microscope slide) and the spheroplasts tend to stick to it. Since picking the spheroplasts from the bottom by a patch pipette to form a giga-ohm seal risks breaking the tip of the pipette, it is better to catch them as they float above the bottom before settling down. One usually has approximately 2 min to manipulate the patch pipette to catch a spheroplast on the tip of the pipette by applying a slight suction before the spheroplasts attach to the bottom of the chamber. Plastic recording chambers can also be used (e.g., small cell culture dishes) that help overcome the problem of spheroplasts sticking to the chamber floor, as this does not occur in such chambers.
20. Laplace's law, which states that tension in a spherical object is equal to the product of the applied pressure times half the radius of the object ($T = p \times r/2$, measured in N/m), can be used to calculate the tension in a patch membrane so long as the membrane patch is visible. Visualizing the membrane patch can typically be done with differential interference contrast (DIC) microscopy or by fluorescence microscopy if the lipid

membrane is stained with a suitable fluorescent dye (Fig. 2b). High magnification 60× or 100× objectives are required. Once an image of the patch inside the pipette is digitally collected, it is a simple matter then to fit a circle to it and calculate its radius (Fig. 2d). Using a pressure recording device, the pressure within the patch pipette can also be recorded (Fig. 2c). From these values the tension in the membrane can be calculated using the aforementioned formula.

Acknowledgments

We wish to thank Maryrose Constantine as well as Drs. Anna Kloda, Evgeny Petrov, and Andrew Battle for critical reading and constructive comments on a draft of the manuscript. This work was supported by the Australian Research Council, the National Health and Medical Research Council of Australia, and Yamada Science Foundation.

References

1. Kubalski A, Martinac B (eds) (2005) Bacterial ion channels and their eukaryotic homologues. ASM, Washington, DC
2. Hamill OP et al (1981) Improved patch-clamp techniques for high-resolution current recording from cells and cell-free membrane patches. Pflugers Arch 391(2):85–100
3. Ruthe HJ, Adler J (1985) Fusion of bacterial spheroplasts by electric fields. Biochim Biophys Acta 819(1):105–113
4. Martinac B et al (1987) Pressure-sensitive ion channel in Escherichia coli. Proc Natl Acad Sci USA 84(8):2297–2301
5. Martinac B, Saimi Y, Kung C (2008) Ion channels in microbes. Physiol Rev 88(4):1449–1490
6. Blount P et al (1999) Mechanosensitive channels of bacteria. Methods Enzymol 294: 458–482
7. Martinac B (2011) Bacterial mechanosensitive channels as a paradigm for mechanosensory transduction. Cell Physiol Biochem 28(6): 1051–1060
8. Hamill OP, Martinac B (2001) Molecular basis of mechanotransduction in living cells. Physiol Rev 81(2):685–740
9. Nguyen T et al (2005) The effects of parabens on the mechanosensitive channels of *E. coli*. Eur Biophys J 34(5):389–395
10. Levina N et al (1999) Protection of Escherichia coli cells against extreme turgor by activation of MscS and MscL mechanosensitive channels: identification of genes required for MscS activity. EMBO J 18(7):1730–1737
11. Sukharev SI et al (1993) Two types of mechanosensitive channels in the Escherichia coli cell envelope: solubilization and functional reconstitution. Biophys J 65(1):177–183
12. Sukharev SI, Blount P, Martinac B, Kung C (1994) Functional reconstitution as an assay for biochemical isolation of channel proteins: application to the molecular identification of a bacterial mechanosensitive channel. Methods 6:51–59
13. Häse CC, Le Dain AC, Martinac B (1995) Purification and functional reconstitution of the recombinant large mechanosensitive ion channel (MscL) of Escherichia coli. J Biol Chem 270(31):18329–18334
14. Edwards MD et al (2005) Pivotal role of the glycine-rich TM3 helix in gating the MscS mechanosensitive channel. Nat Struct Mol Biol 12(2):113–119
15. Kloda A, Martinac B (2001) Structural and functional differences between two homologous mechanosensitive channels of Methanococcus jannaschii. EMBO J 20(8):1888–1896
16. Schumann U, Edwards MD, Rasmussen T, Bartlett W, van West P, Booth IR (2010) YbdG in Escherichia coli is a threshold-setting mechanosensitive channel with MscM activity. Proc Natl Acad Sci USA 107:12664–12669

17. Blount P et al (1996) Single residue substitutions that change the gating properties of a mechanosensitive channel in Escherichia coli. Proc Natl Acad Sci USA 93(21): 11652–11657
18. Saimi Y, Loukin SH, Zhou X-L, Martinac B, Kung C (1999) Ion channels in microbes. Methods Enzymol 294:507–524
19. Cortes DM, Cuello LG, Perozo E (2001) Molecular architecture of full-length KcsA: role of cytoplasmic domains in ion permeation and activation gating. J Gen Physiol 117(2): 165–180
20. Cordero-Morales JF, Cuello LG, Perozo E (2006) Voltage-dependent gating at the KcsA selectivity filter. Nat Struct Mol Biol 13(4): 319–322
21. Blount P, Sukharev SI, Moe PC, Martinac B, Kung C (1999) Mechanosensitive channels in bacteria. Methods Enzymol 294:458–482
22. Santos JS et al (2006) Molecular template for a voltage sensor in a novel K+ channel. I. Identification and functional characterization of KvLm, a voltage-gated K+ channel from Listeria monocytogenes. J Gen Physiol 128(3):283–292
23. Vásquez V et al (2008) A structural mechanism for MscS gating in lipid bilayers. Science 321(5893):1210–1214
24. Sukharev SI, Martinac B, Arshavsky VY, Kung C (1993) Two types of mechanosensitive channels in the E. coli cell envelope: solubilization and functional reconstitution. Biophys J 65: 177–183
25. Moe PC, Blount P, Kung C (1998) Functional and structural conservation in the mechanosensitive channel MscL implicates elements crucial for mechanosensation. Mol Microbiol 28(3):583–592
26. Börngen K et al (2010) The properties and contribution of the Corynebacterium glutamicum MscS variant to fine-tuning of osmotic adaptation. Biochim Biophys Acta 1798(11): 2141–2149
27. Kloda A, Martinac B (2001) Molecular identification of a mechanosensitive channel in archaea. Biophys J 80(1):229–240
28. Petrov E, Palanivelu D, Rohde PR, Constantine M, Minor D, Martinac B (2011) Patch-clamp characterization of the MscS-like mechanosensitive channel from Silicibacter pomeroyi. Biophysical Society Meeting Abstracts. Biophys J Suppl 100(3):281a, Abstract, 1532-Pos
29. Li Y et al (2002) Ionic regulation of MscK, a mechanosensitive channel from Escherichia coli. EMBO J 21(20):5323–5330
30. Kuo MM et al (2008) The desensitization gating of the MthK K+ channel is governed by its cytoplasmic amino terminus. PLoS Biol 6(10):e223
31. Sakmann B, Neher E (1995) Single-channel recording. Plenum, New York
32. Sokabe M, Sachs F, Jing ZQ (1991) Quantitative video microscopy of patch clamped membranes stress, strain, capacitance, and stretch channel activation. Biophys J 59(3):722–728
33. Yoshimura K et al (1999) Hydrophilicity of a single residue within MscL correlates with increased channel mechanosensitivity. Biophys J 77(4):1960–1972
34. Haswell ES, Phillips R, Rees DC (2011) Mechanosensitive channels: what can they do and how do they do it? Structure 19(10): 1356–1369

Chapter 24

Electrophysiological Characterization of Bacterial Pore-Forming Proteins in Planar Lipid Bilayers

Owen S. Mapingire, Beau Wager, and Anne H. Delcour

Abstract

Together with patch-clamp, the planar lipid bilayer technique is one of the electrophysiological approaches used to study the biophysical properties of bacterial pore-forming proteins. Electrophysiological studies have provided important insight into the mechanistic details underlying the function of this class of proteins. Although there are different apparatus designs and variations to the process of obtaining channel recordings, the general architecture of a planar lipid bilayer setup involves two compartments filled with an ionic solution and separated by a septum with a micro-aperture, where a phospholipid bilayer is formed, and an amplifier used to clamp the membrane potential and record currents. Bacterial outer membrane porins and translocons, among others, can be reconstituted in this bilayer and their electrophysiology probed in different physicochemical conditions or through functional assays with substrates or potential modulators. This chapter describes specifically the reconstitution of detergent purified outer membrane pore-forming proteins into artificial lipid membranes using a laboratory customized planar lipid bilayer apparatus and the subsequent recording of channel activity under voltage clamp.

Key words: Channel, Porin, Planar lipid bilayer, Reconstitution, Voltage, Current, Membrane

1. Introduction

Bacterial pore-forming proteins represent a class of proteins where electrophysiology has been applied to study the movement of ions, solutes, or substrates across a biological membrane. As an electrophysiology method, the planar lipid bilayer technique, also known as the black lipid membrane technique (BLM), offers a unique alternative to probe the mechanistic details of how these pore-forming transporters work. While some techniques such as liposome swelling assays and dye leakage assays are useful in demonstrating general pore function, electrophysiology offers the unique benefit of real time single-molecule analysis in mapping out the intrinsic operations of a pore-forming protein. It reveals that

Anne H. Delcour (ed.), *Bacterial Cell Surfaces: Methods and Protocols*, Methods in Molecular Biology, vol. 966,
DOI 10.1007/978-1-62703-245-2_24, © Springer Science+Business Media New York 2013

channels are dynamic entities, oscillating between closed and open states, and characterized by a trademark kinetic signature under specific conditions. The technique also allows structure function relationship and mechanistic studies by investigating how the channel signature is affected by mutation, physicochemical conditions, protein substrates, ligands, or any other factors.

This technique's usefulness is aptly demonstrated in Gram-negative bacteria, where the outer membrane porins serve as conduits for ions, antibiotics, nutrient molecules, and other hydrophilic solutes (1). For example, planar lipid bilayer studies have elucidated the mechanism of selectivity and specific substrate transport (2–4), and uncovered other important phenomena such as modulation of porins by voltage, pH, and ligands (5–8). In addition, through the contribution of planar lipid bilayer electrophysiology, the molecular mechanism for the interaction of some types of antibiotics with general diffusion porins has been uncovered (9, 10).

Pore-forming proteins of bacterial origins, such as colicins and toxins, have also been extensively studied with planar lipid bilayer electrophysiology (11). This approach is providing valuable information on mechanisms of channel formation, interactions with lipids, channel blocking, conformational changes, and substrate translocation (12–15). The combination of planar lipid bilayer electrophysiology and streptavidin–biotin biochemistry is even providing molecular details on the translocation process, for example in the anthrax toxin channel (16, 17). A similar strategy has advanced our understanding of conformational changes leading to channel formation, for example in colicin Ia (18).

While most work has centered on porins and secreted pore-forming proteins, to date the utilization of the planar lipid bilayer technique has been extended to investigate even some of the more noncanonical channel forming proteins such as translocons. Several types of secretion systems have been uncovered so far in Gram-negative bacteria (19). These systems are macromolecular complexes of many different proteins that assemble to form a secretion apparatus across the bacterial cell envelope. All secretion systems require a designated translocon that forms a pore in the outer membrane or the target membrane to shuttle its protein substrates (20, 21). Although the physiological role of the translocon is not to provide a path for ion flux, the use of electrophysiology to assay translocon function and/or assembly, and possibly also to monitor the actual translocation process, has been reported in many different cases. For example, planar lipid bilayer electrophysiology has shed insight into the process of pore opening of TolC, a promiscuous *E. coli* outer membrane protein that serves as the translocon of the type 1 secretion system and a member of a tripartite drug efflux pump (22–24). Other examples include studies on the secretion systems of uropathogenic *E. coli* (UPEC) (25), *Bordetella pertussis* (26, 27), and *Hemophilus influenzae* (28), all involved in the translocation of cell surface appendages or adhesins that act as critical virulence factors. In all these cases, electrophysiology has demonstrated the dynamic nature of the whole protein, and has provided insight in the structural components that underlie or modulate the basic opening and closing activity of these pores.

Several protocols have been reported in the literature for electrophysiological investigations using planar lipid bilayers (29, 30). For many of them, the actual bilayer formation requires the "painting" of a hole in a partition with a mixture of lipids and organic solvents. In this chapter, we describe an adaption of the Montal and Mueller technique (31), which achieves the formation of essentially solvent-free bilayers. The major difference with other methods is that the lipids are first allowed to form a monolayer at the air–water interface, prior to being apposed in a bilayer over the hole in the partition. Since the formation of a stable bilayer is critical to the experiments, we will highlight factors that affect the success of forming stable lipid bilayer membranes. As such, it is critical to optimize the conditions of bilayer formation and be able to comprehensively trouble shoot problems. This chapter also addresses some of the common challenges that typically arise in planar lipid bilayer experiments and suggests systematic ways to address them. Since our method utilizes a customized chamber for the experiments (Fig. 1), some of the challenges and solutions presented here might not be relevant to other setups, commercial or otherwise. However, we have attempted to be as comprehensive as possible in outlining the common themes to general planar lipid bilayer electrophysiology experiments.

2. Materials

All solutions should be freshly made using deionized water and filter-sterilized. Solutions can be stored in the fridge at 4°C and used repeatedly. All procedures described in this protocol are performed at room temperature. It is recommended to retrieve the solution from the fridge an hour prior to use so it can attain the ambient temperature. The following protocol is based on the use of an Axopatch 200B amplifier from Molecular Devices and a custom-made chamber. There are commercially available bilayer chambers and equipment alternatives to those mentioned here.

2.1. Electrophysiology Rig

1. Vibration isolation table.
2. Circular level.
3. Faraday cage with a gate for easy access.
4. Bilayer headstage: CV203BU (Molecular Devices).
5. Amplifier: Axopatch 200B (Molecular Devices).

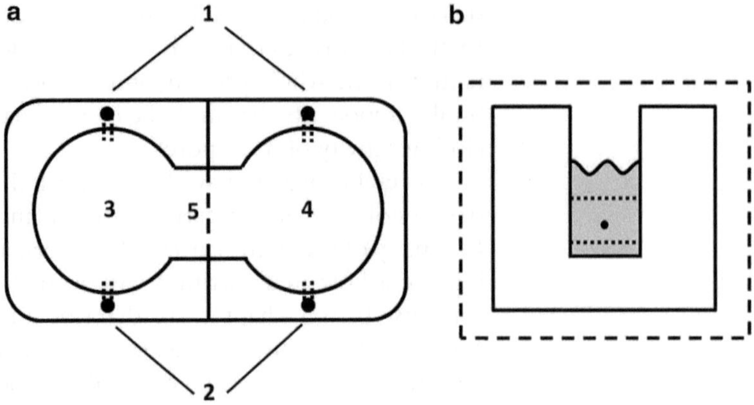

Fig. 1. Cartoon representations of the planar lipid bilayer chamber. (**a**) *A top view* illustration of the planar lipid bilayer chamber. The numbers represent the following: (1) entrances to the two conduits drilled in the Teflon where syringe tubing is inserted (syringe conduits); (2) entrances to the two conduits where the electrodes are inserted (electrode conduits); (3) *cis* chamber-half; (4) *trans* chamber-half; (5) PTFE film that separates the two chamber halves and their respective buffers. (**b**) *A side view* of one half-chamber filled with buffer. The *dashed square* depicts the PTFE film covering the open side of the half-chamber. The volume filled in *gray* color represents the buffer in the half-chamber. The *dotted lines* mark the ideal range of positions for the hole in relation to the half-chamber and buffer surface.

6. Oscilloscope.
7. Analog/digital converter: ITC-18 Computer Interface.
8. Computer capable of connection with analog/digital converter.
9. Electrophysiology magnetic stirrer.
10. Micro magnetic stirring bars (5 mm).
11. Ag/AgCl pellet electrodes.
12. Two 1 mL syringes.
13. At least 10 cm in length of plastic tubing to be connected to the syringes.
14. Grounding cables with banana plug or alligator clips.
15. Soldering gun and solder for making grounding cables and electrodes, if required.
16. Data acquisition program.
17. Data analysis program.

2.2. Preparation of the Chamber

1. A custom designed Teflon chamber (Fig. 1).
2. A metal casing designed to house the chamber and secure it tightly.
3. A roll of polytetrafluoroethylene (PTFE) film of 10 μm thickness (Goodfellow).
4. A dissecting microscope.

5. High Frequency Spark Tester PPM MK3 (Buckleys Ltd.).
6. Stands and clamps to hold the spark tester and the sandwiched PTFE square.
7. Vacuum grease.
8. Sterile cotton swabs.
9. Capillary tube dropper with 1 mm diameter capillary tubes.
10. Pretreatment solution: 1% hexadecane in pentane (HPLC High purity solvent grade). Transfer 495 μL of pentane into a 1.5 dram glass vial. Add 5 μL of hexadecane into the same vial and mix. Store at 4°C until use.

2.3. Electrophysiology

1. Chamber Buffer: 1 M KCl, 5 mM HEPES, pH 7.2. Weigh 74.55 g of KCl; 1.19 g HEPES, and place it in a beaker (see Note 1). Measure out 900 mL water and add to beaker. Thoroughly dissolve this mixture on a magnetic stir plate using a stir bar. With a pH meter, measure the pH of the solution and adjust to pH 7.2 using KOH (see Note 2). Make up the solution to 1 L, filter-sterilize and store at 4°C until use.
2. Lipid solution: 5 mg/mL "azolectin" in pentane. "Azolectin" is a specific preparation of phosphatidylcholine from Sigma-Aldrich (1,2-Diacyl-*sn*-glycero-3-phosphocholine Type II-S). Weigh 5 mg of azolectin granules and transfer to a 1.5 dram glass vial (see Notes 3 and 4). Add 1 mL of pentane to vial and stir gently until granules are dissolved (see Note 5). Keep lipid solution on ice when in use and store at 4°C. Use within 1 week.
3. Purified protein sample.

3. Methods

3.1. Building the Rig

These steps only need to be performed before the first experiment. Once all of the equipment is properly connected and positioned, they can remain indefinitely.

1. Start by finding an isolated location to place the vibration isolation table, where traffic and disturbances are minimized. The table should be leveled for all experiments and should be floated (see Note 6).
2. Surround the vibration isolation table with a Faraday cage with openings large enough for the bilayer headstage to pass through inside the Faraday cage.
3. Nearby the vibration isolation table, connect the oscilloscope to the amplifier. Connect the amplifier to the analog/digital converter. Connect the analog/digital converter to a computer installed with the electrophysiology data acquisition software (see Note 7).

4. Place the magnetic stirrer on the vibration isolation table. Keep the magnetic stirrer controller outside of the Faraday cage.
5. Connect the bilayer headstage to the amplifier and move it into the Faraday cage through an opening. Position the headstage above the magnetic stirrer.
6. Connect tubing to two 1 mL syringes and position these above the magnetic stirrer (see Note 8).
7. Attach the signal and ground electrodes to the headstage.
8. Ground the vibration isolation table and the magnetic stirrer using grounding cables (see Note 7).

3.2. Preparation of the Chamber

3.2.1. Preparation of PTFE Sheet Squares

1. Cut squares of the 10 μm PTFE sheet roughly 1 cm larger than the width and height of the bilayer chamber. Wear gloves while handling the PTFE sheets and cut the squares on a sterile surface that is free of dust. Place the PTFE squares into a large Petri plate or other storage item capable of protection from dust and moisture (see Note 9).
2. Sandwich a PTFE square between two plastic or glass panels with holes, such that only the center of the PTFE square is accessible.
3. Structurally secure the spark tester such that it is horizontally leveled. Clamp the PTFE sandwich such that the accessible region of the PTFE film is ~2 cm away from the barrel tip of the spark tester. Position a grounded needle on the opposite side of the PTFE film ~1 cm away. The path from the spark tester barrel through the PTFE film to the grounded needle should be a straight line.
4. Set the spark tester at 45 kV and gently squeeze the trigger to fire a single spark arc through the PTFE sheet, forming a hole of between 50 and 100 μm.
5. Inspect the PTFE film square using the dissecting microscope to verify that one, and only one, hole has formed in the square, and that the hole has the appropriate size (see Note 10).

3.2.2. Assembling the Bilayer Chamber

1. With a clean cotton swab apply moderate amount of vacuum grease to the chamber face where the PTFE film will be placed on both chamber halves (see Fig. 1b). Avoid the inner edge (see Note 11) and ensure a thin layer of grease fully surrounds the outer edges of the rim surface.
2. Take the punctured PTFE film square and carefully place it on one greased chamber half. Place the hole centered just above the floor of the half-chamber and take care that the hole does not make contact with grease (see Fig. 1b, Note 12). The PTFE film should remain taut across the half-chamber face.

3. Carefully press the second half-chamber on top of the first one. Make sure the chamber halves are aligned and adhere together. Do not press too hard as this may cause movement of the PTFE film and the location of the hole.
4. Secure the vacuum grease seal by placing the adhered chamber halves in a metal casing and by clamping the metal casing tightly. This is now a complete bilayer chamber (see Note 13).
5. Place the chamber on the dissecting microscope at an angle such that the hole in the PTFE film can be observed. Verify that the hole is not covered by vacuum grease or is otherwise inaccessible.

3.2.3. Pretreatment of the Hole

This procedure coats the PTFE film hole with a hydrophobic surface critical for the formation of a stable lipid bilayer.

1. Dip a capillary tube connected to a dropper into the vial of pretreatment solution. The solution should enter the tube via capillary action alone.
2. While observing the chamber under a dissecting microscope, focus on the PTFE film hole and position the tip of the capillary tube in close proximity to the hole (see Note 14).
3. Apply one or two drops of the pretreatment solution to the film. Observe the solution spread across the PTFE film, enter the hole, and spread across the opposite side of the PTFE film.
4. Allow 5 min for the pretreatment solution to set before continuing.

3.3. Electrophysiology

3.3.1. Forming a Stable Lipid Bilayer

1. Turn on the Axopatch 200B amplifier at least 1 h prior to use.
2. Set the Axopatch 200B to these settings:

   Zap: 0.5 ms.

   Pipette Capacitance Comp.: Minimum (fully counterclockwise).

   Series Resistance Comp. % Prediction: 0%, switch OFF.

   Series Resistance Comp. % Correction: 0%, switch OFF.

   Series Resistance Comp. Lag: 1 μs.

   Whole Cell Cap.: 0 pF.

   Series Resistance: 0 MΩ.

   Holding Command 0 mV, ×1, switch OFF.

   Seal Test: switch OFF.

   Meter: I.

   Mode: V-Clamp.

   Config: Whole Cell ($\beta = 1$).

   Output Gain: $\alpha = 1$.

   Lowpass Bessel Filter: 1 kHz.

   Leak Subtraction: ∞ MΩ, switch OFF.

3. Move the chamber to the vibration isolation table and place it on top of the magnetic stirrer. Ground the chamber with a grounding cable.
4. Add 1.5 mL of chamber buffer to each half-chamber.
5. Add 5 μL of the lipid solution to the center of the buffer surface in each half-chamber.
6. Turn off the amplifier while introducing the electrodes into the chamber in order to avoid excessive amount of noise. Insert the electrodes into the electrode conduits (see Fig. 1a), such that they reach the bottom of the chamber, with the Ag/AgCl pellet fully submerged in solution. Place the signal electrode on the side selected for protein addition. This will be defined as the *cis* side. Place the ground electrode on the opposite side, which shall be defined as the *trans* side. In this configuration, the *trans* side is grounded and the desired voltage is applied to the *cis* side relative to the *trans* side.
7. Insert the syringe tubing into the two syringe conduits.
8. Turn on the oscilloscope. Move the position knob of the oscilloscope to set the trace to coincide with the ground level.
9. Turn on the amplifier. The introduction of the electrodes into the buffer and the communication of the buffers from the two chamber halves will create so-called junction potentials that lead to a deflection of the current signal trace away from the zero current level (see Note 15). With time the amplifier will electronically compensate for these junction potentials.
10. Wait for 5 min while monitoring the current signal either on the oscilloscope or with the data acquisition software and ensure that it equilibrates to near zero current level with only mild fluctuations. If this does not happen, proceed to manually move the current signal near the zero current level using the Pipette Offset dial.
11. Lower the chamber buffer in one half-chamber below the PTFE film hole using one of the syringes. The current signal will move exactly to the zero current level with no fluctuations when the buffer reaches below the hole, because the electrical circuit is interrupted due to the lack of communication of the ionic solutions through the hole. Quickly, raise the chamber buffer back above the hole and to its original volume.
12. If a stable planar lipid bilayer has formed, the current signal will remain at the zero current level. If a stable bilayer does not form, or ruptures after formation, both indicated by rapid fluctuation of the current signal and return to a large current value, repeat step 11 (see Note 16).
13. Adjust the Output Gain dial to $\alpha = 5$. This will give a total amplification $\alpha \times \beta$ of 5 pA/mV.

14. To test the strength of the lipid bilayer, apply a steady voltage (≥50 mV) by adjusting the Holding Command to the voltage required and flipping the switch to either (+) or (−). The voltage being applied can be monitored directly by adjusting the Meter switch to V_m (see Note 17).
15. Continue applying voltage for at least 3 min before switching it off. If the bilayer ruptures, repeat step 11.
16. If the bilayer remains stable, verify that multiple bilayers have not formed by breaking the bilayer using the Zap trigger on the amplifier. When the bilayer breaks, the current signal trace will jump out of range on the oscilloscope. Return to step 11 to reform the membrane and continue by disregarding step 16 and proceeding to Subheading 3.3.2. If the signal remains at the zero current level, this indicates that the bilayer did not break, and it is likely that multiple bilayers have formed. To break a multiple bilayer, use plastic Pasteur pipette to blow buffer at the PTFE film. If this fails proceed to remaking a new chamber. If this continues to be a problem over a day or two, make new buffers and new lipid solutions.

3.3.2. Recording Channel Activity

1. Add the micro stir bars to the *cis* half of the bilayer chamber.
2. Turn on the magnetic stirrer and verify that the stir bar is spinning steadily in the center of the *cis* half-chamber (see Note 18).
3. Start recording with the data acquisition software on the computer.
4. Add the protein sample to the *cis* half-chamber (see Notes 19 and 20).
5. Apply a voltage to the chamber.
6. Observe the insertion of a protein channel, indicated by deflections in the current signal, using either the oscilloscope or the data acquisition program (see Fig. 2 and Notes 21 and 22).
7. Turn the magnetic stirrer off after the desired number of insertions occurs.
8. Continue electrophysiological recordings using the data acquisition program in a manner consistent with the type of experiment being run.

3.4. Cleaning Up

1. Turn off the amplifier and oscilloscope.
2. Remove stir bars, electrodes, syringe tubing, and the chamber grounding cable.
3. Discard chamber buffer and wash the inside of the chamber thoroughly with distilled water followed by ethanol. Plastic dispensing bottles can be used to gently clean the PTFE film.
4. Allow the chamber to air-dry and return to Subheading 3.2.3 to begin a new experiment (see Note 23).

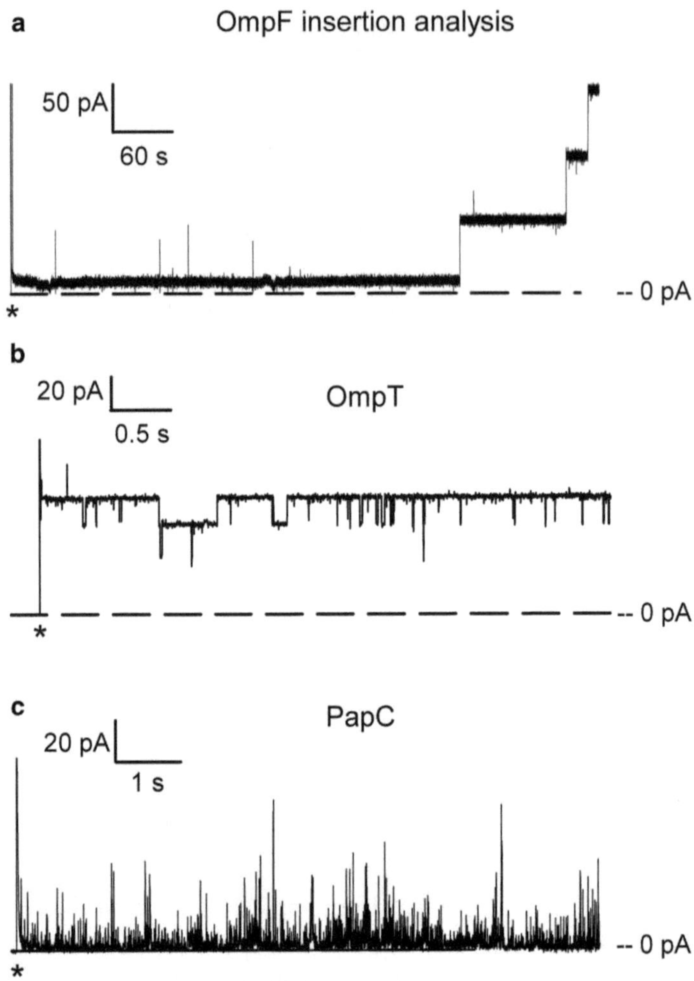

Fig. 2. Examples of typical traces obtained from bacterial pore-forming proteins. (**a**) Example of consecutive insertions of open channels. The trace is a 10 min recording obtained at +50 mV after the addition of purified *E. coli* OmpF porin. When voltage is applied (at the time marked by *asterisk*) the current signal remains near the 0 pA level until the first insertion occurs. Since the protein exists predominantly in an open state, when the first insertion occurs, ions pass immediately through the inserted open channel, and the signal jumps to a higher current level where it remains until another insertion occurs. Each stepwise deflection to a higher current level is due to channel insertion. The size of the deflections reflects whether a single protein (a trimer in the case of OmpF) inserts, or whether many proteins insert simultaneously. In this case, the three observed upward deflections were due to the insertions of a single trimer. By convention, channel opening (increase in current) is shown by upward deflections at positive voltages, and downward deflections at negative voltages. (**b**) Example of a recording from an inserted channel with high open probability (i.e., an open channel). This trace was obtained at +50 mV after the addition of purified *V. cholerae* OmpT porin, and it typifies the behavior of bacterial porins. When voltage is applied (at the time marked by *asterisk*) the current signal shifts to a higher level. This is the full open level and marks the highest amount of current that can pass through this trimeric protein at any time. However, channels are dynamic and spontaneously oscillate between open and closed states. On this trace at positive voltage, closing events are seen as *downward* transitions, and opening events as *upward* transitions. Note that the consecutive closures of two channels will appear as step-wise transitions of

4. Notes

1. The chamber buffer's ionic strength can be varied. We have used buffers with ionic strengths ranging from 50 mM to 3 M KCl. Although a 1 M salt buffer does not represent physiological conditions, it is often used in the initial characterization of new pore-forming proteins because of many advantages such as good current signal strength and signal-to-noise ratio, ease in forming stable bilayers, and promotion of channel insertions. In our experience, proteins tend to insert less readily in buffers with ionic concentrations lower than 1 M. Other salts, such as NaCl, can also be used.
2. The pH of the chamber buffer solution can be varied. However, basic pH tends to present challenges in forming a stable bilayer. Lipid bilayers are equally stable in acidic and neutral pH conditions.
3. Check the quality of the azolectin granules before dissolving them. On delivery of a fresh batch, it is critical to check the color and texture of the granules. Ideal granules should be bright yellow and soft. Dark and hard granules should be discarded. Azolectin is a rough mix of 24% phosphatidylcholine, 39% phosphatidylethanolamine, 19% phosphatidylserine, 6% phosphatic acid, 1% phosphatidylinositol, 9% lyso derivatives, and 2% other phospholipids (32).
4. If a bilayer made of a single type of lipid is required, it is suggested to use a 1,2 diphytanoyl-*sn*-glycero-3-phosphocholine (DPhPC) (Avanti Polar Lipids). A substantial amount of planar lipid bilayer electrophysiology has been done with azolectin and DPhPC. If necessary, it can be mixed with other types of 1,2 diphytanoyl-*sn*-glycero-3-phospholipids.
5. The quick evaporation of pentane at the air–water interface in the planar lipid bilayer chamber is required to allow the formation of a solvent-free lipid bilayer. Alternatively, hexane, decane, or squalene can be used as lipid solvents. Consideration should be made on the choice of lipid solvent, as evidence exists that it can alter channel behavior in some cases (33).

Fig. 2. (continued) identical sizes. (**c**) Example of a recording from an inserted channel with low open probability. This trace was obtained at +50 mV after the addition of purified *E. coli* usher PapC, and typifies the behavior of closed pore proteins such as translocons. Closed pores have a low open probability and the current signal will stay mostly at a level close to 0 pA but shifts to higher current levels as opening transitions. When voltage is applied (at the time marked by *asterisk*), the current signal does not shift far from 0 pA, and often dwells at this low current level during the trace because the channel spends most of its time in a closed state. Channel openings randomly occur and are seen as upward deflections of the current traces. For this particular channel, openings tend to be very brief and appear as flickering events.

6. To check that the table is floating properly, apply gentle pressure on the table. The table should slightly depress and recover to the original level. If it feels stiff to the pressure, this means that the table is not floating and adjustments should be made according to the vibration isolation table manual. A table that is not properly floated will lead to a lot of noise in the traces, and difficulty in forming the bilayer, since it will not compensate for vibrations introduced during manipulations.
7. In case of difficulties with connecting the different electronic instruments to each other and grounding the setup, consult some tutorials available on the web, in particular the "Axon Guide," published by Axon Instruments, and the Resource Center from Molecular Devices. Note that these tutorials tend to be focused on patch-clamping techniques, but many of the same concepts apply in planar lipid bilayer electrophysiology. It is advisable to perform a functional checkout of new equipment, especially the amplifier. The Axopatch 200B comes with a detailed manual that should be followed when performing the functional checkout.
8. It is most effective to have a rod or plate to securely clamp both the headstage and the syringes above the magnetic stirrer. Movement or vibrations of either the headstage or the syringes can raise electronic noise during experiments.
9. It is recommended to make numerous pre-cut squares and store them in a clean Petri dish. These can be retrieved when needed in future experiments. On handling, use tweezers as they tend to stick to gloves. Use an air duster to eliminate dust or other physical particles that can attach to the PTFE film.
10. The hole formed by the spark tester should be a regular circular shape with an ideal diameter between 50 and 100 μm. The spark gun trigger is sensitive, and the duration of the trigger controls to some extent the diameter of the hole. One has to use trial and error to gauge how long the trigger should be pulled to obtain a hole of the desired size (measured using a dissecting microscope with a stage micrometer). However, if the duration of pulling the trigger is too long, burning of the hole or the formation of multiple holes can occur. The size and shape of the hole affects the properties of lipid bilayers formed on the hole. Discard the square and restart if the hole is not round, the hole has radiating tears, or there are multiple holes.
11. If grease is smeared too close to the inner edge of the chamber, some of it might dissolve in the experimental buffer and potentially contaminate the solution and plug the hole.
12. The position of the hole on the PTFE square is vital to bilayer formation. During the buffer lowering and raising step, a planar lipid bilayer is formed when the air–water interface of the buffer sweeps over the hole. Having the hole too low can result

in disruption of the newly formed bilayer upon raising the solution. Placing it too high or close to the air–buffer interface might not allow proper layering of the lipids within the hole when the solution is lowered.

13. The grease helps create the seal between the chamber halves, but mechanical force needs to be applied to tightly adhere the two pieces together. If not properly sealed, buffer leakage will occur. Buffer leakage is usually not large scale enough to notice visually, but it will result in an open circuit and make electrophysiology data recording impossible. To diagnose a suspected leak with the oscilloscope, remove the buffer from one half-chamber until the buffer surface is below the PTFE hole. If a leak is present, the current signal line will continue to fluctuate instead of returning to the zero current level during this step. If a leak is confirmed it is necessary to make a new chamber.
14. Locating the hole within the chamber under the dissecting microscope requires patience. The task becomes easier with experience. It is easiest to view the hole while the chamber is tilted at a 30–45 ° angle on the stage.
15. After turning on the amplifier prior to protein addition and in preparation to forming the bilayer (equilibration step), the current signal trace should never be fixed at the zero current level on the oscilloscope, because a large amount of current is passing through the open hole. The current signal should be out of range or fluctuating to or away from the zero current level. A fixed stationary line at the zero current level means there is blockage at the hole resulting in a strong electrical resistance. This blockage may be a contaminating foreign particle attached to the PTFE film or a bubble that moved into the hole during buffer addition. The blockage may be removed by rapid and repeated blowing of the buffer near the film with a plastic Pasteur pipette. If you are unable to remove the blockage, clean the chamber, and restart at Subheading 3.2.3.
16. Formation of a stable bilayer is critical when performing planar lipid bilayer experiments. When having limited success forming stable bilayers you can add more lipids or empty the chamber and repeat the hole pretreatment process.
17. When applying voltage to the bilayer, the current signal should not move from the zero current level by more than a few pA, because no protein channel has yet been reconstituted in the bilayer. If the current signal moves away from the zero current level by more than a few pA, either the bilayer has not fully formed over the hole or there are contaminating proteins (possibly from a prior experiment) in the chamber that have inserted into the bilayer. This is also the case if one observes a patterned behavior indicative of a channel activity. In this case, the chamber needs to be cleaned thoroughly.

18. Breakage of bilayers can occur after the stir bar has just started spinning. It is important to choose an appropriate stirrer speed that allows the stir bar to generate sufficient turbulence to aid protein insertion, but not strong enough to destabilize the bilayer. If the chamber and stirrer are aligned properly the bar will spin smoothly in the compartment center and not cause a disruptive force. If the stir bar is disrupting the bilayer, the problem will appear as frequent spiky noise events on the recording.
19. The amount of protein to be added to the *cis* half-chamber has to be determined by trial and error, and depends on the purpose of the experiment, i.e., the need for single channel or multiple channels. A pure protein sample is typically first diluted in a buffer solution containing the same concentration of detergent as used for the purification. Insertion rates vary among pore-forming proteins. In our experience, porins, in particular *E. coli* OmpF, insert with ease, and nanogram quantities added to the *cis* half-chamber are usually sufficient. On the contrary, translocons, which are often larger molecular complexes than porins, tend to require much larger amounts, in the order of a few micrograms.
20. There are different ways of preparing the protein samples that might affect the success rate of insertions. The age of the protein preparation and the number of freeze-thaw cycles are an important consideration. It is advisable to aliquot the purified protein such that it undergoes only one freeze-thaw cycle before being used in the planar lipid bilayer setup. The type of detergent and the extent of the dilution (which impacts the protein to detergent ratio) are also factors. The lipid composition of the bilayer may also impact the ability of the protein to insert.
21. The insertion of a porin trimer is seen as a large current deflection to a steady level, because the three channels insert in an open state (see Fig. 2). Insertion of closed channels may not be seen readily per se (sometimes a small deflection of the baseline level occurs), but evidence that a channel has inserted is given by the channel activity, i.e., the succession of opening and closing transitions. Note that membrane disruptions can sometimes appear as channel activity. But they will be more random, they have low reproducibility, and they typically show a range of current sizes. They may also be associated with a drift in the baseline level.
22. The number of insertions into a bilayer can be controlled by changing the amount of protein added to the *cis* half-chamber. If the objective is to perform single channel analysis, switching off the stirrer after the first insertion can help reduce the chance of more insertions. Low insertion rates can be overcome with higher protein concentration, a larger hole size, and longer periods of time waiting for channel insertion.

Typically, insertions occur within the first 5 min. If this is not the case, add more protein to the *cis* half-chamber (or to both chamber halves if protein orientation is not an issue in the experimental design), or increase the membrane potential. If still no insertion occurs by 10 min, break the membrane using the Zap trigger on the amplifier, and form a new bilayer in the presence of the already added protein. If the problem persists, consider changing conditions such as lipids, detergents, dilution factors, and buffer composition.

23. The chamber apparatus can be reused with the same PTFE film, but it is recommended only when using the same protein. After cleaning at the end of an experiment, the hole can be pretreated again and another experiment started. If there is need to work with another protein, dismantle the chamber completely, dispose of the PTFE film, bath-sonicate the chamber halves in a 50:1 water:alcohol mixture for at least 30 min. Some chemicals can stick to the Teflon walls of the chamber, resulting in contamination of future experiments if the chamber is not sonicated. If working with viscous substances like polyethylene glycols (PEGs) in the chamber, sonicate the chamber each time before starting a new experiment.

References

1. Nikaido H (2003) Molecular basis of bacterial outer membrane permeability revisited. Microbiol Mol Biol Rev 67:593–656
2. Dargent B, Hofmann W, Pattus F, Rosenbusch JP (1986) The selectivity filter of voltage-dependent channels formed by phosphoporin (PhoE protein) from *E. coli.* EMBO J 5:773–778
3. Kullman L, Winterhalter M, Bezrukov SM (2002) Transport of maltodextrins through maltoporin: a single-channel study. Biophys J 82:803–812
4. Saint N, Lou KL, Widmer C, Luckey M, Schirmer T, Rosenbusch JP (1996) Structural and functional characterization of OmpF porin mutants selected for larger pore size. II. Functional characterization. J Biol Chem 271:20676–20680
5. Van Gelder P, Saint N, Phale P, Eppens EF, Prilipov A, van Boxtel R, Rosenbusch JP, Tommassen J (1997) Voltage sensing in the PhoE and OmpF outer membrane porins of *Escherichia coli*: role of charged residues. J Mol Biol 269:468–472
6. Nestorovich EM, Rostovtseva TK, Bezrukov SM (2003) Residue ionization and ion transport through OmpF channels. Biophys J 85: 3718–3729
7. Baslé A, Iyer R, Delcour AH (2004) Subconductance states in OmpF gating. Biochim Biophys Acta 1664:100–107
8. Delcour AH (2003) Solute uptake through general porins. Front Biosci 8:d1055–d1071
9. Danelon C, Nestorovich EM, Winterhalter M, Ceccarelli M, Bezrukov SM (2006) Interaction of zwitterionic penicillins with the OmpF channel facilitates their translocation. Biophys J 90:1617–1627
10. Nestorovich EM, Danelon C, Winterhalter M, Bezrukov SM (2002) Designed to penetrate: time-resolved interaction of single antibiotic molecules with bacterial pores. Proc Natl Acad Sci U S A 99:9789–9794
11. Parker MW, Feil SC (2005) Pore-forming protein toxins: from structure to function. Prog Biophys Mol Biol 88:91–142
12. Zakharov SD, Kotova EA, Antonenko YN, Cramer WA (2004) On the role of lipid in colicin pore formation. Biochim Biophys Acta 1666:239–249
13. Wu Z, Jakes KS, Samelson-Jones BS, Lai B, Zhao G, London E, Finkelstein A (2006) Protein translocation by bacterial toxin channels: a comparison of diphtheria toxin and colicin Ia. Biophys J 91:3249–3256

14. Karginov VA, Nestorovich EM, Schmidtmann F, Robinson TM, Yohannes A, Fahmi NE, Bezrukov SM, Hecht SM (2007) Inhibition of S. *aureus* alpha-hemolysin and *B. anthracis* lethal toxin by beta-cyclodextrin derivatives. Bioorg Med Chem 15:5424–5431
15. Collier RJ, Young JA (2003) Anthrax toxin. Annu Rev Cell Dev Biol 19:45–70
16. Basilio D, Jennings-Antipov LD, Jakes KS, Finkelstein A (2011) Trapping a translocating protein within the anthrax toxin channel: implications for the secondary structure of permeating proteins. J Gen Physiol 137:343–356
17. Zhang S, Finkelstein A, Collier RJ (2004) Evidence that translocation of anthrax toxin's lethal factor is initiated by entry of its N terminus into the protective antigen channel. Proc Natl Acad Sci U S A 101:16756–16761
18. Qiu XQ, Jakes KS, Kienker PK, Finkelstein A, Slatin SL (1996) Major transmembrane movement associated with colicin Ia channel gating. J Gen Physiol 107:313–328
19. Holland IB (2010) The extraordinary diversity of bacterial protein secretion mechanisms. Methods Mol Biol 619:1–20
20. Karuppiah V, Berry JL, Derrick JP (2011) Outer membrane translocons: structural insights into channel formation. Trends Microbiol 19:40–48
21. Mattei PJ, Faudry E, Job V, Izore T, Attree I, Dessen A (2011) Membrane targeting and pore formation by the type III secretion system translocon. FEBS J 278:414–426
22. Zgurskaya HI, Nikaido H (2000) Multidrug resistance mechanisms: drug efflux across two membranes. Mol Microbiol 37:219–225
23. Andersen C, Hughes C, Koronakis V (2002) Electrophysiological behavior of the TolC channel-tunnel in planar lipid bilayers. J Membr Biol 185:83–92
24. Andersen C, Koronakis E, Bokma E, Eswaran J, Humphreys D, Hughes C, Koronakis V (2002) Transition to the open state of the TolC periplasmic tunnel entrance. Proc Natl Acad Sci U S A 99:11103–11108
25. Mapingire OS, Henderson NS, Duret G, Thanassi DG, Delcour AH (2009) Modulating effects of the plug, helix, and N- and C-terminal domains on channel properties of the PapC usher. J Biol Chem 284:36324–36333
26. Clantin B, Delattre AS, Rucktooa P, Saint N, Meli AC, Locht C, Jacob-Dubuisson F, Villeret V (2007) Structure of the membrane protein FhaC: a member of the Omp85-TpsB transporter superfamily. Science 317:957–961
27. Meli AC, Hodak H, Clantin B, Locht C, Molle G, Jacob-Dubuisson F, Saint N (2006) Channel properties of TpsB transporter FhaC point to two functional domains with a C-terminal protein-conducting pore. J Biol Chem 281:158–166
28. Duret G, Szymanski M, Choi KJ, Yeo HJ, Delcour AH (2008) The TpsB translocator HMW1B of *Haemophilus influenzae* forms a large conductance channel. J Biol Chem 283:15771–15778
29. Van Gelder P, Dumas F, Winterhalter M (2000) Understanding the function of bacterial outer membrane channels by reconstitution into black lipid membranes. Biophys Chem 85:153–167
30. Benz R (1988) Structure and function of porins from gram-negative bacteria. Annu Rev Microbiol 42:359–393
31. Montal M, Mueller P (1972) Formation of bimolecular membranes from lipid monolayers and a study of their electrical properties. Proc Natl Acad Sci U S A 69:3561–3566
32. Papini E, Colonna R, Cusinato F, Montecucco C, Tomasi M, Rappuoli R (1987) Lipid interaction of diphtheria toxin and mutants with altered fragment B. 1. Liposome aggregation and fusion. Eur J Biochem 169:629–635
33. Sobko AA, Kotova EA, Antonenko YN, Zakharov SD, Cramer WA (2006) Lipid dependence of the channel properties of a colicin E1-lipid toroidal pore. J Biol Chem 281: 14408–14416

Index

A

Adhesin ... 3, 5, 6, 38, 39, 41, 46–51, 206, 383
Adhesion ... 3, 4, 47, 48, 78, 100
Agrobacterium tumefaciens ... 188
Alcian blue ... 290, 291, 294, 300, 301
Alkaline hydrolysis ... 292–293, 298–299
Amine coupling ... 136–138, 140–141, 150
AMS labeling ... 325
Amyloid ... 53–73
Analyte ... 135, 136, 144–149, 152–154, 250
Arabinogalactan ... 7, 309, 311, 314, 319–320
ASKA over-expression library ... 337–356
Assembly ... 3, 37, 38, 53–55, 59, 72, 77, 78, 99, 134, 137, 152, 153, 157–171, 181, 223–236, 382
Autotransporters ... 205

B

BAM complex ... 223, 224
β-barrel ... 3, 6, 7, 149, 205, 223, 224, 232, 236, 325
β-galactosidase ... 339, 345, 351, 355, 363
Binding kinetics ... 145, 153
Biofilm ... 53, 55, 59, 66–69, 72, 78, 97, 121, 122, 125–129, 259
Biogenesis ... 1, 3, 9, 37, 38, 54, 197, 223, 224, 259
Biotinylation ... 137, 142–144, 150
Black lipid membrane ... 381
Bligh–Dyer extraction ... 241, 246

C

Capsule ... 2, 8, 20, 21, 109–119, 261, 269, 310, 315
Cell
  envelope ... 2, 5–7, 16–18, 20–23, 26, 29, 134, 158, 169, 188, 225, 228–236, 309–323, 338, 359, 382
  wall ... 16, 18–20, 22, 24, 77, 78, 84, 86, 90, 276, 280, 289–307, 309, 311, 312, 314, 317, 319, 320, 338, 359, 367, 368, 378
Cephalexin ... 368, 371
Channel ... 6, 7, 38, 98, 104, 130, 134, 188, 193, 198, 199, 256, 367–379, 382, 389–391, 393, 394
Chaperone ... 54, 173, 174, 177, 179, 182, 183, 223, 224
Chaperone-subunit complex ... 37–51
Chitosan ... 99–101, 103, 105, 106
Chromatin immunoprecipitation (ChIP) ... 188, 198, 202
Chromatography
  affinity ... 38, 44, 63, 177, 179, 194, 195, 212, 213, 276, 279, 282–283
  gas ... 314–315, 321–322
  gel filtration ... 38, 43, 195, 202
  ion exchange ... 38, 213–214, 276, 279, 283–284
  metal affinity ... 38, 177, 179, 276
  reverse phase ... 110, 111, 113, 166
Coating procedures ... 97
Collision induced dissociation (CID) ... 242, 244, 248–250, 252, 253, 256
Congo red ... 55–56
Crystallography ... 2, 3, 24, 25, 28, 79, 188, 206, 210, 220, 221
Crystal violet ... 5, 6, 59, 66–68
Curli ... 53–72
Cysteine ... 137, 138, 140, 142, 167, 168, 325–335

D

Delipidation ... 316
Density gradient centrifugation ... 190, 195, 198, 224, 261
Disulfide bond ... 136, 166, 325–335, 359
DMB labeling ... 110–113, 115, 116, 118
DNA conjugation ... 187
Drug efflux pump ... 382
Dye leakage assay ... 381

E

Electrophysiology ... 367–379, 381–385, 387–389, 391–393
Escherichia coli ... 2, 17, 38, 53, 109–119, 133, 206, 224, 240, 254, 259, 261, 273, 278, 325, 337–356, 360
Exopolysaccharides ... 5, 121–130

F

Fimbriae ... 25, 37, 79
FLAG ... 190, 192, 194, 195
Fluorescence ... 55, 63, 65, 66, 72, 98, 100, 107, 112, 119, 127, 129, 135, 158, 159, 169–171, 174, 176–178, 181–183, 265, 274, 374, 378
Formaldehyde (FA) ... 188, 192, 193, 199–203, 294, 305, 314

Anne H. Delcour (ed.), *Bacterial Cell Surfaces: Methods and Protocols*, Methods in Molecular Biology, vol. 966, DOI 10.1007/978-1-62703-245-2, © Springer Science+Business Media New York 2013

French press....................................39, 41, 42, 45, 49, 177, 179, 278, 282, 313, 316
Fruiting body..122, 124–125, 128

G

Glucan...310, 311, 322
Glycosyltransferase..273–278, 310
Gold particles....................................79, 82, 87, 88, 90, 92
Gram-negative bacteria................................2–7, 9, 16, 20, 37, 77, 133, 134, 205, 206, 224, 239–256, 259, 273, 274, 359, 382
Gram-positive bacteria..............................2, 5, 6, 9, 77–93, 133, 273, 289, 290, 374

H

Heat modifiability..224, 234, 236
His-tag.................................38, 39, 41, 43–47, 49, 63, 69, 174, 179, 183, 210, 211, 220, 221, 278, 281, 282

I

Image analysis..100, 103, 105, 122
Immunoprecipitation............................192–193, 198–203
Intracellular bacteria.............................295–296, 302–303
iTRAQ...................................158, 159, 161, 166–169, 171

K

Klebsiella pneumoniae...261

L

Large unilamellar vesicles (LUVs).................................181
Library screen...337–356
Lipid A...239–256
Lipid II...273–287
Lipoarabinomannan (LAM)...............................310, 311
Lipomannan..310, 311
Lipopolysaccharide (LPS).......................2, 6, 16, 110, 115, 118, 239, 310, 359, 362
Liposome...173–184
Listeria monocytogenes...................................289–307, 373
Luciferase...339
Luminescence..........339, 345, 346, 348, 349, 351, 352, 356

M

Maleimide.............................138, 142, 151, 327, 329, 334
Mass spectrometry..........................3, 7, 158, 241, 242, 244, 248, 303, 305, 314–315, 320–321, 360–363
Mechanosensitive channels..372
Membrane fusion proteins (MFP)......................134, 137, 149, 150, 152
Microfluidic chamber..99–104
Microscopy
 atomic force microscopy (AFM)............2, 5, 19, 79–80, 82–83, 87–93, 99, 100
 confocal laser scanning microscopy (CLSM)...121–130
 electron microscopy (EM)...............2, 3, 16–18, 23–25, 27–29, 69, 77, 79, 80, 82, 85–88, 90, 98, 188, 193, 197, 198, 202, 224, 260
 fluorescence microscopy...........15, 27, 98, 157–171, 378
 light microscopy......................................2, 16, 98, 282
Motility...1, 2, 8, 97–107
Muramidase....................................78, 80, 83, 86, 90, 277, 280–281, 284–286, 292, 297–298, 303, 304
Muropeptides.......................................276–278, 280–281, 284–287, 306
Mycobacterium tuberculosis......................................309–323
Myxococcus xanthus.............................98, 100, 103, 121–130

N

Neisseria meningitides...224, 261

O

Osmotic shock..360, 362, 363, 365
Outer membrane................................3, 4, 6–9, 16, 20–22, 24–26, 37–51, 54, 63, 133, 134, 188, 205, 206, 211, 223–236, 239–256, 274, 278, 325, 338, 359–363, 365, 375, 376, 382
Outer membrane vesicles.............................5, 164, 259–271

P

Patch-clamp.....................367, 369, 372–375, 377, 378, 392
PCR.......................................193, 198–203, 339–341, 343, 345–348, 352, 353, 373
Penicillin-binding proteins (PBPs)...............................273
Peptidoglycan...........................4–6, 8, 9, 16, 20, 21, 26, 51, 78, 79, 289–292, 294–297, 300–307, 309, 311, 314, 319–320, 323, 359
Peptidoglycan synthesis...273–287
Periplasm.......................5, 6, 29, 37, 38, 40, 46–47, 51, 134, 169, 188, 223, 224, 325–335, 359, 360, 362, 363
Photodissociation..........................239, 242, 245, 250, 257
Pilus...37–51, 77–93, 98, 194, 197
Planar lipid bilayer..381–395
Plug western blot.......................................56, 60, 62–63
Polydimethylsiloxane (PDMS)......................................102
Polyethylene glycol (PEG)............................151, 209, 395
Polymerization...................................20, 55, 58, 63, 65–66, 71, 78, 110, 119, 202, 206, 229, 299, 300, 306, 307
Polymyxin...360, 362–365
Polysaccharides..................3, 8, 69, 109, 110, 290, 309–323
Pore....................................54, 63, 173–184, 295, 381–395
Porin...48, 49, 390, 394
Precipitation
 acetone..84
 ammonium sulfate..211
 TCA...84, 333, 334, 366
PreScission protease...208, 213

Proteinase K 203, 233, 292, 293, 306, 312, 314, 316, 317, 319
Protein purification 71, 137, 207–208, 212–215
Protein translocation 187
Proteomics 157–171, 244, 261, 291
Protoplast 83, 84, 90, 374
Pseudomonas aeruginosa 174, 261, 362
Pyrogallol-red-molybdate-methanol (PRMM) method 169

R

Reconstitution 134, 135, 161, 166
Redox state 325–335
Reporter gene 337–356
Reverse phase evaporation 174

S

Sarkosyl 39, 42, 49
Secretion
- chaperone-usher 37
- two-partner 205
- type III 25, 173, 193
- type IV 25, 187
- type V 205–221

Seeding 55, 58, 65, 66, 72, 206, 218–219
Sensor chip 135, 137–139
Sialic acid 109, 110, 113, 114, 116–119, 129
Silver staining 290, 300, 301
S-layer 2, 8–9, 18–21, 24, 26
Sortase 5, 78, 90, 91, 290
Spheroplast 40, 46, 48, 51, 367–379
Streptavidin 13–139, 142–144, 190, 194, 382
Stress 3, 4, 53, 54, 99, 100, 106, 337–356
Sulforhodamine-B 174, 177, 178, 180, 181, 183, 184
Surface plasmon resonance (SRP) 133–154
Surface proteins 3, 5, 7, 78, 79, 289–307

T

Tangential flow concentrator 262–263, 266
Teichoic acids 4, 5, 273, 289–307
Termed transfer DNA immunoprecipitation (TrIP) 188, 192–193, 198–202
Thin layer chromatography (TLC) 7, 241, 243–245, 247–248, 253
Thioflavin T (ThT) 55, 58, 65
Thiol 136, 137, 140–142, 150, 151, 331
Thiol labeling 327, 329, 334
Thrombin 38–41, 43, 45, 49, 50, 275, 276, 279, 283, 286
Translocator 173–184, 187, 205, 211
Transpeptidase 5, 273–275, 277, 278, 306
Transport 4, 5, 110, 133–137, 145, 223, 242, 326, 381, 382
Tris/sucrose/EDTA extraction 362, 364
Trypsin 161, 166–168, 225, 228–229, 233, 234, 236, 297, 303–305

U

Urea 81, 90, 224–226, 229, 233, 234
Usher 37–51, 391

V

Vibrio cholerae 158, 253, 260

W

Wall techoic acids (WTAs) 5, 290

MIX
Papier aus verantwortungsvollen Quellen
Paper from responsible sources
FSC® C105338

If you have any concerns about our products,
you can contact us on
ProductSafety@springernature.com

In case Publisher is established outside the EU,
the EU authorized representative is:
Springer Nature Customer Service Center GmbH
Europaplatz 3, 69115 Heidelberg, Germany

Printed by Libri Plureos GmbH
in Hamburg, Germany